GEORGE B. THOMAS, Jr.
Massachusetts Institute of Technology

ROSS L. FINNEY
University of Illinois, Urbana-Champaign

CALCULUS AND ANALYTIC GEOMETRY

Fifth Edition

SELF-STUDY MANUAL

by Maurice D. Weir, Naval Postgraduate School

ADDISON-WESLEY PUBLISHING COMPANY
Reading, Massachusetts
Menlo Park, California · London · Amsterdam · Don Mills, Ontario · Sydney

Reproduced by Addison-Wesley from camera-ready copy prepared by the author.

ISBN 0-201-07655-1
ABCDEFGHIJ-AL-79

PREFACE TO THE STUDENT

This study manual has been designed especially for you, the student. It conforms with the fifth edition of CALCULUS, by George B. Thomas and Ross L. Finney. It is intended as a self-study workbook to assist you in mastering the basic ideas in CALCULUS.

The study manual is organized section by section to correspond with the Thomas/Finney text. For each section we specify its main ideas by stating appropriate behavioral OBJECTIVES. Each objective states a particular task for you to perform in order to master that objective. Usually the task requires you to solve a certain type of problem related to the discussion in the text; sometimes the task requires you to demonstrate proficiency with certain key terms or concepts. In every case the objective is highly specific and states the behavior expected of you.

One or more examples follows each objective and illustrates its requirements. Each example is written in a semi-programmed format; that is, the example is only partially worked out, so you must supply some of the intermediate results yourself. Correct answers to each intermediate result are supplied at the bottom of the page. Thus, each example is broken down into a sequence of steps to guide you through the procedures and techniques associated with its solution. Each example has been carefully selected not to repeat any example or problem in the Thomas/Finney text; thus, you retain the full array of the text problems for practice and further development.

We have provided an OBJECTIVE-PROBLEM KEY at the end of each chapter in this manual to assist you in coordinating the textbook problems at the end of each section in Thomas/Finney with the specific objectives of this workbook.

At the end of each chapter there is a SELF-TEST. Each test is followed by complete solutions to all the test problems. The test problems cover the objectives and are similar in scope and difficulty to the examples in this manual and the examples and problems in Thomas/Finney. The test should be useful in preparing for class examinations.

We recommend that this manual be used in the following way:

1. Carefully read the section of Thomas/Finney assigned you by your calculus instructor.

2. Read each objective and work through the associated example(s) in the corresponding section of this manual. You should conceal the answers to the examples at the bottom of the page. Work with pencil and scratch paper as you are guided through each solution, writing in the intermediate results in the blanks provided.

3. After all the blanks for a given problem are filled in, compare your answers with the correct answers at the bottom of the page. If you have difficulty or do not fully understand the answers given, review the material in the textbook or consult your instructor.

4. Solve the homework problems assigned to you by your instructor from this section of Thomas/Finney. If you encounter difficulty in solving a particular problem, look in the OBJECTIVE-PROBLEM KEY at the end of the chapter and determine to which of the objectives (if any) it corresponds. Review the example(s) associated with the objective, paying particular attention to the steps and techniques.

If additional help is required, see your instructor. Make an effort to specify the place where you are having trouble.

5. After you complete a chapter in Thomas/Finney, review the objectives in this manual. Then take the chapter self-test and compare your solutions with those provided. Problems in the self-test sometimes bring together several ideas from the chapter.

We caution you that the objectives given in this manual by no means exhaust all the possible objectives that could be written for a careful study of Thomas/Finney: we have tried to identify the main ones. However, your instructor may have additional requirements. For instance, he or she may want you to be able to prove certain theorems or derive results in the text. We have not stated objectives of this sort. Also, your instructor may consider some objectives far more important than others and not require that you master some objectives at all. So it is imperative that you find out specifically what your instructor considers essential, and study accordingly. This manual should be helpful to you both in identifying the tasks and successfully mastering them. The problems assigned to you by your instructor should help you discover those concepts and applications of calculus that your instructor wishes to stress.

January, 1979

Maurice D. Weir

PREFACE TO THE INSTRUCTOR

Although this manual has been written primarily for the student, we hope it will also be useful to the instructor. For instance, the OBJECTIVE-PROBLEM KEY at the end of each chapter, which coordinates the Thomas/Finney textbook problems with each specific section objective in this manual, should be helpful in selecting problems for student homework assignments. Moreover, we have found the written objectives to be very useful in identifying for the student those concepts and classes of problems that we consider especially important and intend to test. As the course instructor you may have additional objectives that you consider essential, but we hope to have identified at least some of the main ones to make your task of specifying them easier.

Since each example and chapter test problem in this manual does not repeat any examples or problems in the Thomas/Finney text, the manual should provide supplementary material for your course. The manual was not written with any particular instructional mode in mind. However, the written objectives, examples, and chapter tests make the manual easily adaptable to calculus courses using individualized instruction, such as PSI (the Keller Plan).

M.D.W.

TABLE OF CONTENTS

CHAPTER 1 THE RATE OF CHANGE OF A FUNCTION

1-1 Introduction.

OBJECTIVE: Broadly define the mathematics of calculus, and specify its two main classes of problems.

1. Calculus is the mathematics of _______________ and _______________.

2. One class of problems in calculus involves finding the _______________ at which a variable quantity is changing. This branch of calculus is called the _________________ calculus.

3. A second class of problems in calculus involves finding a _____________ when its rate of change is known. This branch is called the _____________________ calculus.

Both branches are important to modern science and engineering.

1-2 Coordinates.

OBJECTIVE: Draw a rectangular coordinate system and plot or locate points within it.

4. Finish labeling the coordinate system at the right, and plot the point

$$P = P(-3,2) .$$

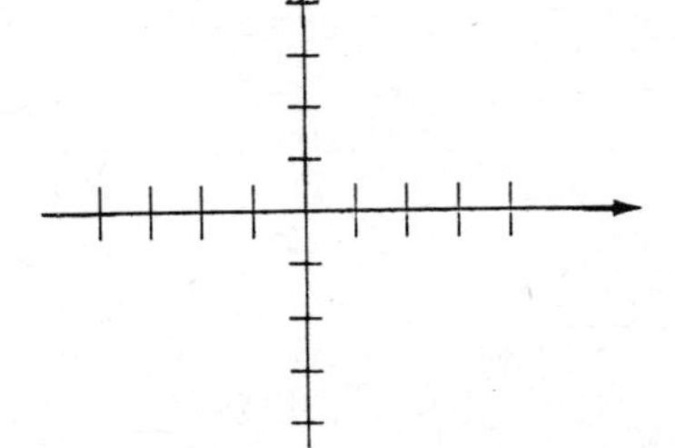

5. Use the same diagram to locate the point Q such that PQ is perpendicular to the x-axis and bisected by it. The coordinates of Q are __________.

6. Use the same diagram to locate the point R such that PR is perpendicular to the y-axis and bisected by it. The coordinates of R are __________.

7. Use the same diagram to locate the point S such that PS is bisected by the origin. The coordinates of S are __________.

---

1. change, motion
2. rate, differential
3. function, integral
4. P(−3,2)
5. Q(−3,−2)
6. R(3,2)
7. S(3,−2)

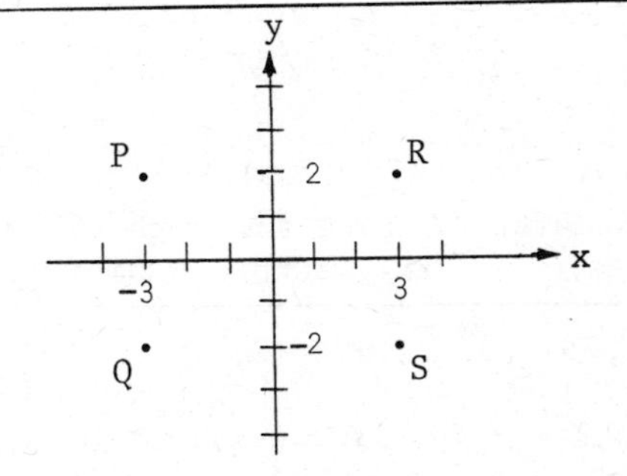

1-3 Increments.

OBJECTIVE A: Given the coordinates of points P and Q in the plane, find the increments Δx and Δy from P to Q and the distance between the points.

8. If a particle starts at P(2,-1) and goes to Q(-7,-3), then its x-coordinate changes by

$$\Delta x = -7 - \_\_\_\_ = \_\_\_\_ .$$

9. Its y-coordinate changes by

$$\Delta y = \_\_\_\_ - (-1) = \_\_\_\_ .$$

10. The distance between P and Q is

$$d = \sqrt{\_\_\_\_ + \_\_\_\_} = \_\_\_\_ .$$

OBJECTIVE B: Given the increments from the point P to the point Q and the coordinates of one of these points, determine the coordinates of the other point.

11. The coordinates of a particle change by $\Delta x = -3$ and $\Delta y = 5$ in moving from P(1,-4) to Q(x,y) . The x-coordinate of Q is given by

$$x = 1 + \_\_\_\_ = \_\_\_\_ .$$

12. The y-coordinate of Q is given by

$$y = \_\_\_\_ + 5 = \_\_\_\_ .$$

OBJECTIVE C: Write an equation of the circle given its radius and the coordinates of its center.

13. An equation of the circle with radius $r = 5$ and center (-1,3) is given by

$$\sqrt{(x - \_\_\_)^2 + (y - \_\_\_)^2} = \_\_\_\_ ,$$

or $\_\_\_\_\_\_\_\_\_\_\_\_\_\_\_\_$.

Remark. A more thorough treatment of the circle will be given in the chapter on Analytic Geometry.

---

8. 2,-9 9. -3,-2 10. $(-9)^2$, $(-2)^2$, $\sqrt{85}$ 11. -3,-2

12. -4,1 13. (-1), 3, 5 or $(x+1)^2 + (y-3)^2 = 25$

1-4 Slope of a Straight Line.

OBJECTIVE A: Define the slope of a straight line, and calculate the slope (if any) of the line determined by two given points.

14. The slope of the line through the points $P_1(x_1,y_1)$ and $P_2(x_2,y_2)$ is given by

$$m = \frac{\text{rise}}{\text{run}} = \underline{\qquad} = \underline{\qquad}, \quad \text{provided that } x_1 \neq x_2 .$$

15. If $x_1 = x_2$, then the line through the points $P_1(x_1,y_1)$ and $P_2(x_2,y_2)$ is a ____________ line. For vertical lines the ________ is not defined.

16. The slope of the line through the points $A\left(-\frac{1}{2}, 1\right)$, $B(0,-2)$ is $m =$ ____.

OBJECTIVE B: Use slopes to determine whether three or more points are collinear (lie on a common straight line).

17. Consider the three points $A(-3,7)$, $B(1,-1)$, $C(2,-3)$. The slope of the line through A and B is

$m_1 =$ _____ .

The slope of the line through B and C is

$m_2 =$ _____ .

Because m_1 and m_2 are ______________, the three points A, B, C are collinear.

OBJECTIVE C: Find the coordinates of the midpoint of the line segment joining two given points.

18. If x_1 and x_2 are two real numbers on the x-axis, the real number located half-way between them is equal to

____________ .

---

14. $\frac{y_2 - y_1}{x_2 - x_1}$, $\frac{y_1 - y_2}{x_1 - x_2}$ 15. vertical, slope 16. -6

17. $m_1 = -2$, $m_2 = -2$, equal 18. $\frac{1}{2}(x_1 + x_2)$

19. If $P_1(x_1,y_1)$ and $P_2(x_2,y_2)$ are two points, the coordinates of the midpoint of the line segment joining them are given by

______________________ .

20. The midpoint of the line segment joining $A(-2,-1)$ to $B(3,5)$ is ________.

1-5 Equations of Straight Lines.

OBJECTIVE A: Write an equation of any vertical line given a point on the line.

21. An equation of the vertical line passing through the point $P(4,-7)$ is __________ .

OBJECTIVE B: Write an equation of any line with given slope and passing through a given point.

22. Using the point-slope equation of the line, we have $y - y_1 = m(x - x_1)$. Thus an equation of the line with slope $m = -2$ through the point $(1,3)$ is given by ______________________ .

23. The line perpendicular to the line in (22) has slope $m = -\frac{1}{-2} = \frac{1}{2}$. Thus an equation of the perpendicular line through $(1,3)$ is ______________.

OBJECTIVE C: Write an equation of any line given two points on the line.

24. Let $P_1(-3,0)$ and $P_2(2,-1)$ be two points on the line L . The slope of L is $m =$ ____ . Thus, an equation of L using P_1 is ________________; using P_2 an equation is ________________ . In either case, solving for y we obtain the equation $y =$ _________ .

---

19. $\left(\frac{x_1+x_2}{2}, \frac{y_1+y_2}{2}\right)$

20. $\left(\frac{1}{2}, 2\right)$

21. $x = 4$

22. $y - 3 = -2(x - 1)$

23. $y - 3 = \frac{1}{2}(x - 1)$

24. $-\frac{1}{5}$, $y - 0 = -\frac{1}{5}(x + 3)$, $y + 1 = -\frac{1}{5}(x - 2)$, $y = -\frac{x+3}{5}$

25. Let $P_1(1,-3)$ and $P_2(1,5)$ be two points on the line L. Since the x-coordinates of the points are the same we conclude that L is a ____________ line and hence has no ________. An equation for L is __________.

OBJECTIVE D: Recognize an equation as representing a line and determine the slope (if any), the x-intercept (if any), and the y-intercept (if any).

26. The equation $3x - 2y = 6$ represents a straight line because it contains only _________ powers of x and y. When $x = 0$, $y =$ _____ which gives the value where the line crosses the y-axis. This is called the ______________. When $y = 0$, $x =$ _____ giving the value where the line crosses the ___________. This is called the x-intercept.

27. The equation $y = 3$ represents a straight line which is parallel to the _________. It is called a ______________ line and has slope $m =$ _______.

28. The equation $xy = 1$ does not represent a straight line because it is not a ___________ equation when the variables x and y are multiplied together.

OBJECTIVE E: Graph any equation representing a line.

29. Graph the line $y = -3x + 1$.

30. Graph the line $\frac{x}{2} - \frac{y}{3} = \frac{1}{2}$.

---

25. vertical, slope, $x = 1$

26. first, -3, y-intercept, 2, x-axis

27. x-axis, horizontal, 0

28. linear

29.

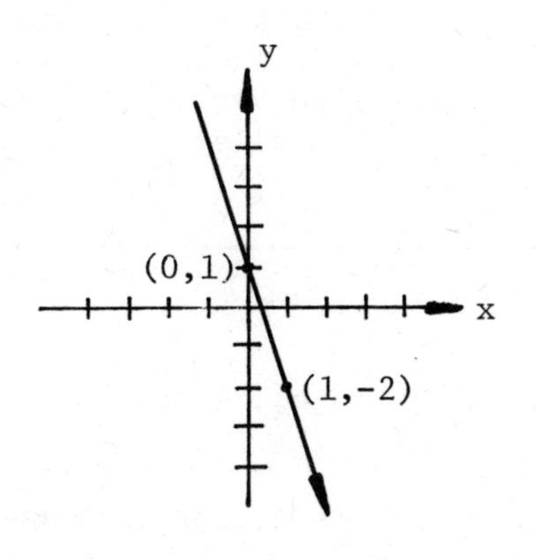

30.

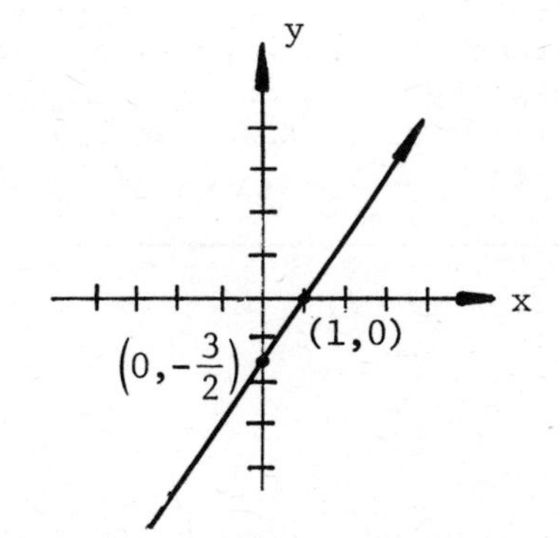

OBJECTIVE F: Find an equation of the line passing through a given point and parallel or perpendicular to a given line.

31. The line containing the point $(-1,2)$ that is parallel to the line $3x - y - 1 = 0$ has slope $m =$ ________. Since the line contains the point $(-1,2)$, its equation in point-slope form is ____________________.

32. The line containing the point $(4,1)$ that is perpendicular to the line $2y - 3x = 5$ has slope $m =$ ________. Since the line contains the point $(4,1)$, its equation in point-slope form is ____________________.

1-6 Functions and Graphs.

In this section of the textbook there are several terms associated with the concept of a "function" with which you will need to become familiar. The following items are designed to assist you in learning the precise mathematical meanings of these various terms.

33. A variable is a symbol such as x, y, t, etc. that may take on any value over a prescribed set. The set of values that a variable may take on is called the ________ of the variable.

34. In applications variables often have domains that are intervals. Symbolically, the open interval $a < x < b$ is given by ________. The closed interval $a \leq x \leq b$ is designated by ________, and the half-open intervals $a < x \leq b$ and $a \leq x < b$ are designated by ________ and ________, respectively.

35. Calculus is concerned with how variables are related. If to each value of the variable x there corresponds a unique value of the variable y, then y is said to be a ________ of x. The key word in this definition of function is ________ : we do not want to input a single value for the variable x with two or more possible outcomes for y. Every function is determined by two things: (1) the ____________ of the first variable x and (2) the ____________ or condition describing how y is obtained from x so that the ordered pairs (x,y) belong to the function. The variable x is called the ______________ variable or ____________ of the function; the second variable y is called the ______________ variable. It is also said that the function ____________ the x-variable to its image y-value. The set of values taken on by the dependent variable y is called the ________ of the function.

---

31. 3, $y - 2 = 3(x + 1)$

32. $-\frac{2}{3}$, $y - 1 = -\frac{2}{3}(x - 4)$

33. domain

34. (a,b), $[a,b]$, $(a,b]$, $[a,b)$

35. function, unique, domain, rule, independent, argument, dependent, maps, range

OBJECTIVE A: Given an equation for a function $y = f(x)$ calculate the value of f at a specified point, find the domain and range of f, and graph f by making a table of pairs.

36. Consider the function $y = -3x + 2$. The domain of the function is the interval _______ . Solving the equation for x, gives $x =$ ___________ so that the variable y may take on any value whatsoever. Thus, the range of the function is the interval _______ . Sketch the graph.

37. Consider the function $y = \dfrac{x^2 - 1}{x+1}$. The function is defined for all values of x except ________; hence the domain consists of the union of the intervals _________ and _________ . When $x \neq -1$,

$$y = \frac{x^2 - 1}{x+1} = \frac{(x-1)(x+1)}{x+1} = \_\_\_\_\_\_\_\_\_\_$$

. Therefore, the range of the function is all real numbers except for $y =$ _________ (because $x \neq -1$), so that the range is the union of the two intervals ________ and ________ . Sketch the graph.

38. The domain of the function $y = -\sqrt{1-x}$ is the interval _________ , since $\sqrt{1-x}$ is defined whenever $1 - x \geq 0$. Squaring both sides and solving the resultant equation for x, we obtain $x =$ ___________ . We see from this last equation that y can take on any value. However, since y is the negative square root, the range is the interval ___________ .

39. If $g(x) = \dfrac{1}{\sqrt{x-2}}$, the domain of g is ________ . The value $g(3)$ is ______ ; $g(11)$ is _______ ; $g(a)$ is ________ ; $g(b+2)$ is ________ .

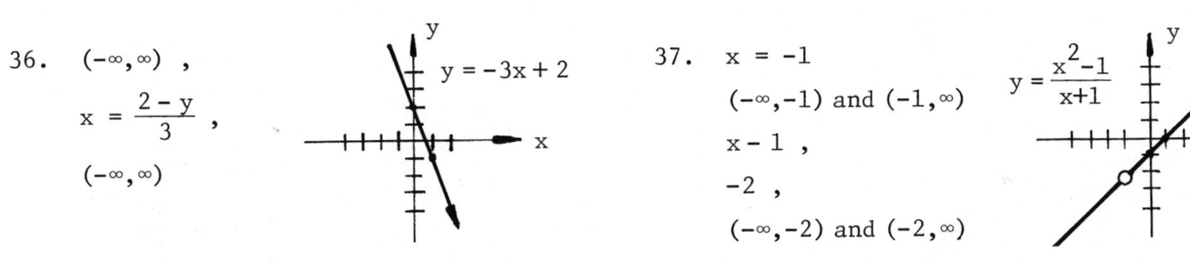

36. $(-\infty, \infty)$, $x = \dfrac{2-y}{3}$, $(-\infty, \infty)$

37. $x = -1$

$(-\infty, -1)$ and $(-1, \infty)$

$x - 1$,

-2,

$(-\infty, -2)$ and $(-2, \infty)$

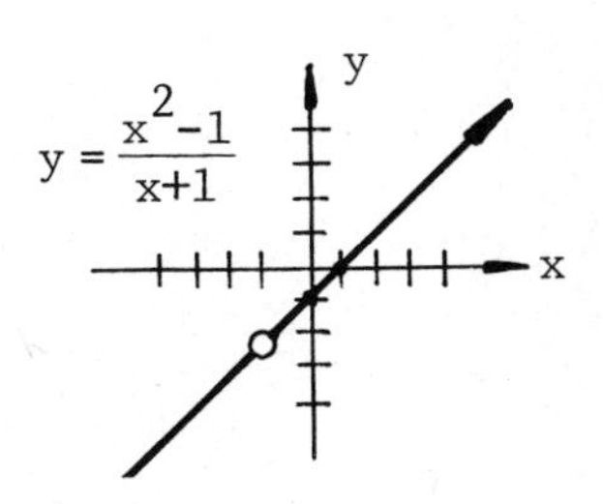

38. $(-\infty, 1]$, $x = 1 - y^2$, $(-\infty, 0]$

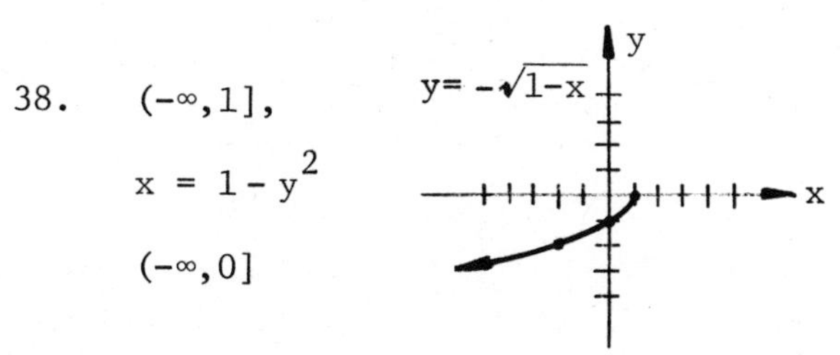

39. $(2, \infty)$, 1, $\dfrac{1}{3}$, $\dfrac{1}{\sqrt{a-2}}$, $\dfrac{1}{\sqrt{b}}$

40. Consider the fucntion $y = |x-1| + 2$. The domain of this function is the interval ________ . If the x values satisfy $x \geq 1$, then $y =$ ________; on the other hand if $x < 1$, then $y =$ ________ . A table of some values for this function is given by (complete the table):

x	-2	-1	0	1	2	3
y	5					

Sketch the graph of the function using the table. From the graph the range of the function is evidently the interval ________ .

41. Consider the function $y = [x-1] + 2$, where $[x-1]$ denotes the greatest integer in ________ . The domain of this function is the interval ________ . A table of some of the values for this function is given by (complete the table):

x	-2.0	-1.5	-1.0	-.5	0	.5	1.0	1.5	2.0	2.5
y	-1.0									

Sketch a graph of the function using the table. The range of this function is not an interval, but the set of numbers ________________ .

OBJECTIVE B: Define absolute value and describe the domain of an absolute value inequality without using absolute value symbols.

42. The absolute value function assigns to the number x the number ________ . Thus, if $x \geq 0$ so that x is nonnegative, $|x|$ is the number ________ ; but if $x < 0$ is negative, then $|x|$ is the number ______ . For instance, $|4| =$ ______ and $|-4| = -($ ______ $) =$ ______ . Thus, the absolute value function is never negative. Its domain is the interval ______ and its range is the interval ______ .

---

40. $(-\infty,\infty)$,

$(x-1) + 2 = x+1$,

$-(x-1) + 2 = 3-x$,

x	-2	-1	0	1	2	3
y	5	4	3	2	3	4

$y = |x-1| + 2$

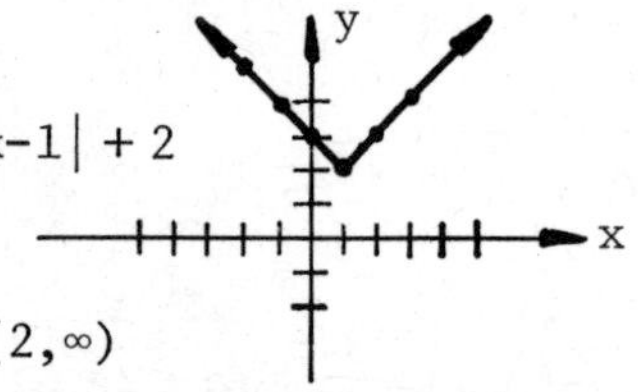

range: $[2,\infty)$

41. $x-1$,

$(-\infty,\infty)$,

x	-2.0	-1.5	-1.0	-.5	0
y	-1.0	-1.0	0	0	1.0

x	.5	1.0	1.5	2.0	2.5
y	1.0	2.0	2.0	3.0	3.0

$y = [x-1] + 2$

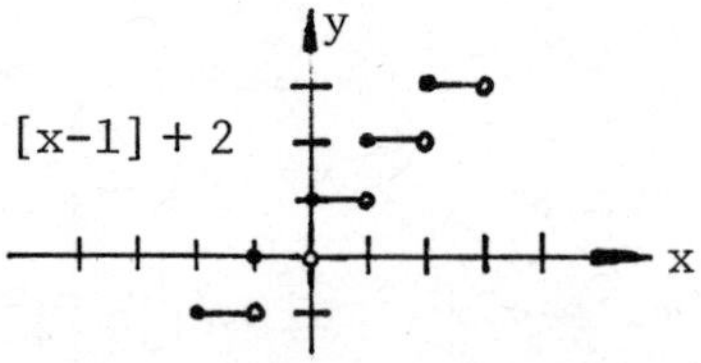

range: $\{\ldots,-2,-1,0,1,2,3,\ldots\}$

42. $\sqrt{x^2}$, x , $-x$, 4 , -4 , 4 , $(-\infty,\infty)$, $[0,\infty)$

43. The number ________ = ________ measures the distance between x and a. If r is a positive real number, then $|x-a| < r$ is equivalent to the inequality __________________ . Thus, x must lie within the interval ______________ .

44. If $|x+2| \leq 7$, then $|x-(-2)| \leq$ _____ . This is equivalent to the inequality _________________ . Therefore x must lie within the closed interval _________ .

45. If $|x| > r > 0$, then x must lie within the union of the intervals _______ and _______ . For instance, $|x| > 7$ implies $x <$ _____ or $x >$ _____ .

46. The relationship between absolute value, addition, and multiplication is given by the two equations

$$|a+b| \leq \_\_\_\_\_\_\_\_\_\_\_\_$$

$$|ab| = \_\_\_\_\_\_\_\_\_\_\_\_ .$$

OBJECTIVE C: Convert radian measure to degree measure, and vice versa.

47. The radian measure of $180°$ is _____ units. Here the symbol _____ represents a real number. This real number corresponds to the length that is subtended by an _____ of a circle of radius 1 with central angle _____. Therefre, $1°$ corresponds to _______ radians, and 1 radian corresponds to _______ degrees.

48. Converting from degree to radian measure, $60°$ = ________ radians, $-45°$ = ________ radians, and $72°$ = ________ radians.

49. Converting from radian to degree measure, $\frac{\pi}{6}$ radians = _______ degrees, $-\frac{3\pi}{2}$ radians = _______ degrees, and 2 radians = _______ degrees.

Remark. Whenever you encounter $\sin 2$, for instance, you must think the sine of 2 radians not the sine of 2 degrees. The latter is written $\sin 2°$.

---

43. $|x-a| = |a-x|$, $-r < x-a < r$ or $a-r < x < a+r$, $(a-r, a+r)$

44. 7, $-7 \leq x+2 \leq 7$, $[-9,5]$

45. $(-\infty,-r)$ and (r,∞), -7, 7

46. $|a| + |b|$, $|a||b|$

47. π, π, arc, $180°$, $\frac{\pi}{180}$, $\frac{180}{\pi}$

48. $\frac{\pi}{3}$, $-\frac{\pi}{4}$, $\frac{2\pi}{5}$

49. 30, -270, $\left(\frac{360}{\pi}\right)$

OBJECTIVE D: Given two functions f and g, write an expression for their composite $f(g(x))$.

50. If $f(x) = 5x + 2$ and $g(x) = x^2$, then a formula for $f(g(x))$ is obtained as follows:

$$f(g(x)) = f(x^2) = \_\_\_\_\_\_\_\_ .$$

The domain of $y = f(g(x))$ is all values of x in the domain of g such that $f(g(x))$ is defined. This is the interval ______ .

51. If $f(x) = \sqrt{x - 1}$ and $g(x) = x + 1$, then

$$f(g(x)) = \_\_\_\_\_\_ .$$

The domain of the composite is all values of x in the domain of g such that $f(g(x))$ is defined. This is the interval ______ .

52. Let $f(x) = x^2$ and $g(x) = \sqrt{x - 1}$. Then $f(g(x)) =$ ______ . The domain of g is the set of all real numbers x satisfying ______ . Thus, the domain of the composite $y = f(g(x))$ is the interval ______ .

OBJECTIVE E: Find two functions f and g that will produce a given composite function h such that $h(x) = f(g(x))$.

53. Consider $h(x) = \sin (x^2 - 1)$. If we let $f(x) = \sin x$ and $u = g(x) =$ ______ , then

$$h(x) = \_\_\_\_\_\_ = f(x^2 - 1) = \_\_\_\_\_\_ .$$

54. Consider $h(x) = |x + 2|$. Then for $f(x) =$ ______ and $u = g(x) =$ ______ ,

$$h(x) = \_\_\_\_\_\_ = f(x + 2) = \_\_\_\_\_\_ .$$

55. If $h(x) = \sqrt{x^5 + 2x^3 - 1}$, then for $f(x) = \sqrt{x}$ and

$u = g(x) =$ ______ , it is true that $h(x) = f(g(x))$.

---

50. $5x^2 + 2$, $(-\infty, \infty)$

51. $\sqrt{x}$, $[0, \infty)$

52. $x - 1$, $x \geq 1$, $[1, \infty)$

53. $x^2 - 1$, $f(g(x))$, $\sin (x^2 - 1)$

54. $|x|$, $x + 2$, $f(g(x))$, $|x + 2|$

55. $x^5 + 2x^3 - 1$

1-7 Slope of a Curve.

OBJECTIVE: Given a curve $y = f(x)$, find the slope of the curve at a point (x,y) on the curve as the limit of slopes of secants to the curve through the point.

56. Consider the curve $y = 6 - 4x - x^2$. If $P(x_1,y_1)$ is a point on the curve and $Q(x_2,y_2)$ is another point on the curve with

$$\Delta x = x_2 - x_1 \quad \text{and} \quad \Delta y = y_2 - y_1 ,$$

then $x_2 =$ __________ and $y_2 =$ __________ . Since the point $Q(x_2,y_2)$ is on the curve,

$$y_2 = 6 - 4x_2 - x_2^2 = \text{\_\_\_\_\_\_\_\_\_\_\_\_\_\_\_\_\_\_\_\_}$$

$$= \text{\_\_\_\_\_\_\_\_\_\_\_\_\_\_\_\_\_\_\_\_} .$$

Since P is on the curve, its coordinates also satisfy the equation:

$$y_1 = \text{\_\_\_\_\_\_\_\_\_\_} .$$

Thus, $\Delta y = y_2 - y_1 =$ _______________ .

Division of both sides by Δx gives,

$$m_{sec} = \frac{\Delta y}{\Delta x} = \text{\_\_\_\_\_\_\_\_\_\_\_\_} .$$

The slope of the tangent to the curve at the point (x_1,y_1) is the __________ of m_{sec} as Δx approaches ________ . Hence, this limit equals __________ . Since (x_1,y_1) is an arbitrary point on the curve, deleting the subscript 1 gives $m =$ __________ , the slope at any point $P(x,y)$.

---

56. $x_1 + \Delta x$, $y_1 + \Delta y$, $6 - 4(x_1 + \Delta x) - (x_1 + \Delta x)^2$,

$6 - 4x_1 - 4\Delta x - x_1^2 - 2x_1\Delta x - (\Delta x)^2$, $6 - 4x_1 - x_1^2$, $-4\Delta x - 2x_1\Delta x - (\Delta x)^2$,

$-4 - 2x_1 - \Delta x$, limit , zero , $-4 - 2x_1$, $m = -4 - 2x$

1-8 Derivative of a Function.

OBJECTIVE A: For a given function f, find the derivative $f'(x)$ by applying the definition.

57. The definition of the derivative of f at the point x_1 is

$$f'(x_1) = \underline{\hspace{6cm}}$$

whenever this limit exists. The set of all pairs of numbers $(x, f'(x))$ is called the ________ or ____________ function. The domain of f' is a subset of the domain of f consisting of all numbers in the domain of f at which the ________ exists.

58. For the function $f(x) = (x+1)^2$,

STEP 1. Form $f(x+\Delta x) = (x+\Delta x+1)^2$

$$= \underline{\hspace{6cm}}$$

and $f(x) = \underline{\hspace{4cm}}$.

STEP 2. Subtract $f(x)$ from $f(x+\Delta x)$:

$$f(x+\Delta x) - f(x) = \underline{\hspace{5cm}}.$$

STEP 3. Divide by Δx:

$$\frac{f(x+\Delta x) - f(x)}{\Delta x} = \underline{\hspace{4cm}}.$$

STEP 4. Take the limit as $\Delta x \to 0$:

$$f'(x) = \lim_{\Delta x \to 0} \frac{f(x+\Delta x) - f(x)}{\Delta x} = \underline{\hspace{2cm}}.$$

59. For the function $f(x) = \dfrac{1}{\sqrt{x-1}}$,

STEP 1. Form $f(x+\Delta x) = \underline{\hspace{3cm}}$ and $f(x) = \underline{\hspace{3cm}}$.

STEP 2. Subtracting $f(x)$ from $f(x+\Delta x)$, and

STEP 3. Dividing by Δx gives,

$$\frac{f(x+\Delta x) - f(x)}{\Delta x} = \frac{\underline{\hspace{4cm}}}{\Delta x\sqrt{x-1}\,\sqrt{x+\Delta x-1}}.$$

---

57. $\lim_{\Delta x \to 0} \dfrac{f(x_1+\Delta x) - f(x_1)}{\Delta x}$, derived, derivative, limit

58. $x^2 + 2x\Delta x + (\Delta x)^2 + 2x + 2\Delta x + 1$, $x^2 + 2x + 1$, $2x\Delta x + (\Delta x)^2 + 2\Delta x$, $2x + \Delta x + 2$, $2x + 2$

STEP 4. To calculate the limit as $\Delta x \to 0$ we observe that both the numerator and the denominator in the previous expression tend to zero as $\Delta x \to 0$. In an attempt to avoid division by zero we rationalize the numerator obtaining:

$$\frac{f(x+\Delta x) - f(x)}{\Delta x} = \frac{\_\_\_\_\_\_\_\_\_\_\_\_}{\Delta x\sqrt{x-1}\sqrt{x+\Delta x-1}} \cdot \frac{(\qquad\qquad)}{\sqrt{x-1} + \sqrt{x+\Delta x-1}}$$

$$= \frac{\_\_\_\_\_\_\_\_\_\_\_\_}{\left(\Delta x\sqrt{x-1}\sqrt{x+\Delta x-1}\right)\left(\sqrt{x-1} + \sqrt{x+\Delta x-1}\right)}$$

$$= \frac{\_\_\_\_\_\_\_\_\_\_\_\_}{\left(\sqrt{x-1}\sqrt{x+\Delta x-1}\right)\left(\sqrt{x-1} + \sqrt{x+\Delta x-1}\right)}$$

Thus, as $\Delta x \to 0$,

$$f'(x) = \lim_{\Delta x \to 0} \frac{f(x+\Delta x) - f(x)}{\Delta x} = \_\_\_\_\_\_\_\_\_\_\_\_$$

$$= -\frac{1}{2}(x-1)^{\_\_\_\_} .$$

OBJECTIVE B: Write an equation of the tangent line to the curve $y = f(x)$ at a specified value $x = a$.

60. To find an equation of the tangent line to the curve $f(x) = \frac{1}{\sqrt{x-1}}$ when $x = 2$ we first calculate the slope m. By definition, $m =$ __________ . From our calculation in the previous Problem 59, that slope has the value ________ . The point on the curve corresponding to $x = 2$ has coordinates ________ . Therefore, the point-slope form gives an equation of the tangent line as ____________________ .

---

59. $\frac{1}{\sqrt{x+\Delta x-1}}$, $\frac{1}{\sqrt{x-1}}$, $\sqrt{x-1} - \sqrt{x+\Delta x-1}$,

$\sqrt{x-1} + \sqrt{x+\Delta x-1} \cdot \left(\sqrt{x-1} + \sqrt{x+\Delta x-1}\right)$, $(x-1) - (x+\Delta x-1)$,

-1, $\frac{-1}{\sqrt{x-1}\sqrt{x-1}\left(2\sqrt{x-1}\right)}$, $-\frac{3}{2}$

60. $f'(2)$, $-\frac{1}{2}$, $(2,1)$, $y-1 = -\frac{1}{2}(x-2)$

1-9 Velocity and Rates.

OBJECTIVE: Given a functional relationship $y = f(x)$ between two variables x and y, use the four step differentiation process to calculate the average rate of change and the instantaneous rate of change of y with respect to x.

61. Every derivative may be interpreted as the instantaneous rate of change of one variable per unit change in the other. If $y = f(x)$, then

$$\frac{\Delta y}{\Delta x} = \_\_\_\_\_\_\_\_\_\_\_\_$$

is interpreted as the __________ rate of change of y by a change of one unit in _____ . Passage to the limit as $\Delta x \to 0$ gives

$$\lim_{\Delta x \to 0} \frac{\Delta y}{\Delta x} = \_\_\_\_\_\_$$

as the ______________ rate of change of _____ with respect to _____ .

62. The derivative $f'(x)$ multiplied by Δx gives the change that would occur in _____ if the point (x,y) were to move along the __________ line to the curve $y = f(x)$ instead of moving along the ________ itself. This is expressed by the identity

$$f'(x) \cdot \Delta x = \_\_\_\_\_\_\_\_ .$$

63. Using the differentiation process for the function $f(x) = ax^3 + bx^2 + cx + d$, where a, b, c, d are constants, we obtain the following:

STEP 1. $f(x + \Delta x) = \_\_\_\_\_\_\_\_\_\_\_\_\_\_\_\_\_\_\_\_ .$

STEP 2. $f(x + \Delta x) - f(x) = \_\_\_\_\_\_\_\_\_\_\_\_\_\_\_\_\_\_\_\_ .$

STEP 3. $\dfrac{f(x + \Delta x) - f(x)}{\Delta x} = \_\_\_\_\_\_\_\_\_\_\_\_\_\_\_\_\_\_\_\_ .$

STEP 4. $f'(x) = \lim_{\Delta x \to 0} \dfrac{f(x + \Delta x) - f(x)}{\Delta x} = \_\_\_\_\_\_\_\_\_\_\_\_\_\_\_\_\_\_\_\_ .$

We will use this result in the following application.

---

61. $\dfrac{f(x + \Delta x) - f(x)}{\Delta x}$, average, x, $f'(x)$, instantaneous, y, x

62. y, tangent, curve, Δy_{tan}

63. $a[x^3 + 3x^2\Delta x + 3x(\Delta x)^2 + (\Delta x)^3] + b[x^2 + 2x\Delta x + (\Delta x)^2] + c(x + \Delta x) + d$,

$a[3x^2\Delta x + 3x(\Delta x)^2 + (\Delta x)^3] + b[2x\Delta x + (\Delta x)^2] + c\Delta x$,

$a[3x^2 + 3x\Delta x + (\Delta x)^2] + b[2x + \Delta x] + c$, $3ax^2 + 2bx + c$

64. Suppose the law of motion of a particle is given by

$$s = t^3 - 6t^2 + 2 \quad .$$

Then the instantaneous velocity is given by

$$v = \frac{ds}{dt} = \_\_\_\_\_\_\_\_\_\_ \quad .$$

When $t = 2.3$ sec , the velocity of the particle is $v(2.3) =$ __________ . If our coordinate axis of motion is such that the positive direction is to the right (which is conventional), the interpretation of this negative velocity means that the particle is moving to the ________ . When $t = 4$ sec , the velocity of the particle is ________ and the particle is at rest. When $t = 4.5$ sec , the velocity of the particle is __________ and the particle is moving to the ________ .

65. Consider the equilateral triangle pictured to the right. By the Pythagorean theorem $s^2 =$ ______________ or, solving for h , $h =$ __________ . Then, the area of the triangle is given by

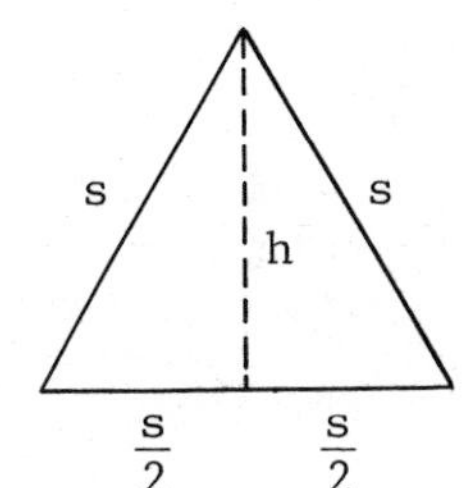

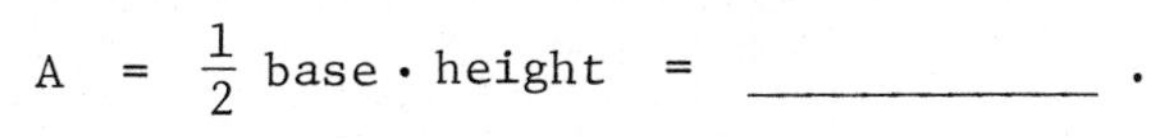

$$A = \frac{1}{2} \text{ base} \cdot \text{height} = \_\_\_\_\_\_\_\_\_\_ \quad .$$

The average rate rate of change of area with respect to side length is

$$\frac{\Delta A}{\Delta s} = \frac{\_\_\_\_\_\_\_\_\_\_\_\_\_\_\_\_}{\Delta s} = \_\_\_\_\_\_\_\_\_\_\_\_ \quad .$$

Taking the limit as Δs tends to zero gives,

$$\frac{dA}{ds} = \lim_{\Delta s \to 0} \frac{\Delta A}{\Delta s} = \_\_\_\_\_\_\_\_\_\_ ,$$

the ______________ rate of change of area with respect to ________ length for an equilateral triangle.

66. Suppose it costs $C(x)$ thousand dollars per year to produce x thousand gallons of antifreeze, where $C(x)$ is given by the table

x	.25	.5	.75	1.0	1.25	1.50	1.75	2.0	2.25	2.5
C(x)	5.875	8.5	10.875	13.0	14.875	16.5	17.875	19.0	19.875	20.0

---

64. $3t^2 - 12t$, -11.73 units/sec. , left , 0 , $\frac{27}{4}$ units/sec. , right

65. $s^2 = h^2 + \frac{s^2}{4}$, $h = \frac{\sqrt{3}}{2} s$, $\frac{\sqrt{3}}{4} s^2$, $\frac{\sqrt{3}}{4} (s + \Delta s)^2 - \frac{\sqrt{3}}{4} s^2$, $\frac{\sqrt{3}}{4} (2s + \Delta s)$,

$\frac{\sqrt{3}}{2} s$, instantaneous, side

The __________ cost at any x is the value of the derivative $C'(x)$. Using the table we estimate $C'(1.75)$ as follows:

$$C'(1.75) \approx \frac{\Delta C}{\Delta x} = \frac{\_\_\_\_\_\_\_\_\_\_}{2.0 - 1.75} = \_\_\_\_\_\_ .$$

Here we have estimated the marginal cost by the __________ cost.

1-10 Properties of Limits.

This article is of a more technical nature than the previous ones. It is intended to make the limiting concept for a function precise. The idea of limit lies at the very heart of calculus.

OBJECTIVE A: Write the formal definition of the limit of a function $F(t)$ as t approaches a number c .

67. Let F be a function defined on an open interval containing the point c , except possibly at c itself. Then the limit of F as t approaches c is L , written

______________ ,

if, given any positive number ε , there is a positive number δ such that

______________ holds whenever $0 < |t - c| < \delta$.

68. As an application of the definition, consider the limit of the function $F(t) = 3 - 2t$ as t approaches 5 . The limit is $L = -7$. To show this it is required to establish that: For any positive ε there is a positive number δ such that

______________ when $0 <$ __________ $< \delta$.

Now, $|(3 - 2t) - (-7)| = 2 \cdot$ __________ . Thus,

$2|t - 5| < \varepsilon$ provided $|t - 5| <$ ______ .

Therefore, if $\delta =$ _____ , then

$|(3 - 2t) - (-7)| < \varepsilon$ whenever ______________ .

That is, $\lim_{t \to 5} (3 - 2t) = -7$.

---

66. marginal, $19.0 - 17.875$, 4.5 , average

67. $\lim_{t \to c} F(t) = L$, $|F(t) - L| < \varepsilon$

68. $|(3 - 2t) - (-7)| < \varepsilon$, $|t - 5|$, $|t - 5|$, $\frac{\varepsilon}{2}$, $\frac{\varepsilon}{2}$, $0 < |t - 5| < \delta$

OBJECTIVE B: Evaluate limits $\lim_{t \to c} \frac{f(t)}{g(t)}$ when $f(t)$ and $g(t)$ are polynomials in t.

69. $\lim_{t \to 4} \frac{t^2 + t - 2}{t^2 - 1} = \frac{\_\_\_\_\_\_\_\_\_\_}{\lim_{t \to 4} (t^2 - 1)} = \frac{18}{\_\_\_}$.

70. $\lim_{t \to 1} \frac{t^3 + t^2 - 3t + 1}{t - 1} \neq \frac{\lim_{t \to 1} (t^3 + t^2 - 3t + 1)}{\_\_\_\_\_\_\_\_\_\_}$

because the limit of the denominator is _____ .

However, $\frac{t^3 + t^2 - 3t + 1}{t - 1} = \frac{(t - 1)(\_\_\_\_\_\_\_\_\_\_)}{t - 1}$,

so that $\lim_{t \to 1} \frac{t^3 + t^2 - 3t + 1}{t - 1} = \lim_{t \to 1} \_\_\_\_\_\_\_\_\_\_ = \_\_\_\_\_$.

71. $\lim_{h \to 0} \frac{(1 + h)^3 - 1}{h} = \lim_{h \to 0} \frac{(\_\_\_\_\_\_\_\_\_\_) - 1}{h}$

$= \lim_{h \to 0} \frac{\_\_\_\_\_\_\_\_\_\_}{h}$

$= \lim_{h \to 0} \_\_\_\_\_\_\_\_\_\_ = \_\_\_\_\_$.

72. To find a limit as t approaches ∞, set $t = 1/h$ and let $h \to 0$. Thus,

$\lim_{t \to \infty} \frac{5t^3}{1 + 3t - 2t^3} = \lim_{h \to 0} \frac{\frac{5}{h^3}}{\_\_\_\_\_\_\_\_\_\_}$

$= \lim_{h \to 0} \frac{5}{\_\_\_\_\_\_\_\_\_\_} = \_\_\_\_\_$.

---

69. $\lim_{t \to 4} (t^2 + t - 2)$, 15

70. $\lim_{t \to 1} (t - 1)$, 0 , $t^2 + 2t - 1$, $t^2 + 2t - 1$, 2

71. $1 + 3h + 3h^2 + h^3$, $3h + 3h^2 + h^3$, $3 + 3h + h^2$, 3

72. $1 + \frac{3}{h} - \frac{2}{h^3}$, $h^3 + 3h^2 - 2$, $-\frac{5}{2}$

OBJECTIVE C: Specify the five important limit properties of the two limit theorems stated in the text.

Assuming all limits exist and are finite:

73. $\lim_{t \to c} [f(t) + g(t)] =$ ____________________ .

74. $\lim_{t \to c} [k\, f(t)] =$ ________________ for every number k .

75. $\lim_{t \to c} [f(t) \cdot g(t)] =$ ____________________ .

76. $\lim_{t \to c} \left[\frac{f(t)}{g(t)}\right] = \frac{\qquad\qquad}{\qquad\qquad}$, provided ________________ .

77. If $f(t) \le g(t) \le h(t)$ for all values of t near c , and if

$\lim_{t \to c} f(t) = \lim_{t \to c} h(t) = L$, then ____________________ .

OBJECTIVE D: For elementary functions $y = f(x)$, find the righthand and lefthand limits as x approaches c , and from these determine if $\lim_{x \to c} f(x)$ exists.

78. Consider the function defined by

$$f(x) = \begin{cases} 4 - x^2 & \text{if } x \le -1 \\ 1 + x^2 & \text{if } x > -1 \end{cases}$$

The graph is shown at the right.

$\lim_{x \to -1^+} f(x) = \lim_{x \to -1^+}$ ________ = _____ ,

$\lim_{x \to -1^-} f(x) = \lim_{x \to -1^-}$ ________ = _____ ,

Since $\lim_{x \to -1^+} f(x) \ne \lim_{x \to -1^-} f(x)$, the limit $\lim_{x \to -1} f(x)$ __________ exist.

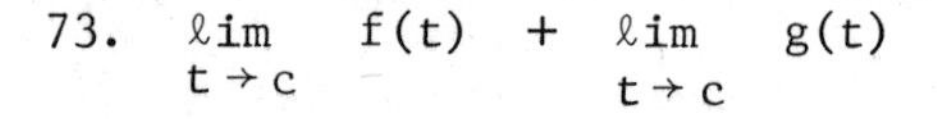

73. $\lim_{t \to c} f(t) + \lim_{t \to c} g(t)$

74. $k \lim_{t \to c} f(t)$

75. $\lim_{t \to c} f(t) \cdot \lim_{t \to c} g(t)$

76. $\dfrac{\lim_{t \to c} f(t)}{\lim_{t \to c} g(t)}$, $\lim_{t \to c} g(t) \ne 0$

77. $\lim_{t \to c} g(t) = L$

78. $1 + x^2$, 2 , $4 - x^2$, 3 , does not

CHAPTER 1 OBJECTIVE - PROBLEM KEY

Objective	Problems in Thomas/Finney Text	Objective	Problems in Thomas/Finney Text
1-2	p. 4, 1-12	1-6 A	p. 27, 1-9,13
1-3 A	p. 7, 1-6	B	p. 27, 14-20
B	p. 7, 11-14	C	p. 27, 24-35
C	p. 7, 7-10	D,E	p. 28, 44-47
1-4 A	p. 10, 1-12	1-7	p. 30, 1-15
B	p. 10, 21-23	1-8 A,B	p. 37, 1-20
C	p. 10, 24	1-9	p. 41, 2-12; p. 42, 18-23
1-5 A	p. 14, 3,5,9	1-10 B	p. 51, 1-8, 14
B,C	p. 14, 1-10	D	p. 51, 9-11
D,E	p. 14, 11-18		
F	p. 14, 20-21		

CHAPTER 1 SELF-TEST

1. For each of the following, draw a pair of coordinate axes, plot the given point, and plot the point meeting the specified requirement giving the coordinates of the second point:

   (a) $P(-3,1)$, and $Q(x,y)$ so that PQ is parallel to the x-axis and bisected by the y-axis;
   (b) $P(2,-2)$, and $R(x,y)$ so that PR is perpendicular to the x-axis and bisected by it;
   (c) $P(-1.3, -0.5)$, and $S(x,y)$ so that PS is bisected by the origin.

2. A particle moves in the plane along a straight line from $P(-3,-1)$ to $Q(7,-3)$. Find the increments Δx and Δy and the distance from P to Q.

3. A particle moves from the point $A(2,-3)$ to the x-axis in such a way that $\Delta y = -6\Delta x$. What are its new coordinates?

4. Write an equation of the circle with center $(1,-3)$ and of radius 4.

5. Determine if the points $A(1,-3)$, $B(-2,9)$, and $C(5,-19)$ are collinear.

6. Find the coordinates of the midpoint of the line segment joining $P(3,-7)$ and $Q(-4,-1)$.

7. Find the slope of the line through the points $(1,4)$ and $(-3,2)$, and write an equation of the line.

8. Determine the slope, the x-intercept, and the y-intercept for each of the following equations:

   (a) $3x + 4y = -1$ (b) $x = 2$ (c) $y = -1$ (d) $x^2 = 2y - 1$

9. Find an equation of the line through the point $(5,-7)$ and perpendicular to the line $2y - x = 8$.

10. Let f be defined by the equation $f(x) = x^2 + 3x - 2$. Find the domain and range of f. Also find the values $f(-2)$, $f(-1)$, $f(0)$, $f(2)$, $f(2b)$ and $f(a+b)$, and sketch the graph of f.

11. Find the domain and range of the function $f(x) = \frac{x^2 - x - 6}{x + 2}$, and sketch the graph.

12. Describe the domain of the following absolute value inequalities without using absolute value symbols.

    (a) $|2x - 3| \leq 5$ (b) $|x - 2| > 4$ (c) $|2 - 3x| < -1$

13. (a) Convert $-211°$ to radians.

    (b) Convert $\frac{121\pi}{360}$ radians to degrees.

14. Let $f(x) = x^2 + 1$ and $g(x) = (x+1)^2$. Find

    (a) $f(x) - g(x)$ (b) $\frac{f(x)}{g(x)}$ (c) $f\big(g(x)\big)$

    (d) $g\big(f(x)\big)$ (e) $g\big(g(x)\big)$ (f) $g(x^2)$

15. Find two functions, f and g, that will produce the given composite $h(x) = f\big(g(x)\big)$. (There is no unique answer.)

    (a) $h(x) = \sqrt{x^2 - 1}$ (b) $h(t) = \left(t + t^{-1}\right)^5$

16. Use the increment method to find the slope of the curve $y = x^3 - 2x + 5$ at a point (x,y) on the curve.

17. Write an equation of the tangent line to the curve in Problem 16 at the point when $x = -2$.

18. Using the fact that for any constant k, if $f(x) = kx^3$, then the derivative $f'(x) = 3kx^2$ holds, calculate the instantaneous rate of change of the volume of a sphere with respect to its radius when the radius is 3 cm.

19. Using the fact that for constants a, b, c and $f(t) = at^2 + bt + c$, then $f'(t) = 2at + b$, find the instantaneous speed of a particle moving along a straight line according to the equation $s = 5t^2 - 3t$ when $t = 2$ sec.

20. Evaluate the following limits.

    (a) $\lim_{t \to 3} \frac{t^2 - 1}{t - 1}$ (b) $\lim_{x \to 2} \frac{2x^2 - 3x - 2}{x - 2}$

    (c) $\lim_{t \to \infty} \frac{t^2}{4 - t^2}$ (d) $\lim_{x \to 1} \frac{3x - 1}{5x^3 - 2x + 1}$

21. Let f be defined by $f(x) = \begin{cases} 2x-3 & \text{if } x \geq 0 \\ -1 & \text{if } x < 0 \end{cases}$

(a) Find $\lim_{x\to 0^+} f(x)$ and $\lim_{x\to 0^-} f(x)$.

(b) Does $\lim_{x\to 0} f(x)$ exist? Justify your answer.

SOLUTIONS TO CHAPTER 1 SELF-TEST

1.

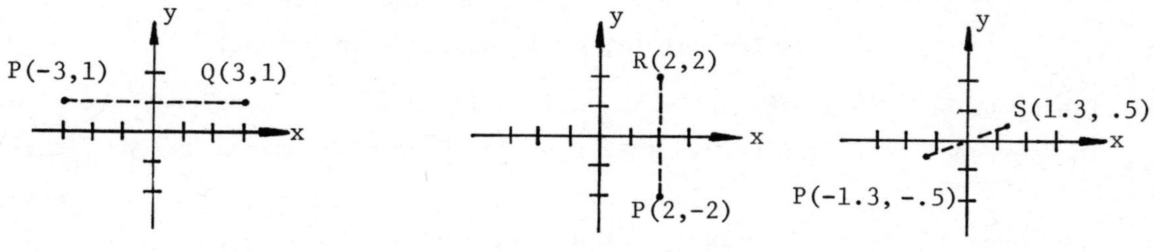

P(-3,1) Q(3,1) y x

(a)

R(2,2) P(2,-2) y x

(b)

S(1.3, .5) P(-1.3, -.5) y x

(c)

2. $\Delta x = 7 - (-3) = 10$, $\Delta y = -3 - (-1) = -2$,

$$d = \sqrt{(\Delta x)^2 + (\Delta y)^2} = \sqrt{104} \approx 10.198$$

3. The new coordinates can be written as $(x,0)$ since the point lies on the x-axis. From $\Delta y = -6\Delta x$, we have $0 - (-3) = -6(x-2)$ or, solving, $x = 3/2$. Thus $\left(\frac{3}{2}, 0\right)$ gives the coordinates of the new position of the particle.

4. $(x-1)^2 + (y+3)^2 = 16$ or $x^2 + y^2 - 2x + 6y - 6 = 0$.

5. The slope of AB is $m_1 = \frac{9-(-3)}{-2-1} = -4$ and the slope of BC is $m_2 = \frac{-19-9}{5-(-2)} = -4$. Since these slopes are equal, the three points do lie on a common straight line.

6. The coordinates of the midpoint are given by

$$\left(\frac{3-4}{2}, \frac{-7-1}{2}\right) \quad \text{or} \quad \left(-\frac{1}{2}, -4\right).$$

7. $m = \frac{2-4}{-3-1} = \frac{1}{2}$ is the slope, and $y - 4 = \frac{1}{2}(x-1)$ or $2y - x = 7$ is an equation of the line.

8. (a) Solving algebraically for y, $y = -\frac{3}{4}x - \frac{1}{4}$. Thus, the slope is $m = -\frac{3}{4}$, the y-intercept is $b = -\frac{1}{4}$; and when $y = 0$, $x = -\frac{1}{3}$ is the x-intercept.

(b) This is a vertical line so it has no slope and no y-intercept. The x-intercept is 2.

(c) This is a horizontal line. It has slope 0 and y-intercept -1. It has no x-intercept.

(d) Since the variable x is squared, this equation does not represent a straight line. When $x = 0$, $y = \frac{1}{2}$ so the y-intercept is $\frac{1}{2}$. If $y = 0$, $x^2 = -1$ which is impossible, so it has no x-intercept.

9. The slope of $2y - x = 8$ or $y = \frac{1}{2}x + 4$ is $m = \frac{1}{2}$. Therefore, the slope of the perpendicular line is $m' = -2$ and an equation is given by $y + 7 = -2(x-5)$ or $y = -2x + 3$.

10. The function $f(x) = x^2 + 3x - 2$ is defined for all values of x, so the domain is $-\infty < x < \infty$ (all real numbers). Setting $y = x^2 + 3x - 2$ or $y + 2 = x^2 + 3x$ and completing the square on the righthand side gives

$$y + 2 + \frac{9}{4} = \left(x + \frac{3}{2}\right)^2 , \quad \text{or} \quad y + \frac{17}{4} = \left(x + \frac{3}{2}\right)^2 .$$

Thus, $y \geq -\frac{17}{4}$ so the range of f is the interval $\left[-\frac{17}{4}, \infty\right)$.

$f(-2) = -4$, $f(-1) = -4$,

$f(0) = -2$, $f(2) = 8$,

$f(2b) = 4b^2 + 6b - 2$,

$f(a+b) = a^2 + 2ab + b^2 + 3(a+b) - 2$

The graph of f is shown at the right.

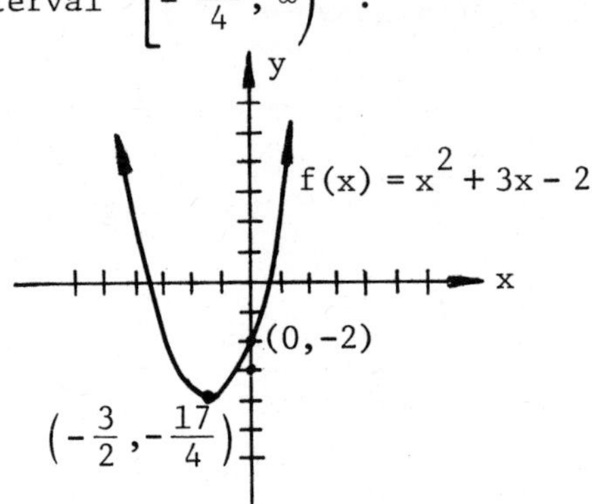

11. $$\frac{x^2 - x - 6}{x+2} = \frac{(x-3)(x+2)}{x+2} .$$

Thus, the domain of f is all real numbers except $x = -2$. Also, for $x \neq -2$, $f(x) = x - 3$. This is a straight line with the point $(-2,-5)$ deleted so the range of f is all real numbers except $y = -5$.

The graph of f is shown at the right.

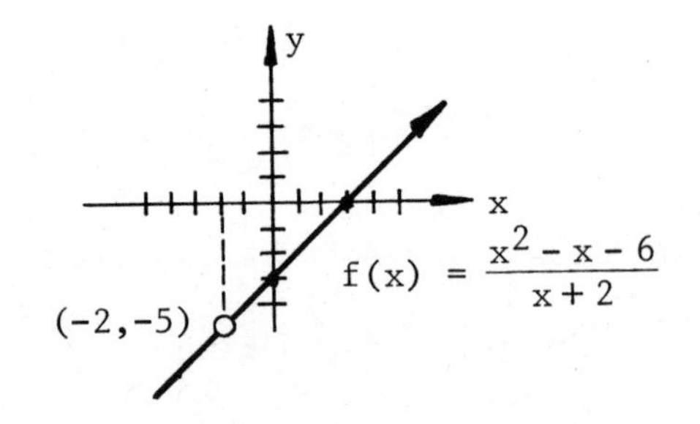

12. (a) $-5 \leq 2x - 3 \leq 5$ so $-1 \leq x \leq 4$

(b) $x - 2 > 4$ or $-(x - 2) > 4$

so $x > 6$ or $-x > 2$;

that is, x lies within the union of the intervals $(-\infty, -2) \cup (6, \infty)$

(c) since for any real number its absolute value is nonnegative, it is impossible for $|2 - 3x|$ to be less than -1 ; thus the domain of x is empty (there is no solution).

13. (a) $-211° = -\frac{211}{180}\pi = -1.1722\bar{2}\pi$ radians

(b) $\frac{121\pi}{360} = \left(\frac{121\pi}{360}\right)\left(\frac{180°}{\pi}\right) = \frac{121°}{2}$ or $60°30'$

14. (a) $f(x) - g(x) = -2x$

(b) $\frac{f(x)}{g(x)} = \frac{x^2 + 1}{(x+1)^2}$

(c) $f\big(g(x)\big) = (x+1)^4 + 1$

(d) $g\big(f(x)\big) = \left(x^2 + 2\right)^2$

(e) $g\big(g(x)\big) = \left(x^2 + 2x + 2\right)^2$

(f) $g(x^2) = \left(x^2 + 1\right)^2$

15. (a) $f(x) = \sqrt{x}$ and $g(x) = x^2 - 1$

(b) $f(t) = t^5$ and $g(t) = t + t^{-1}$

16. STEP 1. $y + \Delta y = (x + \Delta x)^3 - 2(x + \Delta x) + 5$

$= x^3 + 3x^2\Delta x + 3x(\Delta x)^2 + (\Delta x)^3 - 2x - 2\Delta x + 5$

STEP 2. Subtracting $(y + \Delta y) - y$:

$$\Delta y = 3x^2\Delta x + 3x(\Delta x)^2 + (\Delta x)^3 - 2\Delta x$$

STEP 3. Dividing by Δx yields,

$$m_{sec} = \frac{\Delta y}{\Delta x} = 3x^2 + 3x\Delta x + (\Delta x)^2 - 2$$

STEP 4. As Δx tends to zero

$$m_{tan} = \lim_{\Delta x \to 0} \frac{\Delta y}{\Delta x} = 3x^2 - 2 .$$

17. When $x = -2$, $y = (-2)^3 - 2(-2) + 5 = 1$, and from the preceding problem solution, $m = 3(-2)^2 - 2 = 10$. Thus , $(y - 1) = 10(x + 2)$ or $y = 10x + 21$ is an equation of the tangent line.

18. The volume of a sphere is given by $V = \frac{4}{3}\pi r^3$, where r is the radius. We seek the value of $\frac{dV}{dr}$ when $r = 3$. Thus, from the derivative formula, $\frac{dV}{dr} = V'(r) = 4\pi r^2$ so that $V'(3) = 36\pi$.

19. We seek the velocity $v = \frac{ds}{dt}$ when $t = 2$. From the given derivative formula,

$v = \frac{ds}{dt} = 10t - 3$ so that $v(2) = 17$ units/sec.

20. (a) $\lim_{t \to 3} \frac{t^2 - 1}{t - 1} = \frac{9 - 1}{3 - 1} = 4$

(b) $\lim_{x \to 2} \frac{2x^2 - 3x - 2}{x - 2} = \lim_{x \to 2} \frac{(2x+1)(x-2)}{x - 2} = \lim_{x \to 2} 2x + 1 = 5$

(c) $\lim_{t \to \infty} \frac{t^2}{4 - t^2} = \lim_{h \to 0} \frac{\frac{1}{h^2}}{4 - \frac{1}{h^2}} = \lim_{h \to 0} \frac{1}{4h^2 - 1} = \frac{1}{0 - 1} = -1$

(d) $\lim_{x \to 1} \frac{3x - 1}{5x^3 - 2x + 1} = \frac{3 - 1}{5 - 2 + 1} = \frac{1}{2}$

21. (a) From the graph of f shown at the right,

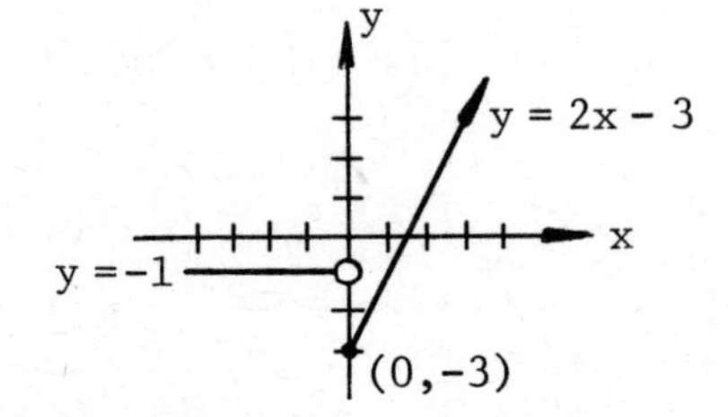

$\lim_{x \to 0^+} f(x) = \lim_{x \to 0^+} (2x - 3) = -3$

$\lim_{x \to 0^-} f(x) = \lim_{x \to 0^-} (-1) = -1$

(b) No, $\lim_{x \to 0} f(x)$ does not exist because the lefthand and righthand limits differ as x tends to zero.

CHAPTER 2 FORMAL DIFFERENTIATION

2-1 Introduction.

1. Consider the absolute value function $f(x) = |x|$. Form the difference quotient

$$\frac{\Delta y}{\Delta x} = \frac{f(0+h) - f(0)}{h} = \underline{\qquad} .$$

Complete the following table showing values of $\Delta y/\Delta x$ for selected values of h :

h	.1	.01	.001	.0001	.00001	-.1	-.01	-.001	-.0001	-.00001
Δy/Δx										

2. It is clear from the table that if $h > 0$, then $\frac{\Delta y}{\Delta x} = \underline{\quad}$; and if $h < 0$, then $\frac{\Delta y}{\Delta x} = \underline{\quad}$. Therefore, the limit $\lim_{h \to 0} \frac{\Delta y}{\Delta x}$ ________ exist at $x = 0$. Thus, though one may be able to calculate (say, on an electronic calculator) successive values of the difference quotient for smaller and smaller values of h , one cannot conclude that the limit exists as $h \to 0$.

2-2 Polynomial Functions and Their Derivatives.

3. If c is a constant and $y = f(x) = c$, then $\frac{dy}{dx} = \underline{\quad}$.

4. If n is any positive integer and $y = f(x) = x^n$, then $\frac{dy}{dx} = \underline{\qquad}$.

5. If $u = f(x)$ is a differentiable function of x , and if $y = cu$ where c is a constant, then

$$\frac{dy}{dx} = \underline{\qquad\qquad} .$$

6. If $u = f(x)$ and $v = g(x)$ are differentiable functions of x , then $y = u + v$ is a __________ function of x , and

$$\frac{dy}{dx} = \underline{\qquad} = \underline{\qquad\qquad} .$$

---

1. $\frac{|h|}{h}$,

h	.1	.01	.001	.0001	.00001	-.1	-.01	-.001	-.0001	-.00001
Δy/Δx	1	1	1	1	1	-1	-1	-1	-1	-1

2. 1 , -1 , does not

3. 0

4. nx^{n-1}

5. $c \frac{du}{dx} = cf'(x)$

6. differentiable , $\frac{du}{dx} + \frac{dv}{dx} = f'(x) + g'(x)$

OBJECTIVE A: Calculate the derivative of any polynomial function.

7. $\frac{d}{dx}\left(3x^2 - 12x + 1\right) = \underline{\qquad\qquad}$.

8. $\frac{d}{dx}\left(\sqrt{3}\,x^4 - \frac{2}{5}x^3 + \frac{1}{3}x^2 - 15x + 109\right) = \underline{\qquad\qquad\qquad\qquad}$.

OBJECTIVE B: Calculate second and higher order derivatives of any polynomial function.

9. If $y = f(x)$, the second derivative of y with respect to x is the derivative of __________ . The second derivative is denoted by ______ or _____ or _______ .

10. In general, the nth derivative of $y = f(x)$ with respect to x is the derivative of ________ , and is denoted by _____ or _____ or ________ .

11. $\frac{d^2}{dx^2}\left(4x^5 - 3x^2 + 2x - 20\right) = \frac{d}{dx}\left(\underline{\qquad\qquad\qquad}\right) = \underline{\qquad\qquad}$.

12. $\frac{d^2}{dx^2}(2x^2 - 1)(x - 3) = \frac{d^2}{dx^2}\left(\underline{\qquad\qquad\qquad}\right) = \frac{d}{dx}\left(\underline{\qquad\qquad\qquad}\right)$

$= \underline{\qquad\qquad\qquad}$.

OBJECTIVE C: If $s = f(t)$ gives the position of a moving body as a function of time t , find and interpret the velocity and acceleration at a specified instant.

13. Suppose a particle is moving along a straight line, negative to the left and positive to the right, according to the law

$$s = t^3 - 3t^2 - 9t + 5 .$$

Then the velocity is given by $\frac{ds}{dt} = \underline{\qquad\qquad} = \underline{\qquad\qquad\qquad}$.
Thus, the velocity is positive when _______ or _______ and the particle is moving to the ________ ; the velocity is __________ when $-1 < t < 3$ so the particle is moving to the ________ .

---

7. $6x - 12$

8. $4\sqrt{3}\,x^3 - \frac{6}{5}x^2 + \frac{2}{3}x - 15$

9. $\frac{dy}{dx} = f'(x)$, $\frac{d^2y}{dx^2}$, y'' , or $f''(x)$

10. $\frac{d^{n-1}y}{dx^{n-1}}$, $\frac{d^ny}{dx^n}$, $y^{(n)}$, or $f^{(n)}(x)$

11. $20x^4 - 6x + 2$, $80x^3 - 6$

12. $2x^3 - 6x^2 - x + 3$, $6x^2 - 12x - 1$, $12x - 12$

13. $3t^2 - 6t - 9 = 3(t - 3)(t + 1)$, $t < -1$ or $t > 3$, right, negative, left

14. The acceleration of the particle is $\frac{d^2s}{dt^2}$ = __________ . When the velocity is zero, t = ______ or ______ and the acceleration has the value ______ or ______ , respectively.

15. Suppose a ball is thrown directly upward with a speed of 96 ft/sec and moves according to the law $y = 96t - 16t^2$, where y is the height in feet above the starting point, and t is the time in seconds after it is thrown. The velocity of the ball at any time t is $v(t) = \frac{dy}{dt}$ = __________ . Hence when t = 2 sec , the velocity of the ball is __________ . Since v(2) is positive the ball is still rising. At its highest point the velocity of the ball is ________ and this occurs when t = _____ seconds. The height corresponding to this time is y = ______ feet and this is the highest point reached. Notice the acceleration is a constant _____ ft/sec^2.

OBJECTIVE D: Find an equation of the tangent line to a curve y = f(x) meeting some specified requirement (such as a condition on the slope).

16. Consider the curve $y = x^3 - 9x^2 + 15x - 5$. The derivative y' gives the value of the ________ of the tangent line at any x . For this particular curve, y' = ________________ = 3(______)(x - 5) . Thus, the tangent line is parallel to the x-axis when x = _____ or x = _____ .

17. When the slope of the tangent line to the above curve equals 15 the value of x is _____ or _____ . The corresponding y values are _____ and _____ , respectively. Equations of the two tangent lines are then given by ________________ and __________________ .

2-3 Rational Functions and Their Derivatives.

OBJECTIVE A: Find the derivative of a product of polynomial functions.

18. If u = f(x) and v = g(x) are differentiable functions of x , then the derivative $\frac{d}{dx}(uv)$ = ______________ .

---

14. $6t - 6$, -1 or 3 , -12 or 12

15. $96 - 32t$, 32 ft/sec, zero, 3, y = 144, -32

16. slope, $3x^2 - 18x + 15$, x - 1, 1, 5

17. 0, 6, -5, -23, $y + 5 = 15x$ and $y + 23 = 15(x - 6)$

18. $u\frac{dv}{dx} + v\frac{du}{dx} = f(x)g'(x) + g(x)f'(x)$

19. If $y = (x^2 - 2)(2x^3 - 5)$, then

$$y' = (x^2 - 2)\frac{d}{dx}(2x^3 - 5) + (2x^3 - 5)\ \_\_\_\_\_\_\_\_\_\_$$

$$= (x^2 - 2)(\ \_\_\_\_\_\_\ ) + (2x^3 - 5)(2x)$$

$$= \_\_\_\_\_\_\_\_\_\_ + 4x^4 - 10x = \_\_\_\_\_\_\_\_\_\_ .$$

20. $\frac{d}{dx}\left[(3x^2 - 2x + 1)(5x - 4)\right]$

$$= (3x^2 - 2x + 1)(\ \_\_\_\_\_\_\ ) + (5x - 4)(\ \_\_\_\_\_\_\ )$$

$$= (15x^2 - 10x + 5) + (\ \_\_\_\_\_\_\_\_\ ) = \_\_\_\_\_\_\_\_\_\_ .$$

OBJECTIVE B: Find the derivative of a quotient of polynomial functions.

21. A __________ function is a quotient of two polynomials, and is generally not itself a polynomial. Every polynomial may be considered to be a rational function with denominator the constant polynomial 1 .

22. If $u = f(x)$ and $v = g(x)$ are differentiable functions of x , then the derivative $\frac{d}{dx}\left(\frac{u}{v}\right) = \_\_\_\_\_\_\_\_\_\_$ when $v \neq 0$.

23. If $y = \frac{3x}{5x^2 - 1}$, then

$$y' = \frac{(5x^2 - 1)(\ \_\_\_\_\ ) - (3x)(\ \_\_\_\_\_\_\ )}{\left(5x^2 - 1\right)^2} = \frac{\_\_\_\_\_\_}{\left(5x^2 - 1\right)^2} .$$

24. $\frac{d}{dx}\left(\frac{1}{x}\right) = \frac{x(\ \_\_\_\_\ ) - 1(\ \_\_\_\_\ )}{x^2} = \_\_\_\_\_\_\_\_ .$

---

19. $\frac{d}{dx}(x^2 - 2)$, $6x^2$, $6x^4 - 12x^2$, $10x^4 - 12x^2 - 10x$

20. 5, $6x - 2$, $30x^2 - 34x + 8$, $45x^2 - 44x + 13$

21. rational

22. $\frac{v\frac{du}{dx} - u\frac{dv}{dx}}{v^2} = \frac{g(x)f'(x) - f(x)g'(x)}{\left[g(x)\right]^2}$

23. 3, $10x$, $-15x^2 - 3$

24. 0, 1, $-\frac{1}{x^2}$

25. $$\frac{d}{dt}\left(\frac{t^2 - 2t + 5}{t^3 + 1}\right) = \frac{(t^3 + 1)(\_\_\_\_\_) - (\_\_\_\_\_\_\_\_)(3t^2)}{\left(t^3 + 1\right)^2}$$

$$= \frac{2(\_\_\_\_\_\_\_\_) - 3t^4 + 6t^3 - 15t^2}{\left(t^3 + 1\right)^2}$$

$$= \frac{(\_\_\_\_\_\_\_\_\_\_\_\_)}{\left(t^3 + 1\right)^2} .$$

OBJECTIVE C: Find the derivative of an integral power of a rational function.

26. If $u = g(x)$ is a differentiable function of x and n is a positive integer, then the derivative

$$\frac{d}{dx}(u^n) = \_\_\_\_\_\_\_\_ .$$

27. The above power rule holds when n is a negative integer at all points x where $g(x)$ is ________ .

28. If $y = \left(5x^3 - x^2 + 7\right)^4$, then

$$y' = 4\left(5x^3 - x^2 + 7\right)^3 \frac{d}{dx}(\_\_\_\_\_\_\_\_) = \_\_\_\_\_\_\_\_\_\_\_\_ .$$

29. $$\frac{d}{dx}\left[(2x - 1)^{-3}\right] = -3(\_\_\_\_\_\_)^{-4} \frac{d}{dx}(\_\_\_\_\_\_\_\_)$$

$$= \_\_\_\_\_\_\_\_ , \text{ if } x \neq \_\_\_\_ .$$

30. Let $y = \frac{2}{(x - 1)^3} + \frac{3}{1 - x^4}$; then $y = 2(x - 1)^{-3} + 3\left(1 - x^4\right)^{-1}$, so that

$$\frac{dy}{dx} = -6(\_\_\_\_\_\_)\frac{d}{dx}(x - 1) - 3\left(1 - x^4\right)^{-2}\frac{d}{dx}(\_\_\_\_\_) = \_\_\_\_\_\_\_\_\_\_ .$$

---

25. $2t - 2$, $t^2 - 2t + 5$, $t^4 - t^3 + t - 1$, $-t^4 + 4t^3 - 15t^2 + 2t - 2$

26. $nu^{n-1}\frac{du}{dx}$

27. not zero

28. $5x^3 - x^2 + 7$, $4\left(5x^3 - x^2 + 7\right)^3(15x^2 - 2x)$

29. $2x - 1$, $2x - 1$, $-6(2x - 1)^{-4}$, $\frac{1}{2}$

30. $(x - 1)^{-4}$, $1 - x^4$, $-6(x - 1)^{-4} + 12x^3\left(1 - x^4\right)^{-2}$

31. $\frac{d}{dx}\left[\left(\frac{x-2}{x+1}\right)^5\right] = 5(\_\_\_\_\_)^4 \frac{d}{dx}(\_\_\_\_\_)$

$= 5\left(\frac{x-2}{x+1}\right)^4 \frac{(x+1)(1) - (\_\_\_\_\_)(1)}{(x+1)^2} = \frac{\_\_\_\_\_}{(x+1)^6}$.

2-4 Inverse Functions and Their Derivatives.

OBJECTIVE A: Find the derivative of $g(x) = x^n$ when n is any rational number $n = p/q$.

32. If $g(x) = x^n$ with $n = 1/m$, where m is a positive odd integer, then $g'(x) = \_\_\_\_\_$ for x satisfying $\_\_\_\_\_$.

33. If $g(x) = x^n$ with $n = 1/m$, where m is a positive even integer, then $g'(x) = \_\_\_\_\_$ for x satisfying $\_\_\_\_\_$.

34. $\frac{d}{dx}\left(x^{1/9}\right) = \_\_\_\_\_$ provided x satisfies $\_\_\_\_\_$.

35. $\frac{d}{dx}\left(x^{1/6}\right) = \_\_\_\_\_$ provided x satisfies $\_\_\_\_\_$.

36. $\frac{d}{dx}\left(x^{3/5}\right) = \_\_\_\_\_$ provided x satisfies $\_\_\_\_\_$.

37. $\frac{d}{dx}\left(x^{-3/4}\right) = \_\_\_\_\_$ provided x satisfies $\_\_\_\_\_$.

38. Let $y = \sqrt{\frac{x+3}{x-3}}$, so $y = u^{1/2}$ where $u = \frac{x+3}{x-3}$. Then

$\frac{dy}{dx} = \frac{1}{2}u^{-1/2}\frac{du}{dx}$ whenever $u > 0$. Thus

$\frac{dy}{dx} = \frac{1}{2}\left(\_\_\_\_\_\right)^{-1/2} \frac{(x-3)(1) - (\_\_\_\_\_)}{(x-3)^2}$

$= \frac{1}{2\sqrt{\frac{x+3}{x-3}}}\ \_\_\_\_\_$ whenever $\frac{x+3}{x-3} > 0$.

---

31. $\frac{x-2}{x+1}$, $\frac{x-2}{x+1}$, $x-2$, $15(x-2)^4$

32. nx^{n-1}, $x \neq 0$

33. nx^{n-1}, $x > 0$

34. $\frac{1}{9}x^{-8/9}$, $x \neq 0$

35. $\frac{1}{6}x^{-5/6}$, $x > 0$

36. $\frac{3}{5}x^{-2/5}$, $x \neq 0$

37. $-\frac{3}{4}x^{-7/4}$, $x > 0$

38. $\frac{x+3}{x-3}$, $(x+3)(1)$, $\frac{-6}{(x-3)^2}$

OBJECTIVE B: Use the rule (Rule 10) to calculate the derivative of the inverse for a specified function.

39. If f and g are inverse functions on suitably restricted domains, then $g(f(x))$ = _____ and $f(g(y))$ = _____ . That is, the composite of g and f or of f and g is the __________ mapping.

40. Given a function $y = f(x)$, to find a formula for the inverse function g , interchange the letters _____ and _____ in the original equation and solve for _____ .

41. If f and g are inverse functions on suitably restricted domains, then $g'(a) = 1/f'(b)$ where a and b are related by ________ and ________ .

42. Let $f(x) = -6x + 2$ and let g denote the inverse of f . We wish to calculate the derivative $g'(14)$. First, $-6x + 2 = 14$ implies $x =$ _____ . Thus, $f(-2) = 14$ so $b =$ _____ and $a =$ _____ in Problem 41. Then,

$f'(x) =$ _____ so that $g'(14) = \dfrac{1}{f'(\_\_\_\_)} =$ _____ .

43. To calculate the inverse of $y = -6x + 2$, interchange the letters x and y obtaining __________ . Solving the resultant equation for y yields ____________ , or $g(x) =$ ___________ is the inverse function of $f(x) = -6x + 2$. Calculating the derivative $g'(14)$ directly from the formula for $g(x)$ gives $-1/6$ as before.

Remark. The advantage of the derivative formula for the inverse function given in Problem 41 (or Rule 10 of the text) is that it provides for the calculation of the derivative $g'(a)$ even though a formula for the inverse function g is not known.

44. Let g be the inverse of $f(x) = x^2 + 4x - 3$ for $x > -2$. To find $g'(-6)$, first set $f(x) = x^2 + 4x - 3$ equal to _____ and solve the quadratic equation yielding $x = -3$ or $x =$ _____ . We reject $x = -3$ because -3 is outside the allowable interval $x > -2$. Thus, $b =$ _____ and $a = f(b) =$ _____ . $\frac{d}{dx}(x^2 + 4x - 3) =$ __________ so $f'(-1) =$ _____ . Thus,

$g'(-6) = \dfrac{1}{f'(\_\_\_\_)} =$ _____ . Notice that we did not need a formula for the inverse function g itself.

---

39. x, y, identity

40. x, y, y

41. $b = g(a)$, $a = f(b)$

42. -2, $b = -2$ and $a = 14$, -6, -2, $-\frac{1}{6}$

43. $x = -6y + 2$, $y = -\frac{1}{6}(x - 2)$, $g(x) = -\frac{1}{6}(x - 2)$

44. -6, -1, -1, -6, $2x + 4$, 2, -1, $\frac{1}{2}$

2-5 Implicit Relations and Their Derivatives.

OBJECTIVE: Compute first and second derivatives by the technique of implicit differentiation.

45. An equation involving the variables x and y is said to determine y ____________ as a function of x, say $y = f(x)$, provided that f satisfies the equation.

46. For instance, consider the equation $x^2 + y^2 = 2$. If we substitute $y = \sqrt{2-x^2}$ into the equation we obtain

$$x^2 + \left(\sqrt{2-x^2}\right)^2 = x^2 + (\_\_\_\_\_\_\_) = \_\_\_\_\_ ,$$

so the equation is satisfied. Similarly, if we substitute $y = -\sqrt{2-x^2}$ into the equation we obtain

$$x^2 + \left(-\sqrt{2-x^2}\right)^2 = x^2 + (\_\_\_\_\_\_\_) = \_\_\_\_\_ ,$$

and the equation is again satisfied. Therefore each of the two functions $y = \sqrt{2-x^2}$ and $y = -\sqrt{2-x^2}$ is defined ____________ by the equation $x^2 + y^2 = 2$.

47. To calculate the derivative dy/dx for $x^2 + y^2 = 2$, differentiate both sides of the equation with respect to x and solve for dy/dx:

$$\frac{d}{dx}(x^2 + y^2) = \frac{d}{dx}(2) \quad \text{or} \quad \frac{d}{dx}(\_\_\_\_) + \frac{d}{dx}(y^2) = \frac{d}{dx}(2) .$$

Thus, $2x + (\_\_\_\_) = 0$ and solving for dy/dx, $\frac{dy}{dx} = \_\_\_\_$.
This derivative is valid whenever y satisfies the condition ______.

48. To calculate d^2y/dx^2 for $x^2 + y^2 = 2$, differentiate both sides of the derivative equation $dy/dx = -x/y$ with respect to x:

$$\frac{d}{dx}\left(\frac{dy}{dx}\right) = \frac{d}{dx}\left(-\frac{x}{y}\right) = -\frac{d}{dx}\left(\frac{x}{y}\right), \quad \text{or}$$

$$\frac{d^2y}{dx^2} = -\left[\frac{(\_\_\_)\frac{dx}{dx} - x(\_\_\_)}{y^2}\right] .$$

Substitution of $-x/y$ for dy/dx in the last equation gives

$$\frac{d^2y}{dx^2} = -\left[\frac{y - x(\_\_\_)}{y^2}\right] = -\frac{\_\_\_\_\_\_}{y^3} = -\frac{\_\_\_}{y^3} .$$

---

45. implicitly

46. $2-x^2$, 2, $2-x^2$, 2, implicitly

47. x^2, $2y\frac{dy}{dx}$, $-\frac{x}{y}$, $y \neq 0$

48. y, $\frac{dy}{dx}$, $-\frac{x}{y}$, x^2+y^2, 2 (because $x^2+y^2 = 2$)

2-6 The Increment of a Function.

OBJECTIVE: Estimate the change Δy produced in a function $y = f(x)$ when x changes by a small amount Δx.

49. An estimate of Δy is given by Δy_{tan} = ________ when x changes by a small amount Δx; this represents the change in y along the __________ line. It is also called the ____________ part of Δy.

50. Suppose we wish to estimate the change in $y = x^3$ when x changes by $\Delta x = 0.1$ at $x = 2$. Now $\Delta y \approx \Delta y_{tan} = \frac{dy}{dx}\Delta x$ = ______ Δx. When $x = 2$ and $\Delta x = 0.1$, $\Delta y_{tan} = 3($ ____ $)^2($ ____ $) = 1.2$. Therefore, since $f(x+\Delta x) = y + \Delta y \approx y + \Delta y_{tan}$, $(2+0.1)^3 \approx 2^3 +$ _____ = _____ .

    The actual value of $(2.1)^3$ is ______ giving an error in our estimate of $\varepsilon\Delta x = \Delta y - \Delta y_{tan}$ = ______ . The positive sign of $\varepsilon\Delta x$ indicates that our estimate 9.2 is too small.

51. To estimate the value of $\sqrt{16.56}$, let $y = \sqrt{x}$, $x = 16$, and $\Delta x = 0.56$. Then $\Delta y_{tan} = ($ _______ $)\,\Delta x$, so when $x = 16$ and $\Delta x = 0.56$, $\Delta y_{tan} = ($ _____ $)(0.56) = .07$. Thus, $\sqrt{16.56} = \sqrt{\_\_\_\_} + .07$ = _______ . (The actual value of $\sqrt{16.56}$ is 4.0694 correct to 5 decimal places, so our estimate is fairly accurate.)

2-7 Newton's Method for Solving Equations.

OBJECTIVE: Use the Newton-Raphson method to estimate the root of an equation $f(x) = 0$ within specified $a \leq x \leq b$.

52. In using the Newton-Raphson method, to go from the nth approximation x_n of the root to the next approximation x_{n+1}, use the formula

    $x_{n+1} = \dfrac{\qquad}{\_\_\_\_\_\_\_\_\_\_}$. This formula fails if the derivative $f'(x_n)$ equals ______ .

---

49. $\frac{dy}{dx}\Delta x$, tangent, principal

50. $3x^2$, 2, 0.1, 1.2, 9.2, 9.261, 0.061

51. $1/2\sqrt{x}$, $\frac{1}{8}$, 16, 4.07

52. $x_n - \dfrac{f(x_n)}{f'(x_n)}$, 0

53. Suppose it is required to find a real root to the equation $f(x) = x^3 + x - 1$. Since $f(0) =$ ____ and $f(1) =$ ____ differ in sign, an unknown root lies somewhere in the interval $0 < x < 1$. As a first guess, choose $x_1 = 0.5$.

To apply the Newton-Raphson formula, we calculate the derivative $f'(x) =$ ________. Then,

$$x_2 = x_1 - \frac{x_1^3 + x_1 - 1}{\rule{2em}{0.4pt}} = \frac{1}{2} - \frac{(1/2)^3 + (1/2) - 1}{\rule{3em}{0.4pt}} = \frac{1}{2} + \frac{\rule{1em}{0.4pt}}{14}$$

$$= \frac{\rule{1em}{0.4pt}}{7} \approx 0.71429 .$$

$$x_3 = x_2 - \frac{\rule{4em}{0.4pt}}{3x_2^2 + 1} = \frac{5}{7} - \frac{\rule{5em}{0.4pt}}{3(5/7)^2 + 1}$$

$$= \frac{5}{7} - \frac{\rule{2em}{0.4pt}}{7 \cdot 124} = \frac{\rule{2em}{0.4pt}}{7 \cdot 124} \approx 0.68318 .$$

With the aid of a hand-held programmable calculator we have computed the following iterations in the same way: $x_4 = 0.68233$ and $x_5 = 0.68233$. Thus, a root to $f(x) = x^3 + x - 1$ is $r = 0.68233$ correct to 5 decimal places. The method is easy, but the arithmetic can be cumbersome without the aid of a calculator.

2-8 Composite Functions and Their Derivatives: The Chain Rule.

OBJECTIVE A: If y is a differentiable function of u and u is a differentiable function of x, use the chain rule to calculate dy/dx.

54. If y is a differentiable function of u and u is a differentiable function of x, then y is a differentiable function of ____, and $\frac{dy}{dx} =$ ____________. This rule is known as the ________ rule for the derivative of a composite function.

55. Consider the chain rule in functional form: let $y = f(u)$ and $u = g(x)$ be differentiable functions. Then the composite $y = F(x) = f(g(x))$ is a differentiable function of ____. When $x = x_o$, let $u = g(x_o) = u_o$. According to the chain rule, the derivative of F evaluated at $x = x_o$ is given by $F'(x_o) =$ ______________. In this equation, $F'(x_o)$

---

53. -1, 1, $3x^2 + 1$, $3x_1^2 + 1$, $3(1/2)^2 + 1$, 3, 5, $x_2^3 + x_2 - 1$, $(5/7)^3 + (5/7) - 1$, 27, 593

54. x, $\frac{dy}{du} \cdot \frac{du}{dx}$, chain

55. x, $f'(u_o) \cdot g'(x_o)$, $(dy/du)_{u_o}$, $g'(x_o)$, x_o, u_o

corresponds to $(dy/dx)_{x_0}$, $f'(u_0)$ corresponds to __________ , and __________ corresponds to $(du/dx)_{x_0}$. It is important that you observe that the derivatives in the chain rule equation $\frac{dy}{dx} = \frac{dy}{du} \cdot \frac{du}{dx}$ are being evaluated at different points: dy/dx and du/dx are evaluated at ____ , whereas dy/du is evaluated at $g(x_0) =$ ____ . Failure to understand this fact can lead to serious misuse of the chain rule equation.

56. To find dy/dx if $y = u^2 - 2u + 3$ and $u = \sqrt{x}$, calculate $dy/du =$ _______ and then substitute $u = \sqrt{x}$ to obtain $(dy/du)_{\sqrt{x}} =$ __________ . According to the chain rule,

$$\frac{dy}{dx} = \frac{dy}{du} \cdot \frac{du}{dx} = \left(2\sqrt{x} - 2\right) \cdot \_\_\_\_\_\_ = \_\_\_\_\_\_\_\_ .$$

57. Suppose $y = z^{-2} + 3z^{-1}$ and $z = x^2 + 1$. Then $dy/dz =$ __________ so that when $z = x^2 + 1$, $(dy/dz)_{x^2+1} =$ ____________________ . Applying the chain rule,

$$\frac{dy}{dx} = \left[-2\left(x^2+1\right)^{-3} - 3\left(x^2+1\right)^{-2}\right] \cdot \_\_\_\_$$

$$= \frac{-2x}{\left(x^2+1\right)^2}\left[\frac{2}{x^2+1} + \_\_\_\right] = \frac{\_\_\_\_\_\_\_\_}{\left(x^2+1\right)^3} .$$

OBJECTIVE B: Given parametric equations $x = f(t)$ and $y = g(t)$, eliminate the parameter t to find an equation in the form $y = F(x)$. Find dy/dx in terms of dy/dt and dx/dt .

58. The equations $x = f(t)$ and $y = g(t)$, which express x and y in terms of t , are called ____________ equations. The variable t is called a ____________ . From the chain rule, the derivative dy/dx is given by $\frac{dy}{dx} =$ ________ .

59. Consider the curve given by the parametric equations $x = t^2$ and $y = t^2 - 2t$. First complete the following table of values for the coordinates of the path:

---

56. $2u - 2$, $2\sqrt{x} - 2$, $\frac{1}{2\sqrt{x}}$, $1 - \frac{1}{\sqrt{x}}$

57. $-2z^{-3} - 3z^{-2}$, $-2\left(x^2+1\right)^{-3} - 3\left(x^2+1\right)^{-2}$, $2x$, 3, $-2x(3x^2 + 5)$

58. parametric, parameter, $\frac{dy/dt}{dx/dt}$

t	-2	-1	0	1	2	3
x						
y						

Plot a graph of the curve in the coordinate system at the right.

Differentiating, $\frac{dx}{dt}$ = ______ and

$\frac{dy}{dt}$ = ________ . Thus $\frac{dy}{dx}$ = ________ .

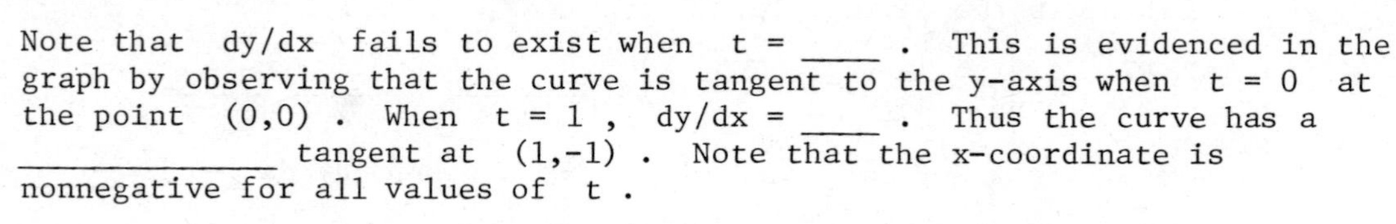

Note that dy/dx fails to exist when t = ____ . This is evidenced in the graph by observing that the curve is tangent to the y-axis when t = 0 at the point (0,0) . When t = 1 , dy/dx = ____ . Thus the curve has a ____________ tangent at (1,-1) . Note that the x-coordinate is nonnegative for all values of t .

60. To eliminate the parameter t in Problem 59, observe that $t = \sqrt{x}$ if $t \geq 0$, and t = ______ if $t < 0$. Thus

$$y = \begin{cases} \_\_\_\_\_\_\_\_\_, & \text{if } t \geq 0 \\ \_\_\_\_\_\_\_\_\_, & \text{if } t < 0 \end{cases}$$

2-9 Brief Review of Trigonometry.

This section of the textbook is intended to give a quick review of some of the basic properties of the trigonometric functions. If you find that a more complete and detailed study is necessary, see Chapters 7-9 in the semi-programmed text ALGEBRA AND TRIGONOMETRY: A FUNCTIONS APPROACH by Keedy and Bittinger, Addison-Wesley Publishing Company, 1978 edition.

Problems 61 - 67 give the most important trigonometric formulas to remember.

61. sin (A + B) = ______________________ . 63. sin (-A) = __________ .

62. cos (A + B) = ______________________ . 64. cos (-A) = __________ .

59.

t	-2	-1	0	1	2	3
x	4	1	0	1	4	9
y	8	3	0	-1	0	3

$\frac{dx}{dt} = 2t, \quad \frac{dy}{dt} = 2(t-1), \quad \frac{dy}{dx} = 1 - \frac{1}{t}$,

0, 0, horizontal

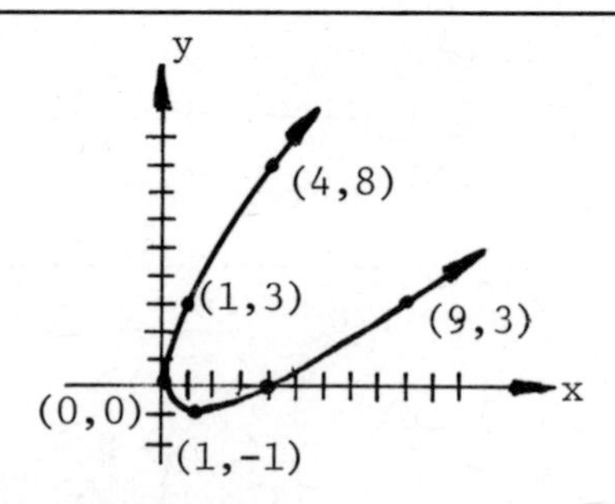

60. $-\sqrt{x}$, $\begin{cases} x - 2\sqrt{x} \\ x + 2\sqrt{x} \end{cases}$

61. sin A cos B + cos A sin B

62. cos A cos B - sin A sin B

63. - sin A

64. cos A

65. Referring to the triangle ABC at the right, the law of cosines states that

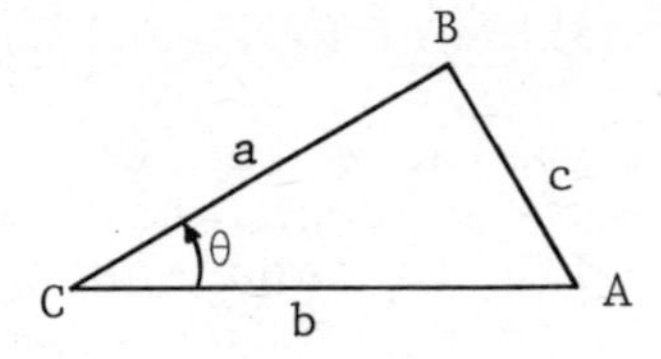

c^2 = 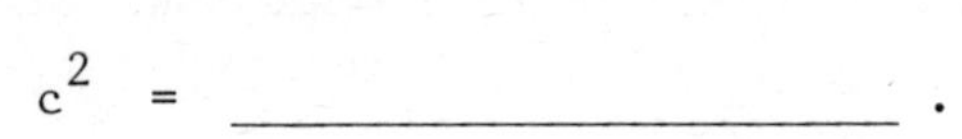 .

66. $\sin^2 \theta + \cos^2 \theta$ = _____ .

67. Dividing both sides of the equation in Problem 66 by $\cos^2 \theta$ gives

$\tan^2 \theta + 1$ = __________ .

OBJECTIVE: Given a general sine function of the form

$$f(x) = A \sin \left[\frac{2\pi}{B}(x - C)\right] + D ,$$

identify the amplitude, period, horizontal shift, and vertical shift. Sketch the graph of the function.

68. Consider the function $f(x) = \frac{1}{2} \sin (3x - 2) + \frac{1}{2}$. Here the amplitude A = _____ . To find the period, we need to write $(3x - 2)$ in the form $\frac{2\pi}{B}(x - C)$, and we proceed as follows:

$$3x - 2 = 3 (\_\_\_\_\_\_\_\_) = \frac{1}{\_\_\_}\left(x - \frac{2}{3}\right) = \frac{2\pi}{\_\_\_}\left(x - \frac{2}{3}\right) .$$

Therefore, the period B = _____ , the horizontal shift C = _____ , and the vertical shift D = _____ . Observe that when $x = C$, the sine of $\frac{2\pi}{B}(x - C)$ is zero. Thus, when C is _____ the graph of f is shifted to the right and when C is negative it is shifted to the _______ . Sketch one cycle of the graph of f in the coordinate system below:

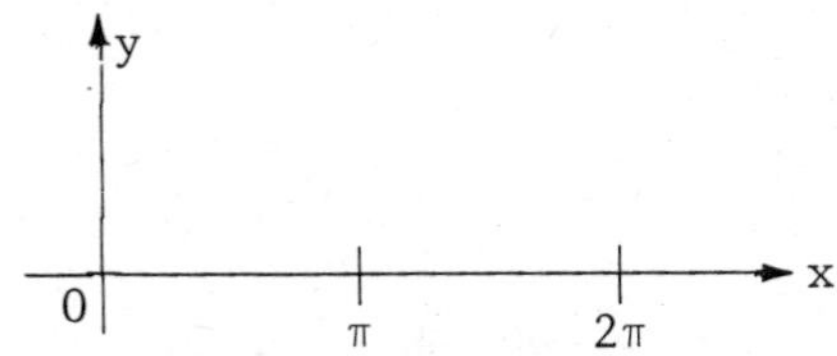

---

65. $a^2 + b^2 - 2ab \cos \theta$

66. 1

67. $\sec^2 \theta$

68. $\frac{1}{2}$, $x - \frac{2}{3}$, $\frac{1}{3}$, $\frac{2\pi}{3}$, $\frac{2\pi}{3}$, $\frac{2}{3}$,

$\frac{1}{2}$, positive, left

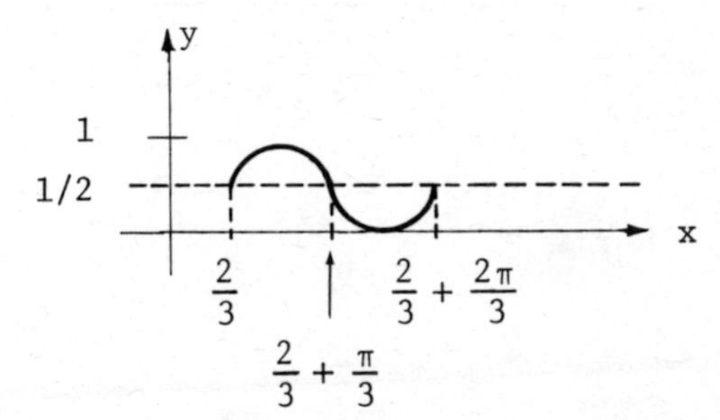

2-10 Differentiation of Sines and Cosines.

OBJECTIVE A: Evaluate limits of trigonometric functions by making use of appropriate trigonometric identities, theorems on limits, and the fact that

$$\lim_{\theta \to 0} \frac{\sin \theta}{\theta} = 1$$

where θ is measured in radians.

69. Consider the limit, $\lim_{x \to 0} \frac{\sin 5x}{2x}$. First, let $u = 5x$. Then $x \to 0$ is equivalent to $u \to 0$, so upon substitution,

$$\lim_{x \to 0} \frac{\sin 5x}{2x} = \lim_{u \to 0} \_\_\_\_\_\_ = \lim_{u \to 0} \frac{5}{2}\left(\_\_\_\_\_\_\right)$$

$$= (\_\_\_\_) \lim_{u \to 0} \frac{\sin u}{u} = \_\_\_\_ .$$

70. Find $\lim_{y \to 0} \frac{\tan 3y}{y}$. Let $u = 3y$. Then $y \to 0$ is equivalent to ______ , so upon substitution, $\lim_{y \to 0} \frac{\tan 3y}{y} = \lim_{u \to 0} \_\_\_\_\_\_$

$$= \lim_{u \to 0} 3 \_\_\_\_\_\_ \cdot \frac{1}{\cos u}$$

$$= 3 \lim_{u \to 0} \left(\_\_\_\_\right) \lim_{u \to 0} \left(\_\_\_\_\right)$$

$$= 3 (\_\_\_\_)(\_\_\_\_) = \_\_\_\_ .$$

71. Find $\lim_{\theta \to \frac{\pi}{2}} \frac{\cos \theta}{\frac{\pi}{2} - \theta}$. Let $u = \frac{\pi}{2} - \theta$ so that $\theta =$ ______ . Then $\theta \to \frac{\pi}{2}$ is equivalent to ______ , so upon substitution

---

69. $\frac{\sin u}{2u/5}$, $\frac{\sin u}{u}$, $\frac{5}{2}$, $\frac{5}{2}$ 70. $u \to 0$, $\frac{\tan u}{u/3}$, $\frac{\sin u}{u}$, $\frac{\sin u}{u}$, $\frac{1}{\cos u}$, 1, 1, 3

71. $\frac{\pi}{2} - u$, $u \to 0$, $\lim_{u \to 0} \frac{\cos\left(\frac{\pi}{2} - u\right)}{u}$, u , 1

$$\lim_{\theta \to \frac{\pi}{2}} \frac{\cos \theta}{\frac{\pi}{2} - \theta} = \underline{\qquad\qquad\qquad} = \lim_{u \to 0} \frac{\sin u}{\underline{\quad}} = \underline{\qquad} .$$

OBJECTIVE B: Calculate the derivatives of functions involving the trigonometric functions, making use of appropriate rules of differentiation and the derivatives of the sine and cosine functions.

72. $\frac{d}{dx} \sin x = \underline{\qquad\qquad}$.

73. $\frac{d}{dx} \cos x = \underline{\qquad\qquad}$.

74. $$\frac{d}{dx} \tan x = \frac{d}{dx} \frac{\sin x}{\underline{\quad}} = \frac{\cos x \cdot \frac{d}{dx}(\underline{\quad}) - \sin x \frac{d}{dx}(\cos x)}{\cos^2 x}$$

$$= \frac{\cos x\,(\underline{\qquad}) - \sin x\,(\underline{\qquad})}{\cos^2 x}$$

$$= \frac{\cos^2 x + \underline{\qquad}}{\cos^2 x} = \underline{\qquad\qquad} .$$

75. $$\frac{d}{dx} \sec x = \frac{d}{dx} \frac{1}{\underline{\qquad}} = -\frac{1}{\cos^2 x} \cdot \frac{d}{dx}(\underline{\quad})$$

$$= \frac{\underline{\qquad}}{\cos^2 x} = \underline{\qquad\qquad\qquad} .$$

Remark. It will be to your advantage in later work to MEMORIZE the derivative formulas derived in Problems 74 and 75.

76. $$\frac{d}{dx} x \tan^2 x = x \frac{d}{dx}(\underline{\qquad}) + \frac{dx}{dx} \tan^2 x = x \cdot 2 \tan x \frac{d}{dx}(\underline{\quad}) + \tan^2 x$$

$$= 2x \underline{\qquad\qquad\qquad} + \tan^2 x$$

$$= \underline{\qquad\qquad\qquad\qquad} .$$

---

72. $\cos x$ 73. $-\sin x$ 74. $\cos x$, $\sin x$, $\cos x$, $-\sin x$, $\sin^2 x$, $\sec^2 x$

75. $\cos x$, $\cos x$, $\sin x$, $\sec x \tan x$

76. $\tan^2 x$, $\tan x$, $\tan x \sec^2 x$, $\tan x\left(2x \sec^2 x + \tan x\right)$

77. $\frac{d}{dx}\sqrt{\frac{1 - \cos x}{1 + \cos x}} = \frac{1}{2}$ ________ $\frac{d}{dx}\left(\frac{1 - \cos x}{1 + \cos x}\right)$

$$= \frac{1}{2}\left(\frac{1 - \cos x}{1 + \cos x}\right)^{-1/2}\left[\frac{(1 + \cos x)(\_\_\_\_\_\_) - (1 - \cos x)(\_\_\_\_\_\_)}{(1 + \cos x)^2}\right]$$

$$= \frac{(1 - \cos x)^{-1/2} \sin x}{\_\_\_\_\_\_\_\_\_\_} .$$

2-11 Continuity.

As was the case with Article 1-10 in Chapter 1, this article is of a more theoretical nature than the previous ones you have been studying in the present chapter. Nevertheless, the idea of continuity lies at the very core of functional behavior, and the properties of continuous functions have a wide range of practical applications. These applications will become increasingly evident as you pursue your studies of calculus and applied mathematics. Here we will concentrate on the meaning and main implications associated with continuity.

OBJECTIVE A: Define precisely what is meant for a function f to be continuous at the point $x = c$.

78. The two conditions that must be satisfied if the function f is to be continuous at the point $x = c$ are that ________ exists and ________________ .

79. A function is continuous over an interval if it is continuous at ________________ within that interval.

80. If a function f is not continuous at the point $x = c$ it is said to be ________________ at c .

---

77. $\left(\frac{1 - \cos x}{1 + \cos x}\right)^{-1/2}$, $\sin x$, $- \sin x$, $(1 + \cos x)^{3/2}$

78. $f(c)$, $\lim_{x \to c} f(x) = f(c)$

79. all points

80. discontinuous

OBJECTIVE B: Given an elementary function $y = f(x)$, determine its points of continuity and discontinuity. Be able to justify your conclusions.

81. Consider $f(x) = \begin{cases} x+4 & \text{if } x < -1 \\ -x & \text{if } x \geq -1 \end{cases}$. Observe that $c = -1$ belongs to the domain of f: $f(-1) = 1$. Does f have a limit as $x \to -1$? To answer that question we calculate the right and lefthand limits:

$$\lim_{x \to -1^-} f(x) = \lim_{x \to -1^-} (\_\_\_\_\_\_) = \_\_\_\_ ,$$

$$\lim_{x \to -1^+} f(x) = \lim_{x \to -1^+} (\_\_\_\_\_\_) = \_\_\_\_ .$$

Since $\lim_{x \to -1^-} f(x) \neq \lim_{x \to -1^+} f(x)$, then $\lim_{x \to -1} f(x)$ ______________.

We conclude that f is ______________ at $x = -1$. Sketch a graph of f.

82. Let $f(x) = \dfrac{x}{x-1}$. Since $x =$ ____ does not belong to the domain of f we conclude that f is ______________ at 1. Also, $\lim_{x \to 1^-} f(x) =$ _____ and $\lim_{x \to 1^+} f(x) =$ _____ so f does not have a finite limit as $x \to 1$. However, as $x \to +\infty$ or $x \to -\infty$, $f(x) \to$ ____. Sketch a graph of f. Observe that f is continuous at all points except $x = 1$.

OBJECTIVE C: Specify the main facts related to continuous functions.

83. If f and g are continuous at c, then $f+g$, $f-g$, and $f \cdot g$ are ______________ at c.

84. If f and g are continuous at c, then $\dfrac{f}{g}$ is ______________ at c provided that ____________.

---

81. $x+4$, 3,
$-x$, 1,
does not exist,
discontinuous

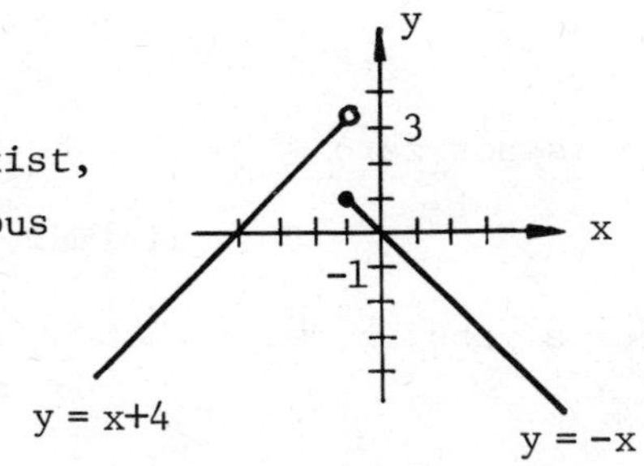

82. 1,
discontinuous,
$-\infty$, $+\infty$, 1

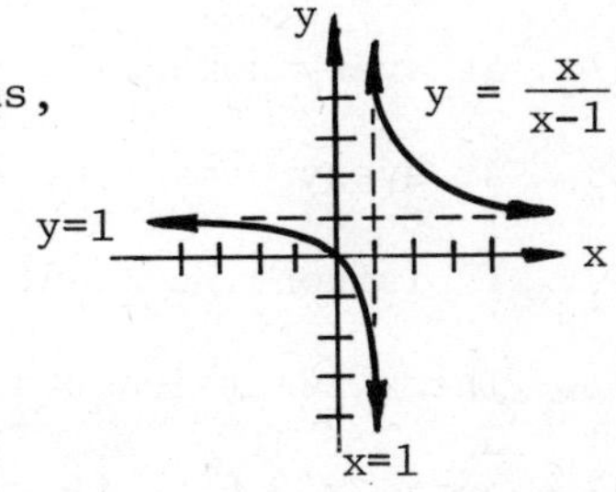

83. continuous

84. continuous, $g(c) \neq 0$

85. If f is continuous at c, and k is any constant, then kf is ______________ at c.

86. Every constant function is continuous ____________________ .

87. Every polynomial function is continuous ____________________ .

88. Every rational function is continuous ____________________________________ ________________________________ .

89. If f is differentiable at $x = c$, then _______________________________ __________________ .

90. Suppose that $f(x)$ is continuous for all x in the closed interval $[a,b]$. Then f has a ___________ value m and a ___________ value M on $[a,b]$.

91. Suppose that $f(x)$ is continuous for all x in the closed interval $[a,b]$, and that N is any number between $f(a)$ and $f(b)$. What is your conclusion? ____ ____ .

OBJECTIVE D: Define uniform continuity, and explain the distinction between uniform continuity and ordinary continuity over an interval.

92. If for each positive number ε, there is a positive number δ such that $|f(x) - f(x')| < \varepsilon$ whenever __________________ and ___________________ , then we say that f is _____________ continuous over the interval $a \leq x \leq b$.

93. Ordinary continuity at a point c means that the bandwidth δ depends not only upon the tolerance ε, but also upon _____ .

94. Uniform continuity means that with any given positive ε we may associate ________ bandwidth δ that works for _______ points in the domain $a \leq x \leq b$.

95. If a function is continuous over an arbitrary domain, is it uniformly continuous?

---

85. continuous

86. at every number

87. at every number

88. at every number at which the denominator is not zero

89. f is continuous at $x = c$

90. minimum, maximum

91. There is at least one number c between a and b such that $f(c) = N$.

92. $a \leq x,\ x' \leq b$, $|x - x'| < \delta$, uniformly

93. c

94. one, all

95. not always

96. If a function is continuous over a closed interval $a \leq x \leq b$, is it uniformly continuous?

97. If a function is uniformly continuous over a domain, is it continuous?

2-12 Differentials.

OBJECTIVE A: Define the differentials dx and dy where $y = F(x)$ is a differentiable function of x, and give a geometric interpretation of the differential dy.

98. If x is the independent variable and $y = F(x)$ is differentiable at x_o, then dx is defined to be an ____________ variable with domain __________ and dy is a ____________ variable satisfying

$$dy = \_\_\_\_\_\_\_\_\_\_\_\_ .$$

Thus the differential dy depends on the value of ______ and the value of _______ .

99. Geometrically, if $y = F(x)$ has a derivative at $x = x_o$, then dx can be interpreted as a _________ in the x-direction, say $x - x_o$, and dy is the corresponding number of units change in the y-direction along __________ ________________________ of the function at the point $(x_o, F'(x_o))$.

100. If $dx \neq 0$, the derivative $F'(x_o) =$ _______ , which is the differential of y divided by ________________________ associated with the point x_o.

101. If t is the independent variable, and if $x = f(t)$ and $y = g(t)$ are differentiable at $t = t_o$, then the differentials of x and y are defined by $dx =$ ____________ and $dy =$ ____________ , where the differential dt is a new independent variable whose domain is $(-\infty, +\infty)$. In this case the derivative dy/dx is still given by

$$\frac{dy}{dx} = \_\_\_\_\_\_\_\_\_\_\_\_\_\_\_\_ = \_\_\_\_\_\_\_\_ .$$

---

96. Yes

97. Yes

98. independent, $(-\infty, +\infty)$, dependent, $F'(x_o)\,dx$, dx, $F'(x_o)$

99. change, the tangent line to the graph

100. $\frac{dy}{dx}$, the differential of x

101. $f'(t_o)\,dt$, $g'(t_o)\,dt$, $\frac{\text{differential of } y}{\text{differential of } x}$, $\frac{g'(t_o)}{f'(t_o)}$

OBJECTIVE B: Given parametric equations $x = f(t)$ and $y = g(t)$ with f and g twice differentiable, find the second derivative d^2y/dx^2 in terms of the parameter t.

102. Let $x = t^2$ and $y = t^2 - 2t$. From Problem 59, $\frac{dx}{dt} = 2t$ and $\frac{dy}{dt} = 2(t-1)$. Thus, $y' = \frac{dy}{dx} =$ __________. Then,

$$\frac{d^2y}{dx^2} = \frac{\_\_\_\_}{dx/dt} = \frac{\_\_\_\_}{2t} = \_\_\_\_\_\_\_\_ .$$

2-13 Formulas for Differentiation Repeated in the Notation of Differentials.

OBJECTIVE A: Given an equation in the variables x and y, calculate the differential dy in terms of dx.

103. If $y = 3x^2 - x^{-1/3} + 4$, then

$dy =$ ______________ dx.

104. Consider $\sin xy + xy^2 = 1$. To calculate dy, first take the differential of each side, equate these, and then solve for dy:

$d\left(\sin xy + xy^2\right) = d(1)$, so

$d(\sin xy) + d($ ____ $) =$ ____ , or

$(\cos xy)\, d($ ____ $) + x\, d($ ____ $) + y^2 d($ ____ $) = 0$

$(\cos xy)($ _______ $) + x($ _______ $) + y^2 dx = 0$

Thus,

$\left(y \cos xy + y^2\right) dx + ($ ______________ $)\, dy = 0$, or

$dy =$ _______________________ .

---

102. $\frac{2(t-1)}{2t}$ or $1 - \frac{1}{t}$, dy'/dt, $1/t^2$, $\frac{1}{2t^3}$

103. $6x + \frac{1}{3} x^{-4/3}$

104. xy^2, 0, xy, y^2, x, $xdy + ydx$, $2ydy$, $x \cos xy + 2xy$, $-\left(\frac{y \cos xy + y^2}{x \cos xy + 2xy}\right) dx$

OBJECTIVE B: Given parametric equations $x = f(t)$ and $y = g(t)$ with f and g twice differentiable, calculate the differentials dx and dy in terms of t and dt. Use these results to calculate dy/dx and d^2y/dx^2.

105. Let $x = \sin 2t$ and $y = \frac{1}{3} \cos 2t$. Then $dx =$ ______________ and $dy =$ ______________. Then, $\frac{dy}{dx} = y' = \frac{\_\_\_\_}{dx/dt} = \frac{-2 \sin 2t}{\_\_\_\_\_\_}$

$= -\frac{1}{3}$ ________ . Also, $\frac{d^2y}{dx^2} = \frac{\_\_\_\_\_\_}{dx/dt} = \frac{\_\_\_\_\_\_}{2 \cos 2t} = -\frac{1}{3}$ ______ .

106. To find an equation of the tangent line to the curve traced out by the point (x,y) in Problem 105 when $t = \frac{\pi}{8}$, we see that

$x\left(\frac{\pi}{8}\right) = \sin$ (____), $y\left(\frac{\pi}{8}\right) = \frac{1}{3} \cos$ (____), and $\frac{dy}{dx}\left(\frac{\pi}{8}\right) =$ ______ ,

or

$x\left(\frac{\pi}{8}\right) =$ ______ , $y\left(\frac{\pi}{8}\right) =$ ______ , and $\frac{dy}{dx}\left(\frac{\pi}{8}\right) =$ _____ .

Therefore, an equation of the tangent line is ______________________ .
Notice that we can easily eliminate the parameter t from Problem 105: $x = \sin 2t$ and $3y = \cos 2t$ so that $x^2 + 9y^2 = 1$. This is an equation of an ellipse.

OBJECTIVE C: Use differentials to obtain reasonable approximations.

107. To find an approximate value for $\sqrt[3]{7.5}$ using differentials, let $y = x^{1/3}$. Then, $dy =$ ____________ . In the problem we want dy when $x = 8$ and $dx = -.5$, so $dy = \frac{1}{3}$ (____)$^{-2/3}$ (____) = _____ . Therefore, the approximation $y + \Delta y \approx y + dy$ gives $\sqrt[3]{7.5} \approx \sqrt[3]{\_\_\_\_} + (\_\_\_\_) =$ ______ .
(If we define $dx = \Delta x$, then $dy = \Delta y_{tan}$ of Article 2-6. Compare the solution method above with the method used in solving Problem 51.)

---

105. $2 \cos 2t\, dt$, $-\frac{2}{3} \sin 2t\, dt$, dy/dt, $6 \cos 2t$, $\tan 2t$, dy'/dt, $-\frac{2}{3} \sec^2 2t$, $\sec^3 2t$

106. $\frac{\pi}{4}$, $\frac{\pi}{4}$, $-\frac{1}{3} \tan \frac{\pi}{4}$, $\frac{\sqrt{2}}{2}$, $\frac{\sqrt{2}}{6}$, $-\frac{1}{3}$, $y - \frac{\sqrt{2}}{6} = -\frac{1}{3}\left(x - \frac{\sqrt{2}}{2}\right)$

107. $\frac{1}{3} x^{-2/3} dx$, 8, -.5, $-\frac{1}{24}$, 8, $-\frac{1}{24}$, $\frac{47}{24} \approx 1.95833$

CHAPTER 2 OBJECTIVE - PROBLEM KEY

Objective	Problems in Thomas/Finney Text	Objective	Problems in Thomas/Finney Text
2-2 A,B	p. 67, 6-15	2-8 A	p. 93, 6-10
C	p. 67, 1-5,16	B	p. 93, 1-5
D	p. 67, 17-19,22,23,25-27	2-9	p. 100, 1-3,8,10
	p. 83, 30-36	2-10 A	p. 107, 1-22
2-3 A	p. 75, 2,5,12,16	B	p. 107, 23-40
B	p. 75, 6,7,9,11,13,17	2-11 A	See page 108
C	p. 75, 3,4,5,8,10,14,15,18	B	p. 117, 2,3,8,10
2-4 A	p. 79, 4,5	C	Theorems 1-5 section 2-11
B	p. 79, 1,2,7	2-12 A	See page 117
2-5	p. 83, 1-9,12,13,15-17,19,21, 22,24-28,29	B	p. 121, 14-20
		2-13 A	p. 121, 1-8
2-6	p. 85, 1-5,6-10	B	p. 121, 14-20
2-7	p. 88, 1-6	C	p. 121, 9-13

CHAPTER 2 SELF-TEST

1. Find $\frac{dy}{dx}$

(a) $y = \left(2x^3 - x^2 + 7x + 3\right)^6$ (b) $y = (x^2 - 9)(3x^5 + 7x)$

(c) $y = \frac{3}{x^4 + 1}$ (d) $y = \frac{x^2 - 1}{3x + 1}$

2. Find $\frac{d^2y}{dx^2}$

(a) $y = \frac{1}{3}x^3 + \frac{1}{2}x^2 - 6x + 8$ (b) $y = (2x^3 - 11)(x^2 - 3)$

3. A particle moves along a horizontal line (positive to the right) according to the law

$$s = t^3 - 6t^2 + 2$$

During which intervals of time is the particle moving to the right and during which is it moving to the left? What is the acceleration and the velocity when $t = 2.3$?

4. Find an equation of the tangent line to the graph of $y = \sqrt{1 - x^2}$ when $x = 1/2$.

5. Find $\frac{dy}{dx}$

(a) $y = x^{4/3} - 5x^{4/5}$ (b) $y = \sqrt{x + \frac{1}{x}}$

(c) $y = \dfrac{1}{2x - \sqrt{x^2 - 1}}$ (d) $y = x \cos(5x - 2)$

6. Let g be the inverse of $f(x) = x^3 + 4x - 5$. Find $g'(-5)$.

7. Find $\dfrac{dy}{dx}$ and $\dfrac{d^2y}{dx^2}$ when $x + y^2 = xy$.

8. Find an equation of the tangent line to the curve $x^3 + 3xy^3 + xy^2 = xy$ at the point $(1,-1)$.

9. Use the tangent line approximation Δy_{tan} (or the differential method) to estimate the value of $\sin 29°$.

10. Beginning with the estimate $x_1 = \frac{\pi}{2}$, apply Newton's method once to calculate a positive solution to the equation $\sin x = \frac{2}{3} x$.

11. Consider the curve given by the parametric equations $x = 3t - 1$ and $y = t^2 - t$. Eliminate the parameter t to find an equation of the form $y = F(x)$ and find dy/dx in terms of dy/dt and dx/dt. For what value of t is $dy/dx = 0$? Sketch the graph of the curve over the interval $-2 \leq t \leq 3$.

12. Identify the amplitude, period, horizontal shift, and vertical shift of the function $f(x) = 2 \sin(3x + 1)$. Sketch one cycle of the graph of the function.

13. Evaluate the following limits:

(a) $\lim\limits_{x \to 0} \dfrac{\sin x^{1/3}}{x^{1/3}}$ (b) $\lim\limits_{x \to \frac{\pi}{2}} \dfrac{\cos x}{\pi - 2x}$

14. Find $\dfrac{dy}{dx}$

(a) $y = \cot 3x$ (b) $y = 3 \sin^2 5x - \sec x$

15. Consider the function $f(x) = \dfrac{x - 1}{x^2 - x}$.

(a) For what values of x is f continuous? Justify your conclusion.

(b) Is f continuous at $x = 1$? If not, what value can be assigned to $f(1)$ so that the resultant function is continuous there?

16. Find the differential dy

(a) $xy^3 + x^2y - 1 = 0$ (b) $y = \sin\sqrt{1 - x^2}$

17. Given the parametric equations $x = t^2 - 1$ and $y = t + 1$

(a) Express dx and dy in terms of t and dt,

(b) Find d^2y/dx^2 in terms of t,

(c) Find an equation of the tangent line to the curve at the point for which $t = 1$.

18. Use differentials to obtain a reasonable approximation to $1/\sqrt{24}$.

SOLUTIONS TO CHAPTER 2 SELF-TEST

1. (a) $\dfrac{dy}{dx} = 6\left(2x^3 - x^2 + 7x + 3\right)^5\left(6x^2 - 2x + 7\right)$

(b) $\dfrac{dy}{dx} = 2x\left(3x^5 + 7x\right) + \left(x^2 - 9\right)\left(15x^4 + 7\right) = 21x^6 - 135x^4 + 21x^2 - 63$

(c) $\dfrac{dy}{dx} = \dfrac{-3(4x^3)}{\left(x^4 + 1\right)^2}$ (d) $\dfrac{dy}{dx} = \dfrac{(3x+1)(2x) - (x^2 - 1)(3)}{(3x+1)^2} = \dfrac{3x^2 + 2x + 3}{(3x+1)^2}$

2. (a) $\dfrac{dy}{dx} = x^2 + x - 6$, $\dfrac{d^2y}{dx^2} = 2x + 1$

(b) $\dfrac{dy}{dx} = 6x^2\left(x^2 - 3\right) + \left(2x^3 - 11\right)(2x) = 10x^4 - 18x^2 - 22x$, $\dfrac{d^2y}{dx^2} = 40x^3 - 36x - 22$

3. $\dfrac{ds}{dt} = 3t^2 - 12t = 3t(t-4)$; $\dfrac{d^2s}{dt^2} = 6t - 12$

The particle is moving to the right when $\dfrac{ds}{dt} > 0$ so $t > 4$ or $t < 0$; it is moving to the left when $0 < t < 4$ and $\dfrac{ds}{dt} < 0$. At $t = 2.3$,

$$\left.\frac{ds}{dt}\right|_{t=2.3} = 3(2.3)^2 - 12(2.3) = -11.73 \text{ , velocity}$$

$$\left.\frac{d^2s}{dt^2}\right|_{t=2.3} = 6(2.3) - 12 = 1.8 \text{ , acceleration}$$

4. $y' = \frac{1}{2}\left(1 - x^2\right)^{-1/2}(-2x) = -x\left(1 - x^2\right)^{-1/2}$, $y\left(\frac{1}{2}\right) = \frac{\sqrt{3}}{2}$ and $y'\left(\frac{1}{2}\right) = -\frac{1}{\sqrt{3}}$

so that $\left(y - \frac{\sqrt{3}}{2}\right) = -\frac{1}{\sqrt{3}}\left(x - \frac{1}{2}\right)$ or $\sqrt{3}\,y + x - 2 = 0$ is an equation of the tangent line.

5. (a) $\dfrac{dy}{dx} = \frac{4}{3}x^{1/3} - 4x^{-1/5}$

(b) $\dfrac{dy}{dx} = \frac{1}{2}\left(x + \frac{1}{x}\right)^{-1/2} \cdot \frac{d}{dx}\left(x + \frac{1}{x}\right) = \frac{1}{2}\left(x + \frac{1}{x}\right)^{-1/2}\left(1 - \frac{1}{x^2}\right)$

(c) $\frac{dy}{dx} = -\left(2x - \sqrt{x^2 - 1}\right)^{-2} \left[2 - \frac{1}{2}\left(x^2 - 1\right)^{-1/2} \cdot 2x\right]$

$= -\left(2x - \sqrt{x^2 - 1}\right)^{-2} \left(2 - \frac{x}{\sqrt{x^2 - 1}}\right)$

(d) $\frac{dy}{dx} = \cos(5x - 2) + x\left[-\sin(5x - 2) \cdot 5\right] = \cos(5x - 2) - 5x \sin(5x - 2)$

6. When $f(x) = -5$, $x^3 + 4x = 0$ or $x(x^2 + 4) = 0$. Thus, $x = 0$. Therefore, $g'(-5) = 1/f'(0)$. Now, $f'(x) = 3x^2 + 4$ so $f'(0) = 4$. Thus, $g'(-5) = 1/4$.

7. Differentiating implicitly, $1 + 2yy' = y + xy'$ or $y' = \frac{y - 1}{2y - x}$.

$$\frac{d^2y}{dx^2} = \frac{(2y - x)(y') - (y - 1)(2y' - 1)}{(2y - x)^2} = \frac{(2 - x)y' + (y - 1)}{(2y - x)^2}$$

$$= \frac{(2 - x)\left(\frac{y - 1}{2y - x}\right) + (y - 1)}{(2y - x)^2} = \frac{2(y - 1)(y - x + 1)}{(2y - x)^3}$$

8. Since $1^3 + 3(1)(-1)^3 + 1(-1)^2 = 1(-1)$ is true, the point $(1,-1)$ is on the curve. Differentiating implicitly, $3x^2 + 3y^3 + 9xy^2y' + y^2 + 2xyy' = y + xy'$, so evaluation at $(1,-1)$ yields $3 - 3 + 9y' + 1 - 2y' = -1 + y'$, or $y' = -\frac{1}{3}$. Thus an equation of the tangent line is given by

$$(y + 1) = -\frac{1}{3}(x - 1) \quad \text{or} \quad x + 3y = -2 .$$

9. The calculation must be done when $y = \sin x$ for x measured in radians. Thus,

$$\sin 29° \approx \sin\frac{\pi}{6} + \Delta y_{tan} ,$$

where $\Delta y_{tan} = dy = \frac{dy}{dx}\Delta x$ when $x = \frac{\pi}{6}$ and $\Delta x = -\frac{\pi}{180}$ radians.

Now, $\left.\frac{dy}{dx}\right|_{\pi/6} = \cos\frac{\pi}{6} = \frac{\sqrt{3}}{2}$ so that $\sin 29° \approx \frac{1}{2} + \left(\frac{\sqrt{3}}{2}\right)\left(-\frac{\pi}{180}\right) \approx .48489$.

10. Let $f(x) = \sin x - \frac{2}{3}x = 0$, $f'(x) = \cos x - \frac{2}{3}$. By Newton's method,

$$x_2 = x_1 - \frac{\sin x_1 - (2/3)x_1}{\cos x_1 - (2/3)} = \frac{\pi}{2} - \frac{1 - \pi/3}{0 - 2/3} = \frac{\pi}{2} + \frac{3}{2}\left(1 - \frac{\pi}{3}\right) = 1.5$$

11. $\frac{dx}{dt} = 3$ and $\frac{dy}{dt} = 2t - 1$, so that $\frac{dy}{dx} = \frac{dy/dt}{dx/dt} = \frac{1}{3}(2t - 1)$. When $t = \frac{1}{2}$, $\frac{dy}{dx} = 0$ which occurs at the point $(x,y) = \left(\frac{1}{2}, -\frac{1}{4}\right)$. To eliminate the parameter t, from the parametric expression for x, $t = \frac{1}{3}(x + 1)$, and substitution into the parametric expression for y gives

$$y = \frac{1}{9}(x+1)^2 - \frac{1}{3}(x+1)$$

or

$$y = \frac{1}{9}(x^2 - x - 2) .$$

A table of coordinate values is as follows:

t	-2	-1	0	1/2	1	2	3
x	-7	-4	-1	1/2	2	5	8
y	6	2	0	-1/4	0	2	6

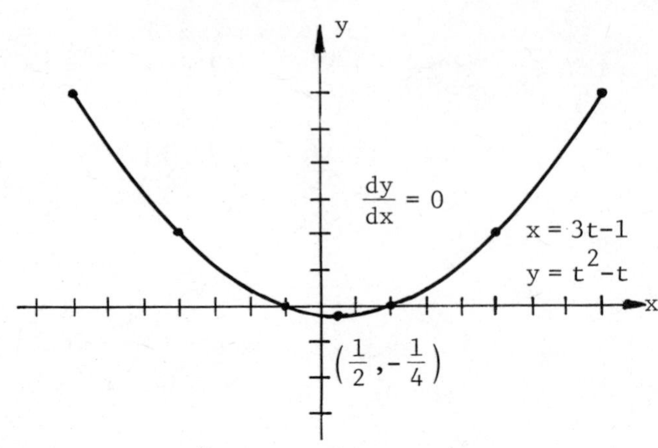

12. $2 \sin(3x+1) = 2 \sin\left[\frac{2\pi}{2\pi/3}\left(x + \frac{1}{3}\right)\right]$

amplitude = 2 ; period = $\frac{2\pi}{3}$;

horizontal shift = $-\frac{1}{3}$ (a shift to the left) ;

vertical shift = 0

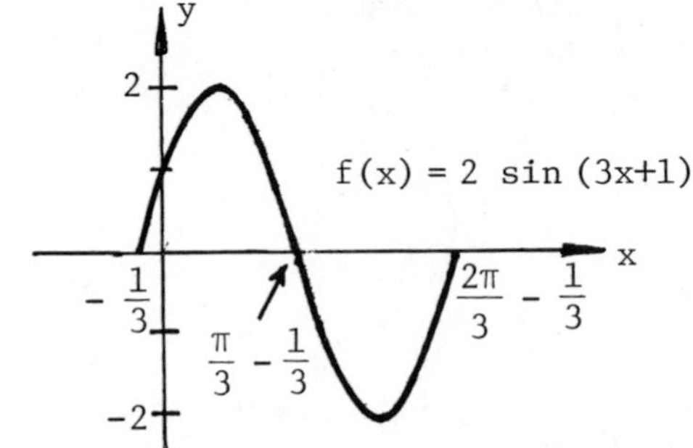

13. (a) Let $u = x^{1/3}$ so that $x \to 0$ is equivalent to $u \to 0$. Thus

$$\lim_{x \to 0} \frac{\sin x^{1/3}}{x^{1/3}} = \lim_{u \to 0} \frac{\sin u}{u} = 1 .$$

(b) $\lim_{x \to \frac{\pi}{2}} \frac{\cos x}{\pi - 2x} = \lim_{x \to \frac{\pi}{2}} \frac{\frac{1}{2}\cos x}{\frac{\pi}{2} - x}$; let $u = \frac{\pi}{2} - x$ so that $x \to \frac{\pi}{2}$ is equivalent to $u \to 0$. Thus,

$$\lim_{x \to \frac{\pi}{2}} \frac{\frac{1}{2}\cos x}{\frac{\pi}{2} - x} = \lim_{u \to 0} \frac{\frac{1}{2}\cos\left(\frac{\pi}{2} - u\right)}{u} = \lim_{u \to 0} \frac{1}{2}\frac{\sin u}{u} = \frac{1}{2} .$$

14. (a) $\frac{dy}{dx} = \frac{d}{dx}\left(\frac{\cos 3x}{\sin 3x}\right) = \frac{\sin 3x\,(-3 \sin 3x) - (\cos 3x)(3 \cos 3x)}{\sin^2 3x}$

$$= \frac{-3 \sin^2 3x - 3 \cos^2 3x}{\sin^2 3x} = -3 \csc^2 3x .$$

(b) $\frac{dy}{dx} = 30 \sin 5x \cos 5x - \sec x \tan x$

15. (a) Since division by zero is never permitted, the points $x = 0$ and $x = 1$ do not belong to the domain of f, and therefore f is discontinuous at those two values; it is continuous for all other values of x.

(b) $f(x) = \frac{x-1}{x^2 - x} = \frac{x-1}{x(x-1)} = \frac{1}{x}$ if $x \neq 1$ and $x \neq 0$. Since

$\lim_{x \to 1} f(x) = \lim_{x \to 1} \frac{1}{x} = 1$, if we specify $f(1) = 1$ the new function so defined is continuous at $x = 1$.

16. (a) $y^3 dx + 3xy^2 dy + 2xydx + x^2 dy = 0$, or $dy = -\left(\frac{y^3 + 2xy}{3xy^2 + x^2}\right) dx$

(b) $dy = y'dx = \frac{-x}{\sqrt{1-x^2}} \cos\sqrt{1-x^2}\, dx$

17. (a) $dx/dt = 2t$ and $dy/dt = 1$, so that $dx = 2t\, dt$ and $dy = dt$.

(b) $\frac{dy}{dx} = \frac{1}{2t}$ so that $\frac{d^2y}{dx^2} = \frac{dy'/dt}{dx/dt} = \frac{-1/2t^2}{2t} = -\frac{1}{4t^3}$

(c) When $t = 1$, $x = 0$, $y = 2$, and $\frac{dy}{dx} = \frac{1}{2}$; thus

$y - 2 = \frac{1}{2}(x - 0)$ or $2y - x = 4$ is an equation of the tangent line.

18. Let $f(x) = x^{-1/2}$ so $f'(x) = -\frac{1}{2} x^{-3/2}$. Set $x = 25$ and $dx = \Delta x = -1$ so that $dy = f'(x)\, dx = -\frac{1}{2}(25)^{-3/2}(-1) = 1/250$. Thus,

$$\frac{1}{\sqrt{24}} \approx \frac{1}{\sqrt{25}} + \frac{1}{250} = \frac{51}{250} = 0.204 .$$

CHAPTER 3 APPLICATIONS

3-1 Sign of the First Derivative. Application to Curve Sketching.

OBJECTIVE: Use the sign of the first derivative dy/dx to determine the values of x where the graph of y versus x is rising and where it is falling.

1. Let $y = f(x)$ be a differentiable function of x. When dy/dx has a ___________ value, the graph of y versus x is rising (to the right). In this case it is also said that the function f is _______________ at that point.

2. When $dy/dx < 0$, the graph of y versus x is ___________ and the function f is _______________ at that point.

3. Let $y = \frac{1}{3}x^3 - x^2 + 2$. Then $y' =$ ________ $= x($_______$)$. The derivative dy/dx is zero when $x =$ ____ or $x =$ ____. Thus, the curve y is rising when $x < 0$, it is falling when x satisfies ______________, and it is rising again when $x >$ ____.
We construct a table of some values for the curve (complete the table):

x	-2	-1	0	1	2	3	4
y							

Sketch the graph in the coordinate system at the right.

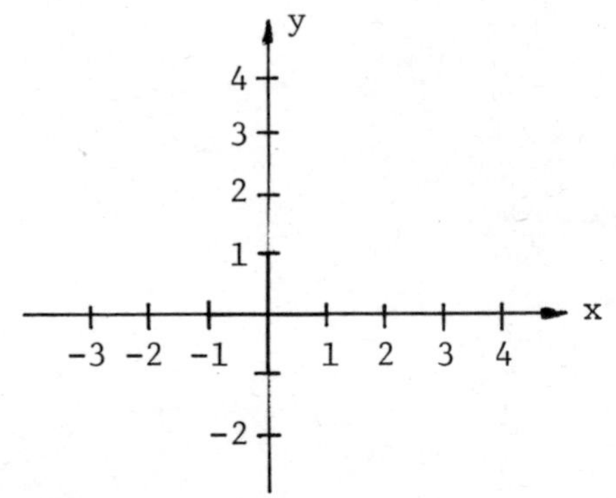

---

1. positive, increasing

2. falling, decreasing

3. $x^2 - 2x$, $x - 2$, 0, 2, $0 < x < 2$, 2

x	-2	-1	0	1	2	3	4
y	-14/3	2/3	2	4/3	2/3	2	4

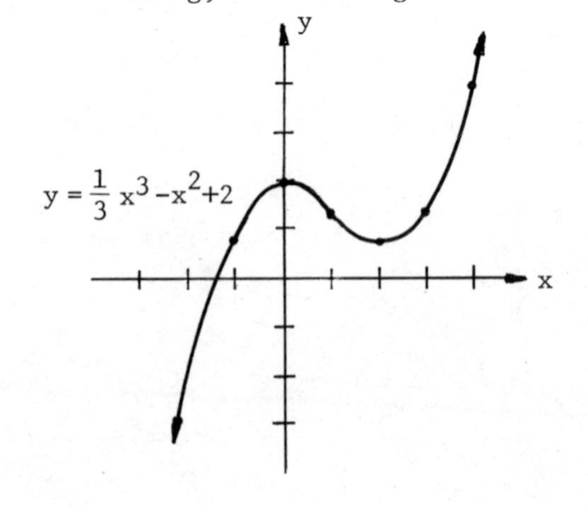

4. The following graph describes a differentiable function $y = f(x)$.

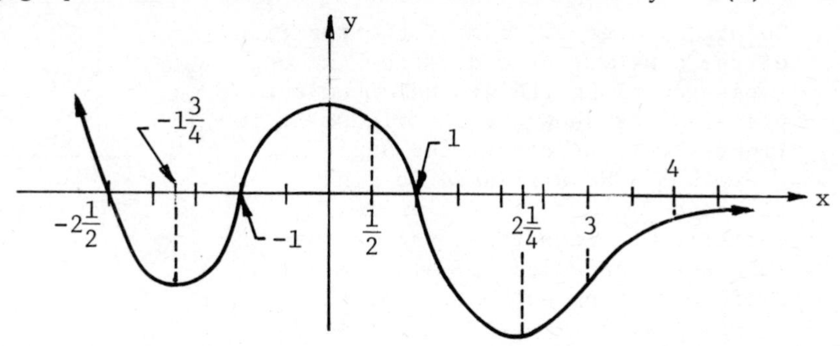

The function f is positive when x belongs to the interval ________ or the interval ________ . There are no points of discontinuity of f . When x = ______ or x = ______ or x = ______ the derivative f' changes sign. The derivative is positive when x belongs to the interval ________ or the interval ________ ; the derivative is ________ when x belongs to $(-\infty, -2\frac{1}{2})$ or ________ . Thus, the curve is ________ for $-1\frac{3}{4} < x < 0$ and $x > 2\frac{1}{4}$, and it is falling for ________ and ________ .

3-2 Related Rates.

In this article we consider problems that ask us to find the rate at which some variable quantity changes when we know the rate at which another quantity related to it changes. Examples abound for problems of this sort. For instance, the rate of production of a certain commodity may depend upon its rate of sales; the rate of increase or decrease in the water level of a dam or reservoir is essential information to a public utility serving the demands of a growing population; the rate at which oil may be spreading on the sea surface from a stricken tanker depends on the rate at which it may be leaking; and so forth.

OBJECTIVE: Solve a related rates problem by
(a) setting up the relationships between the variables of the problem in the form of equations valid for all time values considered,
(b) finding the rate of change of one or more of the variables at a particular instant, given the values of some or all variables and the rates of change of some of them at that instant.

5. A plane flying at 1 mile altitude is 2 miles distant from an observer, measured along the ground, and flying directly away from the observer at 400 mph. How fast is the angle of elevation changing?

---

4. $(-\infty, -2\frac{1}{2})$, $(-1,1)$, $-1\frac{3}{4}$, 0, $2\frac{1}{4}$, $(-1\frac{3}{4}, 0)$, $(2\frac{1}{4}, \infty)$, negative, $(0, 2\frac{1}{4})$, rising, $x < -1\frac{3}{4}$, $0 < x < 2\frac{1}{4}$

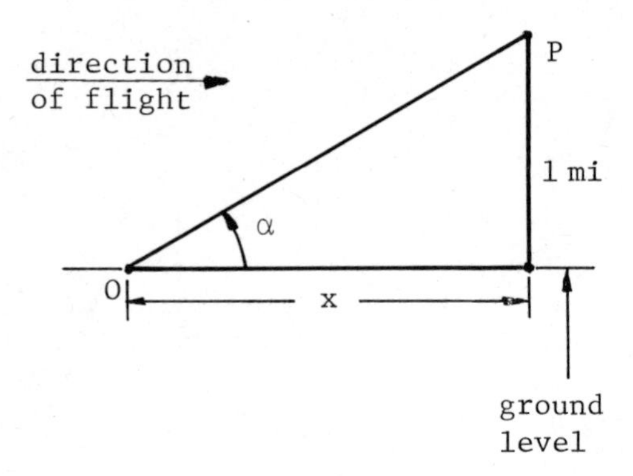

Solution. Let O denote the position of the observer at a distance x units (measured along the ground) from the plane P as shown in the figure at the right. Let α denote the angle of elevation. We are asked to find the rate ________ . The angle α satisfies the equation $\tan\alpha =$ _______ . This equation holds for all time t . Differentiating both sides of the equation with respect to t yields,

$$\sec^2\alpha \cdot (\_\_\_\_\_\_) = -\frac{1}{x^2}\cdot(\_\_\_\_\_\_),$$

or, since $\sec^2\alpha = 1 + \tan^2\alpha =$ ____________ , we solve to find

$$\frac{d\alpha}{dt} = \left(\_\_\_\_\_\_\_\_\_\_\_\_\right)\frac{dx}{dt}.$$

Thus, when $x = 2$ and $dx/dt = 400$, $\frac{d\alpha}{dt} =$ ______ radians per hour,

or _______ rad/sec, or _____ deg/sec. Notice that the angle of elevation is decreasing because $d\alpha/dt$ is ____________ .

6. A trough 10 ft long has a cross section that is an isosceles triangle 3 ft deep and 8 ft across. If water flows in at the rate of 2 cu ft/min, how fast is the surface rising when the water is 2 feet deep?

Solution. A cross section of the trough is shown in the figure at the right. In the figure h denotes the depth of the water and b its width across the trough at any instant t . Thus, b and h are both functions of ____ . At any instant of time the volume of water in the trough is given by the formula,

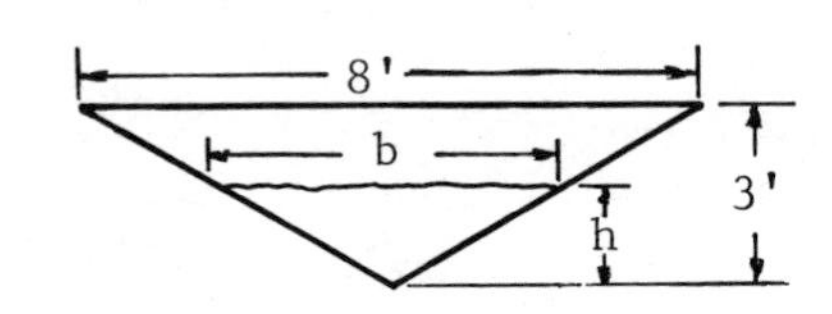

$$V = 10\ (\_\_\_\_\_\_).$$

---

5. $d\alpha/dt$, $1/x$, $d\alpha/dt$, dx/dt, $1 + \frac{1}{x^2}$, $-\frac{1}{x^2}\left(\frac{x^2}{1+x^2}\right)$, -80, $\frac{-80}{3600}$,

(approx) -1.3 , negative

6. t, $\frac{1}{2}bh$, dh/dt, h, $\frac{8}{3}h$, $\frac{40}{3}h^2$, $\frac{80}{3}h\,\frac{dh}{dt}$, $\frac{3}{80}$

We are given the rate $dV/dt = 2$ and we are asked to find the rate ______ when $h = 2$. Since the formula for V involves both the variables b and h we need to write down a formula relating these variables. From the geometry of similar triangles in the figure we have

$$\frac{b}{\_\_} = \frac{8}{3} \quad \text{or,} \quad b = \_\_\_\_\_ .$$

Substitution into the formula for V gives $V = \_\_\_\_\_\_\_$. Differentiation of both sides of this last equation with respect to t yields,

$$\frac{dV}{dt} = \_\_\_\_\_\_\_\_\_\_ .$$

Solving for dh/dt when $dV/dt = 2$ and $h = 2$ gives

$dh/dt = \_\_\_\_\_\_$ ft/min .

7. A walk is perpendicular to a long wall, and a man strolls along it away from the wall at the rate of 3 ft/sec. There is a light 8 ft from the walk and 24 ft from the wall. How fast is his shadow moving along the wall when he is 20 ft from the wall?

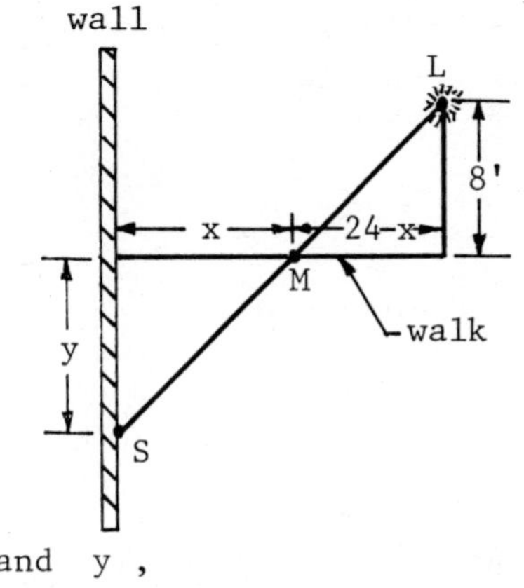

Solution. The situation is pictured in the figure at the right. Here M denotes the position of the man at a distance x units from the wall, S denotes the position of his shadow on the wall at a distance y units from where the wall and the walk intersect, and L denotes the position of the light. We are given that _______ = 3 and are asked to find _______ when $x = 20$. From similar triangles we can establish a relationship between the variables x and y,

$$\frac{x}{y} = \frac{\_\_\_\_\_\_\_\_}{8} , \quad \text{or} \quad 8x = \_\_\_\_\_\_\_\_\_\_\_\_\_ .$$

This equation is valid at any instant of time t. Differentiating both sides with respect to t yields

$$8\frac{dx}{dt} = (\_\_\_\_\_)\frac{dy}{dt} - \_\_\_\_\_ .$$

When $x = 20$, $y = \dfrac{8(20)}{\_\_\_\_\_\_} = \_\_\_\_$, so substitution into the previous derivative equation yields

$$8 \cdot 3 = \_\_\_\_\_\_\_ - 120 , \quad \text{or} \quad \frac{dy}{dt} = \_\_\_\_ \text{ ft/sec} .$$

---

7. dx/dt, dy/dt, $24 - x$, $(24 - x)y$, $24 - x$, $y\dfrac{dx}{dt}$, $24 - 20$, 40, $4\dfrac{dy}{dt}$, 36

3-3 Significance of the Sign of the Second Derivative.

OBJECTIVE: Relate the concavity of a function $y = f(x)$ to the second derivative d^2y/dx^2.

8. If the second derivative d^2y/dx^2 is positive, the y-curve is concave ________ at that point; if d^2y/dx^2 is ________ the curve is concave downward at that point.

9. When a curve is concave upward at a point, locally the curve lies ________ the tangent line; when it is concave ________, locally the curve lies below the tangent line.

10. A point where the curve changes concavity is called a ________, and is characterized by a change in sign of ________.

11. A point of inflection occurs where d^2y/dx^2 is ________ or ________ ________.

12. Does the condition $d^2y/dx^2 = 0$ guarantee a point of inflection? ________

3-4 Curve Plotting.

OBJECTIVE A: Given a function $y = f(x)$, find the intervals of values of x for which the curve is rising, falling, concave upward, and concave downward. Sketch the curve, showing the high turning points, the low turning points, and the points of inflection.

13. Consider the function $f(x) = x^4 - 4x^3 + 10$. We follow the five step procedure outlined in the Thomas/Finney text in order to sketch the graph of $y = f(x)$.

(a) $\frac{dy}{dx} =$ ________ and $\frac{d^2y}{dx^2} =$ ________.

(b) In factored form, $dy/dx = 4x^2(x - 3)$ so that the curve is falling when x belongs to the interval ________ and rising when $x >$ ____. The slope of the curve is zero when $x =$ ____ or $x =$ ____.

---

8. upward, negative

9. above, downward

10. point of inflection, d^2y/dx^2

11. zero, fails to exist

12. No, the function $y = x^4$ affords a counterexample at $x = 0$.

13. (a) $4x^3 - 12x^2$, $12x^2 - 24x$ (b) $(-\infty, 3)$, 3, 0, 3

(c) In factored form, d^2y/dx^2 = ____________ . Thus d^2y/dx^2 is negative when x belongs to the interval ________ and consequently the curve is concave ____________ there. The second derivative is positive when x satisfies ________ or ________ , and the curve is concave ____________ . Therefore, the second derivative changes sign when x = ____ or x = ____ so that these are points of inflection of f .

(d) Complete the following table.

x	y	y'	y"	Conclusions
-2	26	-	+	falling; concave up
-1	15	-	+	falling; concave up
0				
1				
2				
3				
4	10	+	+	rising; concave up

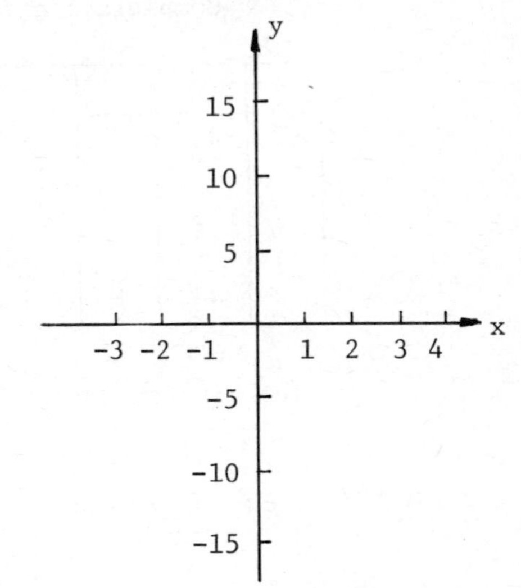

(e) Sketch a smooth curve of y = f(x) in the given coordinate system to the right.

14. Sketch the graph of $y = x^{-2} + 2x$.

Solution. We follow the five steps as in Problem 13.

(a) y' = ____________ and y" = ________ .

(b) In fractional form, $y' = \dfrac{\_\_\_\_\_\_}{x^3}$ so that y' is zero when x = ____ . The curve is falling when x belongs to the interval ______ ; it is rising for x satisfying ________ and ________ .

(c) d^2y/dx^2 is always ____________ so the curve is everywhere concave ____________ . Therefore there are no points of inflection.

---

13. (c) $12x(x-2)$, $(0,2)$, downward, $x < 0$, $x > 2$, upward, 0, 2

(d)

x	y	y'	y"	Conclusions
0	10	0	0	point of inflection
1	7	-	-	falling; concave down
2	-6	-	0	point of inflection
3	-17	0	+	"Holds water"; min.

(e)

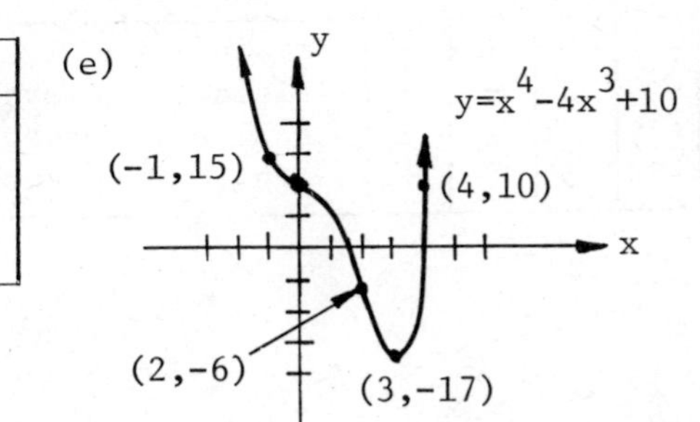

14. (a) $-2x^{-3} + 2$, $6x^{-4}$

(b) $2(x^3 - 1)$, 1, $(0,1)$, $x < 0$, $x > 1$

(c) positive, upward

(d) The curve is discontinuous at $x =$ ____ . For large values of $|x|$, the curve is approximately $y \approx$ _____ . When x is small, the curve is approximately $y \approx$ ______ .

Complete the following table:

x	y	y'	y"	Conclusions
-2	$-3\frac{3}{4}$	+	+	rising; concave up
-1	-1			
$-\frac{1}{2}$	2			
$\frac{1}{2}$	5			
1	3			
2	$4\frac{1}{4}$	+	+	rising; concave up

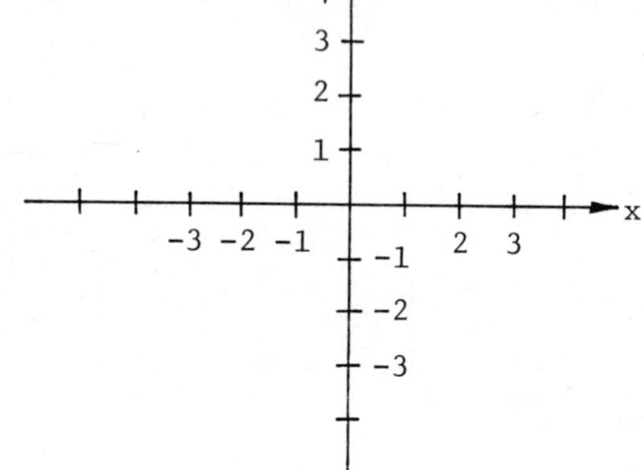

(e) Sketch the graph at the right.

OBJECTIVE B: Given a function $y = f(x)$, locate its vertical tangents, in addition to the information in Objective A above. Sketch the curve showing its relevant features.

15. Sketch the graph of $y = x^{1/3} + x^{2/3}$.

Solution. We follow the five steps as before.

(a) $y' =$ ________________ and $y'' =$ ________________ .

(b) In factored form, $y' = \frac{1}{3}x^{-2/3}$ (__________) . Therefore, $y' = 0$ when $x =$ _____ . We observe that y' fails to exist when $x =$ ____ . However, $dx/dy =$ ______________ which equals zero when $x = 0$. Thus, the tangent to the curve at (0,0) is ____________ .

---

14. (d) 0, 2x, $1/x^2$

x	y	y'	y"	Conclusions
-1	-1	+	+	rising; concave up
$-\frac{1}{2}$	2	+	+	rising; concave up
$\frac{1}{2}$	5	−	+	falling; concave up
1	3	0	+	min.; concave up

(e)

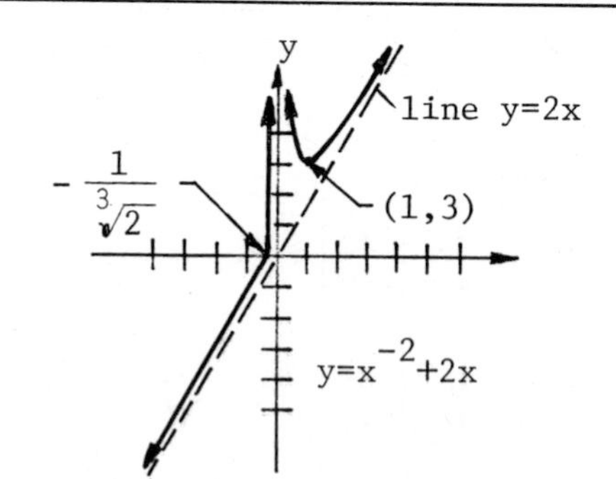

15. (a) $\frac{1}{3}x^{-2/3} + \frac{2}{3}x^{-1/3}$, $-\frac{2}{9}x^{-5/3} - \frac{2}{9}x^{-4/3}$

(b) $1 + 2x^{1/3}$, $-\frac{1}{8}$, 0 , $\frac{3x^{2/3}}{1 + 2x^{1/3}}$, vertical

(c) In factored form,

$$y'' = -\frac{2}{9}x^{-5/3}(\_\_\_\_\_\_\_\_\_\_).$$

Therefore, y'' is positive for ____________ and negative whenever x belongs to the intervals _______ or _______ . Hence the points $x =$ ____ and $x =$ ____ are inflection points.

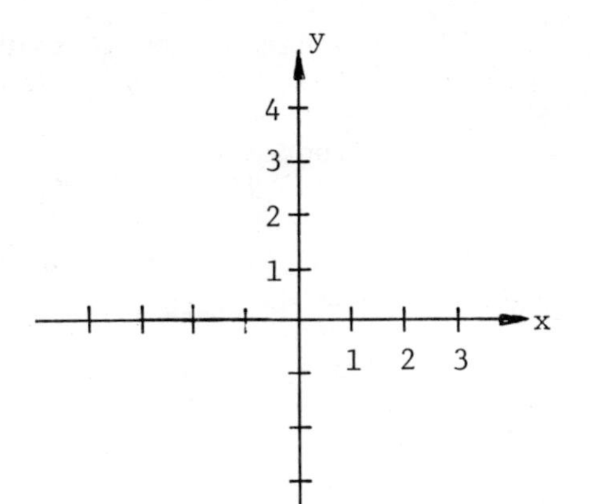

(d) Complete the table.

x	y	y''	y''	Conclusions
-2	1/3	-	-	falling; concave down
-1				
-1/8				
0				
1	2	+	-	rising; concave up

(e) Sketch the graph. Observe that y'' fails to exist at the vertical tangent when $x = 0$. Note, however, that the curve is everywhere continuous.

3-5 Maxima and Minima Theory.

OBJECTIVE A: Define the terms relative maximum, relative minimum, absolute maximum, and absolute minimum.

16. A function f is said to have a relative maximum at $x = a$ if ____________________ for all positive and negative values of h near ________ .

---

15. (d)

x	y	y'	y''	Conclusions
-2	1/3	-	-	falling, concave down
-1	0	-1/3	0	falling, inflection pt.
-1/8	-1/4	0	+	"Holds water"; min.
0	0	+	0	rising; inflection pt.
1	2	+	-	rising; concave down

(e)

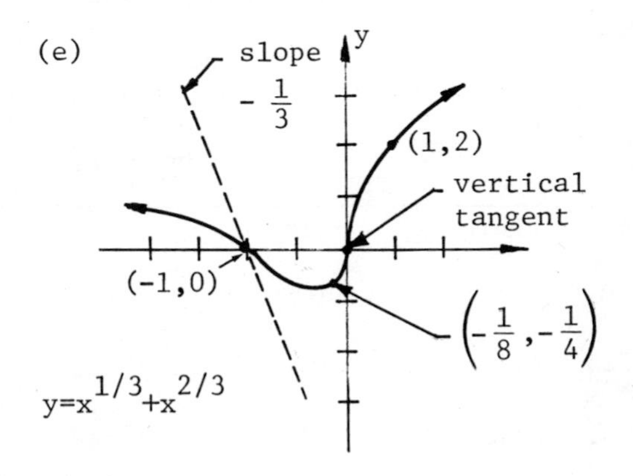

16. $f(a) \geq f(a+h)$, zero

17. A function f is said to have a ___________ maximum over its domain at $x = a$ if $f(a) \geq f(x)$ for all x belonging to the __________ of f .

18. A function f is said to have a ___________ minimum over its domain at $x = a$ if $f(a) \leq f(x)$ for all x close to a .

19. If $f(a) \leq f(x)$ for all x in the domain of f , then f is said to have a ___________ ___________ at $x = a$.

20. Can a relative maximum also be an absolute maximum for a function f ? Can a relative minimum also be an absolute minimum? ____________________

OBJECTIVE B: Interpret correctly the theorem in this article of the text relating relative extrema at an interior point $x = c$ of the domain $a \leq x \leq b$ of a function f and the derivative $f'(c)$.

Answer questions 21 - 24 true or false.

21. If $f'(c) = 0$, then f has either a relative maximum or a relative minimum at the interior point $x = c$. (True or False)

22. If f has a relative minimum at the interior point $x = c$, then $f'(c) = 0$. (True or False)

23. If f has a relative maximum or relative minimum at an endpoint of the interval of definition of the function, then the lefthand (or righthand) tangent must have slope zero there. (True or False)

24. If f has an absolute maximum at an interior point $x = c$ and $f'(c)$ exists as a finite number, then $f'(c)$ is necessarily zero. (True or False)

25. A point on the curve $y = f(x)$ at which $f'(x) = 0$ is called a ___________ point of the curve. Values of x that satisfy the equation $f'(x) = 0$ are called ___________ values of the function f .

26. If f is continuous over the closed interval $a \leq x \leq b$, then every point where f has a (relative or absolute) maximum or minimum must be an ___________ of the interval, a point where f' ________________ , or an ___________ point where f' equals ____ .

---

17. absolute, domain 18. relative 19. absolute minimum

20. yes to both questions 21. False, it could have a point of inflection

22. False, the derivative may fail to exist 23. False

24. True 25. stationary, critical

26. endpoint, does not exist, interior, 0

3-6 Maxima and Minima. Problems.

27. If $\frac{dy}{dx} = 0$ and $\frac{d^2y}{dx^2} > 0$, then y is a ____________ .

28. If $\frac{dy}{dx} = 0$ and $\frac{d^2y}{dx^2} < 0$, then y is a ____________ .

29. If $\frac{dy}{dx} = 0$ and $\frac{d^2y}{dx^2} = 0$, the second derivative test __________ .

30. Another test for a relative maximum at $x = c$ is $f'(c) = 0$ and $f'(x)$ positive for _______ and negative for _______ ; for a relative minimum at $x = c$ the conditions are $f'(c) = 0$, $f'(x)$ ____________ for $x < c$ and ____________ for $x > c$. If dy/dx does not change sign as x advances through c , neither a maximum nor a minimum need occur.

OBJECTIVE: Solve a maxima or minima problem by
(a) writing an equation for the quantity y that is to be a maximum or minimum,
(b) setting up auxiliary equations, if necessary, relating all the variables involved,
(c) locating the points where the derivative of the quantity y is zero or fails to exist, and testing each of these points for a possible maximum or minimum, and
(d) testing the endpoints of the interval over which y is defined (if any exist) for possible extreme values of y .

31. At 9:00 A.M. ship B was 65 miles due east of another ship A. Ship B was then sailing due west at 10 miles per hour, and ship A was sailing due south at 15 miles per hour. If they continue to follow their respective courses, when will they be nearest one another and how near?

Solution. Let A_o and B_o denote the original positions of the ships at 9:00 A.M., and let A and B denote their new positions, respectively, at t hours later. This is pictured in the figure at the right. Let s denote the distance between A and B . The problem is to minimize ____ and to find the time when its minimum occurs.

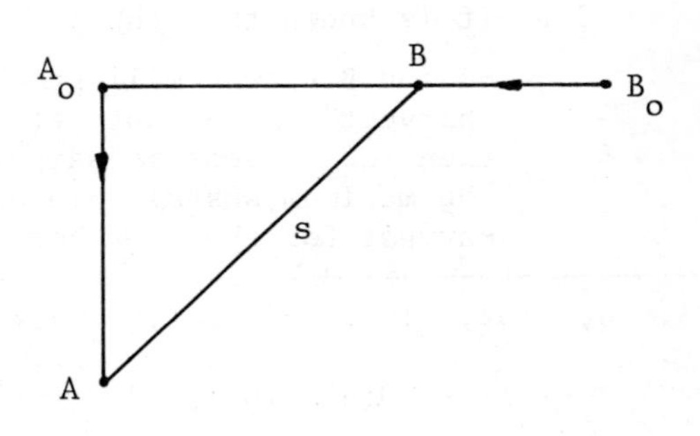

---

27. minimum

28. maximum

29. fails

30. $x < c$, $x > c$, negative, positive

Since rate × time = distance , the distance covered by ship A in t hours is _____ miles, and by ship B _____ miles. The original distance between A and B is given as 65 miles, so the distance between the original position A_o and ship B after t hours is ________ . Fill this information into the figure and then calculate the square of the distance s :

$$s^2 = \_\_\_\_\_\_\_\_\_\_\_\_\_\_\_\_ .$$

Differentiation of both sides of this equation with respect to t gives

$$2s \frac{ds}{dt} = \_\_\_\_\_\_\_\_\_\_\_\_\_\_\_\_ .$$

Thus, $ds/dt = 0$ when $30(15t) - 20(65 - 10t) = 0$ or $t =$ ____ hours. Simplifying ds/dt algebraically, we see that

$$\frac{ds}{dt} = \_\_\_\_\_\_\_\_\_\_\_\_ .$$

Thus, ds/dt is ___________ when $t < 2$ and ___________ when $t > 2$. Therefore, a relative __________ distance occurs for s at $t = 2$ hours. Solving for the distance s after two hours, we find

$$s^2 = (30)^2 + (\_\_\_)^2 \quad \text{or} \quad s = \_\_\_\_\_\_ \text{ miles,}$$

the distance the ships are apart at 11:00 A.M. when they are nearest each other.

32. A company's cost function is $C(x) = 10x + 3$, and its revenue function is $R(x) = 50x - 0.5x^2$, both in thousands of dollars per thousand items. Find the company's maximum profit.

Solution. If $P(x)$ denotes the profit function, then $P(x) = R(x) -$ ______ $=$ _______________ . The maximum profit occurs when $P'(x) =$ ____ , so $P'(x) =$ _________ $= 0$. Thus, $x =$ ____ thousand items. Since $P''(x) =$ ____ is always negative this yields a maximum profit of $P(40) =$ _____ thousand dollars.

33. It is known that the population P for the fur-bearing snowshoe hare in the Hudson Bay area will grow to $f(P) = -0.025P^2 + 4P$ in one year. If they "harvest" the amount $f(P) - P$ so the initial population is not depleted, then the harvest is said to be "sustained." Find the population at which the maximum sustainable harvest occurs, and find the maximum sustainable harvest for the snowshoe hare. Assume P is measured in thousands.

---

31. s, $15t$, $10t$, $65 - 10t$, $(15t)^2 + (65 - 10t)^2$

$30(15t) - 20(65 - 10t)$, 2, $\frac{325t - 650}{s}$, negative,

positive, minimum, 45, $15\sqrt{13}$

A_o B B_o $65 - 10t$ $10t$ $15t$ s A

32. $C(x)$, $40x - 0.5x^2 - 3$, 0, $40 - x$, 40, -1, 797

Solution. The harvest function $H(P) = f(P) - P =$ ________________ . The maximum sustainable harvest occurs when $H'(P) =$ ____ , so $H'(P) =$ ______________ $= 0$, or $P =$ _____ thousand hares. This is the population at which the maximum sustainable harvest occurs, since $H''(P) =$ _______ is always ____________ . The maximum harvest is $H(60) =$ ____ thousand animals.

34. Determine the point on the ellipse $4x^2 + 9y^2 = 36$ that is nearest the origin.

Solution. The problem is to ____________ the distance s from a point (x,y) on the ellipse to the origin. That is, find the minimum of $s =$ ____________ subject to the auxiliary condition that $4x^2 + 9y^2 = 36$. We can just as well minimize $s^2 = S$ since that will also minimize s . Since $S = x^2 + y^2$ is a function of both the variables x and y , we use the equation of the ellipse to eliminate the variable y :

$y^2 = 4 - \frac{4}{9} \cdot x^2$ so that substitution gives

$$S(x) = x^2 + y^2 = \text{\_\_\_\_\_\_\_\_\_\_\_\_} .$$

The minimum distance occurs when $dS/dx =$ ____ , so $(10/9)x = 0$ or $x =$ ____ . Since $S''(x) > 0$ this value of x yields a ___________ . When $x = 0$, $y^2 = 4$ on the ellipse so that $(0,2)$ and $(0,-2)$ are the points on the ellipse that are nearest to the origin.

35. A lighthouse is at a point A , 4 miles offshore from the nearest point 0 of a straight beach; a store is at point B , 4 miles down the beach from 0 . If the lighthouse keeper can row 4 miles/hour and walk 5 miles/hour, find the point C on the beach to which the lighthouse keeper should row to get from the lighthouse to the store in the least possible time.

Solution. The information is sketched in the figure at the right. From the diagram and the Pythagorean theorem, the distance from A to C is _____________ . The total time required to get from A to C to B is

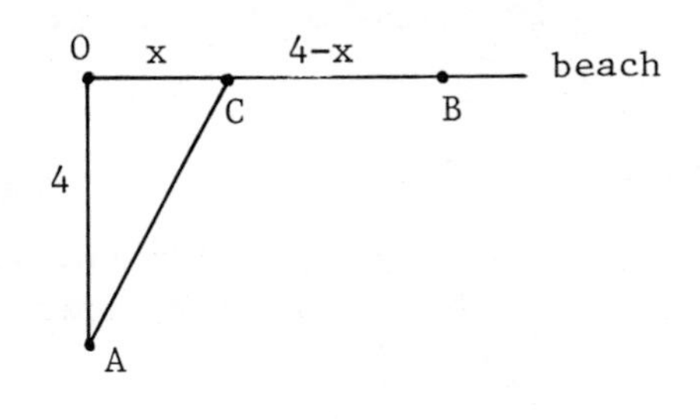

$$T = \frac{\sqrt{16 + x^2}}{4} + \left(\frac{\text{\_\_\_\_}}{5}\right) ,$$

where $0 \leq x \leq 4$.

---

33. $-0.025P^2 + 3P$, 0, $-0.05P + 3$, 60, -0.05, negative, 90

34. minimize, $\sqrt{x^2 + y^2}$, $\frac{5}{9}x^2 + 4$, 0, 0, minimum

35. $\sqrt{16 + x^2}$, $4 - x$, $\frac{1}{8}\left(16 + x^2\right)^{-1/2}(2x) - \frac{1}{5}$, $4\sqrt{16 + x^2}$, $\frac{256}{9}$, $\frac{16}{3}$, endpoints, 1.8, $\sqrt{2}$, 4

The minimum time occurs when $dT/dx = 0$ or,

$$0 = T'(x) = \_\_\_\_\_\_\_\_\_\_\_\_\_\_\_\_ .$$

Simplifying algebraically, $5x =$ ______ or $x^2 =$ ____ or $x =$ ____. However, $x = 16/3$ is outside the allowable range of values $0 \leq x \leq 4$. Therefore, the minimum must be taken on at one of the ________ of the interval. Checking each point, $T(0) =$ ____ hours and $T(4) =$ ____ hours. The smaller of these values occurs when $x =$ ____ so our conclusion is that the lighthouse keeper should row all the way to get to the store in the least possible time.

36. The cost per hour of driving a ship through the water varies approximately as the cube of its speed in the water. Suppose a ship runs into a current of V miles per hour, measured relative to the ocean bottom. Find the total cost for the ship to travel M miles, and find the most economical speed of the ship relative to the ocean bottom.

Solution. Let x denote the speed of the ship relative to the water. Then ______ will be its speed relative to the bottom. The time taken to travel M miles will be ________ . The cost per hour in fuel will be kx^3 for some constant of proportionality k, so the total cost function is given by

$$C(x) = \_\_\_\_\_\_\_\_ .$$

To find the most economical speed, we want to minimize the cost. Now,

$$C'(x) = \_\_\_\_\_\_\_\_\_\_\_\_\_\_\_\_ .$$

The minimum cost occurs when $dC/dx = 0$, or $kMx^2[3(x-V)-x] = 0$. Thus, $x =$ ____ or $x =$ ____ . Since $x = 0$ is ruled out if the ship moves, and since $C(x) \to +\infty$ as $x \to V^+$, we see that $x = 1.5\,V$ must provide the minimum cost.

3-7 Rolle's Theorem.

OBJECTIVE: Apply Rolle's Theorem to show that a given equation $f(x) = 0$ has exactly one root in the specified interval $a \leq x \leq b$.

37. Suppose $y = f(x)$ and its first derivative $f'(x)$ are continuous over $a \leq x \leq b$. If $f(a)$ and $f(b)$ have opposite sign then according to the Intermediate Value Theorem there is at least one point c satisfying $a < c < b$ and $f(c) =$ ____ .

---

36. $x-V$, $M/(x-V)$, $\dfrac{kMx^3}{x-V}$, $\dfrac{(x-V)3kMx^2 - kMx^3}{(x-V)^2}$, 0, 1.5V

37. 0

38. Suppose there is another point d satisfying $a < d < b$ and $f(d) = 0$. Then, according to Rolle's Theorem, there is a point between c and d for which ______ is zero. Thus, if $f'(x)$ is different from zero for all values of x between a and b, there is exactly _____ root to the equation $f(x) = 0$ in the interval ____________.

39. Consider the equation $x^3 + 2x^2 + 5x - 6 = 0$ for $0 \le x \le 5$. When $x = 0$, the value of the left side is _____; and when $x = 5$, the value is ______. These values differ in sign. Calculating the derivative, we have

$$\frac{d}{dx}\left(x^3 + 2x^2 + 5x - 6\right) = \underline{\hspace{4cm}},$$

and this is always ___________ for $0 < x < 5$. Therefore, we conclude from Problems 37 and 38 that there is exactly one root to the equation somewhere between $x =$ ____ and $x =$ ____. We could in fact use Newton's method of Article 2-7 to locate this real root.

3-8 The Mean Value Theorem.

OBJECTIVE A: Given a function $y = f(x)$ satisfying the hypotheses of the Mean Value Theorem for $a \le x \le b$, use the theorem to find a number c satisfying the conclusion of the theorem.

40. The hypotheses of the Mean Value Theorem are that f is ____________ over the closed interval $a \le x \le b$ and _______________ over the open interval ______.

41. The conclusion of the Mean Value Theorem is that there is at least one point c in the open interval ______ satisfying ________________. A geometric interpretation of the conclusion is that the slope of the curve $y = f(x)$ when $x = c$ is the same as the slope of the ________ joining the endpoints $(a, f(a))$ and ________ of the curve.

42. Let $f(x) = 3x^2 + 4x - 3$ over $1 \le x \le 3$. Then $f'(x) =$ _________, so that f and f' satisfy the hypotheses of the Mean Value Theorem. To find a value for c, the equation $f(b) - f(a) = f'(c)(b - a)$ becomes $f(3) - f(1) = f'(c)($ ___ $)$, or $36 - 4 = 2($ ______ $)$. Solving for c gives $c =$ ___.

---

38. $f'(x)$, one, $a \le x \le b$

39. -6, 194, $3x^2 + 4x + 5$, positive, 0, 5

40. continuous, differentiable, $a < x < b$

41. (a,b), $f(b) - f(a) = f'(c)(b - a)$, chord, $(b, f(b))$

42. $6x + 4$, 2, $6c + 4$, 2

43. Does the Mean Value Theorem apply to the function $f(x) = |x|$ in the interval $[-2,1]$?

No, because the derivative $f'(x)$ is not defined for $x =$ ___ so the function f is not ______________ over the open interval ________ as required by the hypotheses.

OBJECTIVE B: Use the approximation $f(b) \approx f(a) + (b-a)f'(a)$ to make reasonable estimates for the value of $y = f(x)$ when $x = b$.

44. Estimate $\sin 29°$.

Here $f(x) = \sin x$, but we must use radian measure. Thus, $b = 29° =$ _____ rad and $a =$ _____ rad . The approximation then becomes,

$$\sin 29° \approx \sin \frac{\pi}{6} + (\_\_\_\_\_\_\_\_\_\_) = \frac{1}{2} - (\_\_\_\_\_\_\_\_\_\_) = .48489 .$$

(The error is less than 10^{-4} from the true value.)

3-9 Indeterminate Forms and l'Hôpital's Rule.

OBJECTIVE A: Evaluate the limits of indeterminate forms using l'Hôpital's rule, whenever applicable.

45. The three hypotheses necessary to apply l'Hôpital's rule

$$\lim_{x \to a} \frac{f(x)}{g(x)} = \lim_{x \to a} \frac{f'(x)}{g'(x)}$$ in the indeterminate case $0/0$ are:

(a) f and g are both ______________ over an open interval I containing the point $x = a$, except possibly at ____ ;

(b) $f(a) = g(a) =$ ____ ; and

(c) $g'(x)$ _____ over I , except possibly when _______ .

46. $\lim_{u \to 1} \frac{u^4 - 1}{\sqrt{u} - 1}$ is of the form _____ . Applying l'Hôpital's rule,

$$\lim_{u \to 1} \frac{u^4 - 1}{\sqrt{u} - 1} = \lim_{u \to 1} \frac{}{\_\_\_\_\_\_\_\_} = \frac{4}{\_\_\_} = \_\_\_ .$$

---

43. 0, differentiable, (-2,1)

44. $\frac{29\pi}{180}$, $\frac{30\pi}{180}$, $-\frac{\pi}{180} \cos \frac{\pi}{6}$, $\frac{\pi}{180}\left(\frac{\sqrt{3}}{2}\right)$

45. differentiable, a, 0, $\neq 0$, $x = a$

46. $0/0$, $\frac{4u^3}{1/2\sqrt{u}}$, $\frac{1}{2}$, 8

47. $\lim\limits_{x\to\infty} \dfrac{\sin \frac{7}{x^2}}{\frac{3}{x^2}}$ is of the form ______ .

Since $d/dx\ (3/x^2) = -6/x^3$ is never ________ , l'Hôpital's rule applies. Thus,

$$\lim_{x\to\infty} \frac{\sin \frac{7}{x^2}}{\frac{3}{x^2}} = \lim_{x\to\infty} \frac{}{\rule{3cm}{0.4pt}} = \lim_{x\to\infty} \left(\rule{1cm}{0.4pt}\right) \cos \frac{7}{x^2} = \rule{1.5cm}{0.4pt} .$$

48. $\lim\limits_{x\to\infty} \dfrac{x^2 - 4x + 200}{5x^4 - 7x^2 + 21}$ is of the form ______ . Applying l'Hôpital's rule,

$$\lim_{x\to\infty} \frac{x^2 - 4x + 200}{5x^4 - 7x^2 + 21} = \lim_{x\to\infty} \frac{}{\rule{3cm}{0.4pt}} \qquad \left[\text{still } \rule{1.5cm}{0.4pt}\right]$$

$$= \lim_{x\to\infty} \frac{2}{60x} = \rule{1cm}{0.4pt} .$$

49. $\lim\limits_{x\to\frac{\pi}{2}^-} (\sec x - \tan x)$ is of the form ______ . We employ trigonometric identities to convert the form to another indeterminate $0/0$ form, and apply l'Hôpital's rule:

$$\lim_{x\to\frac{\pi}{2}^-} (\sec x - \tan x) = \lim_{x\to\frac{\pi}{2}^-} \left(\frac{\rule{2cm}{0.4pt}}{\cos x}\right) \qquad \left[\frac{0}{0}\right]$$

$$= \lim_{x\to\frac{\pi}{2}^-} \frac{}{\rule{3cm}{0.4pt}} = \rule{1cm}{0.4pt} .$$

---

47. $0/0$, zero, $-\dfrac{14}{x^3} \cos \dfrac{7}{x^2} \Big/ -\dfrac{6}{x^3}$, $\dfrac{14}{6}$, $\dfrac{7}{3}$

48. ∞/∞ , $\dfrac{2x - 4}{20x^3 - 14x}$, $\dfrac{\infty}{\infty}$, 0

49. $\infty - \infty$, $1 - \sin x$, $\dfrac{-\cos x}{-\sin x}$, 0

50. $\lim_{x \to \frac{\pi}{2}^+} (\sec x - \tan x)$ is of the form ______ . As in Problem 49,

$$\lim_{x \to \frac{\pi}{2}^+} (\sec x - \tan x) = \lim_{x \to \frac{\pi}{2}^+} \left(\frac{\quad}{\quad}\right) \qquad \left[\frac{0}{0}\right]$$

$$= \lim_{x \to \frac{\pi}{2}^+} \frac{\quad}{\quad} = \underline{\quad} .$$

We conclude that $\lim_{x \to \frac{\pi}{2}} (\sec x - \tan x) = \underline{\quad}$.

51. $\lim_{x \to +\infty} x^2 \sin \frac{1}{x}$ is of the form ______ . We can reduce this to a form 0/0 by algebraic manipulation and then apply l'Hôpital's rule:

$$\lim_{x \to +\infty} x^2 \sin \frac{1}{x} = \lim_{x \to +\infty} \frac{\sin \frac{1}{x}}{\quad} \quad \left[\frac{0}{0}\right] = \lim_{x \to +\infty} \frac{\quad}{\quad}$$

$$= \lim_{x \to +\infty} \left(\quad\right) \cos \frac{1}{x} = \underline{\quad} .$$

There is no finite limit.

52. Which argument is correct, (a) or (b)? Explain.

(a) $$\lim_{x \to 2} \frac{x^3 - x^2 - x - 2}{x^3 - 3x^2 + 3x - 2} = \lim_{x \to 2} \frac{3x^2 - 2x - 1}{3x^2 - 6x + 3} = \lim_{x \to 2} \frac{6x - 2}{6x - 6}$$

$$= \lim_{x \to 2} \frac{6}{6} = 1$$

(b) $$\lim_{x \to 2} \frac{x^3 - x^2 - x - 2}{x^3 - 3x^2 + 3x - 2} = \lim_{x \to 2} \frac{3x^2 - 2x - 1}{3x^2 - 6x + 3} = \frac{12 - 4 - 1}{12 - 12 + 3} = \frac{7}{3} .$$

---

50. $-\infty - (-\infty)$ or $\infty - \infty$ again, $\frac{1 - \sin x}{\cos x}$, $\frac{-\cos x}{-\sin x}$, 0, 0

51. $\infty \cdot 0$, $\frac{1}{x^2}$, $-\frac{1}{x^2} \cos \frac{1}{x} \Big/ -\frac{2}{x^3}$, $\frac{x}{2}$, $+\infty$

52. (b) is correct: l'Hôpital's rule does not apply to the computation of either $\lim_{x \to 2} \frac{3x^2 - 2x - 1}{3x^2 - 6x + 3}$ or $\lim_{x \to 2} \frac{6x - 2}{6x - 6}$ because the numerators and denominators have finite, nonzero limits.

OBJECTIVE B: Given functions f and g that satisfy the hypotheses of the Cauchy Mean Value Theorem for $a \le x \le b$, use the theorem to find values of c satisfying the conclusion of the theorem.

53. Consider $f(x) = \cos x$ and $g(x) = \sin x$ for $0 \le x \le \frac{\pi}{3}$. We observe that f and g are both ____________ for $0 \le x \le \frac{\pi}{3}$, and they are ______________ for $0 < x < \frac{\pi}{3}$. In fact, $f'(x) =$ __________ and $g'(x)$ __________ for $0 < x < \frac{\pi}{3}$. Therefore, the three hypotheses of the Cauchy Mean Value Theorem are satisfied. Translating the conclusion,

$$\frac{f'(c)}{g'(c)} = \frac{f(b) - f(a)}{g(b) - g(a)} \quad \text{becomes} \quad \rule{6cm}{0.4pt},$$

or simplifying, $-\tan c =$ ________________ $=$ _______ , for some c satisfying ____________ . Using trigonometric tables (or a hand-held calculator), c = __________ radians, or c = _____ degrees.

54. Interpreting Problem 53 geometrically, let $x = f(t) = \cos t$ and $y = g(t) = \sin t$ for $0 \le t \le \frac{\pi}{3}$ be parametric equations. Thus, the point (x,y) describes the arc of the unit circle in the plane between the points _______ and _______ . The angle c (in radians) is such that the slope of the tangent line to the arc at the point _______ equals the slope of the _________ joining the two endpoints of the arc. Draw this situation in the coordinate system shown at the right.

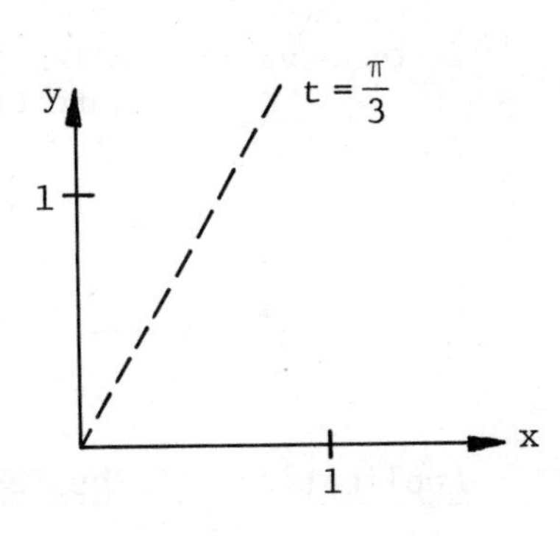

---

53. continuous, differentiable, $-\sin x$, $\cos x$, $\neq 0$, $\dfrac{-\sin c}{\cos c} = \dfrac{\cos\frac{\pi}{3} - \cos 0}{\sin\frac{\pi}{3} - \sin 0}$,

$$\frac{\frac{1}{2} - 1}{\frac{\sqrt{3}}{2} - 0} = \frac{1}{\sqrt{3}}, \quad 0 < c < \frac{\pi}{3}, \quad .524 = \frac{\pi}{6}, \quad 30°$$

54. $(1,0)$, $\left(\frac{1}{2}, \frac{\sqrt{3}}{2}\right)$, $(\cos c, \sin c)$, chord

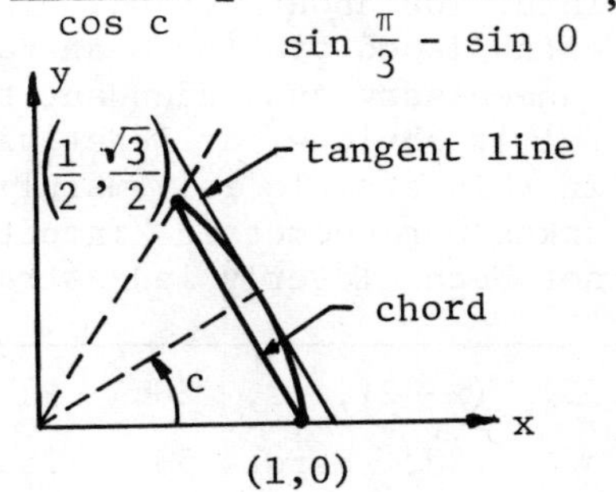

3-10 Extension of the Mean Value Theorem.

OBJECTIVE: Apply the Extended Mean Value Theorem to write any given polynomial $f(x)$ of degree n precisely in the form

$$f(x) = f(a) + f'(a)(x-a) + \frac{f''(a)}{2}(x-a)^2 + \ldots + \frac{f^{(n)}(a)}{n!}(x-a)^n$$

for any specified number a.

55. Consider the polynomial $f(x) = 5x^4 - 3x^3 + x^2 - 7$. Take $a = 2$ so we will express $f(x)$ in powers of _______ rather than powers of x. We need to calculate the first four derivatives of f and evaluate them when $x =$ ____ . Thus,

$f'(x) =$ _______________ so $f'(2) =$ _____ ,

$f''(x) =$ _______________ so $f''(2) =$ _____ ,

$f^{(3)}(x) =$ _______________ so $f^{(3)}(2) =$ _____ ,

$f^{(4)}(x) =$ _______________ so $f^{(4)}(2) =$ _____ .

Observe that all higher derivatives $f^{(n)}(x)$, $n \geq 5$, are identically ________ . Therefore, by the Extended Mean Value Theorem,

$f(x) =$ ____ + ____ $(x-2)$ + ____ $(x-2)^2$ + ____ $(x-2)^3$ + ____ $(x-2)^4$

$=$ ____ .

3-11 Applications of the Mean Value Theorem to Curve Tracing.†

This article of the text is of a more technical nature than the previous ones in this chapter. The discussions of the interpretations of the first and second derivatives in Articles 3-1 and 3-3 relied upon and underscored your geometric intuition about curves. In the present section these same interpretations are established via the Mean Value Theorems. It is very likely that this may seem unnecessary and redundant to you: why re-prove what we have already discovered to be fairly obvious, at least with a little thinking? The point is that the methods used in this article rely mainly on theorems that have already been rigorously proved: it takes our geometric "intuition" out of the picture for the moment to make sure we have not been cleverly led astray. After all, it was "obvious" to not-so-ancient Western

---

55. $(x-2)$, 2, $20x^3 - 9x^2 + 2x$, 128, $60x^2 - 18x + 2$, 206, $120x - 18$, 222, 120, 120, zero, 53, 128, $\frac{206}{2!}$, $\frac{222}{3!}$, $\frac{120}{4!}$,

$f(x) = 53 + 128(x-2) + 103(x-2)^2 + 37(x-2)^3 + 5(x-2)^4$

†The textbook authors note that this section may be omitted without loss of continuity.

civilization that the earth was the center of the universe! Of course, after we establish by rigorous methods that which our geometric intuition seems to be telling us is true, that same intuition is greatly reinforced and made all the more valuable as a tool to aid us further in our mathematical discoveries. And the methods themselves stress the importance and applicability of the Mean Value Theorems.

If we were to write an objective for this section it might be of a vague nature, like "understand or comprehend the arguments used in proving the three theorems." The difficulty is how to measure or test your level of comprehension, and ability to follow the arguments? One possibility is to require that you "be able to reproduce from memory the proofs to these theorems." However, experience has taught that most beginning calculus students struggle painfully with proofs, so such a requirement seems to be not very realistic unless there is a deeper motive. So we forego writing any objectives and merely suggest that you read and study the proofs in this article carefully. You may wish to consult with your calculus instructor to determine if more is to be required of you.

CHAPTER 3 OBJECTIVE - PROBLEM KEY

Objective	Problems in Thomas/Finney Text	Objective	Problems in Thomas/Finney Text
3-1	p. 127, 1-9	3-7	p. 155, 1-3, 6
3-2	p. 131-132, 1-16	3-8 A	p. 159, 1-5
3-3	p. 138, 8,19,21	B	p. 159, 7
3-4 A	p. 137, 1-6, 9-14, 23	3-9 A	p. 165, 1-19
B	p. 138, 15-18	B	p. 165, 20-22
3-6	p. 151-152, 1-35	3-10	p. 167, 3,4

CHAPTER 3 SELF-TEST

In Problems 1 - 3, sketch the curves. Find the intervals of values of x for which the curve is rising, falling, concave upward, and concave downward. Locate all vertical tangents and asymptotes.

1. $y = \dfrac{x}{\sqrt{1+x^2}}$ 2. $y = \dfrac{4x}{x^2+1}$ 3. $y = 1 - (x+1)^{1/3}$

4. A cameraman is televising a 100-yard dash from a position 10 yards from the track in line with the finish line. When the runners are 10 yds. from the finish line, his camera is turning at the rate 3/5 rad/sec. How fast are the runners moving then?

5. A swimming pool is 40 ft long, 20 ft wide, 8 ft deep at the deep end, and 3 ft deep at the shallow end, the bottom being rectangular. If the pool is filled by pumping water into it at the rate of 40 cu. ft/min, how fast is the water level rising when it is

   (a) 3 ft deep at the deep end? (b) 6 ft deep at the deep end?

6. A guy wire is to pass from the top of a pole 36 ft high to an anchorage on the ground 27 ft from the base of the pole. One end of the wire is made fast to the anchorage, and a man climbs the pole with the wire, keeping it taut. If he climbs 2 ft/sec how fast is he paying out the wire when he reaches the top of the pole?

7. Find the absolute maximum and minimum values (if they exist) of $f(x) = x^3 - x^2 - x + 2$ over the interval $0 \leq x < 2$.

8. Suppose a company can sell x items per week at a price $P = 200 - 0.01x$ cents, and that it costs $C = 50x + 20{,}000$ cents to produce the x items. How much should the company charge per item in order to maximize its profits?

9. The weight W (lbs/sec) of flue gas passing up a chimney at different temperatures T is represented by

$$W = A(T - T_0)(1 + \alpha T)^{-2},$$

where A is a positive constant, T the absolute temperature of the hot gases passing up the chimney, T_0 the temperature of the outside air (all in °C), and $\alpha = 1/273$ is the coefficient of expansion of the gas. For a given $T_0 = 15°C$, find the temperature T at which the greatest amount of gas will pass up the chimney.

10. Apply Rolle's Theorem to show that the equation $\cos x = \sqrt{x}$, $x \geq 0$, has exactly one real root.

11. Find all numbers c which satisfy the conclusion of the Mean Value Theorem for $f(x) = 1 + 2x^2$ over $-1 \leq x \leq 1$.

12. Let $f(x) = \frac{1}{x}$. Show that there is no c in the interval $-1 < x < 2$ such that $f'(c) = \frac{f(2) - f(-1)}{2 - (-1)}$. Explain why this does not contradict the Mean Value Theorem.

13. Find a reasonable estimate to $\sqrt[3]{25}$.

Evaluate the limits in Problems 14 - 17.

14. $\lim_{x \to 0} \frac{\sin x - x \cos x}{x^3}$

15. $\lim_{x \to \infty} \frac{x^2 - 5}{2x^2 + 3x}$

16. $\lim_{x \to \infty} \frac{\sin (3/x)}{2/x}$

17. $\lim_{x \to \frac{\pi}{2}^-} \left(x \tan x - \frac{\pi}{2} \sec x\right)$

18. Find all numbers c such that $f(x) = 3x^4$ and $g(x) = x^3 - 2$ satisfy the conclusion of Cauchy's Mean Value Theorem over the interval $[0,1]$.

19. Express the polynomial $f(x) = x^5 - 3x^2 + 1$ in powers of $x + 1$.

SOLUTIONS TO CHAPTER 3 SELF-TEST

1. $y = \dfrac{x}{\left(1+x^2\right)^{1/2}}$

$$y' = \frac{\left(1+x^2\right)^{1/2} - x \cdot \frac{1}{2}\left(1+x^2\right)^{-1/2} 2x}{1+x^2} = \frac{1}{\left(1+x^2\right)^{3/2}}, \text{ and } y'' = \frac{-3x}{\left(1+x^2\right)^{5/2}}.$$

Note that $y = \dfrac{1}{\left(1/x^2 + 1\right)^{1/2}}$ for $x \geq 0$ $\left(\text{since } \sqrt{x^2} = |x|\right)$

and that $y = \dfrac{1}{\left(1/x^2 + 1\right)^{1/2}}$ for $x < 0$. Thus $\lim\limits_{x \to \infty} y = 1$ and $\lim\limits_{x \to -\infty} y = -1$. Hence, the lines $y = 1$ and $y = -1$ are horizontal asymptotes.

Since y' exists for all x and is never zero, there are no critical points. At $x = 0$, $y'' = 0$ so that $x = 0$ is a point of inflection where the graph has slope 1. On $(-\infty, 0)$ $y'' > 0$ and the function is concave upward; on $(0, \infty)$ it is concave downward. Since $y' > 0$ for all x, the graph is everywhere an increasing function of x. This information yields the graph sketched at the right.

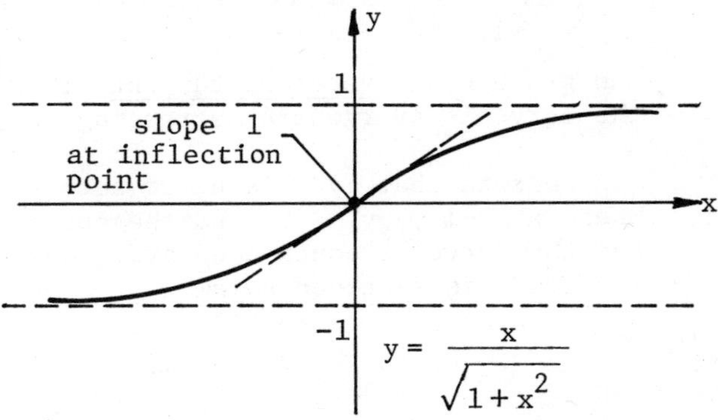

2. Since $\lim\limits_{x \to \pm\infty} y = \lim\limits_{x \to \pm\infty} \dfrac{4/x}{1 + 1/x^2} = 0$, the x-axis is a horizontal asymptote.

Next, $$y' = \frac{4(1-x^2)}{\left(x^2+1\right)^2} \quad \text{and} \quad y'' = \frac{8x(x^2-3)}{\left(x^2+1\right)^3}.$$

Hence, $y' = 0$ implies $x = \pm 1$. Since $y'' > 0$ at $x = -1$ and $y'' < 0$ at $x = 1$, it follows from the second derivative test that $y(-1) = -2$ is a relative minimum and $y(1) = 2$ is a relative maximum.

Next, $y'' = 0$ when $x = 0, -\sqrt{3}$, and $\sqrt{3}$ so that these values for x are points of inflection. Moreover, for

$x < -\sqrt{3}$, $y'' < 0$ and the graph of y is concave downward;

$\sqrt{3} < x < 0$, $y'' > 0$ and the graph of y is concave upward;

$0 < x < \sqrt{3}$, $y'' < 0$ and the graph of y is concave downward,

$x > \sqrt{3}$, $y'' > 0$ and the graph of y is concave upward.

Note that at $x = 0$, $y' = 4$. The graph of y is sketched below (note the symmetry about the origin).

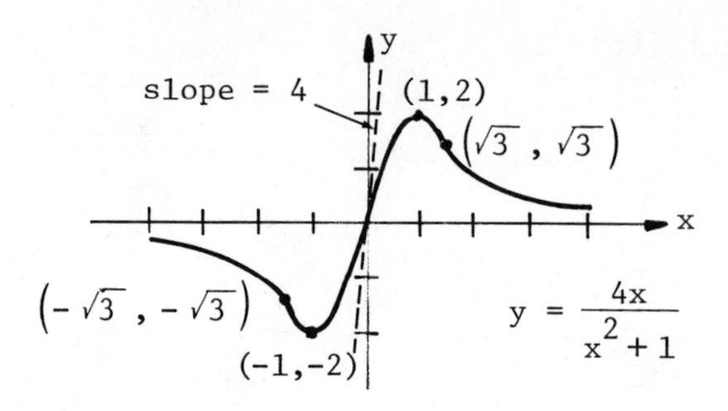

3. $y = 1 - (x+1)^{1/3}$, $y' = -\frac{1}{3}(x+1)^{-2/3}$, and $y'' = \frac{2}{9}(x+1)^{-5/3}$.

The derivative y' does not exist when $x = -1$, although the curve y is continuous at $x = -1$. Since

$$\lim_{x \to -1} \frac{dx}{dy} = \lim_{x \to -1} -3(x+1)^{2/3} = 0 ,$$

the graph has a <u>vertical tangent</u> at $x = -1$. Since $y' < 0$ for all $x \neq -1$, the curve is everywhere decreasing.

We note that y'' is never zero. However, y'' fails to exist at $x = -1$. When $x < -1$, $y'' < 0$ and the curve is concave downward; when $x > -1$, $y'' > 0$ and the curve is concave upward. Therefore, $x = -1$ is a point of inflection. The graph is sketched below.

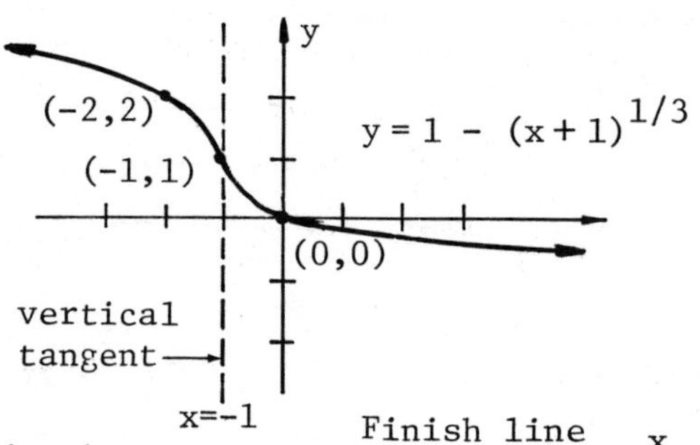

4. The situation is pictured in the figure at the right. Thus,

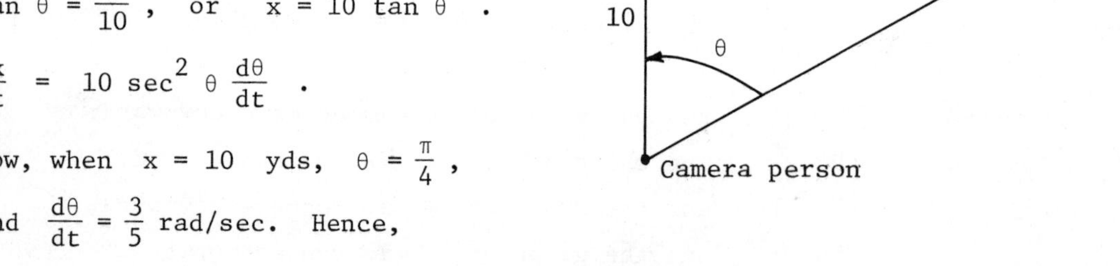

$\tan\theta = \frac{x}{10}$, or $x = 10\tan\theta$.

$$\frac{dx}{dt} = 10\sec^2\theta\,\frac{d\theta}{dt} .$$

Now, when $x = 10$ yds, $\theta = \frac{\pi}{4}$, and $\frac{d\theta}{dt} = \frac{3}{5}$ rad/sec. Hence,

$$\left.\frac{dx}{dt}\right|_{x=10} = 10\left(\sec^2\frac{\pi}{4}\right)\left(\frac{3}{5}\right) = 10\left(\sqrt{2}\right)^2\left(\frac{3}{5}\right) = 12 \text{ yd/sec} .$$

5. A vertical cross-section of the pool is pictured in the figure at the right: y denotes the depth of the water at any time t, and x denotes the horizontal length of the water in the bottom of the pool.

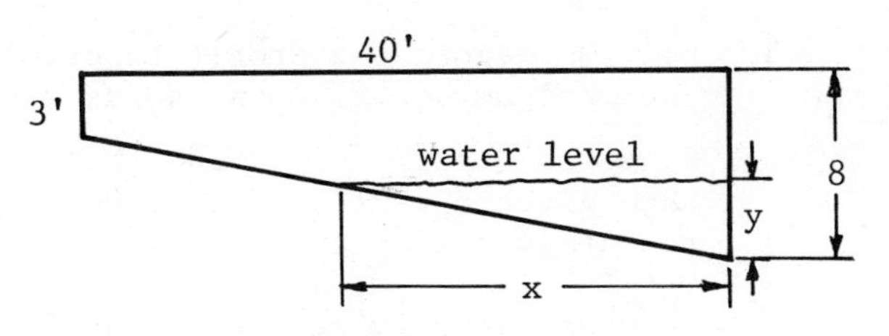

(a) When $y < 5'$, we have from the geometry of similar triangles in the figure that, $\frac{x}{40} = \frac{y}{5}$ or $x = 8y$. The volume of water in the pool is given by $V = \frac{1}{2} x \cdot y \cdot 20 = 80y^2$. Hence $\frac{dV}{dt} = 160y \frac{dy}{dt}$, and since $\frac{dV}{dt} = 40$ is given, solving for dy/dt yields $\left.\frac{dy}{dt}\right|_{y=3} = \frac{40}{160(3)} = \frac{1}{12}$ ft/min.

(b) When $y > 5'$, the total volume of water is given by

$$V = \frac{1}{2}(40)(5)(20) + (40)(20)(y-5) = 800y - 2000 .$$ Hence,

$\frac{dV}{dt} = 800 \frac{dy}{dt}$, and since $\frac{dV}{dt} = 40$ is given, solving for dy/dt yields

$$\left.\frac{dy}{dt}\right|_{y=6} = \frac{40}{800} = \frac{1}{20} \text{ ft/min.}$$

6. The situation is pictured at the right, where h is the height of the man above the ground and ℓ is the length of guy wire played out at any instant of time t. We want to find $\frac{d\ell}{dt}$ when $h = 36$. Now, $\ell^2 = h^2 + (27)^2$, and differentiation with respect to t gives

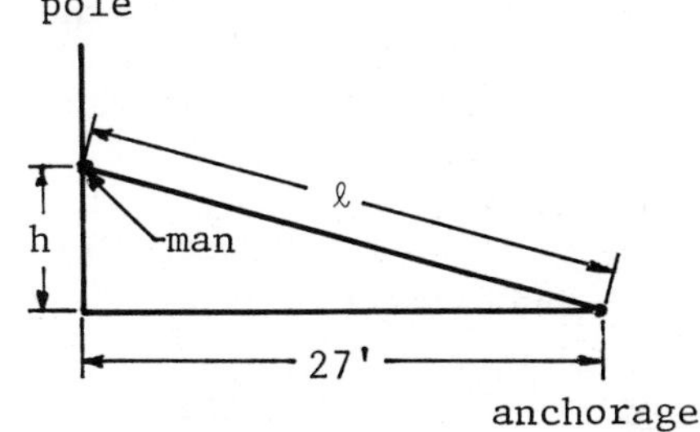

$2\ell \frac{d\ell}{dt} = 2h \frac{dh}{dt}$. When $h = 36$ we have

$$\ell = \sqrt{(36)^2 + (27)^2} = 9\sqrt{4^2 + 3^2} = 45 .$$

Thus, for $dh/dt = 2$,

$$\left.\frac{d\ell}{dt}\right|_{h=36} = \left.\frac{h}{\ell}\frac{dh}{dt}\right|_{h=36} = \frac{36}{45} \cdot 2 = \frac{72}{45} = \frac{8}{5} \text{ ft/sec} .$$

7. $f(x) = x^3 - x^2 - x + 2$ for $0 \leq x < 2$, and

$$f'(x) = 3x^2 - 2x - 1 = (3x+1)(x-1) .$$

Thus, $f'(x) = 0$ implies $x = -\frac{1}{3}$ or $x = 1$. Then $x = 1$ is the only critical point in the interval $[0,2)$. Next note that $f'(x) < 0$ in $[0,1)$ so f is decreasing to the left of $x = 1$, and $f'(x) > 0$ in $(1,2)$ so f is increasing to the right of $x = 1$. Also, $f(0) = 2$, $f(1) = 1$ and $f(2) = 4$. Since $x = 2$ is not in the interval, there is no absolute maximum. The absolute minimum value is $f(1) = 1$ (which is also a relative minimum).

8. Let Q denote the profit function. Then, $Q(x) = xP - C = 150x - 0.01x^2 - 20000$. The maximum occurs when $dQ/dx = 0$, or $150 - .02x = 0$; thus $x = 7500$ items. Since $d^2Q/dx^2 = -.02 < 0$ this provides a maximum profit. The price per item is then given by $P(7500) = 200 - (.01)(7500) = 125$ cents, the price required to obtain the maximum profit $Q(7500) = \$5,425.00$.

9. We want to maximize the weight function W. Now, $W = A(T - T_o)(1 + \alpha T)^{-2}$ and $\frac{dW}{dT} = A(1+\alpha T)^{-2} - 2A\alpha(T - T_o)(1+\alpha T)^{-3}$. Setting $dW/dt = 0$, and simplifying algebraically, gives $(1+\alpha T) - 2\alpha(T - T_o) = 0$, or $T = (1 + 2\alpha T_o)/\alpha$.
Thus, for $T_o = 15°C$ and $\alpha = 1/273$ as given,

$$T = \frac{1}{\alpha} + 2T_o = 273 + 30 = 303°C .$$

Since $\frac{dW}{dT} > 0$ if $T < 303$, and $\frac{dW}{dT} < 0$ if $T > 303$, it is clear that $T = 303°$ provides an absolute maximum for W.

10. Let $f(x) = \cos x - \sqrt{x}$. Since $|\cos x| \le 1$, we see that $f(x) < 0$ if $x > 1$. Thus, the only possible root must lie within the interval $[0,1]$. Now, $f(0) = 1$ and $f\left(\frac{\pi}{2}\right) = -\sqrt{\frac{\pi}{2}}$ so the Intermediate Value Theorem guarantees a root in the interval $\left[0, \frac{\pi}{2}\right]$: we know in fact that the root must lie in $[0,1]$.
Calculating the derivative, $f'(x) = -\sin x - \frac{1}{2\sqrt{x}}$, we see that f' is negative in the interval $(0,1)$. Since f' is different from zero for all values of x between 0 and 1, we conclude there is exactly one real root to the equation $f(x) = 0$ for $x \ge 0$.

11. $f(-1) = 3$ and $f(1) = 3$; $f'(x) = 4x$. Hence $\frac{f(1) - f(-1)}{1 - (-1)} = f'(c)$ translates into $0 = 4c$, or $c = 0$.

12. $\frac{f(2) - f(-1)}{2 - (-1)} = \frac{\frac{1}{2} - (-1)}{3} = \frac{1}{2}$ and $f'(c) = -\frac{1}{c^2}$. Since $-\frac{1}{c^2} = \frac{1}{2}$ is impossible to solve for real values of c, there is no number c in the interval $(-1,2)$ satisfying the conclusion of the Mean Value Theorem. However, this does not contradict the Theorem because $f(x) = 1/x$ is not continuous over the closed interval $[-1,2]$: it fails to be continuous at $x = 0$. Thus the hypotheses of the Theorem are not satisfied.

13. Let $f(x) = x^{1/3}$. Then by the Mean Value Theorem,

$f(25) \approx f(27) + (25 - 27)f'(27)$ or,

$$\sqrt[3]{25} \approx \sqrt[3]{27} + (-2)\frac{1}{3(27)^{2/3}} = 3 - \frac{2}{3 \cdot 9} = \frac{81 - 2}{27} = \frac{79}{27} \approx 2.926 .$$

(A calculator gives $\sqrt[3]{25} \approx 2.924017738$.)

14. $\lim_{x\to 0} \dfrac{\sin x - x\cos x}{x^3}$ is of the form $0/0$. Applying l'Hôpital's rule,

$$\lim_{x\to 0} \frac{\sin x - x\cos x}{x^3} = \lim_{x\to 0} \frac{\cos x - \cos x + x\sin x}{3x^2}$$

$$= \lim_{x\to 0} \frac{\sin x}{3x} = \frac{1}{3} \lim_{x\to 0} \frac{\sin x}{x} = \frac{1}{3}.$$

15. $\lim_{x\to\infty} \dfrac{x^2 - 5}{2x^2 + 3x}$ is of the form ∞/∞. Applying l'Hôpital's rule,

$$\lim_{x\to\infty} \frac{x^2 - 5}{2x^2 + 3x} = \lim_{x\to\infty} \frac{2x}{4x+3} \quad [\text{still } \infty/\infty\ ] = \lim_{x\to\infty} \frac{2}{4} = \frac{1}{2}.$$

16. $\lim_{x\to\infty} \dfrac{\sin(3/x)}{2/x}$ is of the form $0/0$. Applying l'Hôpital's rule,

$$\lim_{x\to\infty} \frac{\sin(3/x)}{2/x} = \lim_{x\to\infty} \frac{(-3/x^2)\cos(3/x)}{-2/x^2} = \lim_{x\to\infty} \frac{3}{2}\cos\frac{3}{x} = \frac{3}{2}\cos 0 = \frac{3}{2}.$$

17. $\lim_{x\to\frac{\pi}{2}^-} \left(x\tan x - \frac{\pi}{2}\sec x\right)$ is of the form $\infty - \infty$. However,

$$\lim_{x\to\frac{\pi}{2}^-} \left(x\tan x - \frac{\pi}{2}\sec x\right) = \lim_{x\to\frac{\pi}{2}^-} \frac{x\sin x - \frac{\pi}{2}}{\cos x}$$ is of the form $0/0$.

Applying l'Hôpital's rule,

$$\lim_{x\to\frac{\pi}{2}^-} \frac{x\sin x - \frac{\pi}{2}}{\cos x} = \lim_{x\to\frac{\pi}{2}^-} \frac{\sin x + x\cos x}{-\sin x} = \frac{1 + \left(\frac{\pi}{2}\right)(0)}{(-1)} = -1.$$

18. The functions f and g satisfy the hypotheses of continuity and differentiability, and $g'(x) = 3x^2 \neq 0$ for $0 < x < 1$. Thus,

$$\frac{f'(c)}{g'(c)} = \frac{f(b) - f(a)}{g(b) - g(a)} \quad \text{translates to} \quad \frac{12c^3}{3c^2} = \frac{3-0}{-1-(-2)}, \quad \text{or} \quad 4c = 3.$$

Hence, $c = 3/4$.

19. Using the Extended Mean Value Theorem with $a = -1$,

$$f(x) = x^5 - 3x^2 + 1 , \qquad f(-1) = -3 ,$$

$$f'(x) = 5x^4 - 6x , \qquad f'(-1) = 11 ,$$

$$f''(x) = 20x^3 - 6 , \qquad f''(-1) = -26 ,$$

$$f^{(3)}(x) = 60x^2 , \qquad f^{(3)}(-1) = 60 ,$$

$$f^{(4)}(x) = 120x , \qquad f^{(4)}(-1) = -120 ,$$

$$f^{(5)}(x) = 120 , \qquad f^{(5)}(-1) = 120 .$$

Thus, $$f(x) = f(-1) + f'(-1)(x+1) + \frac{1}{2} f''(-1)(x+1)^2 + \frac{1}{3!} f^{(3)}(-1)(x+1)^3$$

$$+ \frac{1}{4!} f^{(4)}(-1)(x+1)^4 + \frac{1}{5!} f^{(5)}(-1)(x+1)^5$$

translates into,

$$x^5 - 3x^2 + 1 = -3 + 11(x+1) - 13(x+1)^2 + 10(x+1)^3 - 5(x+1)^4 + (x+1)^5 .$$

CHAPTER 4 INTEGRATION

4-1 Introduction.

The notion of the inverse of an operation implies an "undoing" or reversal of the operation. That is, if we first perform an operation and then perform its inverse, we return to the original state. In arithmetic, for example, subtraction is the inverse operation of addition: if we begin with the number x and add 3 we obtain $x+3$; subtracting 3 from $x+3$ brings us back to the original number x . In Article 2-4 the idea of the inverse of a function was introduced: if two functions f and g are mutually inverse to each other, the composite of f with g or of g with f produces the identity mapping $y = x$. Thus, we saw that the inverse of the cube function $f(x) = x^3$ is the cube root function $g(x) = x^{1/3}$; and the inverse of the square function $f(x) = x^2$ on the domain $x \geq 0$ is the square root function $g(x) = \sqrt{x}$. In the next several sections the idea of "undoing" the differentiation process will be studied. This undoing process is commonly termed indefinite integration or antidifferentiation. The significance of the indefinite integration process will be revealed by an important theorem and by the development further on in this chapter.

4-2 The Indefinite Integral.

Let u and v denote differentiable functions of some independent variable (say of x), and suppose a, n, and C are constants. Then the four basic integration formulas of this section are as follows:

1. $\int du =$ ________ 2. $\int a\, du =$ ________

3. $\int (du + dv) =$ ____________ 4. $\int u^n\, du =$ __________ , $n \neq$ ____

OBJECTIVE A: Find indefinite integrals of elementary functions using the four formulas of this section.

5. $\int (3x^4 - x^{-2} + 5)\, dx = \int 3x^4\, dx -$ ________ $+ \int 5\, dx$

$= 3$ ________ $- \int x^{-2}\, dx + 5 \int dx$

$= 3($ ______ $) - ($ _____ $) + 5\,($ ____ $) + C$

$=$ ________________________ .

---

1. $u + C$ 2. $a \int du$ 3. $\int du + \int dv$ 4. $\frac{u^{n+1}}{n+1} + C$, -1

5. $\int x^{-2}\, dx$, $\int x^4\, dx$, $\frac{1}{5} x^5$, $-x^{-1}$, x , $\frac{3}{5} x^5 + x^{-1} + 5x + C$

6. $\int (5x-3)^9\,dx$.

Let $u = 5x-3$, $du =$ ______ . Thus the integral becomes,

$$\int (5x-3)^9\,dx = \frac{1}{5}\int \underline{\qquad}\,du = \frac{1}{5}\left(\underline{\qquad}\right) + C$$

$$= \frac{1}{50}\left(\underline{\qquad\qquad}\right) + C .$$

7. $\int \frac{x\,dx}{\left(7-x^2\right)^5}$.

Let $u = 7-x^2$. Then $du =$ ________ so $xdx =$ ________ , and

$$\int \frac{x\,dx}{\left(7-x^2\right)^5} = -\frac{1}{2}\int \frac{du}{\underline{\qquad}} = -\frac{1}{2}\left(\underline{\qquad}\right) + C$$

$$= \frac{1}{8}\left(\underline{\qquad\qquad}\right) + C .$$

8. $\int \left(x^{3/2} - 2x^{2/3} + 5\right)\sqrt{x}\,dx = \int \left(x^2 - \underline{\qquad} + 5x^{1/2}\right)dx$

$$= \left(\underline{\quad}\right) - 2\left(\underline{\qquad}\right) + 5\left(\underline{\qquad}\right) + C .$$

$$= \underline{\qquad\qquad\qquad\qquad} .$$

OBJECTIVE B: Solve an elementary differential equation when dy/dx equals an expression in which both x and y may occur, and the variables are separable.

9. The solution technique to solving such an equation is to ____________ the variables so that the differential equation is in the form $g(y)\,dy = f(x)\,dx$, and then ____________ both sides.

10. To solve the differential equation $\frac{dy}{dx} = \frac{y^3}{x^2}$ for $x > 0$ and $y > 0$, separate the variables obtaining the equivalent expression $y^{-3}\,dy =$ _____ .

---

6. $5dx$, u^9, $\frac{u^{10}}{10}$, $(5x-3)^{10}$

7. $-2xdx$, $-\frac{1}{2}\,du$, u^5, $\frac{u^{-4}}{-4}$, $\left(7-x^2\right)^{-4}$

8. $2x^{7/6}$, $\frac{x^3}{3}$, $\frac{x^{13/6}}{13/6}$, $\frac{x^{3/2}}{3/2}$, $\frac{1}{3}x^3 - \frac{12}{13}x^{13/6} + \frac{10}{3}x^{3/2} + C$

9. separate, integrate

10. $x^{-2}\,dx$, $\frac{y^{-2}}{-2}$, $x/2(1-Cx)$

Integration of each side then gives (_____) $= -x^{-1} + C$, or solving for y^2 , $y^2 =$ ____________________ .

11. Let $\frac{dy}{dx} = \sqrt{x} + 2x$, $x \geq 0$. Then $dy = ($ __________ $) dx$, and integration of both sides gives $y = ($ ____________ $) + C$.

12. Let $\frac{ds}{dt} = \left(t - \frac{1}{\sqrt[3]{t}}\right)^2$, $t \neq 0$. Then $ds = \left(t - \frac{1}{\sqrt[3]{t}}\right)^2 dt$ or, expanding the right side algebraically, $ds =$ __________________ .

Integration of both sides gives $s =$ ________________________ .

4-3 Applications of Indefinite Integration.

OBJECTIVE: Solve initial value problems involving differential equations with separable variables.

13. Consider the differential equation $\frac{dy}{dx} = \frac{x^2+1}{\sqrt{y}}$, $y > 0$, subject to the initial condition $y = 1$ when $x = 0$. Writing the equation in differential form gives, $\sqrt{y}\, dy =$ ___________ . Integrating both sides,

$\frac{2}{3} y^{3/2} =$ ________________ . Next we impose the ___________ condition to evaluate the constant of integration: $\frac{2}{3}(1)^{3/2} =$ _______________ ; so $C =$ ____ . Substituting this value of C into the solution of the differential equation gives ____________________ as the solution to the initial value problem.

14. Suppose the acceleration of a moving particle is given by the equation $a = \sqrt{t}$, and that when $t = 0$ it is known that the particle has initial velocity $v = v_0$ and initial position $s = s_0$. Let us find the equation of motion. Since $a = dv/dt$, the original equation can be written in differential form as $dv =$ _______ . Integration then gives $v =$ ______ $+ C_1$. Imposing the intial condition $v = v_0$ when $t = 0$ yields $v_0 =$ ___________; so $C_1 =$ ____ . Hence the velocity equation becomes $v =$ _________________ .

---

11. $\sqrt{x} + 2x$, $\frac{2}{3}x^{3/2} + x^2$

12. $t^2 - 2t^{2/3} + t^{-2/3}$, $\frac{1}{3}t^3 - \frac{6}{5}t^{5/3} + 3t^{1/3} + C$

13. $(x^2+1)\, dx$, $\frac{1}{3}x^3 + x + C$, initial, $\frac{1}{3}(0)^3 + 0 + C$, $\frac{2}{3}$, $\frac{2}{3}y^{3/2} = \frac{1}{3}x^3 + x + \frac{2}{3}$

Now, $v = ds/dt$ so the last equation can be written in differential form as $ds = \left(\frac{2}{3} t^{3/2} + v_o\right) dt$. Integration gives $s =$ __________ $+ C_2$. From the initial condition $s = s_o$ when $t = 0$ it is readily seen that $C_2 =$ ____ . Therefore, the equation of motion is $s =$ __________ valid for all $t \geq 0$.

4-4 Integration of Sines and Cosines.

OBJECTIVE: Find indefinite integrals of elementary functions involving the sine and cosine. Substitution and algebraic manipulation may be required.

15. $\int \sin (3 - 2x)\, dx$.

Let $u = 3 - 2x$. Then $du =$ ______ so $dx = -\frac{1}{2} du$. Thus the integral becomes

$$\int \sin (3 - 2x)\, dx = \int \_\_\_\_\_\_\_\_ = \frac{1}{2} \int \_\_\_\_\_\_$$

$$= \frac{1}{2} \_\_\_\_ + C = \_\_\_\_\_\_\_\_ .$$

16. $\int x^2 \cos (4x^3)\, dx$.

Let $u = 4x^3$. Then $du =$ ______ so $x^2\, dx =$ ______ . Thus the integral becomes,

$$\int x^2 \cos (4x^3)\, dx = \int \cos (4x^3) \cdot x^2 dx = \int \_\_\_\_\_\_ du$$

$$= \frac{1}{12} \_\_\_\_ + C = \_\_\_\_\_\_\_\_ .$$

17. $\int (3 - \sin 2t)^{1/3} \cos 2t\, dt$.

Let $u = 3 - \sin 2t$. Then $du =$ ________ . Substitution into the integral gives,

$$\int (3 - \sin 2t)^{1/3} \cos 2t\, dt = \int \_\_\_\_\_\_ du = \_\_\_\_\_\_\_\_ + C$$

$$= \_\_\_\_\_\_\_\_\_\_ .$$

---

14. $\sqrt{t}\, dt$, $\frac{2}{3} t^{3/2}$, $\frac{2}{3}(0)^{3/2} + C_1$, v_o, $\frac{2}{3} t^{3/2} + v_o$, $\frac{4}{15} t^{5/2} + v_o t$, s_o, $\frac{4}{15} t^{5/2} + v_o t + s_o$

15. $-2dx$, $\sin u \cdot \left(-\frac{1}{2}\right) du$, $-\sin u\, du$, $\cos u$, $\frac{1}{2} \cos (3 - 2x) + C$

16. $12x^2\, dx$, $\frac{1}{12} du$, $\frac{1}{12} \cos u$, $\sin u$, $\frac{1}{12} \sin (4x^3) + C$

17. $-2 \cos 2t\, dt$, $-\frac{1}{2} u^{1/3} du$, $-\frac{3}{8} u^{4/3}$, $-\frac{3}{8} (3 - \sin 2t)^{4/3} + C$

18. $\int \cos \sqrt{x}\, dx$.

We try $\sin \sqrt{x} + C$. Its differential is ______________ . Thus we have the factor $1/2\sqrt{x}$ which we do not want. So let us try the new trial function $2\sqrt{x} \sin \sqrt{x} + C$. The differential of this new function is ______________________ which does have the term $\cos \sqrt{x}$, but also the term $\frac{1}{\sqrt{x}} \sin \sqrt{x}$ that we do not want at all. To eliminate it we want to add to the trial function a term whose differential is $-\frac{1}{\sqrt{x}} \sin \sqrt{x}$. Since it may not be clear what this term is we try adding $\cos \sqrt{x}$. This changes the trial function to __________________________ whose differential is $\left(\frac{1}{2\sqrt{x}} \sin \sqrt{x} + \rule{6cm}{0.4pt}\right)dx$. Thus we see that the term we added needs to be altered by a factor of 2 . We finally try $2 \cos \sqrt{x} + 2\sqrt{x} \sin \sqrt{x} + C$ whose differential is

____ dx which equals

$\cos \sqrt{x}\, dx$. Therefore, $\int \cos \sqrt{x}\, dx =$ ___________________________ .

4-5 Area Under A Curve.

19. Let $y = f(x)$ define a positive continuous function of x on the closed interval $a \le x \le b$. The area under the curve and above the x-axis from $x = a$ to $x = b$ is defined to be the _________ of the sums of the areas of inscribed _____________ as their number ___________ without bound.

OBJECTIVE A: Given an elementary function $y = f(x)$ positive and continuous over the interval $a \le x \le b$, approximate the area under the curve by summing a specified number n of inscribed rectangles of uniform width.

20. Consider $y = \cos x$; $a = 0$, $b = \frac{\pi}{2}$, $n = 5$. The interval $0 \le x \le \pi/2$ is divided into ____ subintervals, each of length $\Delta x = (\quad\quad)/5 =$ ____ $\approx .314$ rads. Thus, the $n - 1 = 4$

---

18. $\frac{1}{2\sqrt{x}} \cos \sqrt{x}\, dx$, $\left(\frac{1}{\sqrt{x}} \sin \sqrt{x} + \cos \sqrt{x}\right) dx$, $\cos \sqrt{x} + 2\sqrt{x} \sin \sqrt{x} + C$,

$\frac{1}{\sqrt{x}} \sin \sqrt{x} + \cos \sqrt{x}$, $-\frac{1}{\sqrt{x}} \sin \sqrt{x} + \frac{1}{\sqrt{x}} \sin \sqrt{x} + 2\sqrt{x} \cos \sqrt{x} \cdot \frac{1}{2\sqrt{x}}$,

$2 \cos \sqrt{x} + 2\sqrt{x} \sin \sqrt{x} + C$

19. limit, rectangles, increases

intermediate points are located at $x_1 = .314$, $x_2 = .628$, $x_3 =$ _____ , and $x_4 =$ _______ . Now, the curve $y = \cos x$ ____________ with x over the interval $0 \le x \le \pi/2$, so the altitude of each inscribed rectangle is the length of its ________ edge. Calculation of the areas of the inscribed rectangles gives (use a calculator or the trigonometric tables for radians in the Appendix of your textbook),

$$\cos(.314) \cdot \Delta x = (0.951)(.314) = 0.299$$

$$\cos(\_\_\_\_) \cdot \Delta x = (0.809)(.314) = \_\_\_\_\_$$

$$\cos(.942) \cdot \Delta x = (\_\_\_\_\_)(.314) = \_\_\_\_\_$$

$$\cos(\_\_\_\_) \cdot \Delta x = (\_\_\_\_\_)(.314) = \_\_\_\_\_$$

$$\cos(\_\_\_\_) \cdot \Delta x = (\_\_\_\_\_)(.314) = \_\_\_\_\_$$

$$\text{Sum} = \_\_\_\_\_ .$$

The estimate is too ________ because we were using inscribed rectangles. In fact, our estimate of 0.835 is about 17 percent too small.

OBJECTIVE B: Interpret and utilize the sigma notation to express or write out sums, and determine (if possible) the value of a sum expressed in sigma notation.

21. Write out the sum $\sum_{k=3}^{7} 2^{k-2}$,

Replace the k in 2^{k-2} by 3 and obtain ____ .
Replace the k in 2^{k-2} by 4 and obtain ____ .
Replace the k in 2^{k-2} by 5 and obtain ____ .
Replace the k in 2^{k-2} by 6 and obtain ____ .
Replace the k in 2^{k-2} by 7 and obtain ____ .

The expanded form is $\sum_{k=3}^{7} 2^{k-2} =$ ________________________ .

This finite sum is equal to _____ .

---

20. 5, $\frac{\pi}{2} - 0$, $\pi/10$, .942, 1.257, decreases, right, .628, 0.254, 0.588, 0.185, 1.257, 0.309, 0.097, $\pi/2$, 0, 0, 0.835, small

21. 2^1, 2^2, 2^3, 2^4, 2^5, $2^1 + 2^2 + 2^3 + 2^4 + 2^5$, 62

22. To express the finite sum $1 + \frac{1}{2} + \frac{1}{4} + \frac{1}{8} + \frac{1}{16}$ in sigma notation, we may observe that the sum can be written as

$$\left(\frac{1}{2}\right)^0 + \left(\frac{1}{2}\right)^1 + \left(\frac{1}{2}\right)^2 + (\_\_)^3 + (\_\_)^4 .$$

The kth term in this expression is $\left(\frac{1}{2}\right)^k$, and we see that k starts at ___ and ends at ___ . Therefore, the required sigma notation is

_________ .

23. To write out the sum $\sum_{k=0}^{3} \frac{(-1)^k}{k!} x^k$, first

recall that $k!$ means $1 \cdot 2 \cdot 3 \cdots k$. Then,

Replace the k in $\frac{(-1)^k}{k!} x^k$ by 0 and obtain ________ = _______ .

Replace the k in $\frac{(-1)^k}{k!} x^k$ by 1 and obtain ________ = _______ .

Replace the k in $\frac{(-1)^k}{k!} x^k$ by 2 and obtain ________ = _______ .

Replace the k in $\frac{(-1)^k}{k!} x^k$ by 3 and obtain ________ = _______ .

The expanded form is $\sum_{k=0}^{3} \frac{(-1)^k}{k!} x^k =$ ______________________________ .

Substitution of $x = 1$ gives the value ____ .

4-6 Computation of Area as Limits.

OBJECTIVE A: Establish elementary arithmetic formulas by the method of mathematical induction. That is, show the formula is correct for $n = 1$; and show that if true for n , the formula is also true for $n + 1$.

24. Consider the formula $2^1 + 2^2 + 2^3 + \cdots + 2^n = 2^{n+1} - 2$. In sigma notation the formula could be written as $\sum_{k=1}^{n}$ ____ = __________ . To

---

22. $\frac{1}{2}$, $\frac{1}{2}$, 0, 4, $\sum_{k=0}^{4} \left(\frac{1}{2}\right)^k$

23. $\frac{(-1)^0}{0!} x^0 = 1$, $\frac{(-1)^1}{1!} x^1 = -x$, $\frac{(-1)^2}{2!} x^2 = \frac{1}{2} x^2$, $\frac{(-1)^3}{3!} x^3 = -\frac{1}{6} x^3$,

$1 - x + \frac{1}{2} x^2 - \frac{1}{6} x^3$, $\frac{1}{3}$

verify the formula for $n = 1$ we need to show that $\sum_{k=1}^{1} 2^k = 2^{1+1} - 2$ or $2^1 = 2^{1+1} - 2$. Is this last statement true?

25. Write the formula for $n = 2$. Is it true? (It is not necessary to verify the formula for $n = 2$ when applying mathematical induction, but this problem gives you some practice in working with the formula.)

26. Suppose now that n is any integer for which the formula in Problem 24 is known to be true (at the moment, n could be the integers ___ and ___). Let us add 2^{n+1} to both sides of the formula,

$$2^1 + 2^2 + 2^3 + \cdots + 2^n + 2^{n+1} = (2^{n+1} - 2) + 2^{n+1}$$

which must also be true for the same n . The right side of this last equation can be written as $(2^{n+1} - 2) + 2^{n+1} = 2(\_\_\_\_\_) - 2 = \_\_\_\_\_\_\_ - 2$.
Thus, $2^1 + 2^2 + \cdots + 2^n + 2^{n+1} = \_\_\_\_\_\_\_\_\_\_\_$. This is just like the original formula in Problem 24 except that n is replaced by ______ .

OBJECTIVE B: Find the area bounded by a curve $y = f(x)$ positive over $a \leq x \leq b$ by finding the limit of the sum of inscribed or circumscribed rectangles, if f is of the form $y = mx^k$ for $k = 0,1,2,3$.

27. To find the area under the graph $y = x^2$, $-1 \leq x \leq 0$ using inscribed rectangles, divide the interval $-1 \leq x \leq 0$ into n subintervals each of equal length $\Delta x =$ _____ by inserting the points

$x_1 = -1 + \Delta x$, $x_2 = -1 + 2\Delta x$, . . . , $x_{n-1} =$ ___________ .

Notice that $y = x^2$ is a ______________ function of x over the interval $-1 \leq x \leq 0$. Thus, the height of the first inscribed rectangle is _________ . The inscribed rectangles have areas

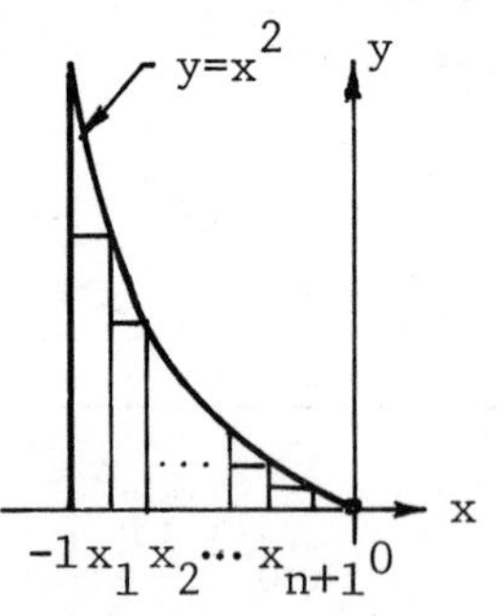

$$f(x_1)\Delta x = (-1 + \Delta x)^2 \cdot \Delta x$$

$$f(x_2)\Delta x = (-1 + 2\Delta x)^2 \cdot \Delta x$$

$$\vdots$$

$$f(x_{n-1})\Delta x = \left(-1 + (n-1)\Delta x\right)^2 \cdot \Delta x$$

$$f(\_\_)\Delta x = \_\_\_\_\_\_\_\_\_\_\_\_\_\_\_\_ ,$$

---

24. $\sum_{k=1}^{n} 2^k = 2^{n+1} - 2$, yes because $2 = 4 - 2$ 25. $2^1 + 2^2 = 2^{2+1} - 2$, yes because $2 + 4 = 8 - 2$

26. 1, 2, 2^{n+1}, 2^{n+2}, $2^{n+2} - 2$ or $2^{(n+1)+1} - 2$, $n + 1$

whose sum is

$$S_n = \left[\left(-1+\frac{1}{n}\right)^2 + \left(-1+\frac{2}{n}\right)^2 + \cdots + \left(-1+\frac{n-1}{n}\right)^2 + 0\right] \cdot \underline{\qquad}$$

Expanding each term on the right side and collecting like terms gives,

$$S_n = \left[(-1)^2(n-1) + 2(-1)\left(\frac{1}{n}+\frac{2}{n}+\cdots+\frac{n-1}{n}\right) + \left(\frac{1^2}{n^2}+\frac{2^2}{n^2}+\cdots+\frac{(n-1)^2}{n^2}\right)\right]\frac{1}{n}$$

$$= \frac{1}{n}\left[(n-1) - \frac{2}{n}\left(\underline{\qquad\qquad\qquad}\right) + \frac{1}{n^2}\left(\underline{\qquad\qquad\qquad}\right)\right]$$

$$= \frac{1}{n}\left[(n-1) - \frac{2}{n}\cdot\left(\underline{\qquad\qquad}\right) + \frac{1}{n^2}\cdot\left(\underline{\qquad\qquad\qquad}\right)\right]$$

$$= \frac{1}{n}\left[(n-1) - \underline{\qquad} + \frac{1}{6}\left(2n - 3 + \frac{1}{n}\right)\right] = \underline{\qquad\qquad\qquad} .$$

The area under the graph is defined to be the limit of ____ as $n \to \infty$. This limit equals ____ .

28. To find the area under the graph $y = 3x^2$, $3 \leq x \leq 4$ using circumscribed rectangles, divide the interval $3 \leq x \leq 4$ into n subintervals each of equal length $\Delta x =$ _____ by inserting the points

$$x_1 = 3+\Delta x, \quad x_2 = 3+2\Delta x, \quad \ldots, \quad x_{n-1} = \underline{\qquad\qquad} .$$

Notice that $y = 3x^2$ is an _____________ function of x over the interval $3 \leq x \leq 4$. Thus, the height of the first circumscribed rectangle is ______; the height of the last circumscribed rectangle is ______ . The circumscribed rectangles have areas

---

27. $1/n$, $-1 + (n-1)\Delta x$, decreasing, $(-1+\Delta x)^2$ or x_1^2, 0, $0 \cdot \Delta x$ or $(-1+n\Delta x)^2 \cdot \Delta x$, $1/n$, $1+2+3+\cdots+(n-1)$, $1^2+2^2+\cdots+(n-1)^2$, $\frac{(n-1)n}{2}$, $\frac{(n-1)n\big(2(n-1)+1\big)}{6}$, $(n-1)$, $\frac{1}{3} - \frac{1}{2n} + \frac{1}{6n^2}$, S_n, $1/3$

28. $1/n$, $3 + (n-1)\Delta x$, increasing, $3x_1^2$ or $3(3+\Delta x)^2$, $3(4)^2$ or $3(3+n\Delta x)^2$, 4, $3(4)^2 \cdot \Delta x$ or $3(3+n\Delta x)^2 \cdot \Delta x$, $3 + \frac{1}{n}$, $\frac{1}{n}+\frac{2}{n}+\cdots+\frac{n}{n}$, $\frac{1^2}{n^2}+\frac{2^2}{n^2}+\cdots+\frac{n^2}{n^2}$, $\frac{n(n+1)}{2}$, $\frac{n(n+1)(2n+1)}{6}$, $2+\frac{3}{n}+\frac{1}{n^2}$, $3\left(3^2+3+\frac{1}{3}\right)$

$$f(x_1)\Delta x = 3(3+\Delta x)^2 \cdot \Delta x$$

$$f(x_2)\Delta x = 3(3+2\Delta x)^2 \cdot \Delta x$$

$$\vdots$$

$$f(\_\_)\Delta x = \_\_\_\_\_\_\_\_ ,$$

whose sum is

$$S_n = \left[3\left(\_\_\_\_\right)^2 + 3\left(3+\frac{2}{n}\right)^2 + \cdots + 3\left(3+\frac{n}{n}\right)^2\right]\frac{1}{n}$$

$$= \frac{3}{n}\left[n3^2 + 6\,(\_\_\_\_\_\_\_\_\_\_\_\_) + (\_\_\_\_\_\_\_\_\_\_\_\_)\right]$$

$$= \frac{3}{n}\left[n3^2 + \frac{6}{n}\cdot\left(\_\_\_\_\_\_\right) + \frac{1}{n^2}\cdot\left(\_\_\_\_\_\_\_\_\right)\right]$$

$$= 3\left[3^2 + 3\left(1+\frac{1}{n}\right) + \frac{1}{6}\left(\_\_\_\_\_\_\_\_\right)\right] .$$

Hence, $\ell im\ S_n$ as $n \to \infty$ equals ____________ = 37 , the area.

4-7 Areas by Calculus.

In Problems 29 - 33, let f be a positive-valued continuous function over the domain $a \leq x \leq b$, and let A_a^x denote the area under the graph and above the x-axis from a to x .

29. A_x^a = ____

30. A_a^a = ____

31. $A_a^c + A_c^b$ = ____

32. $A_a^{x+\Delta x} - A_a^x$ = ____

33. The area function A_a^x satisfies the differential equation __________ , and the initial condition ________ .

34. The notation $g(x)\Big]_c^d$ means __________ .

---

29. $-A_a^x$

30. 0

31. A_a^b

32. $A_x^{x+\Delta x}$

33. $\dfrac{dA_a^x}{dx} = f(x)$, $A_a^a = 0$

34. $g(d) - g(c)$

OBJECTIVE: Find the area under the graph of a positive-valued continuous function $y = f(x)$ over an interval $a \le x \le b$ by integration.

35. To find the area under the graph of $y = \sqrt{9+x}$ for $-9 \le x \le 0$, we first find an ____________ integral for $y = f(x)$:

$$\int \sqrt{9+x}\; dx = \underline{\hspace{3cm}} + C .$$

We then compute the area $A_{-9}^{0} = \underline{\hspace{3cm}}\Big]_{-9}^{0} = \underline{\hspace{1cm}}$.

Notice that the constant C of integration does not enter into the final calculation, as discussed in the Thomas-Finney text.

36. Find the area between the curve $y = \cos \sqrt{x}$, the x-axis, and the lines $x = 0$, $x = \pi^2/4$.

Solution. From Problem 18 of this chapter, the indefinite integral

$$\int \cos \sqrt{x}\; dx = \underline{\hspace{4cm}} .$$

Therefore, the desired area is given by

$$\int \cos \sqrt{x}\; dx\Big]_{\_\_}^{\_\_} = \left(2 \cos \_\_ + 2\left(\frac{\pi}{2}\right) \sin \_\_\right) - \left(2 \cos \_\_ + 2(0) \sin \_\_\right)$$

$$= \underline{\hspace{1.5cm}} .$$

37. Find the area between the curve $y = x^2 + x + 1$, the x-axis, and the lines $x = 2$, $x = 3$.

Solution. The area is

$$A_2^3 = \int (x^2 + x + 1)\; dx\Big]_2^3 = \underline{\hspace{3cm}}\Big]_{\_\_}^{\_\_}$$

$$= \left(\frac{27}{3} + \_\_ + 3\right) - \left(\_\_ + \frac{4}{2} + 2\right) = \_\_ .$$

---

35. indefinite, $\frac{2}{3}(9+x)^{3/2}$, $\frac{2}{3}(9+x)^{3/2}$, 18

36. $2 \cos \sqrt{x} + 2\sqrt{x} \sin \sqrt{x} + C$, $\Big]_0^{\pi^2/4}$, $\frac{\pi}{2}$, $\frac{\pi}{2}$, 0, 0, $\pi - 2$

37. $\frac{1}{3}x^3 + \frac{1}{2}x^2 + x\Big]_2^3$, $\frac{9}{2}$, $\frac{8}{3}$, $\frac{59}{6}$

4-8 The Definite Integral and The Fundamental Theorem of Integral Calculus.

OBJECTIVE A: Write from memory the statement of the Fundamental Theorem of Integral Calculus.

38. The hypotheses of the Fundamental Theorem are:

(a) The function f is ____________ over $a \le x \le b$;

(b) $x_1, x_2, \ldots, x_{n-1}$ are numbers that partition the interval $a \le x \le b$, with ________________________, into n equal subintervals each of length Δx = ________ .

(c) $c_1, c_2, \ldots, c_n$ are n numbers satisfying ______________ for $k = 1, 2, \ldots, n$ and $x_o = a$, $x_n = b$;

(d) S_n = ____________ ; and

(e) $F(x)$ = __________ .

39. The conclusion of the Fundamental Theorem is that as $n \to \infty$,

$\ell im\ S_n$ = ____________ = __________ .

40. The limit, $\ell im \sum f(c_k)\,\Delta x$, is called the ________ ________ of f from a to b. It is a number. This number is denoted by the symbol __________ . The number a in the symbol is called the ________ ________ of integration, and ___ is the upper limit.

41. For k any constant, $\int_a^b kf(x)\,dx$ = ____________ .

42. $\int_a^b \left[f(x) + g(x) \right] dx$ = ______________________ .

---

38. continuous, $a < x_1 < x_2 < \cdots < x_{n-1} < b$, $(b-a)/n$, $x_{k-1} \le c_k \le x_k$,
$\sum_{k=1}^{n} f(c_k)\,\Delta x$, $\int f(x)\,dx$ or is any integral of $f(x)$

39. $\ell im \sum f(c_k)\,\Delta x$, $F(b) - F(a)$

40. definite integral, $\int_a^b f(x)\,dx$, lower limit, b

41. $k \int_a^b f(x)\,dx$

42. $\int_a^b f(x)\,dx + \int_a^b g(x)\,dx$

43. $\int_b^a f(x)\,dx = -$ ______________ .

44. If f is positive-valued and continuous over $a \le x \le b$, then $\int_a^b f(x)\,dx$ is the ________ under the graph of $y = f(x)$, above the __________ from $x = a$ to $x = b$.

45. If f is both positive-valued and negative-valued over $a \le x \le b$, then $\int_a^b f(x)\,dx$ is the algebraic sum of the __________ areas bounded by the graph $y = f(x)$ and the x-axis, from $x = a$ to $x = b$.

46. If $f(x) \ge 0$ for $a \le x \le b$, then $\int_a^b f(x)\,dx$ is ________________ .

47. If $f(x) \le g(x)$ for $a \le x \le b$, then their definite integrals are related by ____________________________ .

OBJECTIVE B: Evaluate definite integrals of elementary continuous functions, using the Fundamental Theorem.

48. Find $\int_{-1}^{2} (x^3 - 2x + 5)\,dx$.

$$\int_{-1}^{2} (x^3 - 2x + 5)\,dx = \underline{\qquad\qquad\qquad}\Big]_{-1}^{2}$$

$$= \left(\frac{1}{4}(2^4) - 2^2 + 5 \cdot 2\right) - \left(\underline{\qquad\qquad\qquad\qquad}\right)$$

$$= 10 - (\underline{\quad}) = \underline{\quad} .$$

---

43. $\int_a^b f(x)\,dx$

44. area, x-axis

45. signed

46. nonnegative or ≥ 0 .

47. $\int_a^b f(x)\,dx \le \int_a^b g(x)\,dx$

48. $\frac{1}{4}x^4 - x^2 + 5x$, $\frac{1}{4}(-1)^4 - (-1)^2 + 5(-1)$, $\frac{-23}{4}$, $\frac{63}{4}$

49. To find $\int_0^1 \frac{x\,dx}{\sqrt{4-x^2}}$, first find $\int \frac{x\,dx}{\sqrt{4-x^2}}$. Using substitution, let $u = 4 - x^2$. Then $du =$ ______ so $x\,dx =$ ______. Substitution into the indefinite integral yields,

$$\int \frac{x\,dx}{\sqrt{4-x^2}} = \underline{\qquad\qquad} = -\frac{1}{2}\left(\underline{\quad}\right) + C = -\sqrt{4-x^2} + C .$$

Thus, $\int_0^1 \frac{x\,dx}{\sqrt{4-x^2}} = \underline{\qquad\qquad}\Big]_0^1 = \underline{\qquad\qquad} \approx 0.268$.

OBJECTIVE C: Calculate the derivative of an integral $\int_a^{b(x)} f(t)\,dt$ with respect to x, when the upper limit of integration is a differentiable function of x. Assume that the integrand f is continuous.

50. If $F(x) = \int_1^x (t^5 - 2t^3 + 1)^4\,dt$, then we may find $F'(x)$ by replacing t by x in the integrand. Thus, $F'(x) =$ ______________ .

51. Let $F(x) = \int_0^{x^2} \sqrt{1+t}\,dt$. If $u = x^2$, then by the chain rule,

$\frac{dF}{dx} = \frac{dF}{du} \cdot \underline{\quad}$. Now, $\frac{dF}{du} = \frac{d}{du}\int_0^u \sqrt{1+t}\,dt =$ ________ . Thus,

$F'(x) = \sqrt{1+u} \cdot \underline{\quad} =$ __________ .

52. Consider $F(x) = \int_{\cos x}^1 \frac{dt}{t^2}$, $0 < x < \frac{\pi}{2}$. Reversing the limits of integration, so that the upper limit is a function of x,

$F(x) = -$ __________ . Let $u = \cos x$. Then,

---

49. $-2x dx$, $-\frac{1}{2}\,du$, $-\frac{1}{2}\int \frac{du}{u^{1/2}}$, $2u^{1/2}$, $-\sqrt{4-x^2}$, $-\sqrt{3} + 2$

50. $(x^5 - 2x^3 + 1)^4$

51. $\frac{du}{dx}$, $\sqrt{1+u}$, $2x$, $2x\sqrt{1+x^2}$

52. $\int_1^{\cos x} \frac{dt}{t^2}$, $\frac{1}{u^2}$, $\cos^2 x$, $\sec x \tan x$

$$F'(x) = \frac{dF}{du}\frac{du}{dx} = -\left(\underline{\quad}\right)(-\sin x) = \frac{\sin x}{\underline{\quad}}$$

$$= \underline{\qquad\qquad} .$$

4-9 Rules for Approximating Integrals.

OBJECTIVE A: Approximate a given definite integral by use of the trapezoidal rule for a specified number n of subdivisions. Estimate the error in this approximation.

53. Let $y = f(x)$ be defined and continuous over the interval $a \leq x \leq b$. Divide the interval $[a,b]$ into n subintervals, each of length $h = (b-a)/n$, by inserting the points $x_1 = a+h$, $x_2 = a+2h$, . . . , $x_{n-1} = a + (n-1)h$. Set $x_o = a$ and $x_n = b$ for convenience in notation. Define $y_k = f(x_k)$ for each $k = 0,1,2,\ldots,n$. Then the trapezoidal approximation for the definite integral is

$$\int_a^b f(x)\,dx = \underline{\qquad\qquad\qquad\qquad} .$$

The error estimate is $E = \underline{\qquad\qquad}$, for some number c satisfying $\underline{\qquad\qquad}$.

54. Use the trapezoidal rule to approximate $\int_0^1 \sqrt{1+x^2}\,dx$, $n = 5$.

<u>Solution</u>. Here, $h = \frac{1-0}{5} = \underline{\quad}$, $x_o = \underline{\quad}$, $x_5 = \underline{\quad}$. The subdivision points are $x_1 = \frac{1}{5}$, $x_2 = \underline{\quad}$, $x_3 = \underline{\quad}$, $x_4 = \underline{\quad}$. The corresponding function values are computed as,

$y_o = \sqrt{1+0^2} = 1$, $y_1 = \sqrt{1+\left(\frac{1}{5}\right)^2} = \underline{\quad} \approx \underline{\qquad}$

$y_2 = \sqrt{1+\left(\frac{2}{5}\right)^2} = \underline{\quad} \approx \underline{\qquad}$, $y_3 \approx \underline{\qquad}$,

$y_4 \approx \underline{\qquad}$, and $y_5 \approx \underline{\qquad}$. Therefore, the trapezoidal

---

53. $\frac{h}{2}(y_o + 2y_1 + 2y_2 + \cdots + 2y_{n-1} + y_n)$, $\frac{b-a}{12} f''(c) \cdot h^2$, $a < c < b$

54. $\frac{1}{5}$, 0, 1, $\frac{2}{5}$, $\frac{3}{5}$, $\frac{4}{5}$, $\frac{\sqrt{26}}{5}$, 1.01980, $\frac{\sqrt{29}}{5}$, 1.07703, 1.16619, 1.28062, 1.41421, 2.15407, 2.33238, 2.56124, 1.41421, 11.50150, 1.15015, 1.15015

approximation is

$$T = \frac{1}{2(5)} \cdot (y_o + 2y_1 + 2y_2 + 2y_3 + 2y_4 + y_5) \text{ , or}$$

$$T = \frac{1}{10} \cdot (1 + 2.03960 + \_\_\_\_\_\_ + \_\_\_\_\_\_ + \_\_\_\_\_\_ + \_\_\_\_\_\_)$$

$$= \frac{1}{10} \cdot (\_\_\_\_\_\_) = \_\_\_\_\_\_ .$$

Therefore,

$$\int_0^1 \sqrt{1+x^2}\, dx \approx \_\_\_\_\_\_ .$$

55. To estimate the error in the approximation of Problem 54, let $f(x) = \sqrt{1+x^2}$. Then, $f'(x) =$ ______ and $f''(x) =$ ______ . Therefore, for $0 \le x \le 1$, we see that

$$|f''(x)| = \frac{1}{\left(1+x^2\right)^{3/2}} < \_\_\_ .$$

Thus, the error E satisfies

$$E = \left|\frac{b-a}{12} f''(c) \cdot h^2\right| < \left| \_\_\_ \cdot \_\_\_ \cdot \_\_\_ \right| \approx \_\_\_\_\_\_ .$$

Then, $\int_0^1 \sqrt{1+x^2}\, dx = 1.15015 \pm E$, or

$$\_\_\_\_\_\_ < \int_0^1 \sqrt{1+x^2}\, dx < \_\_\_\_\_\_ .$$

(In fact, the actual value of the integral to five decimal places is 1.14779. Thus the error is about 2/10 of one percent.)

56. How many subdivisions are required to obtain $\int_0^1 \sqrt{1+x^2}\, dx$ to 5 decimal places of accuracy by the trapezoidal rule?

Solution. From Problem 55, $E = \left|\frac{1}{12} f''(c)\, h^2\right| < \frac{1}{12} h^2$. In order to obtain 5 place accuracy, we need $E < 5 \cdot 10^{-6}$. Thus,

---

55. $x\left(1+x^2\right)^{-1/2}$, $\left(1+x^2\right)^{-3/2}$, 1, $\left|\frac{1}{12} \cdot 1 \cdot \frac{1}{25}\right|$, 0.00333, 1.14682, 1.15348

$\frac{1}{12} h^2 < 5 \cdot 10^{-6}$ implies $h^2 <$ ________ or, since $h = \frac{b-a}{n} =$ ____ ,

$n^2 >$ ________ . Thus, $n >$ ________ $\approx$ _____ , and choosing $n =$ _____ as the number of subdivisions ensures 5 place accuracy. (This is only an upper estimate: fewer subdivisions may work, but there are no guarantees.)

OBJECTIVE B: Approximate a given definite integral by use of Simpson's rule for a specified even number n of subdivisions. Estimate the error in this approximation.

57. Let $y = f(x)$ be defined and continuous over the interval $a \leq x \leq b$. Divide the interval $[a,b]$ into n subintervals, where n is an even number, each of length $h = (b-a)/n$, using the points $x_o = a$, $x_1 = a+h$, $x_2 = a+2h$, . . . , $x_n = a+nh = b$. Define $y = f(x_k)$ for each $k = 0,1,2,\ldots,n$. Then the Simpson approximation for the definite integral is

$$\int_a^b f(x)\, dx = \underline{\hspace{6cm}} .$$

The error estimate is $E =$ ______________ .

58. Simpson's rule is exact if $f(x)$ is ____________________ .

59. Approximate $\int_0^1 \sqrt{1+x^2}\, dx$ by Simpson's rule with $n = 6$.

Solution. Here $h =$ ___ , $x_o =$ ___ , and $x_6 =$ ___ . The subdivision points are,

$x_1 = \frac{1}{6}$, $x_2 =$ ___ , $x_3 =$ ___ , $x_4 =$ ___ , and $x_5 =$ ___ .

---

56. $60 \cdot 10^{-6}$, $\frac{1}{n}$, $\frac{1}{60} \cdot 10^6$, $\frac{1}{2\sqrt{15}} 10^3$, 129, 130

57. $\frac{h}{3}(y_o + 4y_1 + 2y_2 + 4y_3 + \cdots + 2y_{n-2} + 4y_{n-1} + y_n)$, $\frac{b-a}{180} f^{(4)}(c) \cdot h^4$

58. a polynomial of degree < 4

59. $\frac{1}{6}$, 0, 1, $\frac{1}{3}$, $\frac{1}{2}$, $\frac{2}{3}$, $\frac{5}{6}$, $\frac{\sqrt{37}}{6}$, 1.01379, 1.05409, 1.11803, 1.20185, 1.30171, $\sqrt{2}$, 1.41421, 2.10819, 4.47212, 2.40370, 5.20684, 1.41421, 20.66021, 1.14779

The corresponding function values are $y_o = 1$,

$y_1 = \sqrt{1+\left(\frac{1}{6}\right)^2} =$ _____ $\approx$ __________ , $y_2 \approx$ __________ ,

$y_3 \approx$ __________ , $y_4 \approx$ __________ , $y_5 \approx$ __________ , and

$y_6 =$ _____ $\approx$ __________ . Therefore, the Simpson approximation is,

$$S = \frac{1}{6 \cdot 3}(y_o + 4y_1 + 2y_2 + 4y_3 + 2y_4 + 4y_5 + y_6)$$

$$= \frac{1}{18}\Big(1 + 4.05518 + \_\_\_\_\_ + \_\_\_\_\_ + \_\_\_\_\_ + \_\_\_\_\_ + \_\_\_\_\_\Big)$$

$$= \frac{1}{18}(\_\_\_\_\_) = \_\_\_\_\_ .$$

Compare the answer with that found in Problem 54 where we employed the trapezoidal rule.

60. To estimate the error in the approximation in Problem 59, let

$f(x) = \sqrt{1+x^2}$. Then, from Problem 55,

$f''(x) = \left(1+x^2\right)^{-3/2}$, so that $f^{(3)}(x) =$ ______________ and

$f^{(4)}(x) =$ ______________ . Now,

$$\frac{4x^2-1}{\left(1+x^2\right)^{7/2}} < \frac{4x^2}{\left(1+x^2\right)^{7/2}} < \frac{4x^2}{1+x^2} < 4 \quad \text{for} \quad 0 \le x \le 1 .$$

Thus, the error E satisfies

$$E = \left|\frac{b-a}{180} f^{(4)}(c) \cdot h^4\right| < \left|\_\_\_ \cdot \_\_\_ \cdot \_\_\_\right| \approx \_\_\_\_\_ .$$

Therefore, we observe that Simpson's rule, with $n = 6$, provides the value of the integral to at least 3 decimal places of accuracy; with $n = 10$ we will obtain at least 4 place accuracy, according to our error estimates.

---

60. $-3x\left(1+x^2\right)^{-5/2}$, $3\left(4x^2-1\right)\left(1+x^2\right)^{-7/2}$, $\frac{1}{180} \cdot 12 \cdot \frac{1}{6^4}$, 0.00005

4-10 Some Comments on Notation.

61. The symbol ____ is used to refer to a sum of expressions, each of the form $f(x)\Delta x$, extending over a set of subintervals from $x = a$ to $x = b$.

62. The symbol $\int_a^b f(x)\,dx$ is used to denote the limit __________ .

63. If we think of dx as a small increment Δx in x, we may interpret $dA = f(x)dx$ as the area of a small ____________ with base extending along the x-axis from x to ________ , and with altitude ______ . This provides an _________________ to that portion of the area under the curve $y = f(x)$ lying between ___ and _______ . Adding together the areas of all these small rectangles from $x = a$ to $x = b$ and taking the _________ of this sum then gives the exact value of the area under the curve $y = f(x)$ between $x = a$ and $x = b$. This limit is symbolized by

$$A = \_\_\_\_\_\_\_\_\_\_ .$$

4-11 Summary.

In Problems 64 - 66, assume that the Riemann integral of f over the interval $[a,b]$ exists, and is denoted by $R_a^b(f)$.

64. The Riemann integral $R_a^b(f)$ is a __________ .

65. This number has the property that given any $\varepsilon > 0$ there is a positive number $\delta > 0$ such that any sum of the form

$$\sum_{k=1}^{n} f(c_k)\cdot(x_k - x_{k-1}) ,$$

where $a = x_o < x_1 < \cdots < x_n = b$ and $x_{k-1} < c_k < x_k$,

differs in value from the number $R_a^b(f)$ by at most ___ provided that Δx_k = ___________ is no bigger than ___ for each k.

66. $R_a^a(f)$ = ___ and R_b^a = _________ .

---

61. S_a^b

62. $\lim_{\Delta x \to 0} S_a^b$

63. rectangle, $x + \Delta x$, $f(x)$, approximation, x, $x + \Delta x$, limit, $\int_a^b f(x)\,dx$

64. number

65. ε, $x_k - x_{k-1}$, δ

66. 0, $-R_a^b(f)$

OBJECTIVE: Calculate the value of the Riemann integral of a given piecewise continuous function over a specified interval.

67. To find $\int_{-1}^{2} f(x)\,dx$, for $f(x) = \begin{cases} x^2, & -1 \le x \le 1 \\ \sqrt{x-1}, & 1 < x \le 2 \end{cases}$

we have $\int_{-1}^{2} f(x)\,dx = \int_{-1}^{1} x^2\,dx + \rule{3cm}{0.4pt}$

$= \rule{1.5cm}{0.4pt} + \frac{2}{3}(x-1)^{3/2}\Big]_1^2$

$= \rule{0.7cm}{0.4pt} + \rule{0.7cm}{0.4pt} = \rule{0.7cm}{0.4pt}\,.$

CHAPTER 4 OBJECTIVE - PROBLEM KEY

Objective	Problems in Thomas/Finney Text	Objective	Problems in Thomas/Finney Text
4-2 A	p. 181, 14-25	4-6 A	p. 196, 1,6
B	p. 181, 1-13	B	p. 196, 2-5
4-3	p. 184, 1-11, 13-20	4-7	p. 200, 1-14, 16
	p. 187, 23,24	4-8 B	p. 210, 6-16
4-4	p. 187, 1-22	C	p. 210, 18-24
4-5 A	p. 191, 1-5	4-9 A,B	p. 218, 1-9
B	p. 191, 6-11	4-11	p. 223, 2

CHAPTER 4 SELF-TEST

1. Find the following indefinite integrals.

(a) $\int (x-1)(2+x)\,dx$

(b) $\int \sqrt{2x-1}\,dx$

(c) $\int x^2\left(5-3x^3\right)^{-1/2}\,dx$

(d) $\int x^{-1/2}\sin\left(\sqrt{x}-3\right)\,dx$

---

67. $\int_1^2 \sqrt{x-1}\,dx$, $\frac{1}{3}x^3\Big]_{-1}^{1}$, $\frac{2}{3}$, $\frac{2}{3}$, $\frac{4}{3}$

2. Solve the differential equations.

(a) $\frac{dy}{dx} = 4x - 3$ (b) $\frac{dy}{dx} = y^{-2} \sec^2 x$; $y = 3$ if $x = 0$

3. Approximate the area under the curve $y = x^2 - 2x + 4$ between $x = 1$ and $x = 4$ by summing $n = 6$ inscribed rectangles of uniform width.

4. Find the numerical values of each of the following.

(a) $\sum_{k=1}^{4} \frac{1}{2k}$ (b) $\sum_{n=1}^{5} n(n-3)$ (c) $\sum_{k=5}^{6} (2k-1)$

5. Establish the formula $1 + 4 + 7 + \cdots + (3n-2) = \frac{n(3n-1)}{2}$ by mathematical induction.

6. Find the area under the curve $y = 1 - 2x$ over the interval $-2 \leq x \leq 0$ by taking the limit of the sum of circumscribed rectangles.

7. Find the area under the curve $y = x\sqrt{x^2+1}$, above the x-axis, between $x = 1$ and $x = 4$.

8. Evaluate the definite integrals.

(a) $\int_1^4 \frac{(x-2)^2}{\sqrt{x}}\, dx$ (b) $\int_{-2}^{0} x^2(4-x)\, dx$

9. Find $\frac{d}{dx} \int_{1-x^2}^{0} \sqrt[3]{t^2+1}\, dt$.

10. Use the trapezoidal rule to approximate $\int_0^1 \sqrt{1+x^3}\, dx$ with $n = 4$.

11. Use Simpson's rule to approximate $\int_1^2 \frac{dx}{x}$ with $n = 6$. Estimate the error in your approximation.

12. Compute $\int_0^{\pi} f(x)\, dx$ where $f(x) = \begin{cases} \sin x, & 0 \leq x < \frac{\pi}{2} \\ \pi x, & \frac{\pi}{2} \leq x \leq \pi \end{cases}$.

SOLUTIONS TO CHAPTER 4 SELF-TEST

1. (a) $\int (x-1)(2+x)\,dx = \int (x^2+x-2)\,dx = \frac{1}{3}x^3 + \frac{1}{2}x^2 - 2x + C$

(b) $\int \sqrt{2x-1}\,dx = \frac{1}{2}\int \sqrt{u}\,du \quad (u = 2x-1) \quad = \frac{1}{3}(2x-1)^{3/2} + C$

(c) $\int x^2\left(5-3x^3\right)^{-1/2} dx = -\frac{1}{9}\int u^{-1/2}\,du \quad (u = 5-3x^3)$

$$= -\frac{2}{9}\left(5-3x^3\right)^{1/2} + C$$

(d) $\int x^{-1/2}\sin\left(\sqrt{x} - 3\right)dx = 2\int \sin u\,du \quad \left(u = \sqrt{x} - 3\right)$

$$= -2\cos\left(\sqrt{x} - 3\right) + C$$

2. (a) $dy = (4x-3)\,dx$ or $y = \int (4x-3)\,dx = 2x^2 - 3x + C$

(b) $y^2 dy = \sec^2 x\,dx$, so $\int y^2 dy = \int \sec^2 x\,dx$ and then

$\frac{1}{3}y^3 = \tan x + C$. Since $y = 3$ if $x = 0$, $\frac{1}{3}(3)^3 = \tan 0 + C$ or $C = 9$.

Hence, the solution is $\frac{1}{3}y^3 = \tan x + 9$.

3. The partition points are $x_0 = 1$, $x_1 = 3/2$, $x_2 = 2$, $x_3 = 5/2$, $x_4 = 3$, $x_5 = 7/2$, and $x_6 = 4$. Since $dy/dx > 0$ on the interval $1 \le x \le 4$, the curve is increasing so the altitude of each rectangle is its left edge. Thus, the areas of the inscribed rectangles for $y = f(x)$ are,

$$f(1)\,\Delta x = 3 \cdot \frac{1}{2} = \frac{3}{2}$$

$$f\left(\frac{3}{2}\right)\Delta x = \frac{13}{4} \cdot \frac{1}{2} = \frac{13}{8}$$

$$f(2)\,\Delta x = 4 \cdot \frac{1}{2} = 2$$

$$f\left(\frac{5}{2}\right)\Delta x = \frac{21}{4} \cdot \frac{1}{2} = \frac{21}{8}$$

$$f(3)\,\Delta x = 7 \cdot \frac{1}{2} = \frac{7}{2}$$

$$f\left(\frac{7}{2}\right)\Delta x = \frac{37}{4} \cdot \frac{1}{2} = \frac{37}{8}$$

Sum $= \frac{127}{8} = 15.875$, approximate area.

4. (a) $\frac{1}{2} + \frac{1}{4} + \frac{1}{6} + \frac{1}{8} = \frac{25}{24}$

(b) $1(-2) + 2(-1) + 3(0) + 4(1) + 5(2) = 10$

(c) $(2 \cdot 5 - 1) + (2 \cdot 6 - 1) = 9 + 11 = 20$

5. For $n = 1$, $(3 \cdot 1 - 2) = \frac{1(3 \cdot 1 - 1)}{2}$ or, $1 = \frac{2}{2}$ is true. Suppose the formula is true for the known integer n. Adding $3(n+1) - 2 = 3n+1$ to both sides of the formula gives,

$$1 + 4 + 7 + \cdots + (3n-2) + (3n+1) = \frac{n(3n-1)}{2} + (3n+1)$$
$$= \frac{3n^2 - n + 2(3n+1)}{2}$$
$$= \frac{3n^2 + 5n + 2}{2} = \frac{(n+1)(3n+2)}{2},$$

which is exactly like the original formula if we replace n by $n+1$.

6. Subdivision of the interval $-2 \leq x \leq 0$ into n equal pieces each of length $\Delta x = \frac{0 - (-2)}{n} = \frac{2}{n}$ gives the subdivision points

$$x_o = -2, \quad x_1 = -2 + \frac{2}{n}, \quad x_2 = -2 + \frac{4}{n}, \quad \ldots, \quad x_{n-1} = -2 + \frac{2(n-1)}{n}, \quad x_n = 0.$$

Since $y = 1 - 2x$ is decreasing over the interval we take the left edge for the altitude of each circumscribed rectangle. These rectangles have area

$$f(-2)\,\Delta x = (1+4) \cdot \frac{2}{n} = 5 \cdot \frac{2}{n}$$

$$f\left(-2 + \frac{2}{n}\right) \Delta x = \left(1 + 4 - \frac{4}{n}\right) \cdot \frac{2}{n} = \left(5 - \frac{4}{n}\right) \cdot \frac{2}{n}$$

$$f\left(-2 + \frac{4}{n}\right) \Delta x = \left(1 + 4 - \frac{8}{n}\right) \cdot \frac{2}{n} = \left(5 - \frac{8}{n}\right) \cdot \frac{2}{n}$$

$$\vdots$$

$$f\left(-2 + \frac{2n-2}{n}\right) \Delta x = \left(1 + 4 - \frac{4(n-1)}{n}\right) \cdot \frac{2}{n} = \left(5 - \frac{4(n-1)}{n}\right) \cdot \frac{2}{n},$$

whose sum is

$$S_n = \left[5n - \left(\frac{4}{n} + \frac{8}{n} + \cdots + \frac{4(n-1)}{n}\right)\right] \cdot \frac{2}{n}$$
$$= \left[5n - \frac{4}{n}\left(1 + 2 + \cdots + (n-1)\right)\right] \cdot \frac{2}{n}$$
$$= \left[5n - \frac{4}{n} \cdot \frac{(n-1)n}{2}\right] \cdot \frac{2}{n} = 10 - \frac{4(n-1)}{n}.$$

Therefore, $\lim_{n \to \infty} S_n = \lim_{n \to \infty} \left(10 - 4 + \frac{4}{n}\right) = 6.$

7. The area is given by

$$\int_1^4 x\sqrt{x^2+1}\,dx = \frac{1}{3}\left(x^2+1\right)^{3/2}\Big]_1^4 = \frac{1}{3}\left(17^{3/2} - 2^{3/2}\right) \approx 22.42146 .$$

8. (a) $$\int_1^4 \frac{(x-2)^2}{\sqrt{x}}\,dx = \int_1^4 \frac{x^2-4x+4}{\sqrt{x}}\,dx$$

$$= \int_1^4 \left(x^{3/2} - 4x^{1/2} + 4x^{-1/2}\right) dx$$

$$= \frac{2}{5}x^{5/2} - \frac{8}{3}x^{3/2} + 8x^{1/2}\Big]_1^4$$

$$= \left(\frac{2}{5}(32) - \frac{8}{3}(8) + 16\right) - \left(\frac{2}{5} - \frac{8}{3} + 8\right) \approx 1.73333 .$$

(b) $$\int_{-2}^0 x^2(4-x)\,dx = \int_{-2}^0 (4x^2 - x^3)\,dx = \frac{4}{3}x^3 - \frac{1}{4}x^4\Big]_{-2}^0$$

$$= -\left[\frac{4}{3}(-2)^3 - \frac{1}{4}(-2)^4\right] = \frac{44}{3} \approx 14.66667 .$$

9. $$\frac{d}{dx}\int_{1-x^2}^0 \sqrt[3]{t^2+1}\,dt = -\frac{d}{dx}\int_0^{1-x^2} \sqrt[3]{t^2+1}\,dt$$

$$= -\sqrt[3]{\left(1-x^2\right)^2 + 1} \cdot \frac{d}{dx}(1-x^2)$$

$$= 2x\left(x^4 - 2x^2 + 2\right)^{1/3} .$$

10. Subdivision points are $x_o = 0$, $x_1 = 1/4$, $x_2 = 1/2$, $x_3 = 3/4$, $x_4 = 1$ and $h = (1-0)/4 = 1/4$. Then,

$y_o = \sqrt{1+0} = 1$, $y_1 = \sqrt{1 + (1/64)} \approx 1.00778$,

$y_2 = \sqrt{1 + (1/8)} \approx 1.06066$, $y_3 = \sqrt{1 + (27/64)} \approx 1.19242$,

$y_4 = \sqrt{1 + 1} \approx 1.41421$. Thus,

$$\int_0^1 \sqrt{1+x^3}\,dx \approx T = \frac{1}{8}\left[y_o + 2y_1 + 2y_2 + 2y_3 + y_4\right]$$

$$\approx \frac{1}{8}(8.93593) \approx 1.11699 .$$

11. Subdivision points are $x_o = 1$, $x_1 = 7/6$, $x_2 = 4/3$, $x_3 = 3/2$, $x_4 = 5/3$, $x_5 = 11/6$, $x_6 = 2$, and $h = 1/6$. Then, $y_o = 1$, $y_1 = 6/7 \approx .85714$, $y_2 = 3/4 = .75$, $y_3 = 2/3 \approx .66667$, $y_4 = 3/5 = .6$, $y_5 = 6/11 \approx .54545$, $y_6 = 1/2 = .5$. Thus,

$$\int_1^2 \frac{dx}{x} \approx \frac{1}{3 \cdot 6}\left[y_o + 4y_1 + 2y_2 + 4y_3 + 2y_4 + 4y_5 + y_6\right]$$

$$\approx \frac{1}{18}(12.47706) \approx 0.69317 .$$

To estimate the error, $f(x) = \frac{1}{x}$, $f'(x) = -\frac{1}{x^2}$, $f^{(3)}(x) = \frac{2}{x^3}$, and $f^{(4)}(x) = -\frac{6}{x^4}$. Since $\left|-\frac{6}{x^4}\right| < 6$ on $1 \leq x \leq 2$, the error satisfies

$$E = \left|\frac{2-1}{180} f^{(4)}(c) \cdot h^4\right| < \frac{1}{180} \cdot 6 \cdot \frac{1}{1296} \approx .00003 .$$

Therefore, the approximation is accurate to 4 decimal places.

12. $$\int_0^{\pi} f(x)\,dx = \int_0^{\pi/2} \sin x\,dx + \int_{\pi/2}^{\pi} \pi x\,dx$$

$$= -\cos x\Big]_0^{\pi/2} + \frac{1}{2}\pi x^2\Big]_{\pi/2}^{\pi} = 1 + \frac{3\pi^3}{8} \approx 12.62735 .$$

CHAPTER 5 APPLICATIONS OF THE DEFINITE INTEGRAL

5-1 Introduction.

Let us review the central theme of the previous chapter.

1. Let A_a^x denote the area function giving the signed area between the curve of the continuous function $y = f(x)$ and the x-axis from a to any abscissa x. Then the area function is an integral of ________ . That is, it satisfies the differential equation ____________ subject to the initial condition ________ .

2. More generally, this motivated the idea of the definite integral as satisfying the equation

$$\int_a^b f(x)\,dx = \underline{\qquad\qquad} ,$$

where $F(x) = \int f(x)\,dx$ is any integral of f .

3. The connection between the definite integral $\int_a^b f(x)\,dx$ and the sums of the form $S_a^b = \sum_a^b f(x)\,\Delta x$ was revealed to be that the definite integral is the ________ of S_a^b as Δx ________________ , provided that the function f is assumed to be ______________ on $a \le x \le b$.

4. Thus, a sum of the form S_a^b for suitably small Δx could be used as an approximation to the value of the definite integral $\int_a^b f(x)\,dx$. Two numerical methods providing such an approximation studied in the previous chapter were the ______________ rule and ____________ rule.

5-2 Area Between Two Curves.

OBJECTIVE A: Find the area bounded by two given continuous curves $y_1 = f_1(x)$ and $y_2 = f_2(x)$ over an interval $a \le x \le b$. It may be required to calculate a and b .

5. Find the area between the curves $y = 1$ and $y = 1 - x^2$ for $1 \le x \le 4$.

---

1. $f(x)\,dx$, $\dfrac{dA_a^x}{dx} = f(x)$, $A_a^a = 0$

2. $F(b) - F(a)$

3. limit, approaches zero, continuous

4. trapezoidal, Simpson's

Solution. The desired region is shown in the figure at the right. The area is given by

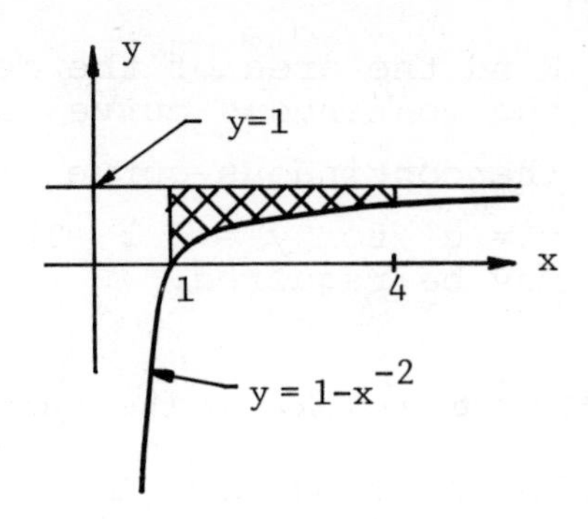

$$A = \int_1^4 (\underline{\qquad\qquad}) \, dx$$

$$= \int_1^4 \underline{\quad} \, dx = \underline{\quad}\Big]_1^4$$

$$= -\frac{1}{4} - (\underline{\quad}) = \underline{\quad} .$$

6. Find the area between the curves $y = \sin x$ and $y = \frac{2x}{\pi}$.

Solution. The region is shown in the figure at the right. We need to determine the x-coordinates of the points A and B of intersection of the two curves. Now, these occur where $\sin x = \frac{2x}{\pi}$;

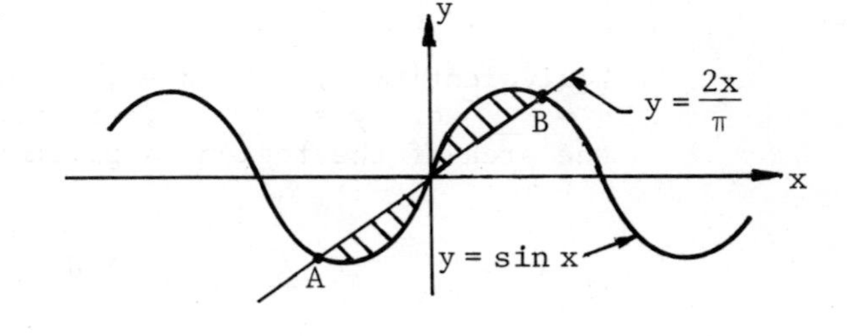

so $x = -\frac{\pi}{2}$, ___ , or ___ .

As x varies from $-\frac{\pi}{2}$ to 0 the curve $y = \sin x$ is ________ the curve $y = 2x/\pi$; however, as x varies from 0 to ___ the situation is reversed. Thus, the area between the two curves is given by

$$A = \int_{-\pi/2}^{0} \left(\frac{2x}{\pi} - \sin x\right) dx + \int_{\underline{\quad}}^{\underline{\quad}} \left(\underline{\qquad\qquad}\right) dx$$

$$= \left(\underline{\qquad\qquad}\right)\Big]_{-\pi/2}^{0} + \left(-\cos x - \frac{x^2}{\pi}\right)\Big]_0^{\pi/2}$$

$$= \left(\underline{\qquad\qquad}\right) + \left(-\cos\frac{\pi}{2} - \frac{\pi^2}{4\pi} + 1\right) = \underline{\qquad} .$$

---

5. $1 - (1 - x^{-2})$, x^{-2}, $-x^{-1}$, -1, $\frac{3}{4}$

6. 0, $\frac{\pi}{2}$, below, $\frac{\pi}{2}$, $\int_0^{\pi/2} \left(\sin x - \frac{2x}{\pi}\right) dx$, $\frac{x^2}{\pi} + \cos x$,

$1 - \frac{\pi^2}{4\pi} - \cos\left(-\frac{\pi}{2}\right)$, $2 - \frac{\pi}{2}$

OBJECTIVE B: Find the area of the region bounded on the right by the continuous curve $x_1 = g_1(y)$ and on the left by the continuous curve $x_2 = g_2(y)$ as y varies from $y = c$ to $y = d$. The calculation of c and d may be required.

7. Find the area bounded by the curves $y^2 = x$ and $y = 6 - x$.

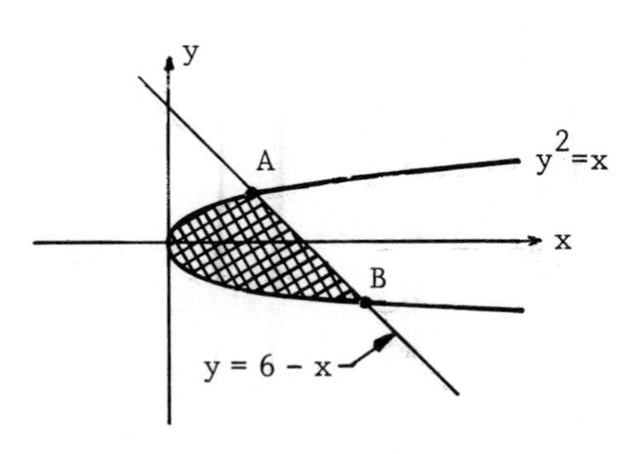

Solution. The first curve is $x = y^2$ and the second is $x = 6 - y$. The region is depicted in the figure at the right. First we will find the y-coordinates of the points A and B of intersection of the two curves. The intersection occurs when $x = y^2$ is the same as $x = 6 - y$ or __________.

Equivalently, $y^2 + y - 6 = 0$, or $y =$ ___ or $y =$ ___. Therefore, the area of the region is given by

$$A = \int_{-3}^{2} (\_\_\_\_\_\_\_\_) \, dy = \_\_\_\_\_\_\_\_\_\_\_\_\Big]_{-3}^{2}$$

$$= \left(12 - 2 - \frac{8}{3}\right) - (\_\_\_\_\_\_\_\_\_\_) = \_\_\_\_\_ \approx \_\_\_\_\_\_\_\_ .$$

5-3 Distance.

OBJECTIVE A: Given a continuous function $v = f(t)$ representing the velocity v of a moving body as a function of time t, find the time intervals in which the velocity is positive and in which it is negative, and calculate the total distance traveled by the body during the specified time interval $t = a$ to $t = b$.

8. Consider the velocity given by $v = \sin(3t - 1)$ for $0 \le t \le 2$. A graph of the velocity curve is shown in the figure on the next page. From our analysis of the sine curve in Article 2-9, we observe that the velocity is of periodicity _____, amplitude ___, and horizontal shift ___.

---

7. $y^2 = 6 - y$, 2, -3, $6 - y - y^2$, $6y - \frac{1}{2}y^2 - \frac{1}{3}y^3$, $-18 - \frac{9}{2} + 9$, $20\frac{5}{6}$,

20.83333

Thus, the velocity v is negative for $0 \leq t < 1/3$, it is positive for ______________, and negative again for ______________.

Therefore, the total distance traveled by the moving body is

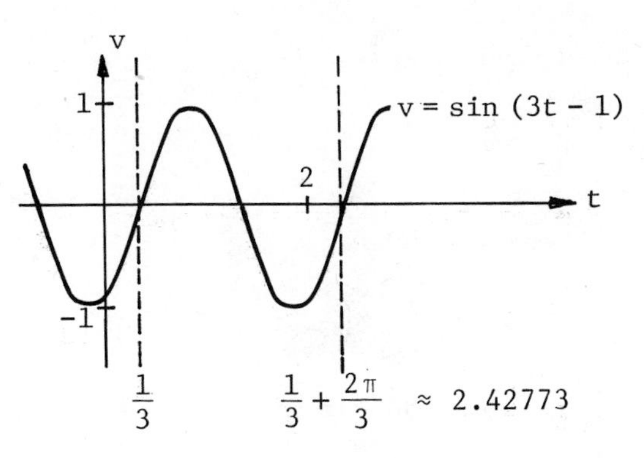

$$s = \int_0^{1/3} -\sin(3t-1)\,dt + \text{\_\_\_\_\_\_\_\_\_\_\_\_\_\_\_\_}$$

$$= \frac{1}{3}\cos(3t-1)\Big]_0^{1/3} + \text{\_\_\_\_\_\_\_\_\_\_\_\_\_\_\_\_}$$

$$= \frac{1}{3}(1 - \cos 1) + \text{\_\_\_\_\_\_\_\_\_\_\_\_}$$

$$= \text{\_\_\_\_\_\_\_\_\_\_\_\_\_\_} \approx 1.24779 .$$

OBJECTIVE B: Given a continuous function $a = f(t)$ representing the acceleration a of a moving body as a function of time t, and given v_o as its velocity at time $t = 0$, find the total distance traveled by the body during the specified time interval $t = 0$ to $t = b$.

9. Suppose the brakes are applied to a car traveling 50 mph and the brakes give the car a constant negative acceleration of 20 ft/sec^2. How far will the car travel before stopping?

Solution. To make the units consistent in the problem we convert miles per hour to feet per second. Since 15 mph is equivalent to 22 ft/sec, it follows that 50 mph is equivalent to _______ ft/sec.

---

8. $\frac{2\pi}{3}$, 1, $\frac{1}{3}$, $\frac{1}{3} < t < \frac{1}{3} + \frac{\pi}{3}$, $\frac{1}{3} + \frac{\pi}{3} < t \leq 2$,

$\int_{\frac{1}{3}}^{\frac{1}{3}+\frac{\pi}{3}} \sin(3t-1)\,dt + \int_{\frac{1}{3}+\frac{\pi}{3}}^{2} -\sin(3t-1)\,dt$,

$-\frac{1}{3}\cos(3t-1)\Big]_{\frac{1}{3}}^{\frac{1}{3}+\frac{\pi}{3}} + \frac{1}{3}\cos(3t-1)\Big]_{\frac{1}{3}+\frac{\pi}{3}}^{2}$, $\frac{2}{3} + \frac{1}{3}(1 + \cos 5)$,

$\frac{4}{3} + \frac{1}{3}\cos 5 - \frac{1}{3}\cos 1$

The problem gives $a =$ ____ , so $v = \int a\,dt =$ _____ $+ C$.

Since $v_o = \frac{220}{3}$ when $t = 0$, it follows that $C =$ _____ . Therefore,

$v =$ ______________ . Next, the total distance traveled after T seconds is given by

$$s = \int_0^T \left(-20t + \frac{220}{3}\right) dt = \_\_\_\_\_\_\_\_\_\_\_\_ .$$

When the car stops, $v =$ ___ , or $-20T + \frac{220}{3} = 0$, and this occurs

at time $T =$ ___ seconds. The total distance traveled during this time is therefore

$$s\Big]_{T=\frac{11}{3}} = -10\left(\_\_\right)^2 + \frac{220}{3}\left(\_\_\right) = \_\_\_\_\_\_ \text{ ft.}$$

5-4 Volumes (Slices).

OBJECTIVE A: Calculate the volume of a solid of revolution generated by the graph $y = f(x)$ of a continuous function over $a \le x \le b$ rotated about the x-axis.

10. Calculate the volume of the solid generated when $y = x^2 - 6x$ is rotated about the x-axis for $0 \le x \le 6$.

Solution. A graph is shown in the figure at the right. Thus, the volume generated is given by

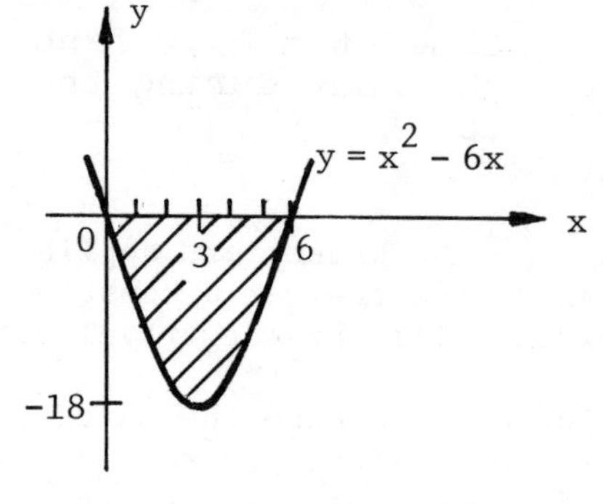

$$V = \_\_\_\_\_\_\_\_\_\_\_\_\_\_$$

$$= \int_0^6 \pi(x^4 - 12x^3 + 36x^2)\,dx$$

$$= \_\_\_\_\_\_\_\_\_\_\_\_\_\_\_\_$$

$$= \pi\left(\_\_\_\_\_\_\_\_\_\_\_\_\_\_\right) - \pi \cdot 0 = \_\_\_\_\_\_ \approx \_\_\_\_\_\_ \text{ cubic units.}$$

---

9. $\frac{50 \cdot 22}{15} = \frac{220}{3}$, -20, $-20t$, $\frac{220}{3}$, $-20t + \frac{220}{3}$, $-10T^2 + \frac{220}{3}T$, 0, $\frac{11}{3}$, $\frac{11}{3}$, $\frac{11}{3}$, $\frac{1210}{9} \approx 134.44$

10. $\int_0^6 \pi\left(x^2 - 6x\right)^2 dx$, $\pi\left[\frac{1}{5}x^5 - 3x^4 + 12x^3\right]_0^6$, $\frac{1}{5}(6)^5 - 3(6)^4 + 12(6)^3$, $\frac{1296\pi}{5}$, 814.3

11. For the volume generated by the graph of $y = |x|$ rotated about the x-axis from $x = -2$ to $x = 1$,

$$V = \int_{-2}^{1} \pi(\underline{\qquad})^2\,dx = \int_{-2}^{1} \underline{\qquad\qquad}$$

$$= \pi\Big[\underline{\qquad}\Big]_{-2}^{1} = \pi(\underline{\qquad\qquad}) = \underline{\qquad} \approx \underline{\qquad\quad} \text{ cubic units.}$$

OBJECTIVE B: Use the method of slicing to calculate the volume of a solid whose base is given and whose cross sectional areas are specified.

12. Consider the solid whose base is a triangle cut from the first quadrant by the line $x + 5y = 5$ and whose cross sections perpendicular to the x-axis are semicircles. A typical cross section of the solid is illustrated at the right. From the equation $x + 5y = 5$ we find that $y = \underline{\qquad\qquad\qquad}$. Since y is the diameter of the semicircle, the cross sectional area given by the area of the semicircle is

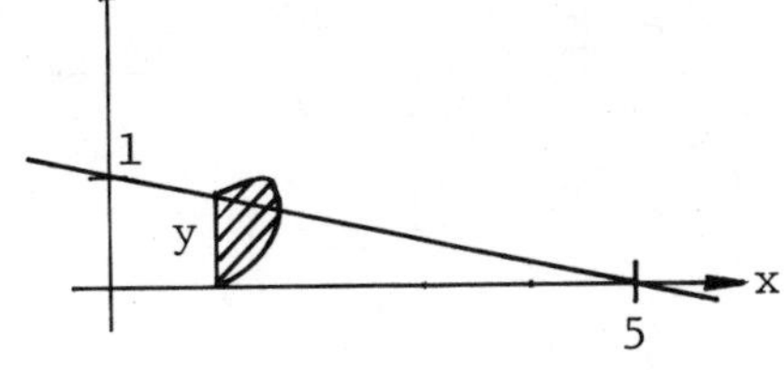

$$A(x) = \pi(\underline{\quad})^2 = \underline{\qquad\quad}.$$

Therefore, the volume of the solid is given by

$$V = \int_0^5 A(x)\,dx = \int_0^5 \frac{1}{4}\pi y^2\,dx = \int_0^5 \frac{\pi}{4}\Big(\underline{\qquad\qquad\qquad}\Big)\,dx$$

$$= \frac{\pi}{4}\left[\frac{x^3}{75} - \frac{x^2}{5} + x\right]_0^5 = \frac{\pi}{4}(\underline{\qquad\qquad\qquad})$$

$$= \underline{\qquad} \approx \underline{\qquad\qquad} \text{ cubic units.}$$

---

11. $|x|$, $\pi x^2\,dx$, $\frac{1}{3}x^3$, $\frac{1}{3} - \left(\frac{-8}{3}\right)$, 3π, 9.42

12. $-\frac{1}{5}x + 1$, $\frac{y}{2}$, $\frac{1}{4}\pi y^2$, $\frac{x^2}{25} - \frac{2x}{5} + 1$, $\frac{5}{3} - 5 + 5$, $\frac{5\pi}{12}$, 1.309

5-5 Volumes (Shells and Washers).

OBJECTIVE A: Use the method of cylindrical shells to find the volume generated when a given planar region is rotated about a specified axis.

13. Consider the volume generated when the region between the curves $y = x^2$ and $y^2 = x$ is rotated about the x-axis. The planar area is shown in the figure at the right. Using the method of shells, we begin with a horizontal strip of width Δy above the x-axis. A typical length of this strip is $x = \sqrt{y}$ minus $x = y^2$, or in terms of y, ________ .

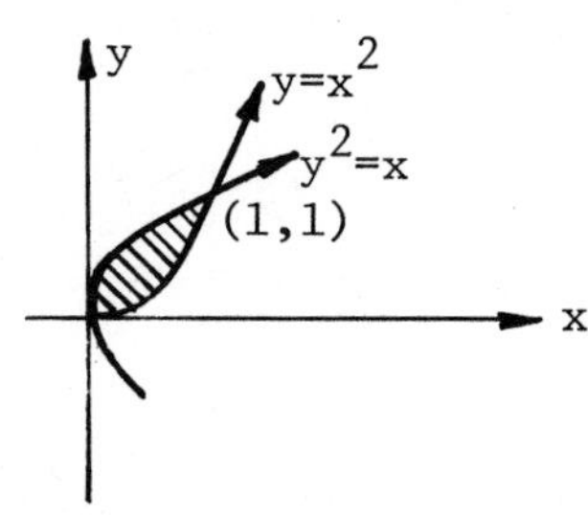

The volume generated by revolving this strip about the x-axis is a hollow cylindrical shell of inner circumference _____ , inner length _________ , and wall thickness ____ . Hence, the total volume is given by the definite integral

$$V = \int_0^1 \text{\_\_\_\_\_\_\_\_\_\_\_\_\_\_} \, dy$$

$$= 2\pi \int_0^1 (\text{\_\_\_\_\_\_\_\_\_\_}) \, dy = 2\pi \Big[\text{\_\_\_\_\_\_\_\_\_\_\_\_\_\_\_\_}\Big]_0^1$$

$$= 2\pi\left(\frac{2}{5} - \frac{1}{4}\right) = \text{\_\_\_\_} \approx 0.94248 \text{ cubic units.}$$

14. Compute the volume generated by rotating about the y-axis the region bounded by $y = 2x$ and $y = x^2 - 2x$.

Solution. A graph of the plane region to be rotated is shown in the figure at the right. We begin with a horizontal strip of width Δx . A typical height of this strip is $y = 2x$ minus $y =$ ________ .

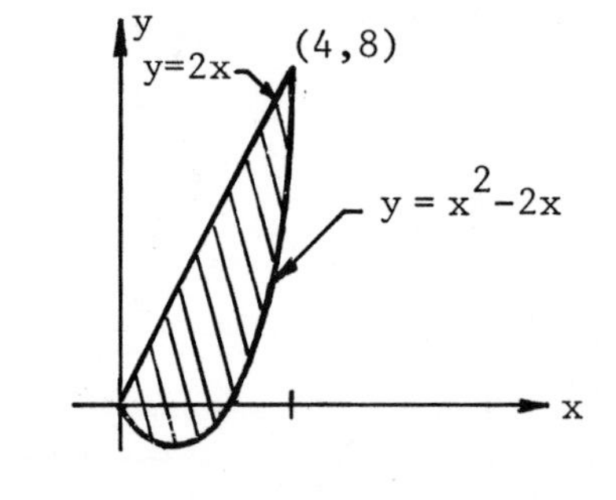

---

13. $\sqrt{y} - y^2$, $2\pi y$, $\sqrt{y} - y^2$, Δy, $2\pi y\left(\sqrt{y} - y^2\right)$, $y^{3/2} - y^3$, $\frac{2}{5} y^{5/2} - \frac{1}{4} y^4$, $\frac{3\pi}{10}$

14. $x^2 - 2x$, $2\pi x$, $2x - (x^2 - 2x)$, Δx, $2\pi x(4x - x^2)$, $4x^2 - x^3$, $\frac{4}{3} x^3 - \frac{1}{4} x^4$, 1, $128\pi/3$

The volume generated by revolving this strip about the y-axis is a hollow cylindrical shell of inner circumference _____ , inner length ___________ , and thickness ___ . Hence, the total volume is given by the definite integral

$$V = \int_0^4 \_\_\_\_\_\_\_\_\_\_\_ \, dx = 2\pi \int_0^4 \_\_\_\_\_\_\_\_\_ \, dx$$

$$= 2\pi \left[\_\_\_\_\_\_\_\_\_\_\_\right]_0^4 = 128\pi\left(\frac{4}{3} - \_\_\_\right)$$

$$= \_\_\_\_\_\_\_ \approx 134.04 \text{ cubic units.}$$

OBJECTIVE B: Use the method of washers to find the volume generated when a given planar region is rotated about a specified axis or line.

15. Let V denote the volume generated when the planar region bounded by the graph of $y^2 = x^2 - 9$, $y = 0$, and $x = 5$ is rotated about the y-axis. The planar region is illustrated in the figure at the right. Now, imagine the solid to be cut into thin slices by planes perpendicular to the y-axis. Each slice is like a thin washer of thickness Δy, inner radius $r_1 = x =$ ________ , and outer radius $r_2 =$ ___ . The area of the face of such a washer is $\pi r_2^2 - \pi r_1^2 = \pi($ ________ $)$. Therefore, the volume of the solid is given by

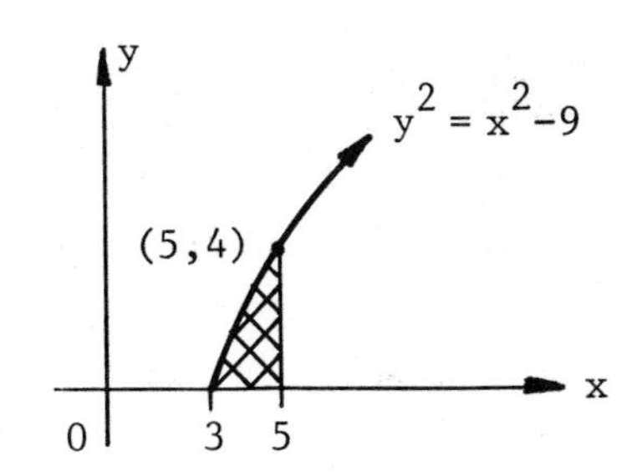

$$V = \int_0^4 \_\_\_\_\_\_\_\_\_\_\_ \, dy = \pi\left[\_\_\_\_\_\_\_\_\_\_\_\right]_0^4 = \_\_\_\_\_\_ \text{ cubic units.}$$

16. Consider the volume generated by rotating the planar region bounded by $y = x^2$ and $y = |x| + 2$ about the y-axis. A sketch of the region is shown on the next page. The two curves intersect at the points where $y = x^2$ is the same as $y = |x| + 2$. For $x > 0$, this occurs when $x^2 = x + 2$ or $x =$ ___ . From symmetry it follows that the points of intersection are _______ and _______ .

---

15. $\sqrt{y^2 + 9}$, 5, $16 - y^2$, $\pi(16 - y^2)$, $16y - \frac{1}{3}y^3$, $\frac{128\pi}{3}$

Imagine the solid to be cut into thin slices by planes perpendicular to the y-axis. As y varies from $y = 0$ to $y = 2$, each slice is a thin washer of thickness Δy, inner radius 0, and outer radius $x =$ _____ . As y varies from $y = 2$ to $y = 4$, each slice is a thin washer of thickness Δy, inner radius $x =$ ______ and outer radius $x = \sqrt{y}$. Therefore, the volume is given by

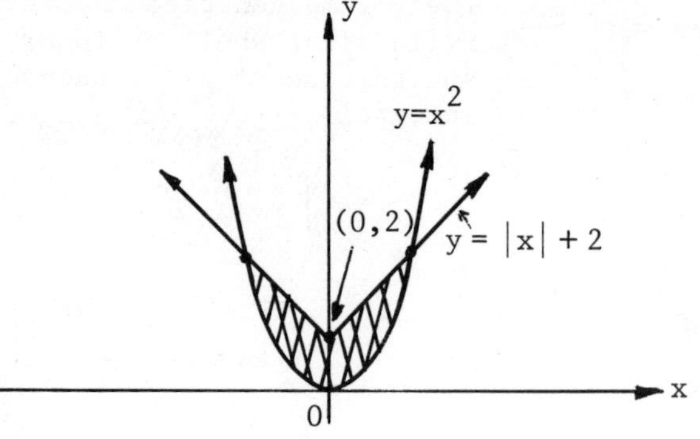

$$V = \int_0^2 \pi(y - \_\_\_)\,dy + \int_2^4 \pi\left[\_\_\_\_\_\_\_\_\_\_\right]dy$$

$$= \frac{\pi y^2}{2}\Big]_0^2 + \pi\Big(\_\_\_\_\_\_\_\_\_\_\Big)\Big]_2^4$$

$$= 2\pi + \pi\left[\left(8 - \frac{8}{3}\right) - (\_\_\_\_\_)\right]$$

$$= \_\_\_\_\_ \approx 16.76 \text{ cubic units.}$$

5-6 Approximations.

17. Roughly speaking, the conclusion of this article in the textbook is that in the approximation

$$\Delta U \approx f(X)\,\Delta x$$

we must include all ________________ terms, but we may omit ______________ terms like $(\Delta x)^2$, $(\Delta x)^3$, or we may omit mixed terms like __________ or $(\Delta x)^2\Delta y$, $(\Delta x)(\Delta y)^2$, and so forth.

18. If each separate piece ΔU is estimated to the degree of accuracy specified in Problem 17 above, then the total quantity U is exactly equal to the limit ________________ , that is the definite integral ______________ .

---

16. 2, $(-2,4)$, $(2,4)$, $\sqrt{y}$, $y-2$, 0, $y-(y-2)^2$, $\frac{y^2}{2} - \frac{(y-2)^3}{3}$, $2-0$, $\frac{16\pi}{3}$

17. first power Δx, higher-power, $(\Delta x)(\Delta y)$

18. $\lim_{\Delta x \to 0} \sum_a^b f(X)\,\Delta x$, $\int_a^b f(x)\,dx$

5-7 Length of a Plane Curve.

OBJECTIVE A: Find the length of a planar curve between two specified points, when the curve is defined by an equation that gives y as a continuously differentiable function of x, or gives x as a continuously differentiable function of y.

19. Consider the curve defined by the function $y = \frac{2}{3}(x-1)^{3/2}$ for $1 \le x \le 5$. We seek the length of the graph from $(1,0)$ to $\left(5, \frac{16}{3}\right)$.

$\frac{dy}{dx} =$ ______, so that $1 + \left(\frac{dy}{dx}\right)^2 =$ ___. The arc length is given by

$$s = \int_1^5 \sqrt{1 + \left(\frac{dy}{dx}\right)^2}\, dx = \int_1^5 \text{\_\_\_\_\_\_}\, dx$$

$$= \text{\_\_\_\_\_\_}\Big]_1^5 = \text{\_\_\_\_\_\_\_\_\_\_} \approx 6.78689 \text{ units.}$$

20. Let the curve be defined by the equation $y^2 = x^3$. To find the arc length from the point $(0,0)$ to the point $(4,8)$, we first calculate dy/dx. Differentiating the equation for the curve implicitly gives

$$2y\frac{dy}{dx} = \text{\_\_\_\_} \quad \text{so that} \quad 1 + \left(\frac{dy}{dx}\right)^2 = 1 + \frac{\text{\_\_\_\_}}{4y^2} = \text{\_\_\_\_\_\_\_\_} .$$

Therefore, the arc length is given by

$$s = \int_{\text{\_\_}}^{\text{\_\_}} \text{\_\_\_\_\_\_\_\_}\, dx = \frac{4}{9} \cdot \frac{2}{3}\left[\text{\_\_\_\_\_\_\_\_\_\_}\right]_0^4$$

$$= \frac{8}{27}\left(\text{\_\_\_\_\_\_\_\_}\right) \approx 9.073 \text{ units.}$$

OBJECTIVE B: Find the length of a curve specified parametrically by continuously differentiable equations $x = f(t)$ and $y = g(t)$ over a given interval $a \le t \le b$.

21. To compute the length of the curve given by $x = t^2 \cos t$ and $y = t^2 \sin t$ for $0 \le t \le 1$, we first calculate dx/dt and dy/dt.

---

19. $\sqrt{x-1}$, x, $\sqrt{x}$, $\frac{2}{3}x^{3/2}$, $\frac{2}{3}(5\sqrt{5} - 1)$

20. $3x^2$, $9x^4$, $1 + \frac{9x}{4}$, $\int_0^4 \sqrt{1 + \frac{9x}{4}}\, dx$, $\left(1 + \frac{9x}{4}\right)^{3/2}$, $10\sqrt{10} - 1$

$$\frac{dx}{dt} = \underline{\qquad\qquad} \; ; \quad \frac{dy}{dt} = \underline{\qquad\qquad} \quad \text{so that}$$

$$\left(\frac{dx}{dt}\right)^2 + \left(\frac{dy}{dt}\right)^2 = t^2\left[(2\cos t - t\sin t)^2 + (\underline{\qquad\qquad})^2\right]$$

$$= t^2\left[4\cos^2 t - 4t\cos t\sin t + t^2\sin^2 t + (\underline{\qquad\qquad\qquad})\right]$$

$$= t^2\left[4 + \underline{\quad}\right].$$

Hence the arc length is given by,

$$s = \int_0^1 \sqrt{\left(\frac{dx}{dt}\right)^2 + \left(\frac{dy}{dt}\right)^2}\, dt = \int_0^1 \underline{\qquad\qquad}\, dt$$

$$= \underline{\qquad\qquad}\Big]_0^1 = \frac{1}{3}(\underline{\qquad\qquad}) \approx 1.0601 \text{ units.}$$

5-8 Area of a Surface of Revolution.

OBJECTIVE A: Find the area of a surface generated by rotating the portion of a given curve $y = f(x)$ between $x = a$ and $x = b$ about a specified axis or line parallel to an axis. The function f is assumed to have a derivative that is continuous on the interval.

22. Find the surface area generated when the arc $y = 2\sqrt{x}$ from $x = 0$ to $x = 8$ is rotated about the x-axis.

Solution. We write $dS = 2\pi y\, ds = 2\pi \cdot 2\sqrt{x} \cdot \underline{\qquad\qquad}\, dx$.

$\frac{dy}{dx} = \underline{\quad}$ so that $\sqrt{1 + \left(\frac{dy}{dx}\right)^2} = \underline{\qquad\qquad}$.

---

21. $2t\cos t - t^2\sin t$, $2t\sin t + t^2\cos t$, $2\sin t + t\cos t$, $4\sin^2 t + 4t\cos t\sin t + t^2\cos^2 t$, t^2, $t\sqrt{4+t^2}$, $\frac{1}{3}\left(4+t^2\right)^{3/2}$, $5\sqrt{5} - 8$

22. $\sqrt{1 + \left(\frac{dy}{dx}\right)^2}$, $x^{-1/2}$, $\sqrt{1+x^{-1}}$, $\int_0^8 4\pi\sqrt{x}\sqrt{1+x^{-1}}\, dx$, $\int_0^8 4\pi\sqrt{x+1}\, dx$, $\frac{2}{3}(x+1)^{3/2}$, $9^{3/2}$, $\frac{208\pi}{3}$

Thus,

$$S = \int_{\_}^{\_} \underline{\qquad\qquad\qquad} dx = \int_{\_}^{\_} \underline{\qquad\qquad} dx$$

$$= 4\pi \left(\underline{\qquad\qquad\qquad} \right) \Big]_0^8 = \frac{8\pi}{3} \left(\underline{\quad} - 1 \right)$$

$$= \underline{\qquad} \approx 217.81709 \text{ sq. units.}$$

23. Find the surface area generated when the arc $y = x^{1/3}$ from $x = 0$ to $x = 1$ is rotated about the y-axis.

Solution. We write $dS = 2\pi \underline{\quad} ds$, since the rotation occurs about the y-axis. Since $x = y^3$, $dx/dy = \underline{\qquad}$ and

$$ds = \sqrt{1 + \left(\frac{dy}{dx}\right)^2} dx = \sqrt{1 + \left(\frac{dx}{dy}\right)^2} dy = \underline{\qquad\qquad} .$$

At $x = 0$, $y = 0$ and at $x = 8$, $y = 2$ so the surface area is

$$S = \int 2\pi x \, ds = 2\pi \int_{\_}^{\_} \underline{\qquad\qquad\qquad} dy$$

$$= 2\pi \left(\frac{1}{36} \cdot \frac{2}{3}\right) \left(\underline{\qquad\qquad\qquad} \right]_{\_}^{\_} = \frac{\pi}{27} \left(\underline{\qquad\qquad\qquad} \right)$$

$$\approx 203.0436 \text{ sq. units.}$$

OBJECTIVE B: Find the area of a surface generated by rotating the arc of a curve specified parametrically by equations $x = f(t)$ and $y = g(t)$ over $a \le t \le b$ about an indicated axis. (Again, f and g are assumed to have continuous derivatives.)

24. Find the surface area generated when the arc $x = 2t$ and $y = \sqrt{2}\, t^2$ from $t = 0$ to $t = 2$ is rotated about the y-axis.

Solution. We write $dS = 2\pi \underline{\quad} ds$, since the rotation occurs about the y-axis. Next, we calculate the derivatives $dx/dt = \underline{\quad}$ and $dy/dt = \underline{\qquad}$ so that the arc length differential is given by

$$ds = \sqrt{\underline{\qquad\qquad\qquad}} \, dt = 2\sqrt{\underline{\qquad\qquad}} \, dt .$$

---

23. x, $3y^2$, $\sqrt{1 + 9y^4}$, $\int_0^2 y^3\sqrt{1 + 9y^4}\, dy$, $\left(1 + 9y^4\right)^{3/2}\Big]_0^2$, $(145)^{3/2} - 1$

Therefore, the surface area is

$$S = 2\pi \int_0^2 \underline{\hspace{3cm}}\, dt = \frac{4\pi}{3}\left(\underline{\hspace{3cm}} \right)\Big]_0^2$$

$$= \frac{4\pi}{3}\left(\underline{\hspace{1.5cm}} \right) = \underline{\hspace{1.5cm}} \approx 108.90854 \text{ sq. units.}$$

5-9 Average Value of a Function.

OBJECTIVE: Calculate the average value of a given continuous function $y = f(x)$ over a specified interval $a \le x \le b$.

25. To find the average value of $y = \sin x - \cos x$ over $0 \le x \le \pi/4$ we have

$$\left(y_{av}\right)_x = \frac{1}{\frac{\pi}{4} - 0} \int_{\_\_}^{\_\_} \left(\underline{\hspace{3cm}} \right) dx$$

$$= \frac{4}{\pi}\left(\underline{\hspace{3cm}} \right)\Big]_{\_\_}^{\_\_}$$

$$= \frac{4}{\pi}\left[\left(-\frac{\sqrt{2}}{2} - \frac{\sqrt{2}}{2}\right) - \left(\underline{\hspace{1.5cm}} \right)\right] = \frac{4}{\pi}\left(\underline{\hspace{1.5cm}} \right) \approx -.52739 .$$

26. Suppose that $y = f(x)$ has a continuous derivative over the interval $a \le x \le b$. Since the slope m of the tangent line at any point is $f'(x)$, the average value of the slope is given by

$$\left(m_{av}\right)_x = \underline{\hspace{3cm}} = \frac{\underline{\hspace{2.5cm}}}{b - a} .$$

Interpreting this last equation geometrically, the average slope of the tangent line over the interval is precisely the slope of the chord joining the points ________ and ________ on the graph of $y = f(x)$. Do you see how this motivates the name of the "Mean Value Theorem" of Article 3-8?

---

24. x, 2, $2\sqrt{2}\,t$, $\left(\frac{dx}{dt}\right)^2 + \left(\frac{dy}{dt}\right)^2$, $2t^2 + 1$, $4t\sqrt{2t^2 + 1}$, $\left(2t^2 + 1\right)^{3/2}$, $27 - 1$, $\frac{104\pi}{3}$

25. $\int_0^{\pi/4} (\sin x - \cos x)\, dx$, $(-\cos x - \sin x)\Big]_0^{\pi/4}$, $(-1 - 0)$, $(1 - \sqrt{2})$

26. $\frac{1}{b - a}\int_a^b f'(x)\, dx$, $f(b) - f(a)$, $\big(a, f(a)\big)$, $\big(b, f(b)\big)$

5-10 Moments and Center of Mass.

OBJECTIVE: Find the center of mass of a thin homogeneous plate covering a given region of the xy-plane.

27. Find the center of mass $(\bar{x},\bar{y})$ of the thin homogeneous triangular plate formed in the first quadrant by the coordinate axes and the line $x + 2y = 2$.

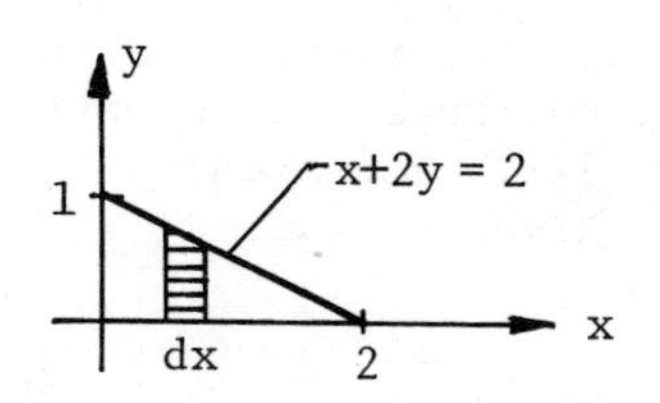

Solution. The plate is depicted in the figure at the right. To find $\bar{x}$, divide the triangle into vertical strips of width dx parallel to the y-axis. A representative strip is shown in the figure and its mass is approximately

$$dm = \delta_2 dA = \delta_2(y \underline{\quad}) = \delta_2(\underline{\qquad}) \, dx .$$

Thus,

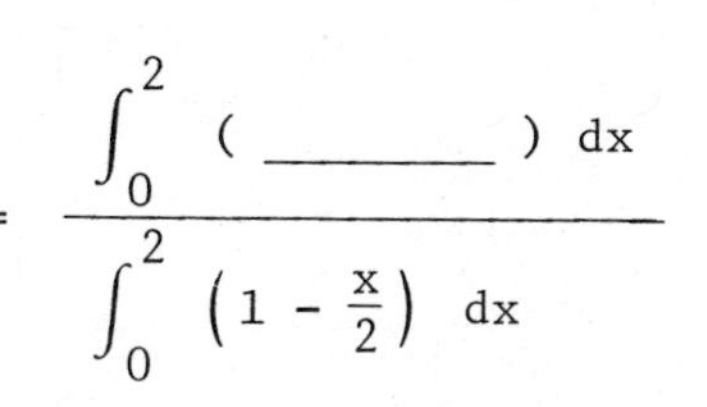

$$\bar{x} = \frac{\int x \, dm}{\int dm} = \frac{\int_0^2 x \cdot (\underline{\qquad}) \, dx}{\int_0^2 \delta_2 \left(1 - \frac{x}{2}\right) dx} = \frac{\int_0^2 (\underline{\qquad}) \, dx}{\int_0^2 \left(1 - \frac{x}{2}\right) dx}$$

$$= \frac{\underline{\qquad}\Big]_0^2}{x - \frac{1}{4}x^2 \Big]_0^2} = \frac{\underline{\quad}}{2 - 1} = \underline{\quad} .$$

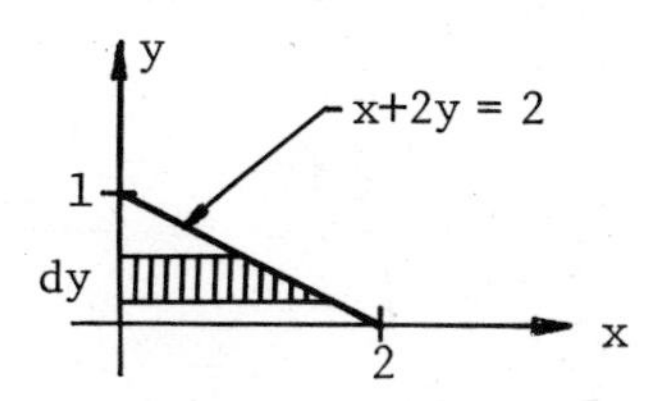

In a similar manner, to find $\bar{y}$, divide the triangle into horizontal strips of width dy. The mass of a representative strip is approximately

$$dm = \delta_2 dA = \delta_2(\underline{\quad} \, dy) = \delta_2(\underline{\qquad}) \, dy .$$

Thus,

$$\bar{y} = \frac{\int y \, dm}{\int} = \frac{\int_0^1 y \cdot (\underline{\qquad}) \, dy}{\int_0^1 \delta_2 (2 - 2y) \, dy} = \frac{\int_0^1 (\underline{\qquad}) \, dy}{2 \int_0^1 (1 - y) \, dy}$$

$$= \frac{\underline{\qquad}\Big]_0^1}{2y - y^2 \Big]_0^1} = \frac{\underline{\quad}}{2 - 1} = \underline{\quad} .$$

---

27. dx, $1 - \frac{x}{2}$, $\delta_2\left(1 - \frac{x}{2}\right) dx$, $x - \frac{1}{2}x^2$, $\frac{1}{2}x^2 - \frac{1}{6}x^3$, $\frac{2}{3}$, $\frac{2}{3}$;

x, $2 - 2y$, $\delta_2(2 - 2y)$, $2(y - y^2)$, $y^2 - \frac{2}{3}y^3$, $1 - \frac{2}{3}$, $\frac{1}{3}$

5-11 Centroid and Center of Gravity.

OBJECTIVE: Find the center of gravity of the planar region bounded by given curves and lines.

28. Find the center of gravity of the region bounded by the parabola $y = x^2$ and the line $y = x$.

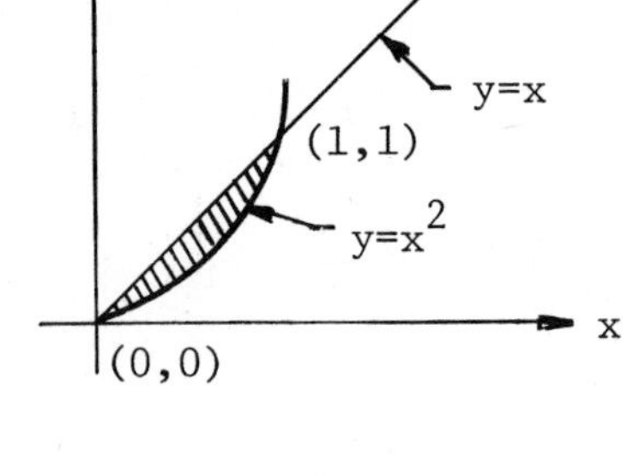

Solution. The region is illustrated at the right. To calculate $\bar{x}$, divide the region into vertical strips of width dx . The area of a typical strip is given by dA = (______) dx so that

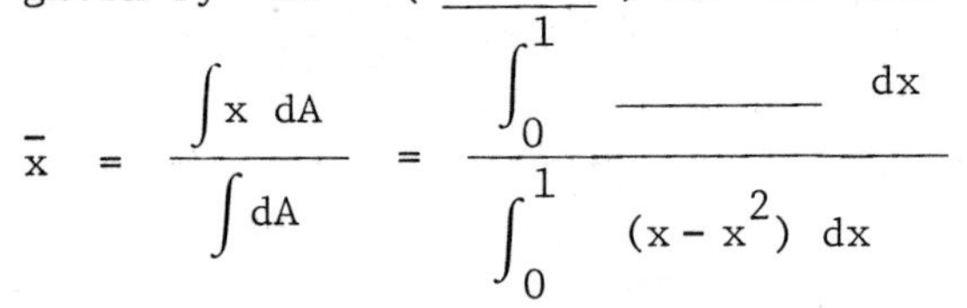

$$\bar{x} = \frac{\int x\, dA}{\int dA} = \frac{\int_0^1 \_\_\_\_\_\_\_\_ \, dx}{\int_0^1 (x - x^2)\, dx}$$

$$= \frac{\left.\_\_\_\_\_\_\_\_\_\_\right]_0^1}{\left.\frac{1}{2}x^2 - \frac{1}{3}x^3\right]_0^1} = \frac{\_\_\_\_\_}{\frac{1}{2} - \frac{1}{3}} = \frac{\_\_\_\_}{\frac{1}{6}} = \_\_\_ .$$

To find $\bar{y}$, divide the region into horizontal strips of width dy . The area of a typical strip is dA = (________) dy so that

$$\bar{y} = \frac{\int y\, dA}{\int dA} = \frac{\_\_\_\_\_\_\_\_\_\_\_\_\_\_\_\_}{\frac{1}{6}} = \frac{\left.\_\_\_\_\_\_\_\_\_\_\_\_\_\right]_0^1}{\frac{1}{6}}$$

$$= 6(\_\_\_\_\_\_) = 6(\_\_\_) = \_\_\_ .$$

5-12 The Theorems of Pappus.

OBJECTIVE: Use Pappus' First Theorem to find the volume of a solid of revolution of a given planar region rotated about a line parallel to one of the coordinate axes.

29. Suppose it is required to find the volume generated by rotating the triangle with vertices (1,0) , (3,-1) , and (3,2) about the y-axis. According to Pappus' Theorem, V = ______ is the area of the region times the distance traveled by its center of gravity. Let the base of the triangle be the line segment joining the vertices (3,-1) and (3,2) .

---

28. $x - x^2$, $x(x - x^2)$, $\frac{1}{3}x^3 - \frac{1}{4}x^4$, $\frac{1}{3} - \frac{1}{4}$, $\frac{1}{12}$, $\frac{1}{2}$, $\sqrt{y} - y$,
$\int_0^1 y\left(\sqrt{y} - y\right) dy$, $\frac{2}{5}y^{5/2} - \frac{1}{3}y^3$, $\frac{2}{5} - \frac{1}{3}$, $\frac{1}{15}$, $\frac{2}{5}$

Then the height of the triangle equals $3 - \_\_\_ = \_\_\_$ units, so the area $A = \frac{1}{2}(2 - \_\_\_)(2) = \_\_\_$ square units. Next, $\bar{x}$ is located 2/3 of the distance along the median from the vertex (1,0). That is, $\bar{x} = 1 + \frac{2}{3}(\_\_\_\_\_) = \_\_\_$. Therefore, $V = 2\pi(\_\_\_)(3) = \_\_\_$ cubic units.

5-13 Hydrostatic Pressure.

OBJECTIVE: Find the total force exerted on a given planar region placed vertically under the surface of an incompressible fluid of constant density w.

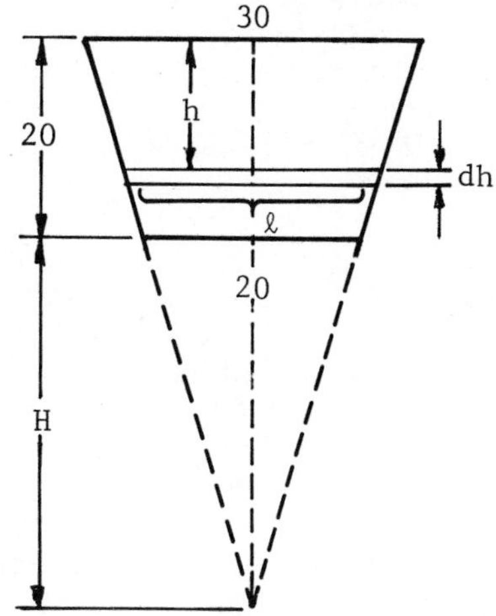

30. Suppose the face of a dam has the shape of an isosceles trapezoid of altitude 20 feet with an upper base of 30 feet and a lower base of 20 feet. A figure illustrating the face of the dam is shown at the right. Let us find the force due to water pressure on the dam when the surface of the water is level with the top of the dam.

From the figure and similar triangles, we see that $\frac{20}{H} = \frac{30}{\_\_\_\_\_}$ or

$30H = \_\_\_\_\_\_\_\_$; thus $H = \_\_\_\_$ feet. Next, consider a horizontal strip of the face of width dh at a depth h below the surface of the water (see figure). Let ℓ denote the horizontal length of this strip.

Then by similar triangles, $\frac{\ell}{60 - h} = \frac{30}{\_\_\_}$, or $\ell = \_\_\_\_\_\_\_\_$.

Therefore, the area of the strip is $dA = \ell dh$, so the total force exerted on the face of the dam is

$$F = \int wh\, dA = \int wh\ell\, dh = \int_0^{20} \_\_\_\_\_\_\_\_\_\_ = \frac{w}{2}\int_0^{20} (\_\_\_\_\_\_)\, dh$$

$$= \frac{w}{2}\Big[\_\_\_\_\_\_\_\_\Big]_0^{20} = 200w\left(30 - \_\_\_\right)$$

$$= \_\_\_\_\_\_ \approx 291{,}667 \text{ pounds using } w = 62.5 .$$

---

29. $2\pi\bar{x}A$, 1, 2, -1, 3, $3-1$, 7/3, 7/3, 14π

30. $20 + H$, $400 + 20H$, 40, 60, $\frac{1}{2}(60 - h)$, $\frac{w}{2} h(60-h)\, dh$, $60h - h^2$, $30h^2 - \frac{1}{3}h^3$, $\frac{20}{3}$, $\frac{14000w}{3}$

5-14 Work.

OBJECTIVE A: For a mechanical spring of given natural length L, if a specified force is required to stretch or compress the spring by a certain given amount, calculate (a) the "spring constant" c and (b) the amount of work done in stretching or compressing the spring from a specified length $L = a$ to a specified length $L = b$.

31. Suppose an unstretched spring is 3 feet long. When the spring is used to suspend a 4-pound weight it stretches to a length of 5 feet. Therefore, the equation $F = cx$ becomes ____________, or $c =$ ___ is the value of the spring constant. To calculate the work required to stretch the spring from a length of 4 feet to a length of 6 feet, we find

$$W = \int F\,ds = \int_1^{\_\_} (\_\_\_\_)\,dx = \_\_\_ \text{ foot-pounds}.$$

OBJECTIVE B: Find the work done in pumping all the water out of a specified container to a given height above the container.

32. A cylindrical tank of radius 5 feet and height 10 feet stands on a platform so that its bottom is 50 feet above the surface of a lake. Water is pumped directly up from the surface of the lake into the water tank. Find the depth d of the water in the tank when half the necessary work has been done to fill the tank.

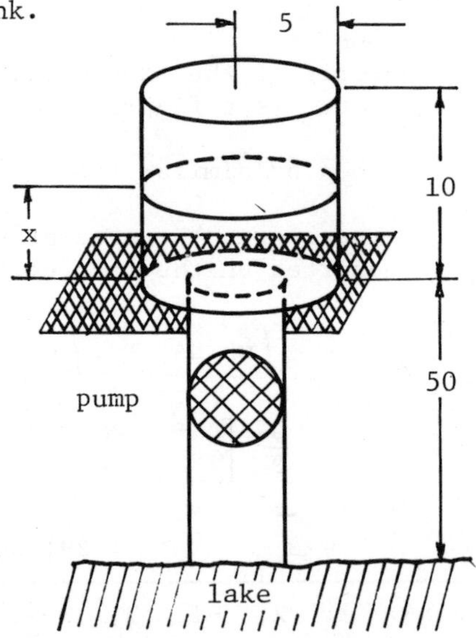

Solution. Let the distance from the bottom of the tank to the surface of the water in the tank be x feet. The situation is depicted in the figure at the right. Consider a slice of water in the tank of thickness dx cut by two planes parallel to the surface of the lake and platform. The volume of this slice is $dV = A\,dx =$ ________. Therefore, its weight is $w\,dV =$ ________, where w is the weight of a cubic foot of water. The total distance this slice is lifted is $50 + x$ feet, where x is its typical height from the bottom of the tank.

---

31. $4 = c(5 - 3)$, 2, $\int_1^3 2x\,dx$, 8

Therefore, the total work required to fill the tank is

$$W_1 = \int \text{distance} \cdot w\, dV = \int_0^{10} (\underline{\qquad\qquad}) \, dx$$

$$= 25\pi w \, (\underline{\qquad\qquad}) \Big]_0^{10} = 25\pi w (\underline{\quad}) \, .$$

On the other hand, the work required to fill the tank to the depth d is given by

$$W_2 = \int_0^d (\underline{\qquad\qquad}) \, dx = 25\pi w (\underline{\qquad\qquad}) \, .$$

We want to find d when $W_2 = \frac{1}{2} W_1$. This translates into

$25\pi w(\underline{\qquad\qquad}) = \frac{1}{2}(25\pi w)(550)$ or $\underline{\qquad\qquad} = 550$.

Solving this quadratic equation for d (which must be positive) gives

$$d = \frac{-100 \pm \sqrt{10000 + \underline{\quad}}}{2} , \text{ or } d \approx \underline{\quad} \text{ feet} .$$

CHAPTER 5 OBJECTIVE - PROBLEM KEY

Objective	Problems in Thomas/Finney Text
5-2 A	p. 230, 2,5,6,8,9,11
B	p. 230, 1,3,4,6,7,9,11
5-3 A	p. 233, 1-8,15
B	p. 233, 9-13
5-4 A	p. 239, 1-8,10a,12
B	p. 239, 13-17
5-5 A	p. 245, 1,2,4,5,6,9-17
B	p. 245, 3,7,8,19
5-7 A	p. 254, 1-6
B	p. 255, 7-10
5-8 A	p. 260, 1-4,6
B	p. 260, 7-9
5-9	p. 263, 1-7
5-10	p. 270, 1-5
5-11	p. 273, 1-5
5-12	p. 276, 1,4,5
5-13	p. 280, 1-5
5-14 A	p. 284, 1-6
B	p. 285, 13,14

---

32. $25\pi\, dx$, $25\pi w\, dx$, $25\pi w(50 + x)$, $50x + \frac{1}{2}x^2$, 550, $25\pi w(50 + x)$, $50d + \frac{1}{2}d^2$, $50d + \frac{1}{2}d^2$, $100d + d^2$, 2200, 5.227

CHAPTER 5 SELF-TEST

1. Find the area of the planar region bounded by the curves $y = x^3 + 1$ and $y = x^2 + x$.

2. Find the area of the planar region bounded by the curves $x = y^{1/3}$ and $y^2 = x$.

3. A train leaving a railroad station has an acceleration of $a = 0.5 + 0.02t$ ft/sec^2. How far will the train move in the first 20 sec of motion? What is its velocity after 20 seconds?

4. Find the volume generated by rotating the ellipse $\frac{x^2}{a^2} + \frac{y^2}{b^2} = 1$ about the x-axis.

5. The base of a certain solid is the region between the planar curves $y = 4 + x^2$ and $y = 12 - x^2$, and the cross sections by planes perpendicular to the x-axis are circles with diameters extending from one curve to the other. Find the volume of the solid.

6. Find the volume generated when the planar region between the lines $y = x$, $y = 3x$, and $x + y = 4$ is rotated about the x-axis.

7. Find the volume generated when the planar region between the curve $y^2 = 9x$ and the lines $x = 4$, $y = 0$ is rotated about the line $x = 5$.

8. Find the lengths of the following curves.

   (a) $y = \frac{2}{3}x^{3/2} - \frac{1}{2}x^{1/2}$ over $0 \leq x \leq 1$, and

   (b) $x = t^3 + 3t^2$ and $y = t^3 - 3t^2$ for $0 \leq t \leq 2$.

9. Find the surface area generated by revolving about the x-axis the arc

$$y = \frac{1}{2\sqrt{2}} x\sqrt{x - x^2}, \quad 0 \leq x \leq 1.$$

10. Find the area of the surface generated by rotating the cardiod $x = 2\cos\theta - \cos 2\theta$, $y = 2\sin\theta - \sin 2\theta$, for $0 \leq \theta \leq \pi$, about the x-axis. First sketch the curve.

11. Find the average value of the function $y = x^2 - x + 1$ over the interval $0 \leq x \leq 2$.

12. Find the center of gravity of the planar region bounded by the curves $y = x + 2$, $y = 2x$, and the y-axis.

13. A water pipe is in the shape of a cylinder placed horizontally in the ground. Its radius is 4 inches. If the pipe is half full of water, find the force on the face of the vertical circular plate that closes off the pipe. Assume the density of water is $w = 62.5$ pounds per cubic foot.

14. If a force of 90 pounds stretches a 10-foot spring by 1 foot, find the work done in stretching the spring from 10 feet to 15 feet.

15. A conical tank is 16 feet across the top, and 12 feet deep. It contains water to a depth of 8 feet. Find the work required to pump all the water to a height of 2 feet above the top of the tank.

SOLUTIONS TO CHAPTER 5 SELF-TEST

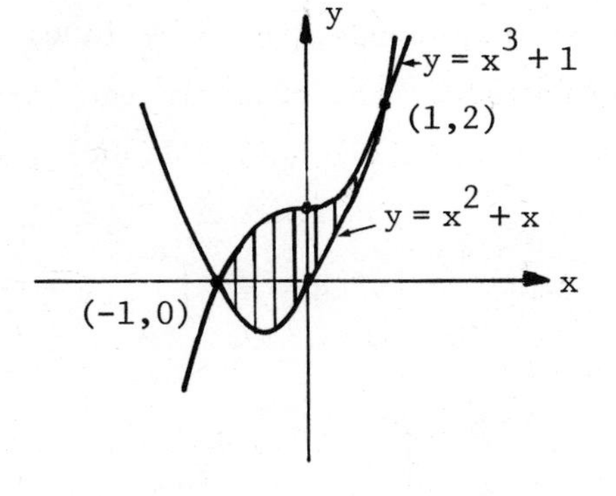

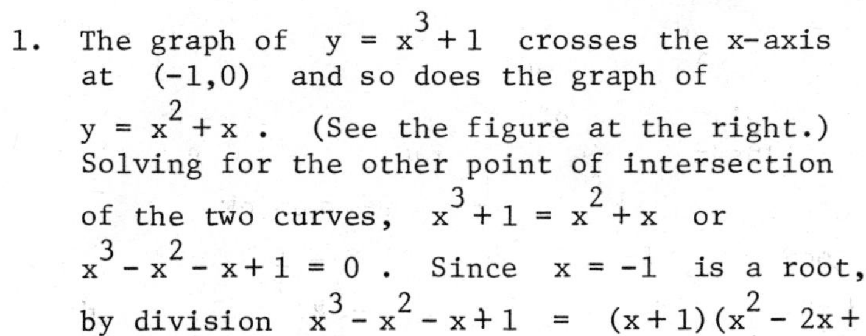

1. The graph of $y = x^3 + 1$ crosses the x-axis at $(-1,0)$ and so does the graph of $y = x^2 + x$. (See the figure at the right.) Solving for the other point of intersection of the two curves, $x^3 + 1 = x^2 + x$ or $x^3 - x^2 - x + 1 = 0$. Since $x = -1$ is a root, by division $x^3 - x^2 - x + 1 = (x+1)(x^2 - 2x + 1) = 0$ or $(x+1)(x-1)^2 = 0$. Thus, the other point of intersection is $(1,2)$. The area between the curves is then given by

$$A = \int_{-1}^{1} \left[(x^3 + 1) - (x^2 + x)\right] dx = \left(\frac{1}{4}x^4 + x - \frac{1}{3}x^3 - \frac{1}{2}x^2\right)\Big]_{-1}^{1}$$

$$= \frac{4}{3} \text{ square units.}$$

2. The planar region is shown at the right. The points of intersection of the two graphs are $(0,0)$ and $(1,1)$. Thus the area is given by

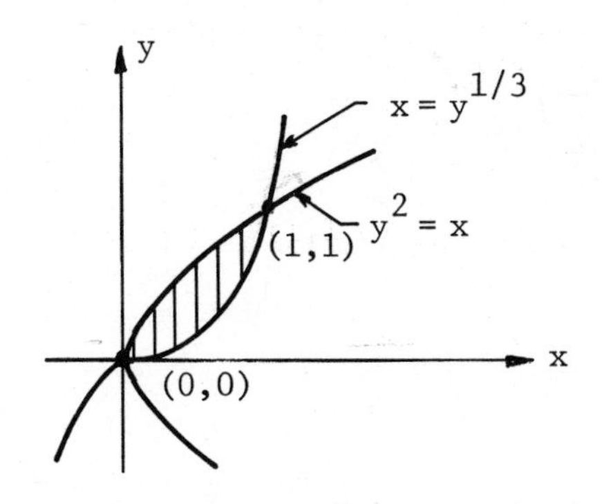

$$A = \int_0^1 \left(y^{1/3} - y^2\right) dy$$

$$= \frac{3}{4}y^{4/3} - \frac{1}{3}y^3\Big]_0^1 = \frac{5}{12} \text{ square units.}$$

Alternatively,

$$A = \int_0^1 \left(\sqrt{x} - x^3\right) dx$$

$$= \frac{2}{3}x^{3/2} - \frac{1}{4}x^4\Big]_0^1 = \frac{5}{12} \text{ square units.}$$

3. Since $v = \int a\,dt$ we have $v = 0.5t + 0.01t^2 + C_1$. At $t = 0$ the train is at rest, so $v = 0$; hence $C_1 = 0$. Next, $s = \int v\,dt$ or $s = 0.25t^2 + \frac{1}{300}t^3 + C_2$. At $t = 0$, $s = 0$ so that $C_2 = 0$. Thus, when $t = 20$ seconds, $s = \frac{1}{4}(400) + \frac{1}{300}(8000) = 126\ 2/3$ feet, the distance traveled by the train in the first 20 seconds. Its velocity at that time is $v = (0.5)(20) + (0.01)(400) = 14$ ft/sec .

4. $$V = \int_{-a}^{a} \pi y^2\,dx = \int_{-a}^{a} \pi \cdot \frac{b^2}{a^2}(a^2 - x^2)\,dx$$

$$= \frac{\pi b^2}{a^2}\left[a^2 x - \frac{1}{3}x^3\right]_{-a}^{a} = \frac{4}{3}\pi b^2 a .$$

This generalizes the formula for the volume of a sphere, when $a = b = r$ is its radius.

5. The two curves intersect when

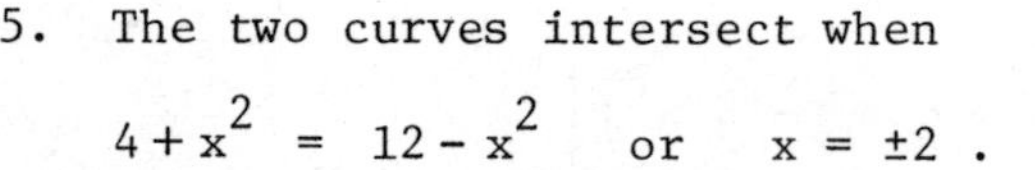

$4 + x^2 = 12 - x^2$ or $x = \pm 2$.

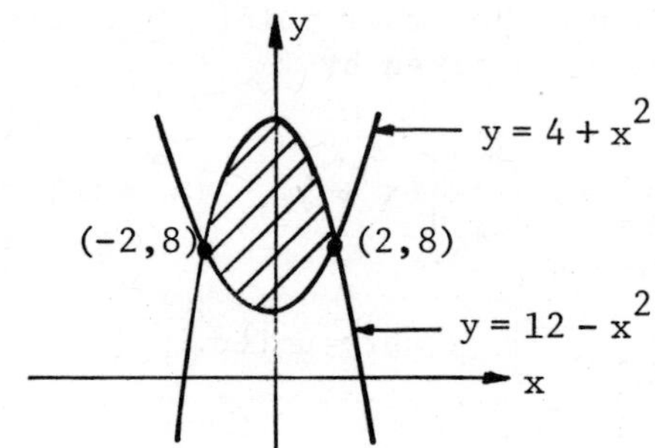

(See the figure at the right depicting the base of the solid.)

The cross-sectional area of a typical slice is

$$\pi\left(\frac{12 - x^2 - (4 + x^2)}{2}\right)^2 = \pi\left(4 - x^2\right)^2 = \pi(16 - 8x^2 + x^4) .$$

Thus, the volume is given by

$$V = \int A(x)\,dx = \int_{-2}^{2} \pi(16 - 8x^2 + x^4)\,dx$$

$$= \pi\left[16x - \frac{8}{3}x^3 + \frac{1}{5}x^5\right]_{-2}^{2} = \frac{512\pi}{15} \approx 107.2 \text{ cubic units.}$$

6. The planar region to be rotated about x-axis is shown in the figure at the right. We slice the region into horizontal slices and use the method of cylindrical shells. Thus, from the figure we see that

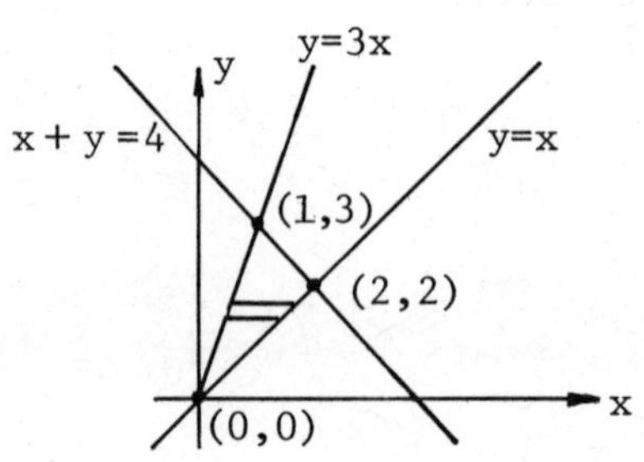

$$V = \int_0^2 2\pi y\left(y - \frac{y}{3}\right) dy + \int_2^3 2\pi y\left(4 - y - \frac{y}{3}\right) dy$$

$$= 2\pi\left[\frac{2}{9}y^3\right]_0^2 + 2\pi\left[2y^2 - \frac{4}{9}y^3\right]_2^3 = \frac{20\pi}{3} \approx 20.9 \text{ cubic units.}$$

7. The planar region to be rotated is shown in the figure at the right. We slice the region into horizontal slices and use the method of washers. Thus, from the figure we find the volume is,

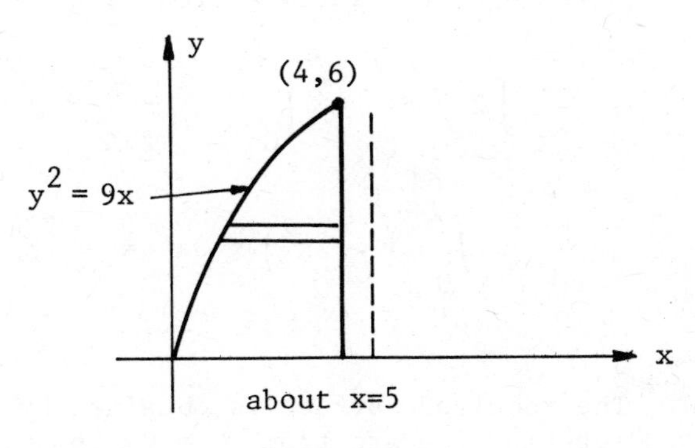

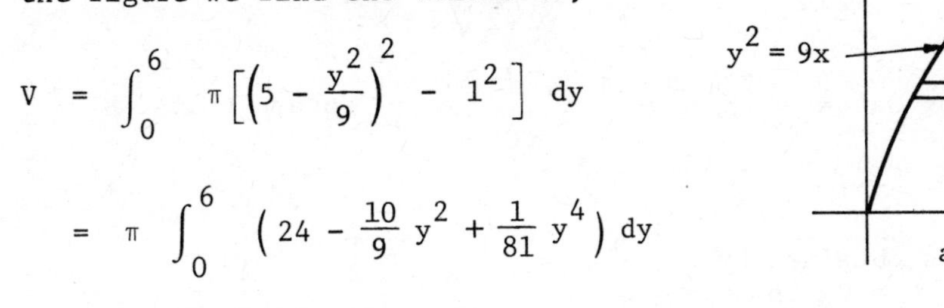

$$V = \int_0^6 \pi\left[\left(5 - \frac{y^2}{9}\right)^2 - 1^2\right] dy$$

$$= \pi\int_0^6 \left(24 - \frac{10}{9}y^2 + \frac{1}{81}y^4\right) dy$$

$$= \pi\left[24y - \frac{10}{27}y^3 + \frac{1}{405}y^5\right]_0^6 = 83.2\pi \approx 261.4 \text{ cubic units.}$$

8. (a) $y = \frac{2}{3}x^{3/2} - \frac{1}{2}x^{1/2}$ and $\frac{dy}{dx} = x^{1/2} - \frac{1}{4}x^{-1/2}$

$$1 + \left(\frac{dy}{dx}\right)^2 = \left(x - \frac{1}{2} + \frac{1}{16}x^{-1}\right) + 1 = \left(x^{1/2} + \frac{1}{4}x^{-1/2}\right)^2,$$

so the arc length is given by

$$s = \int_0^1 \left(x^{1/2} + \frac{1}{4}x^{-1/2}\right) dx = \frac{2}{3}x^{3/2} + \frac{1}{2}x^{1/2}\Big]_0^1 = \frac{7}{6} \text{ units.}$$

(b) $\frac{dx}{dt} = 3t^2 + 6t$ and $\frac{dy}{dt} = 3t^2 - 6t$ so that

$$\left(\frac{dx}{dt}\right)^2 + \left(\frac{dy}{dt}\right)^2 = \left(9t^4 + 36t^3 + 36t^2\right) + \left(9t^4 - 36t^3 + 36t^2\right) = 18t^2\left(t^2 + 4\right).$$

Thus, the arc length is given by

$$s = \int_0^2 3\sqrt{2}\, t\sqrt{t^2 + 4}\, dt = 3\sqrt{2}\left(\frac{1}{3}\right)\left(t^2 + 4\right)^{3/2}\Big]_0^2$$

$$= 8\left(4 - \sqrt{2}\right) \approx 20.7 \text{ units.}$$

9. Notice that $8y^2 = x^2 - x^4$. Differentiating implicitly,

$$16y\left(\frac{dy}{dx}\right) = 2x - 4x^3 \quad \text{or} \quad \frac{dy}{dx} = \frac{x - 2x^3}{8y} . \quad \text{Thus,}$$

$$1 + \left(\frac{dy}{dx}\right)^2 = 1 + \frac{\left(x - 2x^3\right)^2}{64y^2} = 1 + \frac{\left(x - 2x^3\right)^2}{8\left(x^2 - x^4\right)} = \frac{\left(3 - 2x^2\right)^2}{8\left(1 - x^2\right)} .$$

Therefore, the surface area is given by

$$S = \int 2\pi y \; ds = \int_0^1 2\pi \frac{x}{2\sqrt{2}} \sqrt{1 - x^2} \cdot \frac{3 - 2x^2}{2\sqrt{2}\sqrt{1 - x^2}} \; dx$$

$$= \frac{\pi}{4} \int_0^1 \left(3 - 2x^2\right) x \; dx = \frac{\pi}{4}\left[\frac{3}{2}x^2 - \frac{1}{2}x^4\right]_0^1 = \frac{\pi}{4} \text{ square units.}$$

10. The required surface is obtained by rotating the arc from $\theta = 0$ to $\theta = \pi$ about the x-axis. The arc is shown in the figure at the right.

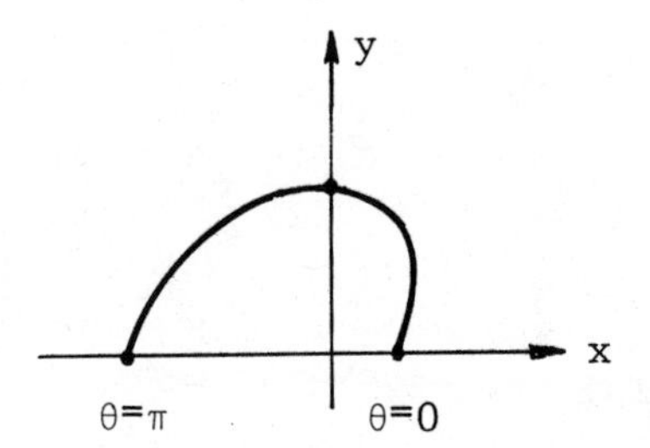

$$\frac{dx}{d\theta} = -2 \sin \theta + 2 \sin 2\theta ,$$

$$\frac{dy}{d\theta} = 2 \cos \theta - 2 \cos 2\theta \quad \text{so that}$$

$$\left(\frac{dx}{d\theta}\right)^2 + \left(\frac{dy}{d\theta}\right)^2 = \left(4 \sin^2 \theta - 8 \sin \theta \sin 2\theta + 4 \sin^2 2\theta\right) + \left(4 \cos^2 \theta - 8 \cos \theta \cos 2\theta + 4 \cos^2 2\theta\right)$$

$$= 8(1 - \sin \theta \sin 2\theta - \cos \theta \cos 2\theta)$$

$$= 8(1 - 2 \cos \theta \sin^2 \theta - \cos \theta \cos 2\theta)$$

$$= 8\left(1 - \cos \theta (2 \sin^2 \theta + \cos 2\theta)\right)$$

$$= 8(1 - \cos \theta)$$

Therefore, the surface area is given by

$$S = \int_0^\pi 2\pi(2 \sin \theta - \sin 2\theta) \cdot 2\sqrt{2}\sqrt{1 - \cos \theta} \; d\theta$$

$$= \int_0^\pi 8\sqrt{2}\,\pi \sin \theta (1 - \cos \theta)^{3/2} \; d\theta$$

$$= \frac{16\sqrt{2}}{5} \pi(1 - \cos \theta)^{5/2}\Big]_0^\pi = \frac{128\pi}{5} \approx 80.4 \text{ square units.}$$

11. The average value is

$$\frac{1}{2-0}\int_0^2 (x^2 - x + 1)\,dx = \frac{1}{2}\left[\frac{1}{3}x^3 - \frac{1}{2}x^2 + x\right]_0^2 = \frac{4}{3}.$$

12. The planar region is shown at the right. We find that

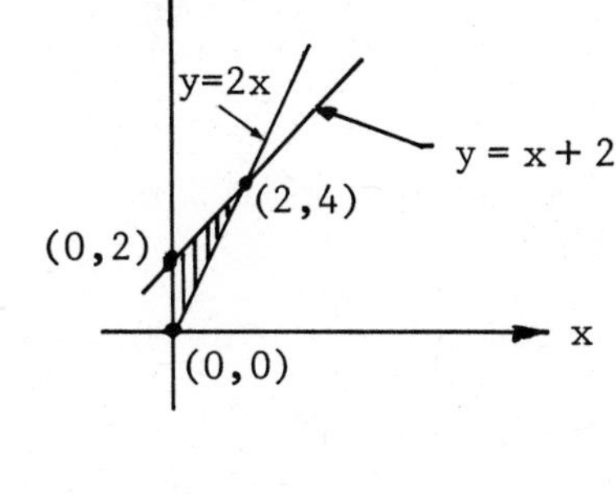

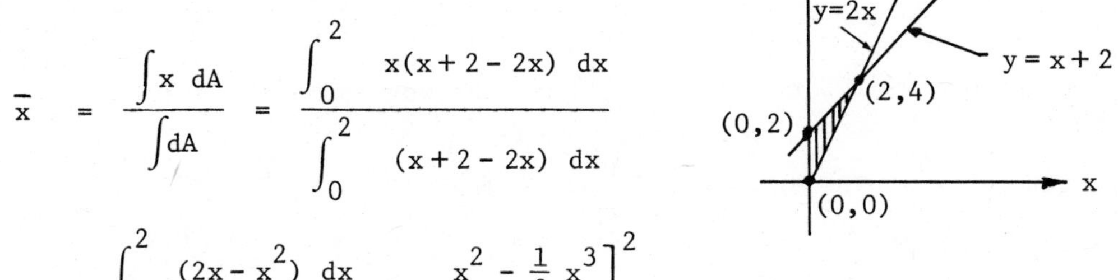

$$\bar{x} = \frac{\int x\,dA}{\int dA} = \frac{\int_0^2 x(x+2-2x)\,dx}{\int_0^2 (x+2-2x)\,dx}$$

$$= \frac{\int_0^2 (2x - x^2)\,dx}{\int_0^2 (2-x)\,dx} = \frac{x^2 - \frac{1}{3}x^3\Big]_0^2}{2x - \frac{1}{2}x^2\Big]_0^2} = \frac{2}{3}, \quad \text{and}$$

$$\bar{y} = \frac{\int y\,dA}{\int dA} = \frac{\int_0^2 y\left(\frac{y}{2}\right)dy + \int_2^4 y\left[\frac{y}{2} - (y-2)\right]dy}{2}$$

$$= \frac{1}{2}\left[\left(\frac{1}{6}y^3\right)\Big]_0^2 + \left(y^2 - \frac{1}{6}y^3\right)\Big]_2^4\right]$$

$$= \frac{1}{2}\left(\frac{4}{3} + 16 - \frac{32}{3} - 4 + \frac{4}{3}\right) = 2.$$

13. A cross section of the half-filled pipe is shown at the right. Here

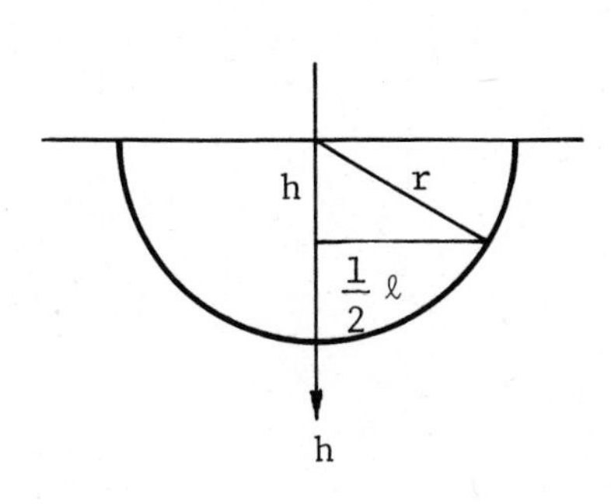

$r = 4$ inches $= \frac{1}{3}$ foot. By the Theorem of Pythagoras,

$$\left(\frac{\ell}{2}\right)^2 + h^2 = r^2 \quad \text{so that}$$

$$\ell = \frac{2}{3}\sqrt{1 - 9h^2}.$$

The force F is given by

$$F = \int wh\, dA = \int_0^{1/3} wh\ell\, dh = \frac{2w}{3} \int_0^{1/3} h\sqrt{1-9h^2}\, dh$$

$$= \left(\frac{2w}{3}\right)\left(\frac{2}{3}\right)\left(-\frac{1}{18}\right)\left(1-9h^2\right)^{3/2}\Big]_0^{1/3} = \frac{2w}{81} \approx 1.54 \text{ pounds} .$$

14. $F = kx$ so $90 = k \cdot 1$ or $k = 90$. Thus, the work done is given by

$$W = \int F\, dx = \int_0^5 90x\, dx = 45x^2\Big]_0^5 = 1125 \text{ ft.-lb.}$$

15. The work done is $W = \int F\, ds$.

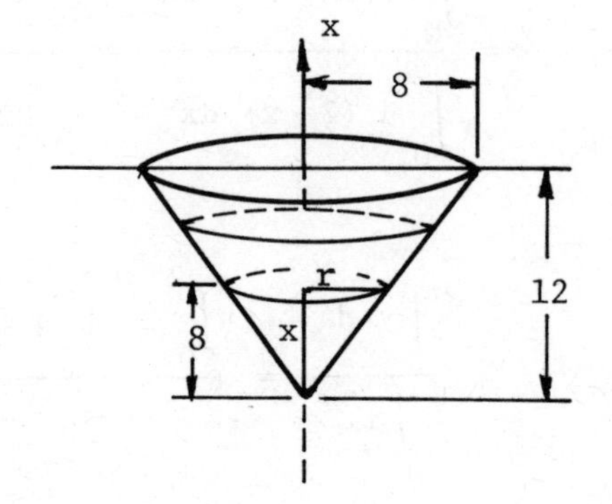

The force is the weight of a typical volume of water. From the figure at the right,

$$\frac{r}{x} = \frac{8}{12} \quad \text{or} \quad r = \frac{2}{3}x .$$

Thus, a typical volume element in a horizontal slice of the water is

$$dV = \pi r^2\, dx = \frac{4\pi}{9} x^2\, dx .$$

The distance this volume element is lifted is $12 - x + 2 = 14 - x$ feet. Thus,

$$W = \int_0^8 w \cdot \frac{4\pi}{9} x^2 (14 - x)\, dx = \frac{4\pi w}{9}\left(\frac{14}{3}x^3 - \frac{1}{4}x^4\right)\Big]_0^8$$

$$= \frac{4\pi w \cdot 8^4}{27} \approx 18.96\pi \text{ ft.-tons} \quad (\text{assuming } w = 62.5) .$$

CHAPTER 6 TRANSCENDENTAL FUNCTIONS

6-1 The Trigonometric Functions.

OBJECTIVE A: Find dy/dx when y is expressed in terms involving trigonometric functions of x.

1. Let $y = \sqrt{1 + 2\tan^2 x}$. Then,

$$\frac{dy}{dx} = \frac{1}{2}\left(1 + 2\tan^2 x\right)^{-1/2} \cdot \frac{d}{dx}\left(\_\_\_\_\_\_\_\_\right)$$

$$= \frac{1}{2}\left(1 + 2\tan^2 x\right)^{-1/2} \cdot 4\tan x \cdot \frac{d}{dx}\left(\_\_\_\_\right)$$

$$= \_\_\_\_\_\_\_\_\_\_\_\_\_\_\_\_\_\_\_\_ .$$

2. Let $y = \sec 2x \csc^2 x$. Then,

$$\frac{dy}{dx} = \frac{d}{dx}(\sec 2x) \cdot \csc^2 x + \sec 2x \cdot \frac{d}{dx}(\_\_\_\_\_\_)$$

$$= \sec 2x \tan 2x \cdot \frac{d}{dx}(\_\_\_)\csc^2 x + \sec 2x \cdot 2\csc x \frac{d}{dx}(\_\_\_\_\_\_)$$

$$= \sec 2x \csc^2 x\,(\_\_\_\_\_\_\_\_\_\_\_\_) .$$

3. If y is expressed implicitly, $2\cot x = \tan y$, differentiation of each side with respect to x gives

$$-2\csc^2 x = \frac{d}{dx}(\_\_\_\_\_\_) = \_\_\_\_\_\_ \frac{d}{dx}(\_\_\_) .$$ Hence,

$$\frac{dy}{dx} = \frac{-2\cos^2 y}{\_\_\_\_\_\_} .$$

---

1. $1 + 2\tan^2 x$, $\tan x$, $2\left(1 + 2\tan^2 x\right)^{-1/2} \tan x \sec^2 x$

2. $\csc^2 x$, $2x$, $\csc x$, $2\tan 2x - \cot x$

3. $\tan y$, $\sec^2 y$, y, $\sin^2 x$

OBJECTIVE B: Find indefinite integrals of simple combinations of trigonometric functions.

4. $\int \left(\frac{4}{\cos^2 x} + \frac{7}{x^2} \right) dx = \int \left(4 \_\_\_\_ + 7x^{-2} \right) dx$

$= 4 \int \_\_\_\_ \, dx + 7 \int x^{-2} \, dx = \_\_\_\_\_\_\_\_ + C.$

5. $\int \csc^2 (1-3x) \, dx = \int \_\_\_\_\_\_ \, du \qquad (u = 1-3x)$

$= \frac{1}{3} \int \_\_\_\_\_\_ = \_\_\_\_\_\_ + C = \_\_\_\_\_\_\_\_ .$

6. $\int \csc^7 x \cot x \, dx$. Let $u = \csc x$, then $du =$ $-\csc x \cot x$ so the integral becomes

$\int \csc^7 x \cot x \, dx = \int -u^6 \, du = -\frac{1}{7} u^7 + C = -\frac{1}{7} \csc^7 x + C .$

OBJECTIVE C: Use l'Hôpital's rule to find limits of indeterminate forms involving trigonometric functions.

7. $\lim_{x \to 0} \frac{\tan x - x}{x - \sin x} = \lim_{x \to 0} \frac{\_\_\_\_\_\_}{1 - \cos x} = \lim_{x \to 0} \frac{\_\_\_\_\_\_}{\sin x}$

$= \lim_{x \to 0} \_\_\_\_\_\_ \cdot \frac{1}{\cos x} = (\_\_) \cdot 1 = \_\_ .$

6-2 The Inverse Trigonometric Functions.

OBJECTIVE A: Find the values of inverse trigonometric functions at selected points without the use of tables or a calculator.

8. $y = \sin^{-1} x$ is equivalent to $\_\_\_\_\_\_$, where $\_\_ \le x \le \_\_$ and $\_\_ \le y \le \_\_$.

---

4. $\sec^2 x$, $\sec^2 x$, $4 \tan x - 7x^{-1}$

5. $-\frac{1}{3} \csc^2 u$, $-\csc^2 u \, du$, $\frac{1}{3} \cot u$, $\frac{1}{3} \cot (1-3x) + C$

6. $-\csc x \cot x$, $-u^6$, $-\frac{1}{7} u^7$, $-\frac{1}{7} \csc^7 x + C$

7. $\sec^2 x - 1$, $2 \sec^2 x \tan x$, $2 \sec^2 x$, 2, 2

8. $x = \sin y$, -1, 1, $-\pi/2$, $\pi/2$

9. $y = \tan^{-1} x$ is equivalent to ________, where ___ $< x <$ ___ and ___ $< y <$ ___ .

10. Let $y = \sin^{-1}\left(-\frac{\sqrt{2}}{2}\right)$; then $\sin y =$ _____ , so $y =$ ___ . That is, $\sin^{-1}\left(-\frac{\sqrt{2}}{2}\right) =$ ___ .

11. If $\alpha = \tan^{-1}\left(-\frac{\sqrt{3}}{3}\right)$, then $\tan \alpha =$ ____ , so $\alpha =$ ____ . Hence, $\sin \alpha =$ ____ and $\cos \alpha =$ ____ .

OBJECTIVE B: Simplify expressions involving inverse trigonometric functions and trigonometric functions.

12. To find $\sin\left(\cos^{-1}\left(-\frac{\sqrt{3}}{2}\right)\right)$, let $y = \cos^{-1}\left(-\frac{\sqrt{3}}{2}\right)$. Then, $\cos y =$ ____ so $y =$ ____ . Hence, $\sin y =$ ____ . Alternatively, since $\sin^2 y + \cos^2 y = 1$ holds,

$$\sin\left(\cos^{-1}\left(-\frac{\sqrt{3}}{2}\right)\right) = \sqrt{1 - \cos^2\left(\cos^{-1}\left(-\frac{\sqrt{3}}{2}\right)\right)}$$

$$= \left[1 - \cos^2\left(\cos^{-1}\left(-\frac{\sqrt{3}}{2}\right)\right)\right]^{1/2} = \sqrt{1 - \_\_\_} = \_\_\_ .$$

6-3 Derivatives of the Inverse Trigonometric Functions.

OBJECTIVE A: Differentiate functions whose expressions involve inverse trigonometric functions.

13. $\frac{d}{dx} \sin^{-1} u =$ __________ .

14. $\frac{d}{dx} \tan^{-1} u =$ __________ .

15. $\frac{d}{dx} \sec^{-1} u =$ __________ .

---

9. $x = \tan y$, $-\infty$, ∞, $-\frac{\pi}{2}$, $\frac{\pi}{2}$

10. $-\frac{\sqrt{2}}{2}$, $-\frac{\pi}{4}$, $-\frac{\pi}{4}$

11. $-\frac{\sqrt{3}}{3}$, $-\frac{\pi}{6}$, $-\frac{1}{2}$, $\frac{\sqrt{3}}{2}$

12. $-\frac{\sqrt{3}}{2}$, $\frac{5\pi}{6}$, $\frac{1}{2}$, $\frac{3}{4}$, $\frac{1}{2}$

13. $\frac{du/dx}{\sqrt{1-u^2}}$

14. $\frac{du/dx}{1+u^2}$

15. $\frac{du/dx}{|u|\sqrt{u^2-1}}$

16. $\frac{d}{dx}\left(\sin^{-1}\frac{x}{5}\right)^2 = \left(2\sin^{-1}\frac{x}{5}\right)\left(\frac{1}{\sqrt{1-(x/5)^2}}\right)\frac{d}{dx}(\_\_\_)$

$= \left(2\sin^{-1}\frac{x}{5}\right)\left(\frac{5}{\sqrt{25-x^2}}\right)(\_\_\_) = \_\_\_\_\_\_\_\_\_\_\_\_\_\_.$

17. $\frac{d}{dx}\sec^{-1}\frac{1}{x} = \frac{1}{\frac{1}{|x|}\sqrt{\_\_\_ - 1}}\cdot\left(\_\_\_\right) = \frac{-1}{\sqrt{\_\_\_}}.$

18. $\frac{d}{dx}\tan^{-1}\sqrt{x-1} = \frac{1}{1+(x-1)}\frac{d}{dx}(\_\_\_\_\_) = \_\_\_\_\_\_\_\_\_\_.$

19. Differentiating $\tan^{-1}\frac{x}{y} = \frac{1}{2}$ implicitly, we find

$0 = \frac{d}{dx}\tan^{-1}\frac{x}{y} = \frac{1}{1+(x/y)^2}\cdot\frac{d}{dx}\left(\_\_\_\right) = \frac{y^2}{x^2+y^2}\left(\_\_\_\_\_\_\_\_\_\_\right)$

$= \frac{y - \_\_\_\_\_}{x^2+y^2}$; thus $\frac{dy}{dx} = \_\_\_$.

OBJECTIVE B: Evaluate definite integrals of functions whose antiderivatives involve inverse trigonometric functions.

20. $\int_0^1 \frac{x\,dx}{\sqrt{1-x^4}}$. Let $u = x^2$, then $du = \_\_\_\_\_$ so that $dx = \_\_\_\_\_$,

and the indefinite integral becomes

$\int \frac{x\,dx}{\sqrt{1-x^4}} = \frac{1}{2}\int\frac{du}{\_\_\_\_\_} = \_\_\_\_\_\_\_\_\_\_ + C$.

---

16. $\frac{x}{5}$, $\frac{1}{5}$, $\frac{2}{\sqrt{25-x^2}}\sin^{-1}\frac{x}{5}$

17. $\frac{1}{x^2}$, $-\frac{1}{x^2}$, $1-x^2$

18. $\sqrt{x-1}$, $\frac{1}{2x\sqrt{x-1}}$

19. $\frac{x}{y}$, $\frac{y - x\frac{dy}{dx}}{y^2}$, $x\frac{dy}{dx}$, $\frac{y}{x}$

20. $2x\,dx$, $\frac{1}{2}du$, $\sqrt{1-u^2}$, $\frac{1}{2}\sin^{-1}u$, $\frac{1}{2}\sin^{-1}x^2$, $\sin^{-1}0$, $\frac{\pi}{4}$

Thus,

$$\int_0^1 \frac{x\,dx}{\sqrt{1-x^4}} = \underline{\qquad\qquad}\Big]_0^1 = \frac{1}{2}\left(\sin^{-1} 1 - \underline{\qquad\qquad}\right) = \underline{\quad} .$$

21. $\displaystyle\int_0^{\pi/2} \frac{\cos x\,dx}{1+\sin^2 x}$. Let $u = \sin x$, then $du = \underline{\qquad\qquad}$, and the indefinite integral becomes

$$\int \frac{\cos x\,dx}{1+\sin^2 x} = \int \frac{du}{\underline{\qquad\qquad}} = \underline{\qquad\qquad} + C .$$ Thus,

$$\int_0^{\pi/2} \frac{\cos x\,dx}{1+\sin^2 x} = \underline{\qquad\qquad\qquad}\Big]_0^{\pi/2} = \tan^{-1} 1 - \underline{\qquad\qquad}$$

$$= \underline{\quad} - 0 = \underline{\quad} .$$

OBJECTIVE C: Use l'Hôpital's rule to find limits of indeterminate forms involving inverse trigonometric functions.

22. $\displaystyle\lim_{x\to 0} \frac{\sin^{-1} x}{\sin x}$ is of the form $0/0$, so l'Hôpital's rule applies. Thus,

$$\lim_{x\to 0} \frac{\sin^{-1} x}{\sin x} = \lim_{x\to 0} \frac{\underline{\qquad\qquad}}{\cos x} = \frac{\underline{\quad}}{1} = \underline{\quad} .$$

23. $$\lim_{x\to +\infty} \frac{\frac{\pi}{2} - \tan^{-1} x}{1/x} = \lim_{t\to 0^+} \frac{\frac{\pi}{2} - \tan^{-1}\frac{1}{t}}{\underline{\quad}}$$

$$= \lim_{t\to 0^+} \left(\underline{\qquad\qquad}\right)\frac{d}{dt}\left(\frac{1}{t}\right) = \lim_{t\to 0^+} \left(-\frac{t^2}{t^2+1}\right)\left(\underline{\quad}\right)$$

$$= \lim_{t\to 0^+} \underline{\qquad\qquad} = \underline{\quad} .$$

---

21. $\cos x\,dx$, $1+u^2$, $\tan^{-1} u$, $\tan^{-1}(\sin x)$, $\tan^{-1} 0$, $\frac{\pi}{4}$, $\frac{\pi}{4}$

22. $\frac{1}{\sqrt{1-x^2}}$, 1, 1

23. t, $-\frac{1}{1+(1/t)^2}$, $-\frac{1}{t^2}$, $\frac{1}{t^2+1}$, 1

6-4 The Natural Logarithm.

OBJECTIVE: Use the trapezoidal rule to calculate natural logarithms of numbers between 0 and 10 .

24. The natural logarithm is defined by $\ln x$ = ______________ for x satisfying ______ .

25. By the Fundamental Theorem of integral calculus, $\frac{d}{dx}\ln x$ = ___ , so the natural logarithm is a continuous function because it is ________________ .

26. To find $\ln 2.5$, we approximate $\int_1^{2.5} \frac{1}{x}\,dx$ by the trapezoidal rule.

Let us take $n = 6$ subdivisions. Hence, $h = (2.5 - 1)/6$ = ___ .
Therefore, the subdivision points are $x_0 = 1$, $x_1 = \frac{5}{4}$, x_2 = ___ , x_3 = ___ , x_4 = ___ , x_5 = ___ , and $x_6 = \frac{5}{2}$. Since $y = \frac{1}{x}$, the trapezoidal approximation gives

$$T = \frac{h}{2}\left(\frac{1}{x_0} + \frac{2}{x_1} + \frac{2}{x_2} + \frac{2}{x_3} + \frac{2}{x_4} + \frac{2}{x_5} + \frac{1}{x_6}\right)$$

$$= \frac{1}{8}\left(\underline{\hspace{8cm}}\right)$$

$$\approx \frac{\underline{\hspace{2cm}}}{8} \approx \underline{\hspace{2cm}} .$$ Hence, $\ln 2.5 \approx$ ________ .

27. Let us estimate the error in the approximation. For $f(x) = \frac{1}{x}$, the error is given by

$$E = \left|\frac{b-a}{12} f''(c)\cdot h^2\right| = \left|\frac{1.5}{12}\left(\underline{\hspace{1cm}}\right)\frac{1}{16}\right|$$

$$= \left|\frac{1}{\underline{\hspace{1cm}}}\right| < \underline{\hspace{1cm}}$$ since $1 < c < 2.5$.

Therefore, because $\frac{1}{64} \approx 0.0156$ the trapezoidal approximation is accurate to at least one decimal place. (The actual error is, in fact, about 43×10^{-4}.)

---

24. $\int_1^x \frac{1}{t}\,dt$, $x > 0$

25. $\frac{1}{x}$, differentiable

26. $\frac{1}{4}$, $\frac{3}{2}$, $\frac{7}{4}$, 2, $\frac{9}{4}$, $\left(1 + \frac{8}{5} + \frac{4}{3} + \frac{8}{7} + 1 + \frac{8}{9} + \frac{2}{5}\right)$, 7.3651, .9206, .9206

27. $-2/c^3$, $64c^3$, 1/64

6-5 The Derivative of $\ln x$.

OBJECTIVE A: Differentiate functions whose expressions involve the natural logarithm function.

28. $\frac{d}{dx} \ln \left(5 + 2x^3\right)^4 = \frac{1}{\left(5+2x^3\right)^4} \frac{d}{dx} \left(\underline{\qquad\qquad}\right)$

$= \frac{1}{\left(5+2x^3\right)^4} \left[\underline{\qquad\qquad}\right] \frac{d}{dx} (5+2x^3)$

$= \frac{\underline{\qquad\qquad}}{\left(5+2x^3\right)^4} = \underline{\qquad\qquad}$.

29. $\frac{d}{dx} \left[\ln (\sin x)\right]^2 = \underline{\qquad\qquad} \frac{d}{dx} \ln (\sin x)$

$= 2 \ln (\sin x) \cdot \underline{\qquad} \cdot \frac{d}{dx} (\sin x)$

$= 2 \csc x \ln (\sin x) \cdot \underline{\qquad} = \underline{\qquad\qquad}$.

30. $\frac{d}{dx} x^2 \ln \sqrt{x} = 2x \ln \sqrt{x} + \underline{\quad} \frac{d}{dx} \ln \sqrt{x}$

$= 2x \ln \sqrt{x} + \underline{\qquad} \frac{d}{dx} \sqrt{x}$

$= 2x \ln \sqrt{x} + \underline{\qquad\qquad} = \frac{x}{2} \left(\underline{\qquad\qquad} \right)$.

OBJECTIVE B: Integrate functions whose antiderivatives involve the natural logarithm function.

31. $\int \frac{x\,dx}{x^2+4}$

Let $u = x^2 + 4$. Then $du = \underline{\quad}$ so $x\,dx = \underline{\qquad}$.

---

28. $\left(5+2x^3\right)^4$, $4\left(5+2x^3\right)^3$, $24x^2\left(5+2x^3\right)^3$, $\frac{24x^2}{5+2x^3}$

29. $2 \ln (\sin x)$, $\frac{1}{\sin x}$, $\cos x$, $2 \cot x \ln (\sin x)$

30. x^2, $x^2 \cdot \frac{1}{\sqrt{x}}$, $x^2 \cdot \frac{1}{\sqrt{x}} \cdot \frac{1}{2\sqrt{x}}$, $2 \ln x + 1$ or, $4 \ln \sqrt{x} + 1$

Thus the integral becomes

$$\int \frac{x\,dx}{x^2+4} = \int \frac{du}{\_\_\_} = \_\_\_\_\_\_\_\_ + C = \_\_\_\_\_\_\_\_\_\_\_\_ .$$

32. $\int \frac{3x+1}{x}\,dx = \int \left(3 + \_\_\right) dx = \int 3\,dx + \_\_\_\_$

$$= 3x + \_\_\_\_\_\_ + C .$$

33. $\int \frac{dx}{x \ln \sqrt{x}}$

Let $u = \ln \sqrt{x}$. Then $\frac{du}{dx} = \_\_\_\_ \frac{d}{dx}\left(\_\_\_\right) = \_\_\_\_$. Hence,

$2\,du = \_\_\_\,dx$. Thus the integral becomes

$$\int \frac{dx}{x \ln \sqrt{x}} = \int \frac{2\,du}{\_\_} = \_\_\_\_\_\_\_\_ + C = \_\_\_\_\_\_\_\_\_\_\_\_ .$$

OBJECTIVE C: Use l'Hôpital's rule to find limits of indeterminate forms involving the natural logarithm function.

34. $\lim\limits_{x \to \infty} \frac{\ln\left(\frac{1+x}{x}\right)}{1/x} = \lim\limits_{x \to \infty} \frac{\ln\left(1 + \frac{1}{x}\right)}{1/x} = \lim\limits_{t \to 0} \frac{\ln(1+t)}{\_\_\_}$

$$= \lim_{t \to 0} \frac{\_\_\_\_\_\_\_\_}{1} = \frac{1}{1 + \_\_} = \_\_\_ .$$

35. $\lim\limits_{x \to 0^+} x^2 \ln x$

As $x \to 0^+$, $x^2 \to 0$ and $\ln x \to \_\_\_\_$ so this is an indeterminate form $0 \cdot \infty$.

Writing the limit as $\lim\limits_{x \to 0^+} x^2 \ln x = \lim\limits_{x \to 0^+} \frac{\ln x}{\_\_\_\_}$ we see this form

is now of the type ∞/∞ . Therefore l'Hôpital's rule applies:

---

31. $2x\,dx$, $\frac{1}{2}\,du$, $2u$, $\frac{1}{2} \ln |u|$, $\frac{1}{2} \ln (x^2+4) + C$

32. $\frac{1}{x}$, $\int \frac{dx}{x}$, $\ln |x|$

33. $\frac{1}{\sqrt{x}}$, $\sqrt{x}$, $\frac{1}{2x}$, $\frac{1}{x}$, u, $2 \ln |u|$, $2 \ln\left(\ln \sqrt{x}\right) + C$

34. t, $1/(1+t)$, 0, 1

35. $-\infty$, $1/x^2$, $\frac{1/x}{-2/x^3}$, $-\frac{1}{2}x^2$, 0, 0

$$\lim_{x \to 0^+} \frac{\ln x}{1/x^2} = \lim_{x \to 0^+} \underline{\qquad} = \lim_{x \to 0^+} \underline{\qquad} = \underline{\quad} .$$ Hence,

$$\lim_{x \to 0^+} x^2 \ln x = \underline{\quad} .$$

36. $\lim_{\theta \to \frac{\pi}{2}^-} \dfrac{\ln (\tan \theta)}{\sec \theta}$

As $\theta \to \frac{\pi}{2}^-$, $\tan \theta \to \infty$ and $\sec \theta \to \underline{\quad}$. Hence $\ln (\tan \theta) \to \underline{\quad}$ and this is an indeterminate form of type $\underline{\qquad}$. Therefore, l'Hôpital's rule applies:

$$\lim_{\theta \to \frac{\pi}{2}^-} \frac{\ln (\tan \theta)}{\sec \theta} = \lim_{\theta \to \frac{\pi}{2}^-} \frac{\underline{\qquad\qquad}}{\sec \theta \tan \theta} = \lim_{\theta \to \frac{\pi}{2}^-} \frac{\underline{\qquad}}{\sin^2 \theta} = \underline{\quad} .$$

6-6 Properties of Natural Logarithms.

OBJECTIVE: Use the three properties of the natural logarithm to rewrite a logarithmic expression as a sum, difference, or multiple of logarithms.

37. $\ln ax = \underline{\qquad\qquad}$ for $a > 0$ and $x > 0$.

38. $\ln \frac{x}{a} = \underline{\qquad\qquad}$ for $a > 0$ and $x > 0$.

39. $\ln x^n = \underline{\qquad}$ for $x > 0$ and n rational.

40. $\ln \sqrt[3]{\dfrac{x^2}{a^4}} = \ln \left(\dfrac{x^2}{a^4}\right)^{\underline{\quad}} = (\underline{\quad}) \ln \left(\dfrac{x^2}{a^4}\right) = \frac{1}{3} \left(\ln x^2 - \underline{\qquad}\right)$

$= \frac{2}{3} \ln x - \underline{\qquad}$.

41. $\ln\left(b^3 \sqrt{x}\right) = \ln b^3 + \underline{\qquad} = 3 \underline{\qquad} + \ln x^{1/2} = \underline{\qquad\qquad}$.

---

36. ∞, ∞, ∞/∞, $\sec^2 \theta/\tan \theta$, $\cos \theta$, 0

37. $\ln a + \ln x$

38. $\ln x - \ln a$

39. $n \ln x$

40. $\frac{1}{3}$, $\frac{1}{3}$, $\ln a^4$, $\frac{4}{3} \ln a$

41. $\ln \sqrt{x}$, $\ln b$, $3 \ln b + \frac{1}{2} \ln x$

42. $\ln(x^2+2x+1) = \ln(x+1)^{\_\_} =$ ________ .

6-7 Graph of $y = \ln x$.

OBJECTIVE: Summarize the characteristics of the graph of $y = \ln x$, and graph functions involving the natural logarithm.

43. The domain of $y = \ln x$ is the set ______ , and its range is the set ________ .

44. The graph of $y = \ln x$ is increasing ________________ . It is concave downward __________ .

45. Since $y = \ln x$ is differentiable for $x > 0$, it is a ________ function of x .

46. $\lim_{x\to\infty} \ln x =$ ____ and $\lim_{x\to 0^+} \ln x =$ ____ .

47. Consider the curve $y = x - \ln x$. The derivative $\frac{dy}{dx} =$ ______ , so that $\frac{dy}{dx} = 0$ implies $\frac{1}{x} =$ ___ or $x =$ ___ . Notice that the domain of y is the set ______ . The second derivative $\frac{d^2y}{dx^2} =$ ___ is always positive. Therefore, the critical point $x = 1$ gives a relative ______ value of $y(1) =$ ___ . As $x \to 0$, $y \to$ ___ . To examine the curve as $x \to \infty$, notice that

$$\lim_{x\to\infty}\left(\frac{x - \ln x}{x}\right) = 1 - \lim_{x\to\infty}\frac{\ln x}{x} = 1 - \lim_{x\to\infty}\frac{\_\_}{\_\_} = \_\_\_$$ by

l'Hôpital's rule. Hence, for large values of x, the ratio $\frac{x - \ln x}{x}$ is approximately ___ or $x - \ln x \approx$ ___ . Sketch a graph of y at the right.

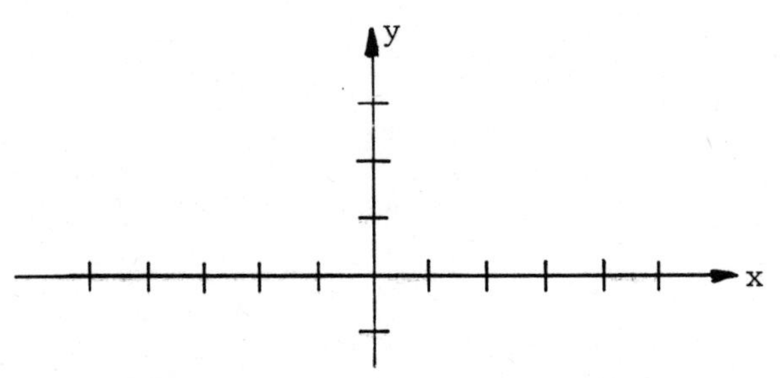

---

42. 2, $2 \ln(x+1)$

43. $x > 0$, $-\infty < y < +\infty$

44. for all x in the domain, for all $x > 0$

45. continuous

46. $+\infty$, $-\infty$

47. $1 - \frac{1}{x}$, 1, 1, $x > 0$, $\frac{1}{x^2}$, minimum, 1, $+\infty$, $\frac{1}{x}$, 1, 1, x ,

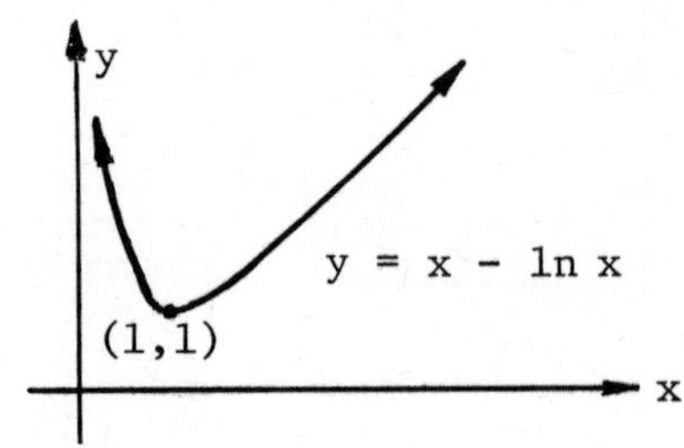

6-8 The Exponential Function.

OBJECTIVE A: Use the equivalent equations $y = e^x$ and $x = \ln y$ to simplify logarithms of exponentials, and exponentials of logarithms.

48. The equation $y = e^{\ln x}$ is equivalent to $\ln y =$ ______ . Since the logarithm is one - one , the last equation is equivalent to ______ ; that is, $e^{\ln x} =$ ___ . In other words, the exponential "undoes" the natural logarithm.

49. The equation $y = \ln (e^x)$ is equivalent to $e^y =$ ____ . Since the exponential function is one - one, the last equation is equivalent to ______ ; that is, $\ln (e^x) =$ ___ .

50. $e^{-2 \ln (x+1)} = e^{\ln \text{\_\_\_\_\_\_\_\_}} =$ __________ by Problem 48.

51. $\ln\left(\sqrt{x}\; e^{2-x}\right) = \ln \sqrt{x} +$ __________ $=$ ______________ .

OBJECTIVE B: Differentiate functions whose expressions involve exponential functions.

52. $\dfrac{d}{dx} e^{\sqrt{1-x^2}} = e^{\sqrt{1-x^2}} \cdot \dfrac{d}{dx}(\text{\_\_\_\_\_\_\_\_})$

$= e^{\sqrt{1-x^2}} \cdot \dfrac{1}{2\sqrt{1-x^2}} \cdot \dfrac{d}{dx}(\text{\_\_\_\_\_\_})$

$=$ ____________________ .

53. $\dfrac{d}{dx} e^{x \ln x} =$ ________ $\cdot \dfrac{d}{dx}(x \ln x) = e^{x \ln x}$ (__________) .

54. $\dfrac{d}{dx} \ln (e^x + 1) = \dfrac{1}{e^x+1} \cdot \dfrac{d}{dx}$ (______) $=$ ______ .

---

48. $\ln x$, $y = x$, x

49. e^x, $y = x$, x

50. $(x+1)^{-2}$, $(x+1)^{-2}$

51. $\ln (e^{2-x})$, $\ln \sqrt{x} + (2-x)$

52. $\sqrt{1-x^2}$, $1-x^2$, $\dfrac{-x}{\sqrt{1-x^2}} e^{\sqrt{1-x^2}}$

53. $e^{x \ln x}$, $\ln x + 1$

54. $e^x + 1$, $\dfrac{e^x}{e^x+1}$

55. $\frac{d}{dx} \sin \sqrt{e^x} = \cos \sqrt{e^x} \cdot \frac{d}{dx}(\_\_\_)$

$= \cos \sqrt{e^x} \cdot \left(\frac{1}{2\sqrt{e^x}}\right) \cdot \frac{d}{dx}(\_\_)$

$=$ ________________ .

OBJECTIVE C: Integrate functions whose antiderivatives involve exponential functions.

56. $\int x^2 e^{-x^3} dx$

Let $u = -x^3$. Then $du =$ ______ so that $x^2\,dx =$ ______ . Thus the integral becomes,

$\int x^2 e^{-x^3} dx = \int \_\_\_\_ du = \_\_\_\_ + C =$ ________ .

57. $\int (e^x - 2)^4 e^x\, dx$

Let $u = e^x - 2$. Then $du =$ ______ and the integral becomes,

$\int (e^x - 2)^4 e^x\, dx = \int \_\_\_ du = \_\_\_\_ + C =$ ________ .

58. $\int \frac{e^x}{1 + e^{2x}} dx$

Let $u = e^x$. Then $du =$ ______ and the integral becomes,

$\int \frac{e^x dx}{1 + e^{2x}} = \int \frac{du}{\_\_\_\_} = \_\_\_\_ + C =$ ________ .

---

55. $\sqrt{e^x}$, e^x, $\frac{1}{2}\sqrt{e^x} \cos \sqrt{e^x}$

56. $-3x^2\,dx$, $-\frac{1}{3}\,du$, $-\frac{1}{3}e^u$, $-\frac{1}{3}e^u$, $-\frac{1}{3}e^{-x^3} + C$

57. $e^x\,dx$, u^4, $\frac{1}{5}u^5$, $\frac{1}{5}(e^x - 2)^5 + C$

58. $e^x\,dx$, $1 + u^2$, $\tan^{-1} u$, $\tan^{-1}(e^x) + C$

OBJECTIVE D: Use l'Hôpital's rule to find limits of indeterminate forms involving exponential functions.

59. $\lim_{x\to\infty} xe^{-x} = \lim_{x\to\infty} \frac{x}{e^x}$ is an indeterminate form of type _____ so l'Hôpital's rule applies. Thus, $\lim_{x\to\infty} \frac{x}{e^x} = \lim_{x\to\infty} \frac{\quad}{\_\_\_\_} = \_\_\_\_$.

60. $\lim_{x\to 0} \frac{e^{2x} - 1}{x}$ is an indeterminate form of type ____ so l'Hôpital's rule applies. Thus, $\lim_{x\to 0} \frac{e^{2x} - 1}{x} = \lim_{x\to 0} \frac{\_\_\_\_}{1} = \_\_\_\_$.

OBJECTIVE E: Graph functions whose expressions contain exponential functions.

61. Consider the function $y = xe^x$. The domain of this function is the set ______________ . To find the critical points: $dy/dx = 0$ implies $e^x +$ _____ $= 0$, or __________ $= 0$. Hence, $x =$ ___ is the only critical point. $\frac{d^2y}{dx^2} = e^x$ (_____) which is positive when $x = -1$; thus $x = -1$ gives a relative ___________ value of $y =$ _____ . $\frac{d^2y}{dx^2} = 0$ gives a point of inflection when $x =$ ____ . For $x < -2$ the curve is concave ____________ , and for $x > -2$ it is concave __________ .
By l'Hôpital's rule,

$$\lim_{x\to-\infty} xe^x = \lim_{x\to-\infty} \frac{x}{e^{-x}} = \lim_{x\to-\infty} \frac{1}{\_\_\_\_} = \_\_\_\_ .$$

Hence, the x-axis is a horizontal asymptote. As $x \to +\infty$, $y \to$ ____ . Sketch the graph in the coordinate system at the right.

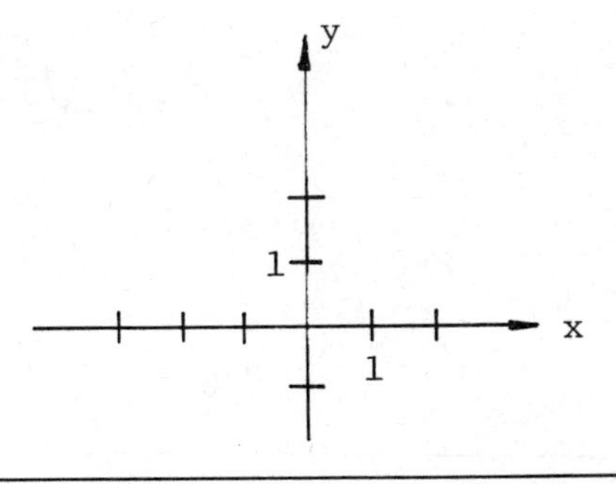

---

59. ∞/∞, $\frac{1}{e^x}$, 0 60. 0/0, $2e^{2x}$, 2

61. $-\infty < x < \infty$, xe^x, $e^x(1+x)$, -1, $x+2$, minimum, $1/e$, -2, downward, upward, $-e^{-x}$, 0, $+\infty$

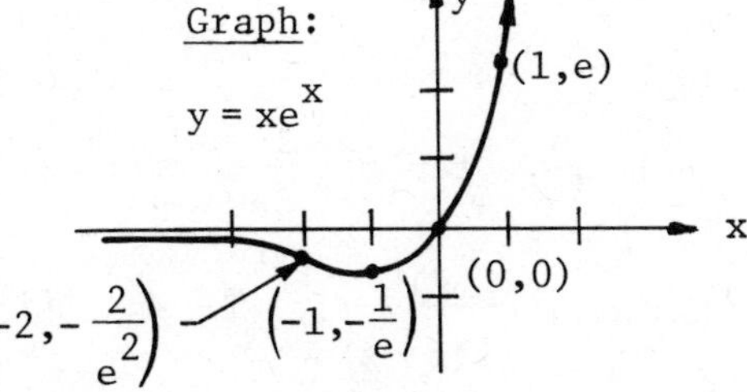

6-9 The Function a^u.

62. The function $y = a^u$ is defined by $a^u =$ ________ and it is well-defined whenever ________ .

63. The derivative of a^u, where u is a differentiable function of x, is given by $\frac{d}{dx} a^u =$ ______________ .

OBJECTIVE A: Use the method of logarithmic differentiation to calculate derivatives.

64. Find dy/dx if $y = x^{\tan x}$, $x > 0$.

Solution.

$$\ln y = \ln\left(x^{\tan x}\right) = \underline{\hspace{3cm}} \text{ so that}$$

$$\frac{1}{y}\frac{dy}{dx} = \sec^2 x \ln x + \underline{\hspace{3cm}}, \text{ or}$$

$$\frac{dy}{dx} = x^{\tan x}\left(\underline{\hspace{5cm}}\right).$$

65. Find dy/dx if $y = \sqrt{\frac{1-x}{1+x}}$, $-1 < x < 1$.

Solution.

$$\ln y = \ln\sqrt{\frac{1-x}{1+x}} = \frac{1}{2}\ln(1-x) - \underline{\hspace{3cm}} \text{ so that}$$

$$\frac{1}{y}\frac{dy}{dx} = -\frac{1}{2(1-x)} - \underline{\hspace{2.5cm}} = \frac{-(1+x) - (\underline{\hspace{1.5cm}})}{2(1-x)(1+x)}$$

$$\frac{dy}{dx} = -y(1-x)^{-1} \cdot \underline{\hspace{2.5cm}} = \underline{\hspace{5cm}}.$$

---

62. $e^{u \ln a}$, $a > 0$

63. $a^u \cdot \frac{du}{dx} \cdot \ln a$

64. $\tan x \cdot \ln x$, $\tan x \cdot \frac{1}{x}$, $\sec^2 x \ln x + \frac{1}{x}\tan x$

65. $\frac{1}{2}\ln(1+x)$, $\frac{1}{2(1+x)}$, $1-x$, $(1+x)^{-1}$, $-(1-x)^{-1/2}(1+x)^{-3/2}$

66. Find dy/dx if $y = \left(x^r\right)^x$, $x > 0$.

Solution.

$$\ln y = \ln\left(x^r\right)^x = \_\_\_\_\_\_\_\_ = \_\_\_\_\_\_\_\_ .$$

$$\frac{1}{y}\frac{dy}{dx} = r \ln x + \_\_\_\_\_\_\_\_ = r\ (\_\_\_\_\_\_\_\_\_\_)$$ so that

$$\frac{dy}{dx} = \_\_\_\_\_\_\_\_\_\_\_\_\_\_\_\_ .$$

OBJECTIVE B: Integrate functions whose antiderivatives involve an exponential function a^u .

67. $\int \frac{dx}{2^x} = \int 2^{-x}\, dx$

Let $u = -x$ so that $du = \_\_\_\_\_$, and the integral becomes

$$\int 2^{-x}\, dx = \int \_\_\_\_\_\ du = \_\_\_\_\_\_\_\_\_\_ + C = \_\_\_\_\_\_\_\_\_\_\_\_\_ .$$

68. $\int_1^2 x\, 10^{x^2-1}\, dx$

Let $u = x^2 - 1$. Then $du = \_\_\_\_\_\_$ so that $x\,dx = \_\_\_\_\_$.

$$\int x\, 10^{x^2-1}\, dx = \int \_\_\_\_\_\ du = \_\_\_\_\_\_\_\_ + C .$$ Hence,

$$\int_1^2 x\, 10^{x^2-1}\, dx = \_\_\_\_\_\_\_\_\_\_\_\_\_\Big]_1^2 = \frac{1}{2\ln 10}\left(\_\_\_\_\_\_\_\right)$$

$$= \_\_\_\_\_\_\_\_ .$$

---

66. $x \ln x^r$, $rx \ln x$, $rx \cdot \frac{1}{x}$, $\ln x + 1$, $r\left(x^r\right)^x (\ln x + 1)$

67. $-dx$, -2^u, $-\frac{1}{\ln 2} 2^u$, $-\frac{1}{2^x \ln 2} + C$

68. $2x\ dx$, $\frac{1}{2}\ du$, $\frac{1}{2}10^u$, $\frac{1}{2\ln 10}10^u$, $\frac{1}{2\ln 10}10^{x^2-1}$, $10^3 - 1$, $\frac{999}{2\ln 10}$

OBJECTIVE C: Graph functions whose expressions involve an exponential function a^u .

69. Consider the curve $y = (0.2)^{x+1} + 2$.

$\frac{dy}{dx}$ = ________________ $\frac{d}{dx}(x+1)$ = ________________ .

Therefore, dy/dx is of constant ____________ sign, since $\ln(.2) < 0$, and the curve is everywhere ______________ . Calculation of the second derivative gives $\frac{d^2y}{dx^2}$ = ________________________ which is of constant ____________ sign; hence the curve is everywhere concave __________ .

As $x \to \infty$, $(.2)^{x+1} \to$ ___ so $y \to$ ___ .

As $x \to -\infty$, $(.2)^{x+1} \to$ ___ so $y \to$ ___ .

Finally, the points (-1, ___) , (0, ___) , (1, ___) lie on the curve. You can now sketch the graph of the curve in the coordinate system provided at the right.

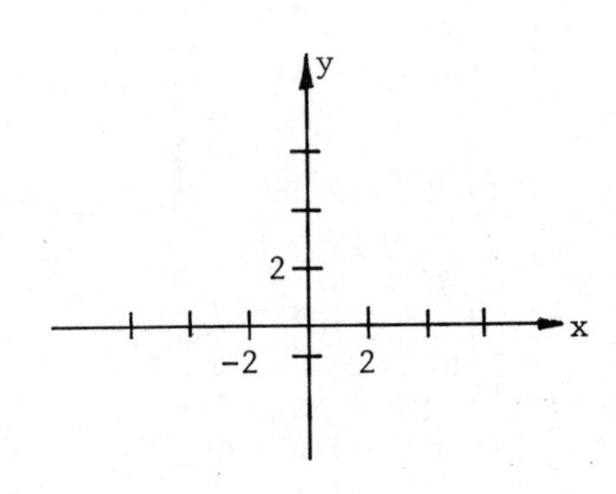

6-10 The Function $\log_a u$.

OBJECTIVE A: Use the definition of $\log_a u$ to evaluate simple expressions.

70. If $a > 0$ and $a \neq 1$, then $y = \log_a u$ is defined and equivalent to a^y = ___ whenever u is positive. If a or u is negative, $\log_a u$ is ____________ .

---

69. $(0.2)^{x+1} \ln(.2)$, $(0.2)^{x+1} \ln(.2)$, negative, decreasing, $(0.2)^{x+1} [\ln(.2)]^2$, positive, upward, 0, 2, $+\infty$, $+\infty$, 3, 2.2, 2.04

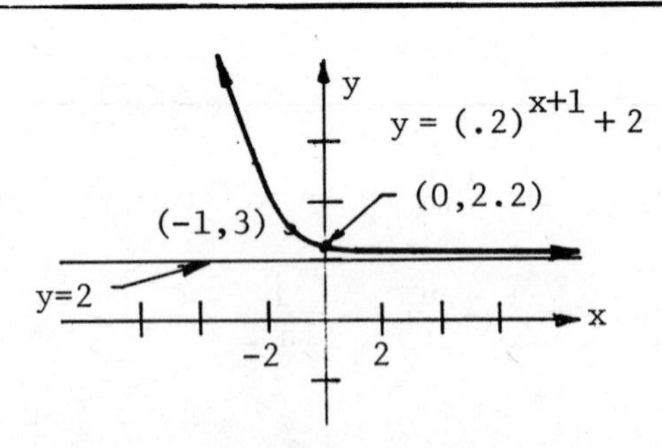

70. u, undefined

71. In terms of natural logarithms, $\log_a u$ = ________ .

72. $\log_8 4$

If $y = \log_8 4$, then 8^y = ___ or $2^{\_\_}$ = 2^2 .

Thus, $3y$ = ___ or y = ___ . Therefore, $\log_8 4$ = ___ .

73. $\log_{.75} \frac{27}{64}$

If $y = \log_{.75} \frac{27}{64}$, then $\frac{27}{64}$ = _____ . Now, $\left(\frac{3}{4}\right)^3$ = _____

so that $\log_{.75} \frac{27}{64}$ = ___ .

OBJECTIVE B: Solve exponential and logarithmic equations.

74. $5^{x^2} = 7^x$

Taking the natural logarithm of both sides gives x^2 (_____) = x ln 7 or x (_______________) = 0 . Hence, x = 0 and x = ________ both solve the equation.

75. $\log_7 \left(x^2 - 6x\right) = 1$ is equivalent to $x^2 - 6x$ = ____ or ,

0 = ____________ = (x - 7)(______) . Therefore, x = ____ and

x = ___ solve the logarithmic equation.

76. $5^{\log_5 2}$ = ___ because $y = 5^u$ and $y = \log_5 u$ are __________ functions of each other.

77. In general, $a^{\log_a u}$ = ___ and $\log_a a^x$ = ___ .

---

71. $\frac{\ln u}{\ln a}$

72. 4, 3y, 2, 2/3, $\frac{2}{3}$

73. $\left(\frac{3}{4}\right)^y$, $\frac{27}{64}$, 3

74. ln 5, x ln 5 - ln 7, $\frac{\ln 7}{\ln 5}$

75. 7^1, $x^2 - 6x - 7$, x + 1, -1, 7

76. 2, inverse

77. u, x

OBJECTIVE C: Differentiate functions involving a logarithmic function $\log_a u$.

78. $\frac{d}{dx} \log_a u = \frac{d}{dx} \frac{\_\_\_\_}{\ln a} = \_\_\_\_\_\_ \cdot \frac{du}{dx}$.

79. $\frac{d}{dx} \log_{10}\left(x^2 - e^x\right) = \frac{1}{\_\_\_\_\_\_\_\_} \frac{d}{dx}\left(x^2 - e^x\right) = \_\_\_\_\_\_\_\_$.

80. Find dy/dx if $y = \left(1 + \sqrt{x}\right)^{\log_2 x}$.

Solution.

$\ln y = \ln\left(1 + \sqrt{x}\right)^{\log_2 x} = \_\_\_\_\_\_\_\_$.

$\frac{1}{y}\frac{dy}{dx} = \_\_\_\_\_\_ \ln\left(1 + \sqrt{x}\right) + \log_2 x \cdot \frac{1}{1 + \sqrt{x}} \cdot \_\_\_\_$

$\frac{dy}{dx} = \left(1 + \sqrt{x}\right)^{\log_2 x}\left[\_\_\_\_\_\_\_\_\_\_\_\_\right]$.

6-11 Compound Interest and Exponential Growth.

OBJECTIVE: Solve exponential growth and decay problems:

$$\frac{dx}{dt} = kx \quad \text{with} \quad x = x_0 \quad \text{when} \quad t = 0 .$$

81. The half-life of a radioactive substance is the length of time it takes for half of a given amount of the substance to disintegrate through radiation. The half-life of the carbon isotope C^{14} is about 5700 years.

---

78. $\ln u$, $\frac{1}{u \ln a}$

79. $\left(x^2 - e^x\right) \ln 10$, $\frac{2x - e^x}{\left(x^2 - e^x\right) \ln 10}$

80. $\log_2 x \cdot \ln\left(1 + \sqrt{x}\right)$, $\frac{1}{x \ln 2}$, $\frac{1}{2\sqrt{x}}$,

$\frac{1}{x \ln 2} \ln\left(1 + \sqrt{x}\right) + \frac{\ln x}{2 \ln 2 \cdot \sqrt{x}\left(1 + \sqrt{x}\right)}$

82. Assume that the amount x of C^{14} present in a dead organism decays exponentially from the time of death. Then, $x = x_0 e^{kt}$, where x_0 is the original amount present. To find the constant k in the case of carbon C^{14},

$$\left(\frac{1}{2}\right) x_0 = x_0 e^{\_\_\_\_ k} \quad \text{or} \quad \ln \frac{1}{2} = \_\_\_\_\_ .$$

Thus, $k = \dfrac{-\ln 2}{\_\_\_\_} \approx -1.22 \times 10^{-4}$.

Suppose we want to determine the percentage of C^{14} present after 10,000 years. Then, the percentage is given by the ratio

$$\frac{x_0 e^{k \cdot 10^4}}{x_0} \approx e^{\_\_\_\_\_\_\_\_} \approx 0.2964 ,$$

or approximately 29.64 percent of the amount of C^{14} remains after 10,000 years.

83. The "1470" skull found in Kenya by Richard Leakey is reputed to be 2,500,000 years old. The percentage of C^{14} remaining is given by

$$\frac{x_0 e^{k \cdot 2.5 \times 10^6}}{x_0} \approx e^{\_\_\_\_} .$$

However, if $x < -21$ then $e^x < 10^{-9}$ so the percentage of C^{14} left in the skull would be negligible. The current reliable limit for C^{14} dating is about 40,000 years, so another method for dating the skull had to be found.

---

82. 5700, 5700k, 5700, $-\dfrac{\ln 2}{57} \times 10^2$

83. -304

CHAPTER 6 OBJECTIVE - PROBLEM KEY

Objective		Problems in Thomas/Finney Text	Objective		Problems in Thomas/Finney Text
6-1	A	p. 294, 13-29, 31, 32, implicit (30, 33)	6-6		p. 309, 1-10
	B	p. 294, 40-47	6-7		p. 311, 1-3, 6
	C	p. 294, 50-53	6-8	A	p. 315, 1
6-2	A	p. 299, 1-5		B	p. 315, 2-16, 18, 20, implicit (17, 19)
	B	p. 299, 8		C	p. 315, 22-30
6-3	A	p. 302, 5-14, 25		D	p. 315, 21
	B	p. 302, 16-21, 25		E	p. 316, 31-34, 41
	C	p. 303, 23	6-9	A	p. 320, 1-8
6-4		p. 304, 1-4, 7		B	p. 320, 11-15
6-5	A	p. 307, 1-18		C	p. 321, 16-18
	B	p. 307, 19-33	6-10	A	p. 323, 1, 2, 6
	C	p. 307, 34		B	p. 323, 3, 4
			6-11		p. 327, 1-6

CHAPTER 6 SELF-TEST

In Problems 1 - 10 calculate the derivative dy/dx .

1. $y = \tan\left(\cos\frac{2}{x}\right)$

2. $y = x^2 e^{-1/x}$

3. $y = x(\ln x)^2$

4. $y = \cos^{-1}\left(\frac{1-x^2}{1+x^2}\right)$, $x > 0$

5. $y = 5^x \log_5 x$

6. $y = 2^{\sin^{-1} x}$

7. $y = 2x\tan^{-1} 2x - \ln\sqrt{1+4x^2}$

8. $y = \dfrac{\sec 3\sqrt{x}}{\sqrt{x}}$

9. $y = (\ln x)^{\sqrt{x}}$, $x > 0$

10. $e^y + e^x = e^{x+y}$

In Problems 11 - 17 calculate the indicated integrals.

11. $\int \sec^2(3x+5)\,dx$

12. $\int_{-5/4}^{5/4} \dfrac{dx}{25+16x^2}$

13. $\int \left(1-e^{-x}\right)^2 dx$

14. $\int \dfrac{1-2\cos x}{\sin^2 x}\,dx$

15. $\int \frac{(\ln 2x)^3}{x}\, dx$

16. $\int \frac{e^{-x}}{\sqrt{1 - e^{-2x}}}\, dx$

17. $\int_1^e \frac{2^{\ln x}}{3x}\, dx$

In Problems 18 - 21 evaluate the limits using l'Hôpital's rule.

18. $\lim_{x \to 0} \csc x \sin^{-1} x$

19. $\lim_{x \to 1} x^{1/(1-x)}$

20. $\lim_{x \to 0^+} \frac{e^x - \cos x}{x \sin x}$

21. $\lim_{x \to 0} \frac{4^x - 2^x}{x}$

22. Determine the following values.

(a) $\sin^{-1}\left(-\frac{\sqrt{3}}{2}\right)$

(b) $\tan^{-1} \frac{\sqrt{3}}{3}$

(c) $\tan^{-1}\left(\cos \frac{\pi}{2}\right)$

(d) $\cos^{-1}\left(\sin \frac{\pi}{6}\right)$

23. Simplify the expression $\frac{\log_3 243}{\log_2 \sqrt[4]{64} + \log_8 8^{-10}}$

24. Let $f(x) = \log_a x$, $f(5) = 1.46$, $f(2) = 0.63$, $f(7) = 1.77$.
Use the properties of logarithms to find,

(a) $f(10)$ (b) $f(49)$ (c) $f\left(\frac{5}{7}\right)$ (d) $f(1.4)$

25. Solve the following equations.

(a) $3^{-8x+6} = 27^{-x-8}$

(b) $\log_5 (5x - 1) = -2$

(c) $e^x = 10^{x+1}$

26. Graph the curve $y = \frac{2}{1 + 3e^{-2x}}$.

27. Sketch the graph of $y = \frac{\ln x}{x^2}$.

28. Suppose that the number of bacteria in a yeast culture grows at a rate proportional to the number present. If the population of a colony of yeast bacteria doubles in one hour, find the number of bacteria present at the end of $3\frac{1}{2}$ hours.

SOLUTIONS TO CHAPTER 6 SELF-TEST

1. $\frac{dy}{dx} = \sec^2\left(\cos\frac{2}{x}\right)\cdot\left(-\sin\frac{2}{x}\right)\cdot\left(-\frac{2}{x^2}\right)$

2. $\frac{dy}{dx} = 2xe^{-1/x} + x^2e^{-1/x}\cdot\left(\frac{1}{x^2}\right) = e^{-1/x}(2x+1)$

3. $\frac{dy}{dx} = (\ln x)^2 + 2x\ln x\cdot\frac{1}{x} = \ln x\,(2+\ln x)$

4. $$\frac{dy}{dx} = \frac{-1}{\sqrt{1-\left(\frac{1-x^2}{1+x^2}\right)^2}}\cdot\left[\frac{\left(1+x^2\right)(-2x)-\left(1-x^2\right)(2x)}{\left(1+x^2\right)^2}\right]$$

$$= \frac{-\left(1+x^2\right)}{\sqrt{\left(1+x^2\right)^2-\left(1-x^2\right)^2}}\cdot\frac{-4x}{\left(1+x^2\right)^2} = \frac{4x}{\left(1+x^2\right)\sqrt{4x^2}}$$

$$= \frac{2x}{\left(1+x^2\right)|x|} = \frac{2}{1+x^2}$$

5. $\frac{dy}{dx} = 5^x\ln 5\cdot\log_5 x + 5^x\cdot\frac{1}{x\ln 5} = 5^x\left(\ln x + \frac{1}{x\ln 5}\right)$

6. $\frac{dy}{dx} = 2^{\sin^{-1}x}(\ln 2)\frac{1}{\sqrt{1-x^2}}$

7. $\frac{dy}{dx} = 2\tan^{-1}2x + 2x\cdot\frac{1}{1+4x^2}\cdot 2 - \frac{1}{\sqrt{1+4x^2}}\cdot\frac{8x}{2\sqrt{1+4x^2}} = 2\tan^{-1}2x$

8. $$\frac{dy}{dx} = \frac{1}{\sqrt{x}}\left(\sec 3\sqrt{x}\tan 3\sqrt{x}\right)\cdot\frac{3}{2\sqrt{x}} + \left(-\frac{1}{2}x^{-3/2}\right)\sec 3\sqrt{x}$$

$$= \frac{\sec 3\sqrt{x}}{2x\sqrt{x}}\left(3\sqrt{x}\tan 3\sqrt{x} - 1\right)$$

9. $y = (\ln x)^{\sqrt{x}}$ gives $\ln y = \sqrt{x}\,\ln x$

$$\frac{1}{y}\frac{dy}{dx} = \frac{1}{2\sqrt{x}}\ln x + \sqrt{x}\cdot\frac{1}{x} = \frac{1}{2\sqrt{x}}(\ln x + 2)$$

$$\frac{dy}{dx} = \frac{1}{2\sqrt{x}}(\ln x + 2)(\ln x)^{\sqrt{x}}$$

10. $e^y \frac{dy}{dx} + e^x = e^{x+y}\frac{d}{dx}(x+y) = e^{x+y}\left(1 + \frac{dy}{dx}\right)$

Hence, $\left(e^y - e^{x+y}\right)\frac{dy}{dx} = e^{x+y} - e^x$, or from the original expression, $-e^x \frac{dy}{dx} = e^y$. Thus, $\frac{dy}{dx} = -e^{y-x}$.

11. $u = 3x+5$, $du = 3\,dx$

$$\int \sec^2(3x+5)\,dx = \frac{1}{3}\int \sec^2 u\,du = \frac{1}{3}\tan(3x+5) + C$$

12. $\int \frac{dx}{25+16x^2} = \frac{1}{25}\int \frac{dx}{1+\left(\frac{4}{5}x\right)^2} = \frac{1}{25}\cdot\frac{5}{4}\tan^{-1}\frac{4}{5}x + C$

$$\int_{-5/4}^{5/4} \frac{dx}{25+16x^2} = \frac{1}{20}\left(\tan^{-1}1 - \tan^{-1}(-1)\right) = \frac{\pi}{40}$$

13. $\int \left(1-e^{-x}\right)^2 dx = \int \left(1 - 2e^{-x} + e^{-2x}\right)dx = x + 2e^{-x} - \frac{1}{2}e^{-2x} + C$

14. $\int \frac{1 - 2\cos x}{\sin^2 x}\,dx = \int \left(\csc^2 x - \frac{2\cos x}{\sin^2 x}\right)dx$

$$= -\cot x - \int \frac{2\,du}{u^2} \qquad (u = \sin x)$$

$$= -\cot x + \frac{2}{\sin x} + C = -\cot x + 2\csc x + C$$

15. $u = \ln 2x\,, \quad du = \frac{1}{2x} \cdot 2\,dx = \frac{1}{x}\,dx$

$$\int \frac{(\ln 2x)^3}{x}\,dx = \int u^3\,du = \frac{1}{4}(\ln 2x)^4 + C$$

16. $u = e^{-x}\,, \quad du = -e^{-x}\,dx$

$$\int \frac{-du}{\sqrt{1-u^2}} = \cos^{-1}\left(e^{-x}\right) + C$$

17. $u = \ln x\,, \quad du = \frac{1}{x}\,dx$

$$\int \frac{2^{\ln x}}{3x}\,dx = \frac{1}{3}\int 2^u\,du = \frac{1}{3\ln 2}\,2^{\ln x} + C$$

$$\int_1^e \frac{2^{\ln x}}{3x}\,dx = \frac{1}{3\ln 2}\left(2^{\ln e} - 1\right) = \frac{1}{3\ln 2}$$

18. $\lim\limits_{x\to 0} \csc x \sin^{-1} x = \lim\limits_{x\to 0} \frac{\sin^{-1} x}{\sin x}$ (type 0/0)

$$= \lim_{x\to 0} \frac{1/\sqrt{1-x^2}}{\cos x} = \frac{1}{1} = 1$$

19. Let $y = x^{1/(1-x)}$ so $\ln y = \frac{\ln x}{1-x}$

$\lim\limits_{x\to 1} \ln y = \lim\limits_{x\to 1} \frac{\ln x}{1-x}$ (type 0/0) $= \lim\limits_{x\to 1} \frac{1/x}{-1} = -1\,.$

By the continuity of the natural logarithm, $\lim\limits_{x\to 1} x^{1/(1-x)} = e^{-1}\,.$

20. $\lim\limits_{x\to 0^+} \frac{e^x - \cos x}{x \sin x}$ (type 0/0) $= \lim\limits_{x\to 0^+} \frac{e^x + \sin x}{x\cos x + \sin x} = +\infty\,.$

21. $\lim\limits_{x\to 0} \frac{4^x - 2^x}{x}$ (type 0/0) $= \lim\limits_{x\to 0} \frac{4^x \ln 4 - 2^x \ln 2}{1}$

$$= \ln 4 - \ln 2 = \ln 2\,.$$

22. (a) $-\frac{\pi}{3}$ (b) $\frac{\pi}{6}$ (c) 0 (d) $\frac{\pi}{3}$

23. $$\frac{\log_3 243}{\log_2 \sqrt[4]{64} + \log_8 8^{-10}} = \frac{5}{\frac{1}{4}(6) - 10(1)} = -\frac{10}{17}$$

24. (a) $\log_a 10 = \log_a 2 + \log_a 5 = 2.09$

(b) $\log_a 49 = 2 \log_a 7 = 3.54$

(c) $\log_a \frac{5}{7} = \log_a 5 - \log_a 7 = -0.31$

(d) $\log_a 1.4 = \log_a 14 - \log_a 10$

$= \log_a 2 + \log_a 7 - \log_a 5 - \log_a 2 = 0.31$

25. (a) $3^{-8x+6} = 3^{-3x-24}$ or $-8x+6 = -3x-24$ or $x = 6$.

(b) $5^{-2} = 5x-1$ or $x = 26/125$.

(c) $x = (x+1) \ln 10$ or $x = \frac{\ln 10}{1 - \ln 10} \approx -1.768$.

26. $f(x) = \frac{2}{1 + 3e^{-2x}}$, $f'(x) = \frac{12e^{-2x}}{\left(1 + 3e^{-2x}\right)^2}$,

$$f''(x) = \frac{24e^{-2x}\left(-1 + 3e^{-2x}\right)}{\left(1 + 3e^{-2x}\right)^3}$$

$f' > 0$ for every x ;

$f''(x) = 0$ when $x = \frac{1}{2} \ln 3 \approx 0.55$

$\lim_{x \to \infty} f(x) = 2$ and $\lim_{x \to -\infty} f(x) = 0$

The graph is sketched at the right.

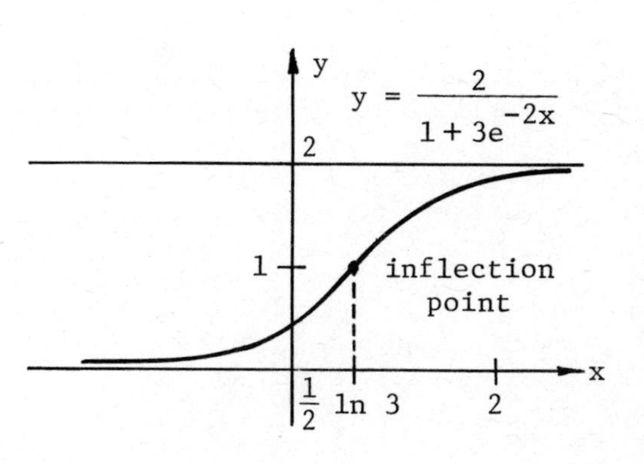

27. $f(x) = \frac{\ln x}{x^2}$,

$f'(x) = \frac{1 - 2\ln x}{x^3}$ so

$f'(x) = 0$ implies $\ln x = \frac{1}{2}$ or $x = \sqrt{e}$

$f''(x) = \frac{-5 + 6\ln x}{x^4}$ so

$f''(x) = 0$ implies $x = e^{5/6} \approx 2.3$

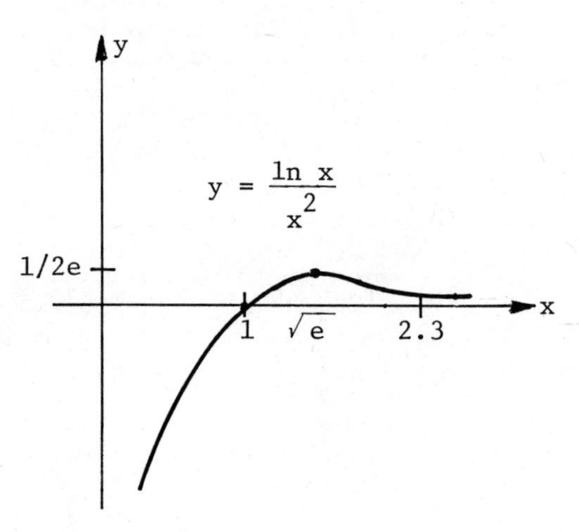

$\lim_{x\to\infty} f(x) = \lim_{x\to\infty} \frac{1/x}{2x} = 0$ and

$\lim_{x\to 0^+} f(x) = -\infty$. The graph is sketched at the right.

28. Let x denote the number of bacteria present at any time t. Then,

$\frac{dx}{dt} = kx$ or $x = Ce^{kt}$, for some constant C.

If x_0 is the initial number of bacteria at $t = 0$, then

$C = x_0$, so $x = x_0 e^{kt}$.

When $t = 1$, $x = 2x_0$ so that $2 = e^k$ or $k = \ln 2$. Therefore,

$$x = x_0 e^{t \ln 2}.$$

Finally, when $t = 3.5$, $x = x_0 e^{3.5 \ln 2} \approx 11.31\, x_0$.

Thus, there are 11.31 times the initial number of bacteria present at the end of 3½ hours.

CHAPTER 7 METHODS OF INTEGRATION

7-1 Basic Formulas.

OBJECTIVE: Evaluate indefinite integrals by reducing the integrands to standard forms through simple substitution.

1. $\int x(a+x)^{1/3}\, dx$, where a is constant.

Let $u = a+x$. Then $du =$ ___ and $x =$ ______ . Thus,

$$\int x(a+x)^{1/3}\, dx = \int \_\_\_\_\_\_\_\_\_\_ du = \int u^{4/3}\, du - \int \_\_\_\_\_\_ du$$

$$= \_\_\_\_\_\_\_ - \frac{3}{4} au^{4/3} + C$$

$$= \_\_\_\_\_\_\_\_\_\_\_\_\_\_\_\_\_\_\_\_ + C .$$

2. $\int \frac{dx}{x\sqrt{x-5}}$

Let $u = \sqrt{x-5}$. Then $du =$ ________ and $x =$ ______ .

$$\int \frac{dx}{x\sqrt{x-5}} = \int \_\_\_\_\_\_ du = \_\_ \int \frac{du}{\left(\frac{u}{\sqrt{5}}\right)^2 + 1}$$

$$= \_\_\_\_ \tan^{-1} \frac{u}{\sqrt{5}} + C = \_\_\_\_\_\_\_\_\_\_\_\_\_\_\_ + C .$$

3. $\int \frac{\cos x\, dx}{\sqrt{1 - \sin x}}$

Let $u = \sin x$, then $du =$ ________ so that

$$\int \frac{\cos x\, dx}{\sqrt{1 - \sin x}} = \int \_\_\_\_\_\_ du = \int (1-u)^{\_\_\_} du$$

$$= \_\_\_\_\_\_\_\_\_ + C = \_\_\_\_\_\_\_\_\_\_\_\_\_\_\_\_ .$$

---

1. dx, $u-a$, $(u-a)u^{1/3}$, $au^{1/3}$, $\frac{3}{7} u^{7/3}$, $3(a+x)^{4/3}\left[\frac{1}{7}(a+x) - \frac{1}{4}\right]$

2. $\frac{dx}{2\sqrt{x-5}}$, u^2+5, $\frac{2}{u^2+5}$, $\frac{2}{5}$, $\frac{2\sqrt{5}}{5}$, $\frac{2}{\sqrt{5}} \tan^{-1} \sqrt{\frac{x}{5} - 1}$

3. $\cos x\, dx$, $\frac{1}{\sqrt{1-u}}$, $-1/2$, $-2(1-u)^{1/2}$, $-2\sqrt{1 - \sin x} + C$

4. $\int \frac{3^{\tan x}\,dx}{\cos^2 x}$

Let $u = \tan x$. Then $du =$ ________ and the integral is

$$\int \frac{3^{\tan x}\,dx}{\cos^2 x} = \int \sec^2 x \; 3^{\tan x} \; dx = \int \_\_\_\_ \; du$$

$$= \_\_\_\_\_\_ + C = \_\_\_\_\_\_\_\_\_\_ .$$

5. $\int \frac{dx}{x + x \ln^2 x}$

Let $u = \ln x$. Then $du =$ ____ so that

$$\int \frac{dx}{x + x\ln^2 x} = \int \frac{dx}{x\left(\_\_\_\_\_\_\right)} = \int \frac{du}{\_\_\_\_\_\_}$$

$$= \_\_\_\_\_\_ + C = \_\_\_\_\_\_\_\_\_\_ .$$

7-2 Powers of Trigonometric Functions.

OBJECTIVE: Evaluate integrals of powers of trigonometric functions.

6. $\int \sin^5 x\, dx = \int \sin^4 x \cdot \_\_\_\_\_\_ = \int \left(1 - \cos^2 x\right)^{\_} \sin x \; dx$

Let $u =$ ______ so that $du = -\sin x\, dx$. Then,

$$\int \left(1 - \cos^2 x\right)^2 \sin x \; dx = \int \_\_\_\_\_\_\_\_ \; du$$

$$= -\int \left(1 - 2u^2 + \_\_\_\right) du = \_\_\_\_\_\_\_\_ - \frac{1}{5} u^5 + C$$

$$= \_\_\_\_\_\_\_\_\_\_\_\_\_\_\_\_\_\_ .$$

---

4. $\sec^2 x\, dx$, 3^u, $\frac{1}{\ln 3}\, 3^u$, $\frac{1}{\ln 3}\, 3^{\tan x} + C$

5. $\frac{1}{x}\, dx$, $1 + \ln^2 x$, $1 + u^2$, $\tan^{-1} u$, $\tan^{-1}(\ln x) + C$

6. $\sin x\, dx$, 2, $\cos x$, $-\left(1 - u^2\right)^2$, u^4, $-u + \frac{2}{3} u^3$,

$-\cos x + \frac{2}{3}\cos^3 x - \frac{1}{5}\cos^5 x + C$

7. $\int \cos^3 x \sin^2 x \; dx = \int \cos^2 x \sin^2 x$ ________

$= \int (1 - \sin^2 x) \sin^2 x$ ________ $= \int (\sin^2 x -$ ________ $) \cos x \; dx$

$= \int \sin^2 x \cos x \; dx - \int$ ________ dx

$= \int u^2 \, du - \int u^4 \, du$, where $u =$ ______ and $du =$ ________ ,

$=$ ____________ $+ C = \frac{1}{3} \sin^3 x +$ ______________ .

8. $\int \tan^3 x \; dx = \int \tan^2 x \cdot$ ________ $= \int (\sec^2 x - 1)$ ________

$= \int$ ____________ $dx - \int \tan x \; dx$

$=$ ________________________ .

9. $\int \sec^5 2x \tan 2x \; dx$

Let $u = \sec 2x$. Then $du =$ ______________ and the integral becomes

$\int \sec^5 2x \tan 2x \; dx = \int \sec^4 2x \cdot$ ______________

$= \int$ __ $\cdot \frac{1}{2} du =$ ______ $+ C$

$=$ ____________________ .

10. $\int \frac{\sin^5 x \; dx}{\sqrt{\cos x}} = \int \frac{\sin^4 x}{\sqrt{\cos x}} \cdot$ ________

Let $u = \cos x$. Then $du =$ ________ . Thus,

---

7. $\cos x \, dx$, $\cos x \, dx$, $\sin^4 x$, $\sin^4 x \cos x$, $\sin x$, $\cos x \, dx$, $\frac{1}{3} u^3 - \frac{1}{5} u^5$, $-\frac{1}{5} \sin^5 x + C$

8. $\tan x \, dx$, $\tan x \, dx$, $\tan x \sec^2 x$, $\frac{1}{2} \tan^2 x + \ln |\cos x| + C$

9. $2 \sec 2x \tan 2x \, dx$, $\sec 2x \tan 2x \, dx$, u^4, $\frac{1}{10} u^5$, $\frac{1}{10} \sec^5 2x + C$

10. $\sin x \, dx$, $-\sin x \, dx$, 2, 2, $u^{7/2}$, $-2u^{1/2} + \frac{4}{5} u^{5/2} - \frac{2}{9} u^{9/2}$, $-2 \cos^{1/2} x + \frac{4}{5} \cos^{5/2} x - \frac{2}{9} \cos^{9/2} x + C$

$$\int \frac{\sin^5 x \ dx}{\sqrt{\cos x}} = \int \frac{\left(1 - \cos^2 x\right)^{\underline{\quad}}}{\sqrt{\cos x}} \sin x \ dx$$

$$= - \int \frac{\left(1 - u^2\right)^{\underline{\quad}}}{u^{1/2}} du = - \int \left(u^{-1/2} - 2u^{3/2} + \underline{\qquad}\right) du$$

$$= \underline{\qquad\qquad\qquad\qquad} + C$$

$$= \underline{\qquad\qquad\qquad\qquad\qquad\qquad} .$$

11. $\int \frac{\sec x \ dx}{\tan^2 x} = \int \frac{\left(\underline{\qquad}\right) \sec x}{\sin^2 x} dx = \int \frac{\underline{\qquad}}{\sin^2 x} dx .$

Let $u = \sin x$, so that $du = \underline{\qquad\qquad}$ and the integral becomes

$$\int \frac{\sec x \ dx}{\tan^2 x} = \int \frac{\cos x \ dx}{\sin^2 x} = \int \underline{\qquad} du$$

$$= \underline{\quad} + C = \underline{\qquad\qquad\qquad} .$$

7-3 Even Powers of Sines and Cosines.

OBJECTIVE: Evaluate integrals involving even powers of $\sin x$ or $\cos x$.

12. $\int \sin^2 (1 - 3x) \ dx$

Since $\sin^2 \theta = \frac{1}{2} (1 - \cos 2\theta)$, we have

$$\int \sin^2 (1 - 3x) \ dx = \frac{1}{2} \int (\underline{\qquad\qquad\qquad\qquad}) \ dx$$

$$= \frac{1}{2} \int dx - \frac{1}{2} \int \underline{\qquad\qquad\qquad\qquad}$$

$$= \frac{1}{2} x - \frac{1}{2} \int \cos u \cdot \underline{\qquad\qquad} , \quad u = 2 - 6x$$

$$= \frac{1}{2} x + \underline{\qquad\qquad} + C$$

$$= \underline{\qquad\qquad\qquad\qquad\qquad\qquad} .$$

---

11. $\cos^2 x$, $\cos x$, $\cos x \ dx$, u^{-2}, $-\frac{1}{u}$, $- \csc x + C$

12. $1 - \cos 2(1 - 3x)$, $\cos (2 - 6x) \ dx$, $- \frac{1}{6} du$, $\frac{1}{12} \sin u$, $\frac{1}{2} x + \frac{1}{12} \sin (2 - 6x) + C$

13. $\int \frac{dx}{\cos^4 x}$

$$\int \frac{dx}{\cos^4 x} = \int \sec^4 x \; dx = \int \sec^2 x \cdot (________) \; dx$$

$$= \int \left(1 + \tan^2 x\right) __________$$

Let $u = \tan x$. Then $du = __________$ so that

$$\int \frac{dx}{\cos^4 x} = \int ______ \; du = __________ + C$$

$$= ____________________ .$$

14. $\int \frac{dx}{\sin^2 x \cos^2 x} = \int \csc^2 x \; __________$

$$= \int \csc^2 x \; (\tan^{__} x + 1) \; dx$$

$$= \int (______ + \csc^2 x) \; dx$$

$$= \tan x + (_______) + C .$$

7-4 Integrals Involving $\sqrt{a^2 - u^2}$, $\sqrt{a^2 + u^2}$, $\sqrt{u^2 - a^2}$, $a^2 + u^2$, $a^2 - u^2$.

OBJECTIVE A: Find indefinite integrals of integrands involving the radicals $\sqrt{a^2 - u^2}$, $\sqrt{a^2 + u^2}$, $\sqrt{u^2 - a^2}$, and the binomials $a^2 + u^2$, $a^2 - u^2$.

15. $\int \frac{x^3 \, dx}{\sqrt{4 - x^2}}$

Let $x = 2 \sin u$ so that $dx = __________$, and the integral becomes

---

13. $\sec^2 x$, $\sec^2 dx$, $\sec^2 dx$, $1 + u^2$, $u + \frac{1}{3} u^3$, $\tan x + \frac{1}{3} \tan^3 x + C$

14. $\sec^2 x \, dx$, 2, $\sec^2 x$, $- \cot x$

15. $2 \cos u \, du$, $8 \sin^3 u$, $8 \sin^3 u \cdot 2 \cos u$, $\cos u$, $\sin u \, du$, $\frac{1}{3} \cos^3 u$,

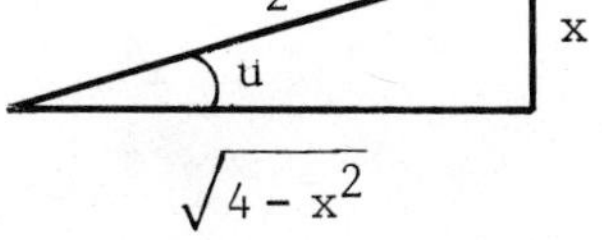

, $\frac{1}{2} \sqrt{4 - x^2}$, $\frac{1}{3} \left(4 - x^2\right)^{3/2}$

$$\int \frac{x^3\,dx}{\sqrt{4-x^2}} = \int \frac{______ \cdot 2\cos u \; du}{\sqrt{4-4\sin^2 u}} = \int \frac{________}{\sqrt{4\cos^2 u}}\,du$$

$$= \pm 8 \int \sin^3 u \; du \quad (\pm \text{ depends on sign of } ______)$$

$$= 8 \int \sin^3 u \; du \; , \quad \text{using only the principal value of } u = \sin^{-1} \frac{x}{2}$$

$$= 8 \int \left(1 - \cos^2 u\right) ______$$

$$= 8\left(-\cos u + ______\right) + C$$

For the substitution $x = 2 \sin u$, finish labeling the diagram at the right. Then, from the diagram,

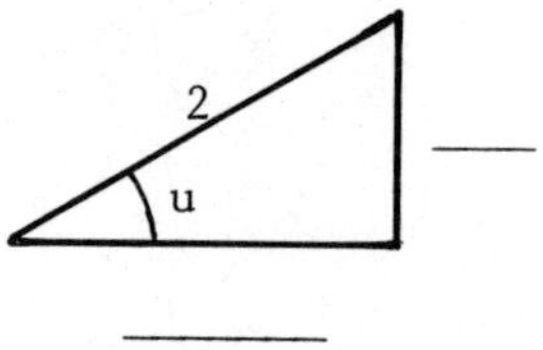

$\cos u =$ ______ , and

substitution gives

$$\int \frac{x^3\,dx}{\sqrt{4-x^2}} = -4\sqrt{4-x^2} + ______ + C .$$

16. $\displaystyle\int \frac{dx}{\left(x^2-9\right)^{3/2}}$

Let $x = 3 \sec u$, $dx =$ ______ and the integral becomes

$$\int \frac{dx}{\left(x^2-9\right)^{3/2}} = \int \frac{______}{\left(9\sec^2 u - 9\right)^{3/2}} = \int \frac{3 \sec u \tan u \; du}{27 \; ______}$$

$$= \frac{1}{9} \int \frac{\sec u \; du}{____} = \frac{1}{9} \int \frac{\cos u \; du}{____} = ______ + C .$$

For the substitution $x = 3 \sec u$, finish labeling the diagram at the right. Then, from the diagram,

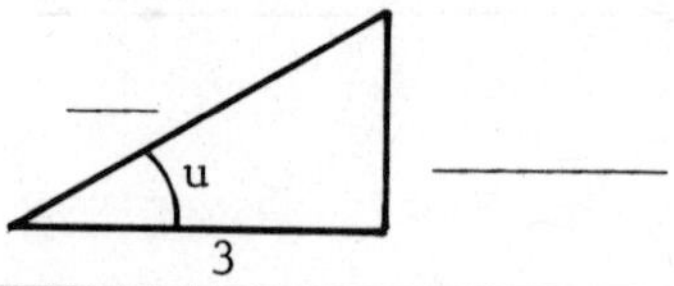

$\sin u =$ ______ ,

---

16. $3 \sec u \tan u \; du$, $3 \sec u \tan u \; du$, $\tan^3 u$, $\tan^2 u$,

$\sin^2 u$, $-\dfrac{1}{9 \sin u}$,

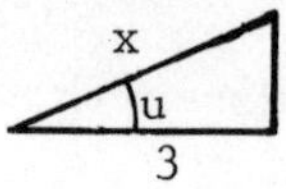

$\sqrt{x^2-9}$, $\dfrac{\sqrt{x^2-9}}{x}$, $\dfrac{-x}{9\sqrt{x^2-9}} + C$

and substitution gives

$$\int \frac{dx}{\left(x^2 - 9\right)^{3/2}} = \underline{\hspace{4cm}} .$$

17. $\displaystyle\int \frac{dx}{x^2\sqrt{x^2 + 16}}$

Let $x = 4 \tan u$, $dx = \underline{\hspace{3cm}}$ and the integral becomes

$$\int \frac{dx}{x^2\sqrt{x^2+16}} = \int \frac{4 \sec^2 u \, du}{\underline{\hspace{4cm}}} = \frac{1}{16} \int \frac{\sec u \, du}{\underline{\hspace{1.5cm}}}$$

$$= \frac{-1}{16} \csc u + C \quad \text{from Problem 11.}$$

For the substitution $x = 4 \tan u$, finish labeling the diagram at the right. Then, from the diagram,

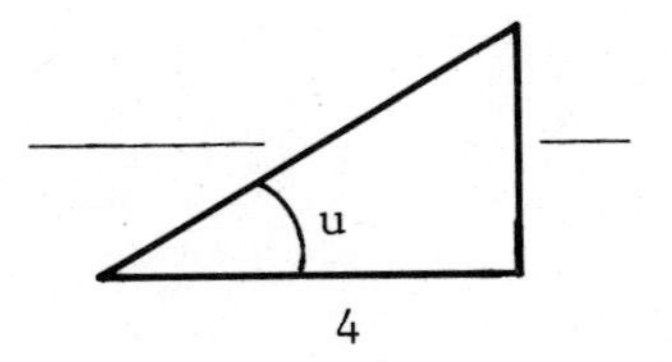

$\csc u = \underline{\hspace{2cm}}$,

and substitution gives

$$\int \frac{dx}{x^2\sqrt{x^2+16}} = \underline{\hspace{4cm}} .$$

OBJECTIVE B: Calculate definite integrals involving $\sqrt{a^2 - u^2}$, $\sqrt{a^2 + u^2}$, $\sqrt{u^2 - a^2}$, $a^2 + u^2$, and $a^2 - u^2$.

18. $\displaystyle\int_4^{4\sqrt{3}} \frac{dx}{x^2\sqrt{x^2+16}}$

As in the previous Problem 17, let $x = 4 \tan u$. When $x = 4$, $\tan u = 1$ or $u = \underline{\hspace{0.6cm}}$; when $x = 4\sqrt{3}$, $\tan u = \sqrt{3}$ or $u = \underline{\hspace{0.6cm}}$.

---

17. $4 \sec^2 u \, du$, $16 \tan^2 u \cdot 4 \sec u$,

$\tan^2 u$,

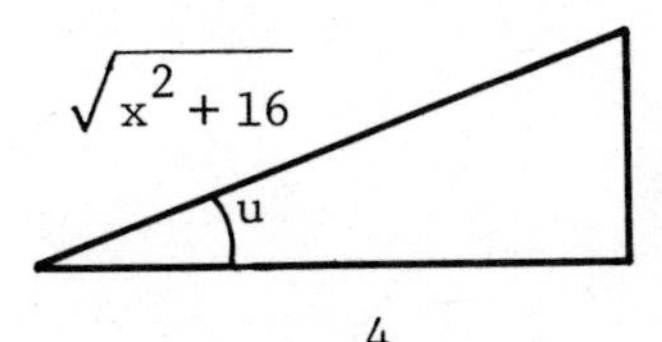

, $\dfrac{\sqrt{x^2+16}}{x}$, $-\dfrac{\sqrt{x^2+16}}{16x} + C$

Thus, upon substitution,

$$\int_{4}^{4\sqrt{3}} \frac{dx}{x^2\sqrt{x^2+16}} = \frac{1}{16}\int_{\_\_\_}^{\_\_\_} \frac{\sec u\ du}{\tan^2 u} \quad \text{(by Problem 17)}$$

$$= -\frac{1}{16} \csc u \Big]_{\_\_\_}^{\_\_\_}$$

$$= -\frac{1}{16}\left(\_\_\_\_\_\_\_\_\_\_\right) \approx 0.016 .$$

7-5 Integrals Involving $ax^2 + bx + c$.

OBJECTIVE: Find indefinite integrals involving quadratics $ax^2 + bx + c$.

19. $\int \frac{(x+1)\ dx}{x^2+x+5}$

The quadratic part may be transformed algebraically as follows:

$$x^2+x+5 = \left(x^2+x+\_\_\_\right) + 5 - \frac{1}{4} = \left(\_\_\_\_\_\right)^2 + \frac{19}{4} .$$

Now, set $u = x + \frac{1}{2}$ and $du = \_\_\_$ so the integral becomes

$$\int \frac{(x+1)\ dx}{x^2+x+5} = \int \frac{(x+1)\ dx}{\left(x+\frac{1}{2}\right)^2 + \frac{19}{4}} = \int \frac{\left(\_\_\_\_\_\right) du}{u^2 + \frac{19}{4}}$$

$$= \int \frac{\_\_\_\_}{u^2 + \frac{19}{4}} + \frac{1}{2}\int \frac{du}{u^2+\frac{19}{4}}$$

---

18. $\frac{\pi}{4}$, $\frac{\pi}{3}$, $\int_{\pi/4}^{\pi/3}$, $\Big]_{\pi/4}^{\pi/3}$, $\frac{2\sqrt{3}}{3} - \sqrt{2}$

19. $\frac{1}{4}$, $x + \frac{1}{2}$, dx, $u + \frac{1}{2}$, $u\,du$, $\frac{1}{2}\,dz$, $\frac{1}{2}\ln|z|$, $u^2 + \frac{19}{4}$,

$\frac{1}{2}\ln\left(x^2+x+5\right) + \frac{1}{\sqrt{19}}\tan^{-1}\frac{2x+1}{\sqrt{19}} + C$

$$= \int \frac{\overline{}}{z} + \frac{1}{2}\int \frac{du}{u^2 + \frac{19}{4}} \qquad \left(z = u^2 + \frac{19}{4}\right)$$

$$= \underline{} + \frac{1}{2}\sqrt{\frac{4}{19}}\ \tan^{-1}\frac{u}{\sqrt{19/4}} + C$$

$$= \frac{1}{2}\ \ln\left(\underline{}\right) + \frac{1}{\sqrt{19}}\ \tan^{-1}\frac{2u}{\sqrt{19}} + C$$

$$= \underline{}\ .$$

7-6 Integration by the Method of Partial Fractions.

OBJECTIVE: Find indefinite integrals of rational functions by the method of partial fractions expansion.

20. The success of separating a rational function $f(x)/g(x)$ into a sum of partial fractions hinges upon two things:
    (1) The degree of $f(x)$ must be ____________________ .
    If this is not the case, one must first perform ______________ , then work with the __________ term.
    (2) The factors of _____ must be known. In practice it may be difficult to perform the factorization.

21. To find A, B, and C in the partial fractions expansion

$$\frac{2x^2 - x + 1}{(x+1)(x-3)(x+2)} = \frac{A}{x+1} + \frac{B}{x-3} + \frac{C}{x+2}$$

by the Heaviside technique, the value of A can be found by covering up the factor _____ in the left side and evaluating the result at $x =$ ____ :

$$A = \frac{2(-1)^2 - (-1) + 1}{\underline{}} = \frac{4}{\underline{}} = \underline{}\ .$$

$$\text{Similarly,}\quad B = \frac{2(3)^2 - (3) + 1}{\underline{}} = \frac{16}{\underline{}} = \underline{}\ ,\ \text{and}$$

$$C = \frac{\underline{}}{(-2+1)(-2-3)} = \frac{\underline{}}{5}\ .$$

---

20. less than the degree of $g(x)$, long division, remainder, $g(x)$
21. $(x+1)$, -1, $(-1-3)(-1+2)$, -4, -1, $(3+1)(3+2)$, 20, $\frac{4}{5}$, $2(-2)^2 - (-2) + 1$, 11

22. $\int \frac{(2x^2 - x + 1)\,dx}{(x+1)(x-3)(x+2)} = \int \frac{-\,dx}{\_\_\_\_} + \frac{4}{5}\int \frac{dx}{\_\_\_\_} + \frac{11}{5}\int \frac{dx}{\_\_\_\_}$

$= \_\_\_\_\_\_\_\_\_\_\_\_\_\_\_\_\_\_\_\_\_\_\_\_ + C .$

23. $\int \frac{4\,dx}{x^3 + 4x}$

To expand $\frac{4}{x^3 + 4x}$ into a sum of partial fractions, first write,

$\frac{4}{x^3 + 4x} = \frac{4}{x(x^2 + 4)} = \frac{A}{x} + \frac{\_\_\_\_}{x^2 + 4}$. Then,

$4 = A(x^2 + 4) + \_\_\_\_\_\_ = (\_\_\_\_)x^2 + Cx + 4A$.

Thus, equating coefficients of like powers of x , $A + B =$ ___ , $C =$ ___ , and $4A =$ ___ . Solving these equations gives $A =$ ___ , $B =$ ___ , and $C =$ ___ . Hence,

$\int \frac{4\,dx}{x^3 + 4x} = \int \frac{dx}{\_\_\_} - \int \frac{x\,dx}{\_\_\_\_}$

$= \ln|x| - \_\_\_\_\_\_\_\_ + C$

$= \_\_\_\_\_\_\_\_\_\_\_\_ .$

24. $\int \frac{dx}{\sin x \cos x}$

Write $\int \frac{dx}{\sin x \cos x} = \int \frac{\sin x\,dx}{\sin^2 x \cos x} = \int \frac{\sin x\,dx}{(1 - \cos^2 x)\cos x}$,

and let $u = \cos x$, $du =$ ________ so that $\int \frac{dx}{\sin x \cos x} = \int \frac{-du}{\_\_\_\_}$.

---

22. $x+1$, $x-3$, $x+2$, $-\ln|x+1| + \frac{4}{5}\ln|x-3| + \frac{11}{5}\ln|x+2|$

23. $Bx+C$, $x(Bx+C)$, $A+B$, 0, 0, 4, 1, -1, 0, x, x^2+4, $\frac{1}{2}\ln(x^2+4)$,

$\ln \frac{|x|}{\sqrt{x^2+4}} + C$

Next, let

$$\frac{1}{(1-u^2)u} = \frac{1}{(1-u)(1+u)u} = \frac{A}{u} + \frac{B}{1-u} + \frac{C}{1+u} .$$

By the Heaviside technique, A = ___ , B = ___ , C = ___ . Thus,

$$\int \frac{-du}{(1-u^2)u} = - \int \frac{du}{\_\_\_} - \frac{1}{2} \int \frac{du}{\_\_\_} + \frac{1}{2} \int \frac{du}{\_\_\_}$$

$$= \underline{\hspace{8cm}} + C'$$

$$= \ln \frac{1}{|u|} \left[\underline{\hspace{4cm}}\right] + C'$$

$$= \ln \frac{\sqrt{1-\cos^2 x}}{\_\_\_\_\_\_} + C' = \underline{\hspace{4cm}} .$$

25. $\int \frac{(x^3 - x + 2)\ dx}{x^2 - x}$

Here the degree of the numerator fails to be less than the degree of the denominator so we must use long division and work with the remainder.

By long division, $\frac{x^3 - x + 2}{x^2 - x} = x + 1 + \_\_\_\_\_\_$. Therefore,

$$\int \frac{(x^3 - x + 2)\ dx}{x^2 - x} = \int (x+1)\ dx + 2 \int \frac{dx}{\_\_\_\_\_\_} .$$

By partial fractions, $\frac{1}{x(x-1)} = \frac{A}{x} + \frac{B}{x-1}$, and the Heaviside technique gives A = ____ and B = ____ .

---

24. $-\sin x\ dx$, $(1-u^2)u$, 1, $\frac{1}{2}$, $-\frac{1}{2}$, u, $1-u$, $1+u$,
$-\ln |u| + \frac{1}{2} \ln |1-u| + \frac{1}{2} \ln |1+u|$, $\left(1-u^2\right)^{1/2}$, $|\cos x|$, $\ln |\tan x| + C'$

25. $\frac{2}{x^2 - x}$, $x(x-1)$, -1, 1, $\frac{1}{2} x^2 + x$, $x-1$, x, $\ln |x-1| - \ln |x|$,
$\frac{x^2}{2} + x + 2 \ln \left|\frac{x-1}{x}\right| + C$

Therefore,

$$\int \frac{(x^3 - x + 2)\ dx}{x^2 - x} = \underline{\hspace{3em}} + 2\int \frac{dx}{\underline{\hspace{2em}}} - 2\int \frac{dx}{\underline{\hspace{2em}}}$$

$$= \frac{x^2}{2} + x + 2(\underline{\hspace{8em}}) + C$$

$$= \underline{\hspace{12em}} .$$

26. $\int \frac{(2x^2 + x + 2)\ dx}{x(x-1)^2}$

By partial fractions,

$$\frac{2x^2 + x + 2}{x(x-1)^2} = \frac{A}{x} + \frac{B}{x-1} + \underline{\hspace{3em}} .$$

Clearing fractions and equating numerators gives,

$$2x^2 + x + 2 = A(x-1)^2 + Bx(x-1) + \underline{\hspace{2em}}$$

$$= (A+B)x^2 + (\underline{\hspace{5em}})x + A$$

Equating coefficients of like powers of x gives the equations

$A + B = 2$, $\underline{\hspace{5em}} = 1$, and $\underline{\hspace{2em}} = 2$.

Solving simultaneously, $A = \underline{\hspace{2em}}$, $B = \underline{\hspace{2em}}$, and $C = 5$. Hence,

$$\int \frac{(2x^2 + x + 2)\ dx}{x(x-1)^2} = 2\int \frac{dx}{\underline{\hspace{1.5em}}} + 5\int \frac{dx}{\underline{\hspace{3em}}}$$

$$= \underline{\hspace{10em}} .$$

7-7 Integration by Parts.

OBJECTIVE: Evaluate integrals by the method of integration by parts whenever feasible.

27. The method of integration by parts is based on the product rule for differentiation. In terms of integrals,

$$\int u\ dv = \underline{\hspace{8em}} .$$

---

26. $\frac{C}{(x-1)^2}$, Cx, $-2A - B + C$, $-2A - B + C$, A, 2, 0, x, $(x-1)^2$, $2 \ln |x| - \frac{5}{x-1} + C$

27. $uv - \int v\ du + C$

28. To employ the method of integration by parts successfully we must be able to integrate the part ____ immediately in order to obtain ___ . Also it is hoped that the integral $\int v\,du$ is simpler than the original integral $\int u\,dv$.

29. $\int \frac{xe^x\,dx}{(1+x)^2}$

Let $u = xe^x$, $du =$ ____________ , $dv = \frac{dx}{(1+x)^2}$, $v =$ ________ . Then,

$$\int \frac{xe^x\,dx}{(1+x)^2} = \frac{-xe^x}{1+x} + \int \_\_\_\_ \,dx + C$$

$$= e^x\left(\frac{-x}{1+x} + \_\_\right) + C = \_\_\_\_\_\_\_\_\_\_ .$$

30. $\int \tan^{-1}\sqrt{x}\,dx$

Let $u = \tan^{-1}\sqrt{x}$, $du =$ ____________ , $dv =$ ____ , $v =$ ___ . Then,

$$\int \tan^{-1}\sqrt{x}\,dx = \_\_\_\_\_\_\_\_\_\_\_\_ - \int \frac{\sqrt{x}\,dx}{2(1+x)} + C .$$

Let $w = \sqrt{x}$ and $dw =$ ______ , and the latter integral becomes

$$\int \frac{\sqrt{x}\,dx}{2(1+x)} = \int \frac{w^2\,dw}{\_\_\_\_\_\_} = \int \left(1 - \_\_\_\_\_\_\right) dw$$

$$= w - \_\_\_\_\_\_\_\_$$

$$= \_\_\_\_\_\_\_\_\_\_\_\_\_\_\_\_\_\_$$

(we may omit the constant of integration here since we already have C)

Putting this together with our previous result, we find

$$\int \tan^{-1}\sqrt{x}\,dx = \_\_\_\_\_\_\_\_\_\_\_\_\_\_\_\_\_\_\_\_\_\_\_\_\_\_\_\_ .$$

---

28. dv, v

29. $e^x(1+x)\,dx$, $-\frac{1}{1+x}$, e^x, 1, $\frac{e^x}{1+x} + C$

30. $\frac{dx}{2\sqrt{x}\,(1+x)}$, dx, x, $x\tan^{-1}\sqrt{x}$, $\frac{dx}{2\sqrt{x}}$, $1+w^2$, $\frac{1}{1+w^2}$, $\tan^{-1} w$, $\sqrt{x} - \tan^{-1}\sqrt{x}$, $(x+1)\tan^{-1}\sqrt{x} - \sqrt{x} + C$

31. $\int_1^e (\ln x)^2 \, dx$

Let $u = (\ln x)^2$, du = __________ , dv = ____ , v = ____ . Then,

$$\int_1^e (\ln x)^2 \, dx = x(\ln x)^2 \Big]_1^e - 2 \int_1^e \_\_\_\_\_\_\_\_\_$$

$$= (e - \_\_\_) - 2 \Big[\_\_\_\_\_\_\_\_\_\_\_\_ \Big]_1^e$$

↑ Example 1, page 363, in the text

$$= e - 2(\_\_\_\_\_\_\_\_\_) = \_\_\_\_\_\_ .$$

32. $\int x^2 \sin x \, dx$

Let $u = x^2$, du = ______ , dv = _________ , v = _________ . Then,

$$\int x^2 \sin x \, dx = \_\_\_\_\_\_\_\_\_\_\_\_ + 2 \int x \cos x \, dx .$$

To determine the latter integral we integrate by parts again: U = x , dU = ___ , dV = __________ , V = _______ . Then,

$$\int x \cos x \, dx = x \sin x - \int \_\_\_\_\_\_\_\_\_ = \_\_\_\_\_\_\_\_\_\_\_\_\_\_\_\_\_\_\_\_ + C$$

Therefore, putting these results together, we find

$$\int x^2 \sin x \, dx = \_\_\_\_\_\_\_\_\_\_\_\_\_\_\_\_\_\_\_\_\_\_\_\_\_\_\_\_\_\_\_\_\_ .$$

33. $\int \frac{\tan^{-1} x \, dx}{x^2}$

Let $u = \tan^{-1} x$, du = __________ , dv = ____ , v = ____ . Then,

---

31. $\frac{2}{x} \ln x \, dx$, dx, x, $\ln x \, dx$, 0, $x \ln x - x$, $(e - e + 1)$, $e - 2$

32. $2x \, dx$, $\sin x \, dx$, $-\cos x$, $-x^2 \cos x$, dx, $\cos x \, dx$, $\sin x$, $\sin x \, dx$, $x \sin x + \cos x$, $(2 - x^2)\cos x + 2x \sin x + C$

33. $\frac{1}{1+x^2} dx$, $\frac{dx}{x^2}$, $-\frac{1}{x}$, $-\frac{\tan^{-1} x}{x}$, $-\frac{\tan^{-1} x}{x}$, $-x \, dx$,
$-\frac{\tan^{-1} x}{x} + \ln |x| - \frac{1}{2} \ln \left(1 + x^2\right) + C$

Then,

$$\int \frac{\tan^{-1} x \, dx}{x^2} = \underline{\hspace{3cm}} + \int \frac{dx}{x(1+x^2)}$$

$$= \underline{\hspace{3cm}} + \int \frac{dx}{x} + \int \frac{\underline{\hspace{1cm}}}{1+x^2}$$

$$= \underline{\hspace{8cm}} .$$

7-8 Integration of Rational Functions of $\sin x$ and $\cos x$, and Other Trigonometric Integrals.

OBJECTIVE A: Integrate rational functions of $\sin x$ and $\cos x$ via the substitution $z = \tan \frac{x}{2}$.

34. Making the substitution $z = \tan \frac{x}{2}$ gives

$\cos x =$ ______ , $\sin x =$ ______ , $x =$ __________ , and $dx =$ ______ .

35. $\displaystyle\int \frac{dx}{\sin x \cos x}$

Using the substitution $z = \tan \frac{x}{2}$,

$$\int \frac{dx}{\sin x \cos x} = \int \frac{1+z^2}{2z} \cdot \frac{1+z^2}{1-z^2} \cdot \frac{2\,dz}{1+z^2} = \int \underline{\hspace{3cm}} \, dz$$

By partial fractions,

$$\int \frac{(1+z^2)\,dz}{z(1-z^2)} = \int \frac{dz}{z} + \int \frac{\underline{\hspace{1cm}}\,dz}{1-z^2} = \ln|z| - \underline{\hspace{3cm}} + C$$

$$= \ln \underline{\hspace{3cm}} + C = \underline{\hspace{5cm}} .$$

---

34. $\dfrac{1-z^2}{1+z^2}$, $\dfrac{2z}{1+z^2}$, $2\tan^{-1} z$, $\dfrac{2\,dz}{1+z^2}$

35. $\dfrac{1+z^2}{z(1-z^2)}$, $2z$, $\ln\left|1-z^2\right|$, $\left|\dfrac{z}{1-z^2}\right|$, $\ln\left|\dfrac{\tan \frac{x}{2}}{1-\tan^2 \frac{x}{2}}\right| + C$

Remark: Using the identity $\tan\frac{x}{2} = \frac{\sin x}{1 + \cos x}$, it can be shown that

$$\frac{\tan\frac{x}{2}}{1 - \tan^2\frac{x}{2}} = \frac{\sin x}{2\cos x} = \frac{1}{2}\tan x . \quad \text{Thus,}$$

$$\ln\left|\frac{\tan\frac{x}{2}}{1 - \tan^2\frac{x}{2}}\right| = \ln|\tan x| - \ln 2 .$$

Now compare the integration method above with that used in Problem 24 of this chapter.

OBJECTIVE B: Find integrals of the functions $\sin mx \sin nx$, $\cos mx \cos nx$, and $\sin mx \cos nx$, where m and n are positive integers.

36. $\int \sin^2 x \cos 3x\, dx$

$$\int \sin^2 x \cos 3x\, dx = \frac{1}{2}\int (1 - \cos 2x) \_\_\_\_\_\_\_\_ dx$$

$$= \frac{1}{2}\int \cos 3x\, dx - \frac{1}{2}\int \_\_\_\_\_\_\_\_\_\_\_\_\_\_\_\_ dx$$

$$= \frac{1}{2}\int \cos 3x\, dx - \frac{1}{4}\int [\_\_\_\_\_\_\_\_\_\_\_\_\_\_\_\_] dx$$

$$= \_\_\_\_\_\_\_\_\_\_\_\_\_\_\_\_\_\_\_\_ - \frac{1}{20}\sin 5x + C .$$

7-9 Further Substitutions.

OBJECTIVE: Calculate integrals by making appropriate substitutions suggested by the form of the integrand. If one substitution fails to simplify the integral, try something else.

37. $\int \frac{1 + x^{1/2}}{1 + x^{1/3}}\, dx$

Here the least common multiple of the denominators of the exponents is 6 .

Hence, take $x = z^6$, $dx = \_\_\_\_\_\_\_\_$.

---

36. $\cos 3x$, $\cos 2x \cos 3x$, $\cos x + \cos 5x$, $\frac{1}{6}\sin 3x - \frac{1}{4}\sin x$

This leads to

$$\int \frac{1 + x^{1/2}}{1 + x^{1/3}}\, dx = \int \frac{1+z^3}{\_\_\_\_\_\_} \cdot 6z^5\, dz = 6 \int \frac{z^8 + z^5}{\_\_\_\_\_\_}\, dz$$

Employing long division in the last integrand gives,

$$\int \frac{1 + x^{1/2}}{1 + x^{1/3}}\, dx = 6 \int \left(z^6 - z^4 + z^3 + z^2 - z - 1 + \_\_\_\_\_\_ \right) dz$$

$$= \frac{6}{7} z^7 - \frac{6}{5} z^5 + \frac{3}{2} z^4 + 2z^3 - 3z^2 - 6z$$

$$+ 6 \int \frac{dz}{z^2+1} + 6 \int \frac{\_\_\_\_\_\_}{z^2+1}$$

$$= \frac{6}{7} z^7 - \frac{6}{5} z^5 + \frac{3}{2} z^4 + 2z^3 - 3z^2 - 6z$$

$$+ \_\_\_\_\_\_\_\_\_\_\_\_\_\_\_\_\_\_ + C$$

$$= \_\_\_\_\_\_\_\_\_\_\_\_\_\_\_\_\_\_\_\_\_\_\_\_$$

$$\_\_\_\_\_\_\_\_\_\_\_\_\_\_\_\_\_\_\_\_ .$$

38. $\displaystyle\int \frac{(1+x)\, dx}{\sqrt{1-x}}$

Let $z^2 = 1 - x$, $-dx = \_\_\_\_\_\_$, and the integral becomes

$$\int \frac{(1+x)\, dx}{\sqrt{1-x}} = \int \frac{\_\_\_\_\_\_}{z} \cdot (-2z\, dz) = -2 \int \_\_\_\_\_\_\, dz$$

$$= -4z + \_\_\_\_\_\_ = \_\_\_\_\_\_\_\_\_\_\_\_ .$$

---

37. $6z^5\, dz$, $1+z^2$, z^2+1, $\dfrac{1+z}{z^2+1}$, $z\, dz$, $6 \tan^{-1} z + 3 \ln (z^2+1)$,

$\frac{6}{7} x^{7/6} - \frac{6}{5} x^{5/6} + \frac{3}{2} x^{2/3} + 2x^{1/2} - 3x^{1/3} - 6x^{1/6} + 6 \tan^{-1} x^{1/6} + 3 \ln\left(x^{1/3} + 1\right) + C$

38. $2z\, dz$, $1 + (1 - z^2)$, $(2 - z^2)$, $\frac{2}{3} z^3 + C$, $-4(1-x)^{1/2} + \frac{2}{3} (1-x)^{3/2} + C$

7-10 Improper Integrals.

39. A definite integral $\int_a^b f(x)\,dx$ is termed improper if,

(a) either limit of integration is ____ or ____ , or

(b) $f(x)$ becomes __________ at some value $x = c$ satisfying __________ .

OBJECTIVE: Determine whether a given improper integral converges or diverges. If convergent, evaluate it.

40. $\int_1^\infty \frac{\ln x}{x^2}\,dx$

$$\int_1^\infty \frac{\ln x}{x^2}\,dx = \lim_{b\to\infty} \int_{\_\_}^{\_\_} \frac{\ln x}{x^2}\,dx$$

To calculate the latter integral, we integrate by parts:
let $u = \ln x$, $du =$ ____ , $dv =$ ____ , $v =$ ____ .

$$\int_1^b \frac{\ln x\,dx}{x^2} = -\frac{\ln x}{x}\Big]_1^b + \int_1^b \_\_\_\_ = \frac{-\ln b}{b} + \_\_\_\_\_ \text{ ; whence}$$

$$\int_1^\infty \frac{\ln x\,dx}{x^2} = \lim_{b\to\infty}\left(\_\_\_\_\_\_\_\_\_\_\right) = \lim_{b\to\infty} \frac{-\ln b}{b} + \_\_\_$$

$$= \lim_{b\to\infty} \frac{\_\_\_}{\_\_\_} + 1 = \_\_\_ .$$

41. $\int_0^1 \ln x\,dx$

Here, $\ln x \to$ ____ as $x \to 0^+$ making the integral improper. Thus,

$$\int_0^1 \ln x\,dx = \lim_{b\to 0^+} \int_{\_\_}^{\_\_} \ln x\,dx .$$

---

39. $+\infty$, $-\infty$, infinite, $a \le c \le b$

40. $\int_1^b$, $\frac{1}{x}\,dx$, $\frac{1}{x^2}\,dx$, $-\frac{1}{x}$, $\frac{dx}{x^2}$, $-\frac{1}{x}\Big]_1^b$, $\frac{-\ln b}{b} - \frac{1}{b} + 1$, 1, $\frac{-1/b}{1}$ (l'Hôpital's rule), 1

41. $-\infty$, $\int_b^1$, $\frac{1}{x}$, dx, x, $x \ln x$, $-b \ln b - 1 + b$, $\frac{1/b}{1/b^2}$ (l'Hôpital's rule), -1

To calculate the latter integral, we integrate by parts:
let $u = \ln x$, $du =$ ____ , $dv =$ ____ , $v =$ ____ .

$$\int_b^1 \ln x\,dx = \underline{\qquad\qquad}\Big]_b^1 - \int_b^1 \quad dx = \underline{\qquad\qquad\qquad} .$$ Therefore,

$$\int_0^1 \ln x\,dx = \lim_{b\to 0^+} (-b \ln b - 1 + b)$$

$$= \lim_{b\to 0^+} \left(\frac{-\ln b}{1/b}\right) - 1$$

$$= \lim_{b\to 0^+} \left(\underline{\qquad}\right) - 1 = \underline{\qquad} .$$

42. $\displaystyle\int_0^\infty \frac{x\,dx}{\sqrt{1+x^3}}$

Observe that $\dfrac{x}{\sqrt{1+x^3}} > \dfrac{1}{x}$ is true whenever $x^2 >$ ________ whenever $x^4 >$ ________ , and the last inequality certainly holds if $x \geq 2$. Thus, we consider the improper integral $\displaystyle\int_2^\infty \frac{dx}{x}$. Now, $\displaystyle\int_2^\infty \frac{dx}{x} = \lim_{b\to\infty} \Big[\ln b - \ln 2 \Big] = +\infty$. Thus, since the original integral satisfies the inequalities,

$$\int_0^\infty \frac{x\,dx}{\sqrt{1+x^3}} \geq \int_2^\infty \frac{x\,dx}{\sqrt{1+x^3}} \geq \int_2^\infty \frac{dx}{x} ,$$

it ____________ to $+\infty$.

43. $\displaystyle\int_0^2 \frac{(x^2 - 3x + 1)\,dx}{x(x-1)^2}$

In this case the integrand becomes infinite when $x =$ ___ and $x =$ ___ , so the integrand is improper.

---

42. $\sqrt{1+x^3}$ (since $x > 0$) , $1+x^3$, diverges

43. 0, 1, $\displaystyle\int_h^2 f(x)\,dx$, Cx, $-2A - B + C$, 1, $-2A - B + C$, 0, -1, $\dfrac{-dx}{(x-1)^2}$, $\dfrac{1}{x-1}$, $\ln x + \dfrac{1}{x-1}$, $\ln x + \dfrac{1}{x-1}$, $\ln x + \dfrac{1}{x-1}$, $+\infty$, $-\infty$, $-\infty$, divergent

To simplify the notation momentarily, let $f(x)$ denote the integrand. Then,

$$\int_0^2 f(x)\,dx = \lim_{b\to 0^+}\int_b^{1/2} f(x)\,dx + \lim_{c\to 1^-}\int_{1/2}^{c} f(x)\,dx$$

$$+ \lim_{h\to 1^+}\int_{\_\_}^{\_\_} \_\_\_\_\_\_\_\_$$

To find the indefinite integral $\int \frac{(x^2-3x+1)\,dx}{x(x-1)^2}$ we expand the integrand by partial fractions:

$$\frac{x^2-3x+1}{x(x-1)^2} = \frac{A}{x} + \frac{B}{x-1} + \frac{C}{(x-1)^2}$$

Therefore,

$$x^2-3x+1 = A(x-1)^2 + Bx(x-1) + \_\_\_\_$$

$$= (A+B)x^2 + (\_\_\_\_\_\_\_\_\_\_)x + A\,.$$

It follows that

$A = 1$, $A+B =$ ___ and __________ $= -3$

Solving simultaneously, $B =$ ___ and $C =$ ___ . Hence,

$$\int \frac{(x^2-3x+1)\,dx}{x(x-1)^2} = \int \frac{dx}{x} + \int \_\_\_\_\_\_\_\_ = \ln|x| + \_\_\_\_\_\_ + C'\,.$$

Therefore,

$$\int_0^2 \frac{(x^2-3x+1)\,dx}{x(x-1)^2} = \lim_{b\to 0^+}\left(\_\_\_\_\_\_\_\_\_\_\_\_\right)\Big]_b^{1/2}$$

$$+ \lim_{c\to 1^-}\left(\_\_\_\_\_\_\_\_\_\_\_\_\right)\Big]_{1/2}^{c} + \lim_{h\to 1^+}\left(\_\_\_\_\_\_\_\_\_\_\_\_\right)\Big]_h^{2}$$

The first limit is ___ , the second limit is ___ , and the third is ___ . Therefore, the improper integral is __________ .

44. The improper integrals in Problems 40 and 41 could be explicitly evaluated with the aid of indefinite integrals. Such is not the case, however, for the improper integral $\int_0^\infty \frac{\sin x\,dx}{1+x^2}$.

---

44. $\int_0^b$, $\tan^{-1} x$, $\frac{\pi}{2}$, convergent

Now, $\int_0^{\infty} \frac{\sin x\ dx}{1+x^2} = \lim_{b\to\infty} \int_{\_\_}^{\_\_} \frac{\sin x\ dx}{1+x^2} \leq \lim_{b\to\infty} \int_0^b \frac{dx}{1+x^2}$

Evaluating the latter integral,

$$\lim_{b\to\infty} \int_0^b \frac{dx}{1+x^2} = \lim_{b\to\infty} \underline{\hspace{3cm}}\Big]_0^b = \underline{\hspace{1cm}} .$$

Therefore, we know that the original integral $\int_0^{\infty} \frac{\sin x\ dx}{1+x^2}$ is ___________.

In order to approximate the value of this integral, let b take on a sufficiently large value, and evaluate numerically (say by Simpson's rule)

$$\int_0^b \frac{\sin x\ dx}{1+x^2} .$$

CHAPTER 7 OBJECTIVE - PROBLEM KEY

Objective	Problems in Thomas/Finney Text	Objective	Problems in Thomas/Finney Text
7-1	p. 335, 1-44, 47	7-6	p. 362, 1-18
7-2	p. 342, 1-24, 28-33	7-7	p. 367, 1-16, 25
7-3	p. 345, 1-10	7-8 A	p. 371, 1-5
7-4 A	p. 352, 2,4,5,8,13-20	B	p. 371, 6-9
B	p. 352, 1,3,6,7,9-12	7-9	p. 374, 1-20
7-5	p. 355, 1-10	7-10	p. 382, 1-19

CHAPTER 7 SELF-TEST

Evaluate the integrals in Problems 1 - 14.

1. $\int \frac{x\ dx}{x^4+1}$

2. $\int \frac{(5x+3)\ dx}{x^3-2x^2-3x}$

3. $\int \frac{\ln x\ dx}{(x+1)^2}$

4. $\int \frac{\sqrt{9-4x^2}\ dx}{x}$

5. $\int \frac{e^{3x/2}\ dx}{1+e^{3x/4}}$

6. $\int \frac{dx}{1-\sin x+\cos x}$

7. $\int \sin^3 x \cos^4 x\ dx$

8. $\int \cot^5 x\ dx$

9. $\displaystyle\int \frac{x\,dx}{x^2+x+1}$

10. $\displaystyle\int \cos^2 x \sin 2x \; dx$

11. $\displaystyle\int_{-\pi/3}^{\pi/4} x \sec^2 x \; dx$

12. $\displaystyle\int_0^1 \frac{(2x^3+x+3)\,dx}{x^2+1}$

13. $\displaystyle\int_5^{5\sqrt{3}} \frac{dx}{x\sqrt{25+x^2}}$

14. $\displaystyle\int_0^a \ln(a^2+x^2)\,dx \quad , \quad a > 0$

In Problems 15 - 18, determine the convergence or divergence of the integral. If the integral is convergent, find its value.

15. $\displaystyle\int_0^{\pi/2} \sec x \tan x \; dx$

16. $\displaystyle\int_{-1}^1 \frac{dx}{x^{2/3}}$

17. $\displaystyle\int_1^{\infty} \frac{\ln x\,dx}{x}$

18. $\displaystyle\int_0^{\infty} \frac{\sin x\,dx}{e^x}$

SOLUTIONS TO CHAPTER 7 SELF-TEST

1. Let $u = x^2$, $du = 2x\,dx$. Then,

$$\int \frac{x\,dx}{x^4+1} = \frac{1}{2}\int \frac{du}{u^2+1} = \frac{1}{2}\tan^{-1} u + C = \frac{1}{2}\tan^{-1} x^2 + C .$$

2. $$\frac{5x+3}{x^3-2x^2-3x} = \frac{5x+3}{x(x+1)(x-3)} = \frac{A}{x} + \frac{B}{x+1} + \frac{C}{x-3}$$

By the Heaviside technique, $A = -1$, $B = -\frac{1}{2}$, $C = \frac{3}{2}$. Thus,

$$\int \frac{(5x+3)\,dx}{x^3-2x^2-3x} = -\int \frac{dx}{x} - \frac{1}{2}\int \frac{dx}{x+1} + \frac{3}{2}\int \frac{dx}{x-3}$$

$$= -\ln|x| - \frac{1}{2}\ln|x+1| + \frac{3}{2}\ln|x-3| + C$$

$$= \ln\sqrt{\frac{(x-3)^3}{x^2(x+1)}} + C$$

3. Let $u = \ln x$, $du = \dfrac{dx}{x}$, $dv = \dfrac{dx}{(x+1)^2}$, $v = \dfrac{-1}{x+1}$; $x \quad 0$

$$\int \frac{\ln x \, dx}{(x+1)^2} = -\frac{\ln x}{x+1} + \int \frac{dx}{x(x+1)}$$

$$= -\frac{\ln x}{x+1} + \int \frac{dx}{x} - \int \frac{dx}{x+1}$$

$$= -\frac{\ln x}{x+1} + \ln x - \ln(x+1) + C$$

$$= -\frac{\ln x}{x+1} + \ln \frac{x}{x+1} + C, \quad x > 0 .$$

4. Let $x = \dfrac{3}{2} \sin u$, $dx = \dfrac{3}{2} \cos u \, du$. Then,

$$\int \frac{\sqrt{9-4x^2}\, dx}{x} = \int \frac{\sqrt{9 - 9\sin^2 u} \cdot \frac{3}{2} \cos u \, du}{\frac{3}{2} \sin u}$$

$$= 3 \int \frac{\cos^2 u \, du}{\sin u} = 3 \int \frac{(1 - \sin^2 u)\, du}{\sin u}$$

$$= 3 \int (\csc u - \sin u)\, du$$

$$= 3 \ln |\csc u - \cot u| + 3 \cos u + C$$

From the substituion $x = \dfrac{3}{2} \sin u$, and the diagram at the right,

$$\csc u = \frac{3}{2x} \quad \text{and} \quad \cot u = \frac{\sqrt{9-4x^2}}{2x} .$$

3/2
x
u
$\frac{1}{2}\sqrt{9-4x^2}$

Thus,

$$\int \frac{\sqrt{9-4x^2}\, dx}{x} = 3 \ln \left| \frac{3}{2x} - \frac{\sqrt{9-4x^2}}{2x} \right| + \sqrt{9-4x^2} + C .$$

5. Let $u = e^x$, $du = e^x \, dx$. Then,

$$\int \frac{e^{3x/2}\, dx}{1 + e^{3x/4}} = \int \frac{e^{x/2} \cdot e^x \, dx}{1 + e^{3x/4}} = \int \frac{u^{1/2}\, du}{1 + u^{3/4}}$$

Next, let $z = u^{1/4}$, $dz = \frac{1}{4} u^{-3/4} du$. Hence, $z^2 = u^{1/2}$, $z^3 = u^{3/4}$, and $du = 4u^{3/4} dz = 4z^3 dz$. Substitution into the last integral gives

$$\int \frac{e^{3x/2} dx}{1 + e^{3x/4}} = \int \frac{z^2 (4z^3 dz)}{1+z^3} = 4 \int \frac{z^5 dz}{z^3+1}$$

$$= 4 \int \left(z^2 - \frac{z^2}{z^3+1} \right) dz$$

$$= \frac{4}{3} z^3 - \frac{4}{3} \ln |z^3+1| + C$$

$$= \frac{4}{3} u^{3/4} - \frac{4}{3} \ln |u^{3/4} + 1| + C$$

$$= \frac{4}{3} e^{3x/4} - \frac{4}{3} \ln \left(e^{3x/4} + 1 \right) + C$$

6. Let $z = \tan \frac{x}{2}$. Then, $\cos x = \frac{1-z^2}{1+z^2}$, $\sin x = \frac{2z}{1+z^2}$, and $dx = \frac{2\,dz}{1+z^2}$.

Substitution gives

$$\int \frac{dx}{1 - \sin x + \cos x} = 2 \int \frac{dz}{\left(1+z^2\right)\left(1 - \frac{2z}{1+z^2} + \frac{1-z^2}{1+z^2}\right)}$$

$$= 2 \int \frac{dz}{1 + z^2 - 2z + 1 - z^2}$$

$$= \int \frac{dz}{1-z} = - \ln |1-z| + C$$

$$= - \ln \left|1 - \tan \frac{x}{2}\right| + C .$$

7. Let $u = \cos x$, $du = - \sin x \, dx$. Then

$$\int \sin^3 x \cos^4 x \, dx = \int \sin^2 x \cos^4 x \sin x \, dx$$

$$= \int (1 - \cos^2 x) \cos^4 x \sin x \, dx$$

$$= \int - (1-u^2) u^4 \, du$$

$$= - \frac{1}{5} u^5 + \frac{1}{7} u^7 + C = - \frac{1}{5} \cos^5 x + \frac{1}{7} \cos^7 x + C .$$

8. Using the trigonometric identity $\cot^2 x = \csc^2 x - 1$,

$$\int \cot^5 x\, dx = \int \cot^3 x (\csc^2 x - 1)\, dx$$

$$= \int \cot^3 x \csc^2 x\, dx - \int \cot^3 x\, dx$$

$$= \int \cot^3 x \csc^2 x\, dx - \int \cot x (\csc^2 x - 1)\, dx$$

$$= \int \cot^3 x \csc^2 x\, dx - \int \cot x \csc^2 x\, dx + \int \cot x\, dx$$

Let $u = \cot x$, $du = -\csc^2 x\, dx$ and we have

$$\int \cot^5 x\, dx = -\frac{1}{4}\cot^4 x + \frac{1}{2}\cot^2 x + \ln|\sin x| + C.$$

9. Completing the square, $x^2 + x + 1 = \left(x + \frac{1}{2}\right)^2 + \frac{3}{4}$. Let $u = x + \frac{1}{2}$, $du = dx$. Then,

$$\int \frac{x\, dx}{x^2 + x + 1} = \int \frac{\left(u - \frac{1}{2}\right) du}{u^2 + \frac{3}{4}}$$

$$= \int \frac{u\, du}{u^2 + \frac{3}{4}} - \frac{1}{2}\int \frac{du}{u^2 + \frac{3}{4}}$$

$$= \frac{1}{2}\ln\left|u^2 + \frac{3}{4}\right| - \frac{1}{2}\cdot\frac{2}{\sqrt{3}}\tan^{-1}\frac{2u}{\sqrt{3}} + C$$

$$= \frac{1}{2}\ln|x^2 + x + 1| - \frac{1}{\sqrt{3}}\tan^{-1}\frac{2x+1}{\sqrt{3}} + C.$$

10. $$\int \cos^2 x \sin 2x\, dx = \frac{1}{2}\int (1 + \cos 2x)\sin 2x\, dx$$

$$= \frac{1}{2}\int \sin 2x\, dx + \frac{1}{2}\int \sin 2x \cos 2x\, dx$$

$$= \frac{1}{2}\int \sin 2x\, dx + \frac{1}{4}\int \sin 4x\, dx$$

$$= -\frac{1}{4}\cos 2x - \frac{1}{16}\cos 4x + C.$$

11. Let $u = x$, $du = dx$, $dv = \sec^2 x\,dx$, $v = \tan x$. Then,

$$\int_{-\pi/3}^{\pi/4} x \sec^2 x\,dx = x \tan x\Big]_{-\pi/3}^{\pi/4} - \int_{-\pi/3}^{\pi/4} \tan x\,dx$$

$$= \frac{\pi}{4}\cdot 1 + \frac{\pi}{3}\cdot\sqrt{3} - \ln|\cos x|\Big]_{-\pi/3}^{\pi/4}$$

$$= \frac{\pi}{4} + \frac{\pi}{\sqrt{3}} - \ln\frac{\sqrt{2}}{2} + \ln\frac{1}{2}$$

$$= \frac{\pi}{4} + \frac{\pi}{\sqrt{3}} - \frac{1}{2}\ln 2 \approx 2.253\,.$$

12. $$\int_0^1 \frac{(2x^3+x+3)\,dx}{x^2+1} = \int_0^1 \left(2x + \frac{3-x}{x^2+1}\right) dx$$

$$= \int_0^1 2x\,dx + 3\int_0^1 \frac{dx}{x^2+1} - \int_0^1 \frac{x\,dx}{x^2+1}$$

$$= x^2 + 3\tan^{-1} x - \frac{1}{2}\ln(x^2+1)\Big]_0^1$$

$$= \left(1 + \frac{3\pi}{4} - \frac{1}{2}\ln 2\right) - (0+0-0) \approx 3.01\,.$$

13. Let $x = 5\tan u$, $dx = 5\sec^2 u\,du$. When $x = 5$, $\tan u = 1$ or $u = \pi/4$; when $x = 5\sqrt{3}$, $\tan u = \sqrt{3}$ or $u = \pi/3$. Thus,

$$\int_5^{5\sqrt{3}} \frac{dx}{x\sqrt{25+x^2}} = \int_{\pi/4}^{\pi/3} \frac{5\sec^2 u\,du}{5\tan u\sqrt{25+25\tan^2 u}}$$

$$= \frac{1}{5}\int_{\pi/4}^{\pi/3} \frac{\sec u\,du}{\tan u} = \frac{1}{5}\int_{\pi/4}^{\pi/3} \csc u\,du$$

$$= \frac{1}{5}\ln\left|\csc u - \cot u\right|\Big]_{\pi/4}^{\pi/3}$$

$$= \frac{1}{5}\left(\ln\left|\frac{2\sqrt{3}}{3} - \frac{\sqrt{3}}{3}\right| - \ln\left|\sqrt{2} - 1\right|\right)$$

$$= \frac{1}{5}\ln\left|\frac{\sqrt{3}}{3\left(\sqrt{2}-1\right)}\right| \approx 0.066\,.$$

14. Let $u = \ln(a^2+x^2)$, $du = \dfrac{2x\,dx}{a^2+x^2}$, $dv = dx$, and $v = x$. Then,

$$\int_0^a \ln(a^2+x^2)\,dx = x\ln(a^2+x^2)\Big]_0^a - \int_0^a \frac{2x^2\,dx}{a^2+x^2}$$

$$= a\ln 2a^2 - 2\int_0^a \left(1 - \frac{a^2}{a^2+x^2}\right)dx$$

$$= a\ln 2a^2 - \left(2x + 2a\tan^{-1}\frac{x}{a}\right)\Big]_0^a$$

$$= a\ln 2a^2 - 2a + \frac{2\pi a}{4}.$$

15. $$\int_0^{\pi/2} \sec x \tan x\,dx = \lim_{b\to\frac{\pi}{2}^-} \int_0^b \sec x\tan x\,dx$$

$$= \lim_{b\to\frac{\pi}{2}^-} \sec x\Big]_0^b = \lim_{b\to\frac{\pi}{2}^-} (\sec b - 1) = \infty.$$

Therefore, the improper interval diverges.

16. $$\int_{-1}^{1} \frac{dx}{x^{2/3}} = \lim_{a\to 0^-} \int_{-1}^{a} \frac{dx}{x^{2/3}} + \lim_{a\to 0^+} \int_a^1 \frac{dx}{x^{2/3}}$$

$$= \lim_{a\to 0^-} 3x^{1/3}\Big]_{-1}^{a} + \lim_{a\to 0^+} 3x^{1/3}\Big]_a^1$$

$$= \lim_{a\to 0^-} \left(3a^{1/3} + 3\right) + \lim_{a\to 0^+} \left(3 - 3a^{1/3}\right) = 6.$$

17. For $x \geq e$, $\ln x \geq 1$. Hence,

$$\int_1^\infty \frac{\ln x\,dx}{x} \geq \int_e^\infty \frac{\ln x\,dx}{x} \geq \int_e^\infty \frac{dx}{x}.$$

Now, $\displaystyle\int_e^\infty \frac{dx}{x} = \lim_{b\to\infty} \ln x\Big]_e^b = \lim_{b\to\infty} \ln b - 1 = \infty.$

Therefore, the integral $\displaystyle\int_1^\infty \frac{\ln x\,dx}{x}$ diverges.

18. $\int_0^\infty \frac{\sin x\, dx}{e^x} = \lim_{b\to\infty} \int_0^b e^{-x} \sin x\, dx$

Let $u = e^{-x}$, $du = -e^{-x}\, dx$, $dv = \sin x\, dx$, $v = -\cos x$.

$$\int_0^b e^{-x} \sin x\, dx = -e^{-x} \cos x \Big]_0^b - \int_0^b e^{-x} \cos x\, dx$$

Let $U = e^{-x}$, $dU = -e^{-x}\, dx$, $dV = \cos x\, dx$, $V = \sin x$, and

$$\int_0^b e^{-x} \cos x\, dx = e^{-x} \sin x \Big]_0^b + \int_0^b e^{-x} \sin x\, dx .$$

Putting these results together,

$$2 \int_0^b e^{-x} \sin x\, dx = -e^{-x} \cos x - e^{-x} \sin x \Big]_0^b$$

$$= -e^{-b} \cos b + 1 - e^{-b} \sin b$$

Now, $\lim_{b\to\infty} \frac{\cos b}{e^b} = \lim_{b\to\infty} \frac{\sin b}{e^b} = 0$. Thus,

$$\int_0^\infty \frac{\sin x\, dx}{e^x} = \lim_{b\to\infty} \int_0^b e^{-x} \sin x\, dx = \frac{1}{2} .$$

CHAPTER 8 PLANE ANALYTIC GEOMETRY

8-1 Curves and Equations.

OBJECTIVE: Given an equation $F(x,y) = 0$, analyze it to investigate properties of symmetry, extent, intercepts, asymptotes, and slope at the intercepts.

1. If the equation of a curve is unaltered when y is replaced by $-y$, that is, if $F(x,y) = F(x,-y)$, then the curve is symmetric with respect to the ________ .

2. A curve is symmetric with respect to the origin if $F(x,y) =$ ____________ . It is symmetric with respect to the __________ if $F(x,y) = F(-x,y)$. Finally, it is symmetric with respect to the __________________ if $F(x,y) = F(y,x)$.

3. The extent of a curve in the x or y direction may be limited by the condition that ____________ numbers do not have ____________________ .

4. The x-intercepts are obtained by setting ________ in the equation and solving for ___ ; the _______________ by setting $x = 0$ and solving for ___ .

5. A fixed line is an asymptote to a curve if the distance between the point $P(x,y)$ on the curve and the line tends to ________ as P moves farther and farther away from ______________ .

6. Consider the curve given by $y^2 = \dfrac{4x}{x^2+1}$. Since y enters into the equation in even powers only, the curve is symmetric with respect to the __________ . It __________ symmetric with respect to the origin or the __________ . Since y^2 is never negative, x can never be ____________ . When $x = 0$, $y =$ ___ and when $y = 0$, $x =$ ___ . Therefore, the intercepts are at the origin. For $y \geq 0$,

$$y = f(x) = \frac{2\sqrt{x}}{\sqrt{x^2+1}} , \text{ and we find}$$

---

1. x-axis
2. $F(-x,-y)$, y-axis, 45° line $y = x$
3. negative, real even roots
4. $y = 0$, x, y-intercepts, y
5. zero, the origin
6. x-axis, is not, y-axis, negative, 0, 0, 0, x-axis, 4, $\dfrac{4(1-x^2)}{(x+1)^2}$, $\dfrac{2(1-x^2)}{y(x^2+1)^2}$, ∞ , vertical, 1, $\left(1, \sqrt{2}\right)$, minimum

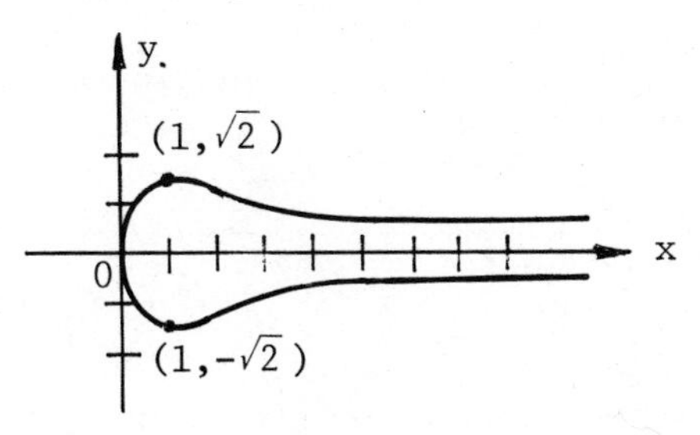

$$\lim_{x \to \infty} f(x) = \lim_{x \to \infty} \frac{\frac{2}{\sqrt{x}}}{\sqrt{1 + 1/x^2}} = \underline{\quad} .$$

Thus the ________ is a horizontal asymptote. Differentiating the original equation implicitly yields,

$$2yy' = \frac{(x^2+1)(\underline{\quad}) - 4x(2x)}{\left(x^2+1\right)^2} = \underline{\qquad\qquad} .$$

Solving for y', $y' = \underline{\qquad\qquad}$. Thus, as $y \to 0$, $y' \to$ ___, and the curve has a __________ tangent at the origin. Also, $y' = 0$ when $x =$ ___, so the point ________ is a maximum point on the curve and the point $\left(1, -\sqrt{2}\right)$ is a __________ point. Sketch the graph.

8-2 Tangents and Normals.

OBJECTIVE A: Find the tangent line and the normal line to a curve
(a) at a given point on the curve or
(b) having some specified slope.

7. To find the tangent to the curve $x^3 + 3xy^3 + xy^2 = xy$ at the point $(1,-1)$ it is first necessary to verify that the point does in fact belong to the curve: $(1)^3 + 3(1)(-1)^3 + (1)(-1)^2 = (1)(-1)$ ___ true, so the point _______ belong to the curve. Next, we differentiate the equation of the curve implicitly: $3x^2 + 3y^3 + 9xy^2y' + y^2 + 2xyy' = \underline{\qquad\qquad}$, or solving for the derivative,

$$y' = \frac{dy}{dx} = \frac{y - 3x^2 - 3y^3 - y^2}{\underline{\qquad\qquad}} .$$

Thus, at the point $(1,-1)$ the slope of the tangent is,

$$\left.\frac{dy}{dx}\right|_{(1,-1)} = \frac{-1 - 3 + 3 - 1}{\underline{\qquad\qquad}} = \underline{\quad} .$$

Therefore, an equation of the tangent line is

$$y + 1 = \underline{\qquad\qquad} \quad \text{or} \quad x + 3y = \underline{\quad} .$$

---

7. is, does, $y + xy'$, $9xy^2 + 2xy - x$, $9 - 2 - 1$, $-\frac{1}{3}$, $-\frac{1}{3}(x-1)$, -2

8. For the slope of the normal line to the curve in Problem 7 at (1,-1) we have, slope of the normal line $= -\left(\overline{\qquad}\right) = \_\_\_$.

Therefore, an equation of the normal line is

$y + 1 = \_\_\_\_\_\_$ or $y - 3x = \_\_\_$.

9. Find the tangent line to the curve $x^3 + y^3 = 9xy$ parallel to the line $y = 0$ when $x \neq 0$.

Solution: The slope of the line $y = 0$ is ___ . Differentiating the equation of the curve implicitly, we find

$3x^2 + 3y^2 \frac{dy}{dx} = \_\_\_\_\_\_\_\_$, or $\frac{dy}{dx} = \_\_\_\_\_\_$. Hence,

$dy/dx = 0$ when $\_\_\_\_\_ = 0$ or $y = \_\_\_$. For $x \neq 0$, substitution of $y = \frac{1}{3}x^2$ into the equation of the curve gives

$$x^3 + \left(\frac{1}{3}x^2\right)^3 = \_\_\_\_ \quad \text{or,} \quad x^3 = \_\_\_\_\_ .$$

Thus, $x = \_\_\_\_\_ \approx 3.78$.

This value yields $y = \frac{1}{3}x^2 \approx 4.76$.

Therefore, an equation of the tangent line is ________ .

A sketch of the curve is shown at the right. The curve is known as the <u>Folium of Descartes</u>.

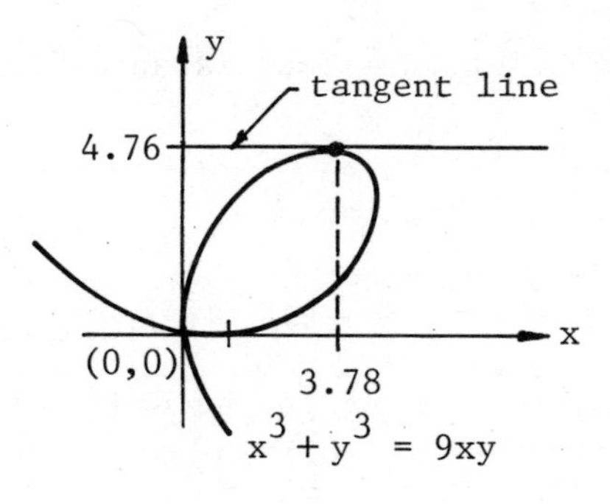

OBJECTIVE B: Find the angle between two given curves at a point where they intersect.

10. Find the angles between the curves $y = x^2$ and $y + x^2 = 2$ at their points of intersection.

Solution. First we calculate the points of intersection of the two curves: $y = x^2$ and $y = 2 - x^2$ intersect when $x^2 = 2 - x^2$ or $x^2 = \_\_\_$.

---

8. $1/(-1/3)$, 3, $3(x-1)$, -4

9. 0, $9x\frac{dy}{dx} + 9y$, $\frac{3y - x^2}{y^2 - 3x}$, $3y - x^2$, $\frac{1}{3}x^2$, $3x^3$, $2(27)$, $3(2)^{1/3}$, $y = 4.76$

10. 1, (1,1) and (-1,1), $-2x$, 2, -2, $\frac{4}{3}$, $\tan^{-1}\frac{4}{3}$, $-\frac{4}{3}$, -0.93 radians

This yields the two points ______ and ______ . For the first curve $y' = 2x$ and for the second curve $y' =$ ____ . The angle ϕ between the two curves at $(1,1)$ satisfies

$$\tan\phi = \frac{m_2 - m_1}{1 + m_2 m_1}, \quad \text{where} \quad m_1 = 2x\Big|_{(1,1)} = \_\_\_ \quad \text{and}$$

$$m_2 = -2x\Big|_{(1,1)} = \_\_\_\_ . \quad \text{Hence} \quad \tan\phi\Big|_{(1,1)} = \_\_\_\_ , \quad \text{so}$$

$\phi =$ __________ $\approx$ 0.93 radians.

Similarly, $\tan\phi\Big|_{(-1,1)} =$ ____ , yielding $\phi =$ ______________ .

8-3 Distance Between Two points; Equations.

OBJECTIVE: Use the distance formula to derive an equation for a set of points $P(x,y)$ that satisfy some specified condition.

11. Find an equation for the points $P(x,y)$ whose distances from $F_1(0,-9)$ and $F_2(0,9)$ have the constant sum 30 .

Solution. The distance from P to F_1 is ________________ , and from P to F_2 is ________________ . Thus, the specified conditions yields the equation

$$\sqrt{x^2 + (y-9)^2} + \_\_\_\_\_\_\_\_\_\_\_\_\_\_\_\_ = 30 .$$

Transposing the second term on the left, to the right side, and squaring both sides gives

$$x^2 + (y-9)^2 = 900 + 60\sqrt{x^2 + (y+9)^2} + \_\_\_\_\_\_\_\_\_\_\_\_\_\_\_$$

or, simplifying algebraically,

$$\sqrt{x^2 + (y+9)^2} = \frac{1}{10}(\_\_\_\_\_\_\_\_\_\_) .$$

Squaring both sides of this last equation we find,

$$x^2 + y^2 + 18y + 81 = \frac{1}{100}\left(\_\_\_\_\_\_\_\_\_\_\_\_\_\_\_\_\_\_\_\_\_\_\_\right),$$

or simplifying algebraically, $100x^2 + 64y^2 =$ _______ . Alternatively,

$$\frac{x^2}{144} + \_\_\_\_\_ = 1 .$$

---

11. $\sqrt{x^2+(y+9)^2}$, $\sqrt{x^2+(y-9)^2}$, $\sqrt{x^2+(y+9)^2}$, $x^2+(y+9)^2$, $-6y-150$, $36y^2 + 1800y + 22500$, 14400, $y^2/225$

12. Find an equation for the points $P(x,y)$ whose distances from the line $y = -3$ are always 4 units less than their distance from the point $F(0,7)$.

Solution. The distance from P to the line is __________ ,

and from P to F is ______________ . Thus, the specified condition

yields the equation $\sqrt{(y+3)^2}$ = ____________________ , or

$\sqrt{x^2 + (y-7)^2}$ = _______________ . Squaring both sides,

$x^2 + y^2 - 14y + 49$ = ___________________________ , or

$x^2 - 20y + 24$ = ______________ .

If $y + 3 \geq 0$, or $y \geq -3$, this equation gives $x^2 - 20y + 24$ = _________ ,
or simplifying algebraically, x^2 = _____ . On the other hand,
if $y + 3 < 0$, or $y < -3$, the equation gives $x^2 - 20y + 24$ = _________
or x^2 = _________ . But for $y < -3$ this last equality would say that x^2 is a negative number. This is clearly impossible so y must be greater than or equal to -3, and a valid equation for the points $P(x,y)$ is

__________ .

8-4 The Circle.

OBJECTIVE A: Find an equation for a circle whose center and radius are known.

13. An equation of the circle with center $C(h,k)$ and radius r is given by

________________________ .

14. An equation of the circle with center $C(-1,3)$ and radius $r = \sqrt{11}$
is ________________________ or $x^2 + y^2 + 2x - 6y +$ (___) = 0 .

---

12. $\sqrt{(y+3)^2} = |y+3|$, $\sqrt{x^2 + (y-7)^2}$, $\sqrt{x^2 + (y-7)^2} - 4$, $\sqrt{(y+3)^2} + 4$,

$(y+3)^2 + 8\sqrt{(y+3)^2} + 16$, $8\sqrt{(y+3)^2}$, $8(y+3)$, $28y$, $-8(y+3)$,

$12y - 48$, $x^2 = 28y$

13. $(x-h)^2 + (y-k)^2 = r^2$

14. $(x+1)^2 + (y-3)^2 = 11$, -1

OBJECTIVE B: Given an equation representing a circle, find the coordinates of its center and the radius.

15. Consider the circle $2x^2 + 2y^2 - 8x + 5y + 8 = 0$. Transposing the constant term and dividing by 2 , we find $x^2 - 4x + (\_\_\_\_\_\_) = -4$.

Completing the squares, $(x-2)^2 + (\_\_\_\_\_\_) = -4 + \frac{25}{16} + \_\_\_$.

Therefore, the center is ________ and the radius is ____ .

OBJECTIVE C: Find an equation of the circle passing through three given noncollinear points.

16. To find the circle determined by the three points A(4,-1), B(-2,1), and C(6,5), substitute the coordinates of each point into the equation $C_1x + C_2y + C_3 = -(x^2+y^2)$. This gives the system of equations:

$4C_1 - C_2 + C_3 = -(16+1)$ from A(4,-1),

____________________ from B(-2,1),

____________________ from C(6,5) .

Solving by the method of elimination: subtraction of the second equation from the first gives ____________ and the second from the third, ____________ . These last two equations may be solved for C_1 and C_2 as follows: 2 times the first added to the second gives $12C_1 + 8C_1 = \_\_\_\_\_\_\_\_$ or $C_1 = \_\_\_$. Substitution of this value of C_1 into the first equation $6C_1 - 2C_2 = -12$ gives $C_2 = -\frac{1}{2}(-12 + \_\_\_)$ or $C_2 = \_\_\_$. Substitution of these two values into the equation $4C_1 - C_2 + C_3 = -17$ gives $C_3 = \_\_\_$. Therefore, the required circle is

________________________ .

---

15. $y^2 + \frac{5y}{2}$, $\left(y + \frac{5}{4}\right)^2$, 4, $\left(2, -\frac{5}{4}\right)$, $\frac{5}{4}$

16. $-2C_1 + C_2 + C_3 = -(4+1)$, $6C_1 + 5C_2 + C_3 = -(36+25)$, $6C_1 - 2C_2 = -12$,

$8C_1 + 4C_2 = -56$, $-24 - 56$, -4, 24, -6, -7,

$x^2 + y^2 - 4x - 6y - 7 = 0$

OBJECTIVE D: Find an equation of a circle given three prescribed conditions on the circle (e.g., two points and a tangent at one of them).

17. Find the circle which is tangent to the line $2x + 3y = 12$ at $(3,2)$ and passes through the point $(6,-1)$

Solution. The normal line to the circle at $(3,2)$ has slope ____ , so an equation of the normal line is $y - 2 =$ ________ , or $3x - 2y =$ ___ .

Given any chord of the circle, the line joining its midpoint to the center is perpendicular to the chord. In particular, the midpoint of the chord joining the two points $(3,2)$ and $(6,-1)$ is ________ . Since the slope of the chord is $-3/3$, an equation of the perpendicular line joining the center $C(x,y)$ and midpoint is $y - \frac{1}{2} =$ ________ or $x - y =$ ___ .

Now, the center of the circle is located at the intersection of the two lines we have constructed: solving simultaneously,

$$\left.\begin{aligned} 3x - 2y &= 5 \\ x - y &= 4 \end{aligned}\right\} \quad \text{yields} \quad x = -3 \quad \text{and} \quad y = \_\_\_ .$$

The radius r is the distance from the center to either one of the two given points, say $(3,2)$; thus,

$$r = \sqrt{36 + \_\_\_} = \_\_\_\_ .$$

Therefore, the required circle is

____________________ .

A diagram illustrating our procedure is given at the right.

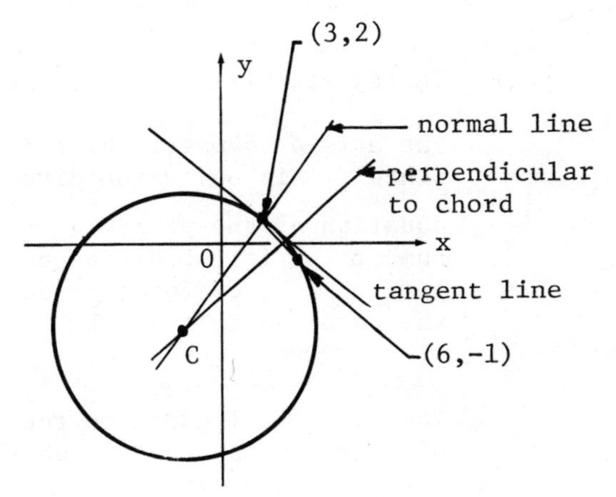

8-5 The Parabola.

OBJECTIVE A: Given the coordinates for the vertex V and the focus F of a parabola, both of which lie along a line parallel to a coordinate axis, find an equation of the parabola and of its directrix. Sketch the graph showing the focus, vertex, and directrix.

18. Vertex at $V(1,2)$ and focus at $F(3,2)$

The axis of symmetry of the parabola is the line containing both V and F , or the line with equation ______ . Since V is to the left of F on the

---

17. $\frac{3}{2}$, $\frac{3}{2}(x-3)$, 5, $\left(\frac{9}{2}, \frac{1}{2}\right)$, $x - \frac{9}{2}$, 4, -7, 81, $\sqrt{117}$, $(x+3)^2 + (y+7)^2 = 117$

axis of symmetry, the parabola will open to the ________ . Thus, an equation of the parabola has the form

____________________ . To calculate the number p , note that it is the distance between the vertex and the focus: thus, p = ___ . Therefore, an equation of the parabola is given by

____________________ .

The directrix is given by ________ .

Graph the equation showing focus, vertex, and directrix.

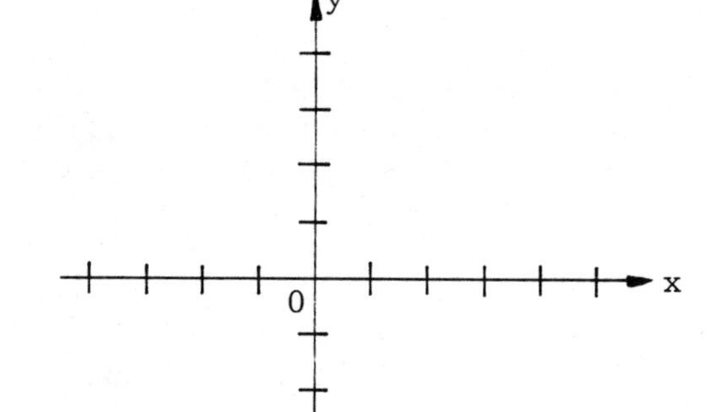

OBJECTIVE B: Given the coordinates for the vertex V , and the directrix L parallel to one of the coordinate axes, find an equation of the parabola so determined and give the coordinates of its focus. Sketch the graph showing the focus, vertex, and directrix.

19. Vertex at V(-1,2) and directrix L: y = 3

The axis of symmetry of the parabola is ________________ to the directrix. Since V is below the directrix, the parabola opens ___________ and so an equation of the parabola has the form ____________________ . The number p is the distance from the vertex to the directrix: thus, p = ___ . Therefore, an equation of the parabola is given by

____________________ .

The focus is located on the axis of symmetry p units from the vertex in the direction which the parabola opens. Thus the focus is _________ . Graph the equation showing the focus, vertex, and directrix.

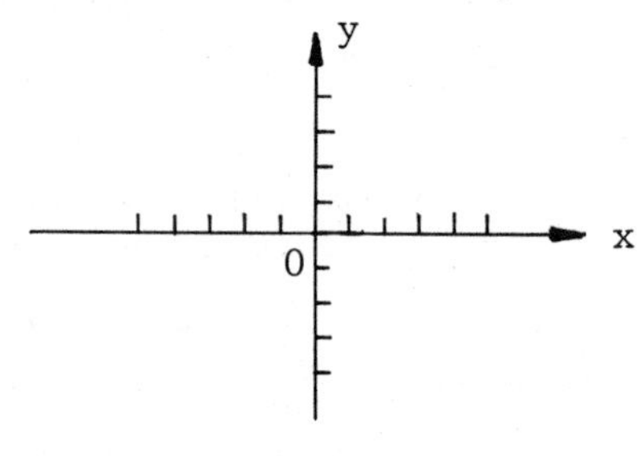

---

18. y = 2, right, $(y-k)^2 = 4p(x-h)$, 2, $(y-2)^2 = 8(x-1)$, x = -1

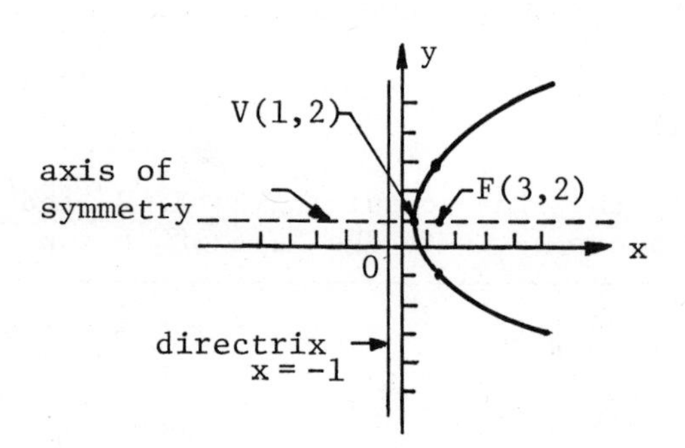

19. perpendicular, downward, $(x-h)^2 = -4p(y-k)$, 1, $(x+1)^2 = -4(y-2)$, F(-1,1)

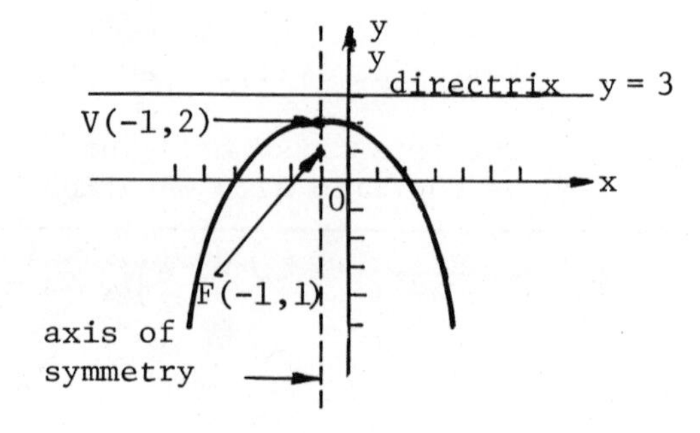

OBJECTIVE C: Given an equation of a parabola, find the vertex, axis of symmetry, focus, and directrix. Sketch the graph showing these features.

20. Consider the parabola $x^2 - 2x - 10y + 6 = 0$. Completing the square in the quadratic terms, _________ $- 10y + 6 = 1$ or _________ $= 10$ (_______) . Therefore, the vertex of the parabola is _________ . The axis of symmetry is _______ because the parabola opens _________ . $4p =$ ___ , so that $p =$ ____ . Therefore, the focus of the parabola is _________ , and the directrix is given by _______ . Sketch the graph showing these features.

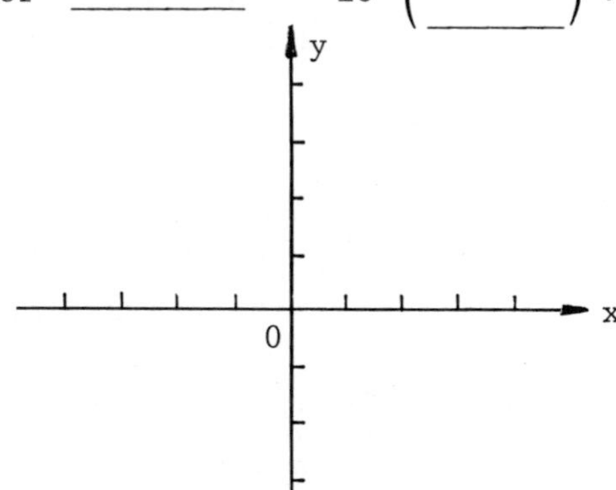

8-6 The Ellipse.

21. In analyzing an ellipse, the intercepts are located on the axes of _________. The major axis is always in the direction of _________ axis length, and the foci always lie on the ________ axis.

22. If we use the letters a, b, and c to represent the lengths of the semimajor axis, semiminor axis, and half-distance between the foci, then it is always true that $c^2 =$ _________ .

23. The eccentricity of an ellipse is the ratio $e =$ _____ . It varies between __________ and measures the degree of departure of the ellipse from circularity: when $e = 0$, the ellipse is a ________ , and when $e =$ ___ it reduces to the line segment F_1F_2 joining the two foci.

---

20. $x^2 - 2x + 1$, $(x - 1)^2$, $y - \frac{1}{2}$,
 $V\left(1, \frac{1}{2}\right)$, $x = 1$, upward,
 10, $\frac{5}{2}$, $F(1,3)$, $y = -2$

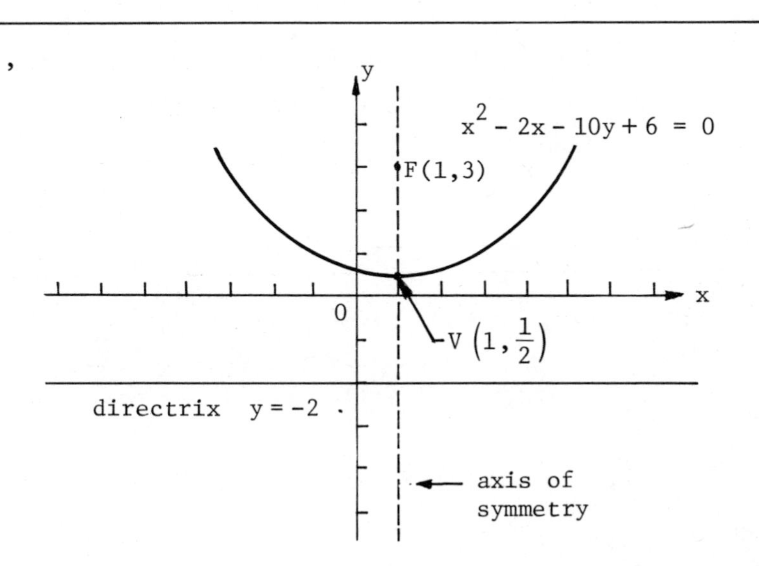

21. symmetry, largest, major

22. $a^2 - b^2$

23. c/a, 0 and 1, circle, 1

OBJECTIVE A: Given an equation of an ellipse, find its center, vertices, and foci. Identify the major and minor axes, and sketch the graph showing all these features.

24. Consider the ellipse given by $9x^2 + y^2 - 18x + 2y + 9 = 0$. Thus, $9(x^2 - 2x) + (\_\_\_\_\_\_) = -9$, so completing the squares in each set of parentheses gives $9(x-1)^2 + \_\_\_\_\_\_ = 1$, or

$$\frac{(x-1)^2}{1/9} + \_\_\_\_\_\_ = 1 .$$

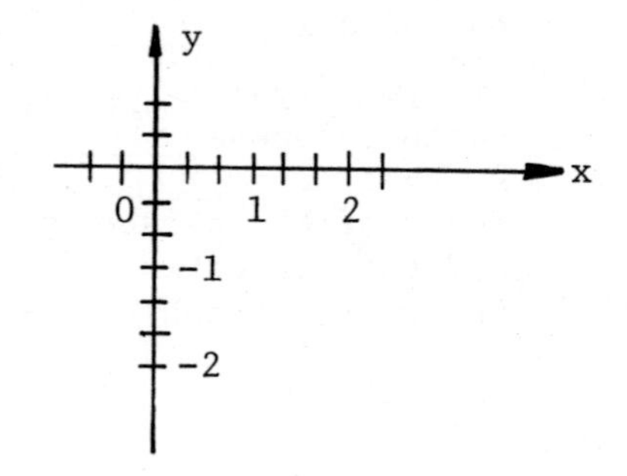

Hence, $c^2 = a^2 - b^2 = \_\_\_\_\_\_$, or $c = \_\_\_\_\_\_$. The center of the ellipse is ______ and the foci are $F_1\left(1, -1 + \frac{2\sqrt{2}}{3}\right)$ and ______ . The length of the semimajor axis is $a = 1$ and the length of the semiminor axis is $b = \_\_\_\_$.
The vertices are the four points $(1,0)$, $\left(\frac{2}{3}, -1\right)$, ______ , and ______ .
The eccentricity of the ellipse is $e = \_\_\_\_\_\_$. Sketch the graph.

OBJECTIVE B: Find an equation of an ellipse having a given center C , focus F , and semimajor axis a . Give the eccentricity of the ellipse and sketch its graph.

25. Center at $C(6,-1)$, focus $F\left(6 + 3\sqrt{3}, -1\right)$, $a = 6$

The two foci are located along the major axis at a distance c units from the center of the ellipse. Thus, $c = \_\_\_\_\_\_$. Since $b^2 = a^2 - c^2$, we find $b^2 = 36 - \_\_\_$ so $b^2 = \_\_\_$. Thus, an equation of the ellipse is given by

____________ .

---

24. $y^2 + 2y$, $(y+1)^2$, $\frac{(y+1)^2}{1}$,
$\frac{8}{9}$, $\frac{2\sqrt{2}}{3}$, $C(1,-1)$, $F_2\left(1, -1 - \frac{2\sqrt{2}}{3}\right)$,
$\frac{1}{3}$, $(1,-2)$, $\left(\frac{4}{3}, -1\right)$, $\frac{2\sqrt{2}}{3}$

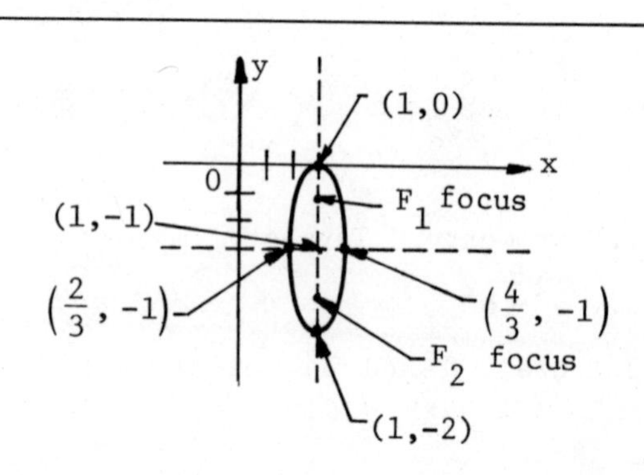

The second focus of the ellipse is  .

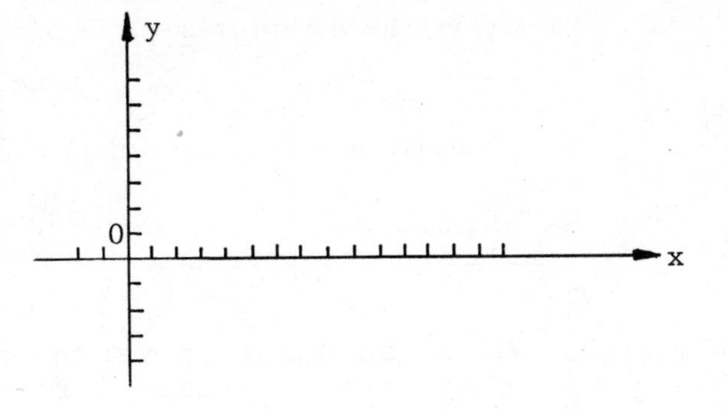

The vertices are the points (6,2), (12,-1), ______ , and ______ . The eccentricity is

e = ______ .

Sketch the graph.

26. To find an equation of the ellipse with center C(-4,-2) , $c = \sqrt{7}$, and one vertex of the major axis at (0,-2) , we note first that the major axis is parallel to the ___ axis. The value of a is the distance between the points C(-4,-2) and ______ , so a = ____ . Therefore, $b^2 = a^2 - c^2$ = ___ - 7 = ___ . Thus, an equation of the ellipse is given by

______________________ .

The eccentricity is e = _____ .

8-7 The Hyperbola.

27. The only differences between the equation of the ellipse and the equation of the hyperbola are the ____________ in the equation of the hyperbola, and the new relation among a, b, and c ; namely, $a^2 - c^2$ = _____ .

28. There is no restriction $a > b$ for the hyperbola as there is for the ellipse. The direction in which the hyperbola opens is controlled by the ________ rather than by the relative ________ of the coefficients of the quadratic terms.

---

25. $3\sqrt{3}$, 27, 9,

$$\frac{(x-6)^2}{36} + \frac{(y+1)^2}{9} = 1 ,$$

$F_2\left(6 - 3\sqrt{3}, -1\right)$, (6,-4),

(0,-1), $\frac{3\sqrt{3}}{6}$

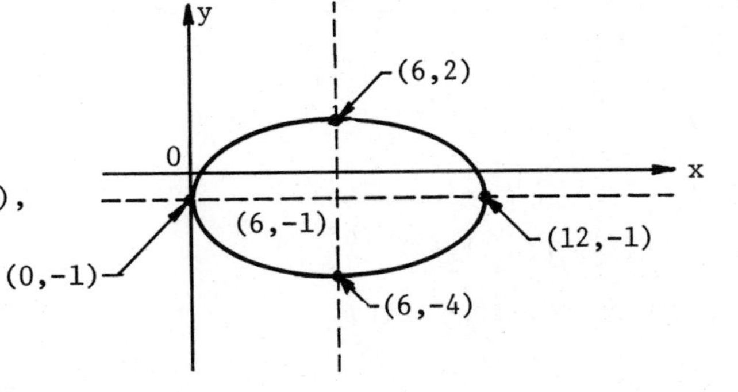

26. x, (0,-2), 4, 16, 9, $\frac{(x+4)^2}{16} + \frac{(y+2)^2}{9} = 1$, $\frac{\sqrt{7}}{4}$

27. minus sign, $-b^2$

28. signs, sizes

29. To obtain the asymptotes for a hyperbola, set the expression ________ or ________ equal to zero (depending on how the hyperbola is defined), and then factor and solve the resulting quadratic equation.

30. The eccentricity for a hyperbola is defined as $e =$ ____ , and is always ________________ one.

OBJECTIVE: Given an equation representing a hyperbola, find the center, vertices, foci, and asymptotes. Sketch a graph showing all these features.

31. Consider the hyperbola given by $\frac{(x-2)^2}{4} - \frac{(y+1)^2}{5} = 1$. Because of the minus sign associated with the y terms, the hyperbola opens ________ and ________. The center of the hyperbola is $C(h,k) =$ ________. Now, $a^2 =$ ___ and $b^2 =$ ___ , so $c^2 = a^2 + b^2 =$ ___ . The coordinates of the foci are $F_1(h-c, k)$ and $F_2(h+c, k)$ or ________ and ________. The coordinates of the vertices are $V_1(h-a, k)$ and $V_2(h+a,k)$ or ________ and ________. Equations may be found for the asymptotes by setting

$$\frac{(x-2)^2}{4} - \frac{(y+1)^2}{5}$$ equal to ______ .

Thus, equations of the asymptotes are

________________ .

The eccentricity is $e =$ ______ .

Sketch the graph of the hyperbola.

---

29. 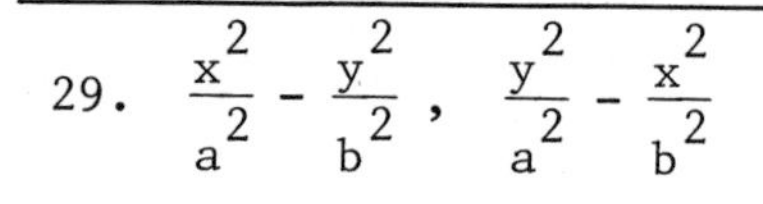
$\frac{x^2}{a^2} - \frac{y^2}{b^2}$, $\frac{y^2}{a^2} - \frac{x^2}{b^2}$

30. c/a, greater than or equal to

31. right, left, C(2,-1), 4, 5, 9, $F_1(1,-1)$, $F_2(5,-1)$, $V_1(0,-1)$, $V_2(4,-1)$, zero,

$$y+1 = \pm\frac{\sqrt{5}}{2}(x-2),$$

$c/a = 3/2$

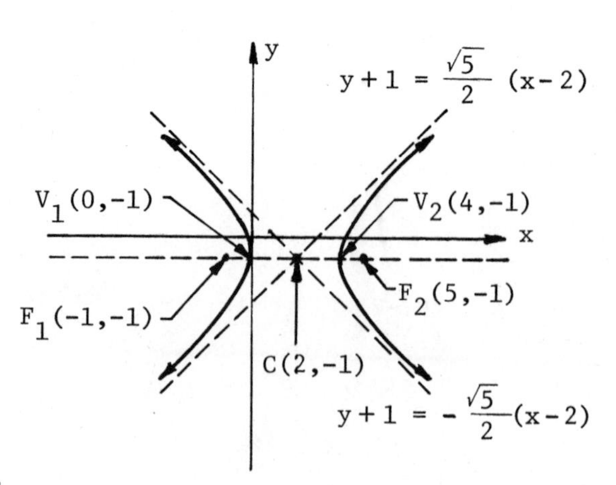

32. Consider the hyperbola $9x^2 - 4y^2 + 18x + 16y + 29 = 0$. Thus,

$9(x^2 + 2x) - 4(\_\_\_\_\_\_) = -29$. Completing the squares in each set of

parentheses gives $9(x+1)^2 - 4(\_\_\_\_\_\_) = -29 + 9 - \_\_\_\_ = \_\_\_\_$, or

$\frac{(y-2)^2}{9} - \frac{\qquad}{\qquad} = 1$.

Because of the minus sign, the hyperbola opens _____ and ______ . The center of the hyperbola is $C(h,k) =$ __________ . The coordinates of the vertices are

$V_1(h, k-a)$ and $V_2(h, k+a)$ or

__________ and __________ . Now,

$c^2 = a^2 + b^2 =$ ____ , and the coordinates

of the foci are $F_1(h, k-c)$ and $F_2(h, k+c)$

or ______________ and ______________ .

Equations for the asymptotes may be found by setting the expression

$\frac{(y-2)^2}{9} - \frac{(x+1)^2}{4}$ equal to ______ . Thus, equations of the asymptotes

are ____________________ . The eccentricity is $e =$ ________ .

Sketch the graph showing the features of the hyperbola.

---

32. $y^2 - 4y$, $(y-2)^2$, 16, -36,

$\frac{(x+1)^2}{4}$, up, down, $C(-1,2)$,

$V_1(-1,-1)$, $V_2(-1,5)$, 13,

$F_1\left(-1, 2 - \sqrt{13}\right)$, $F_2\left(-1, 2 + \sqrt{13}\right)$,

zero, $y - 2 = \pm \frac{3}{2}(x+1)$,

$\frac{c}{a} = \frac{\sqrt{13}}{3}$

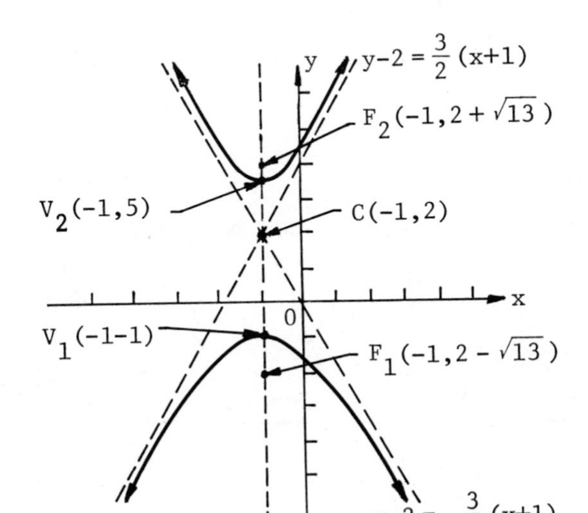

33. Any second degree equation in x and y represents a circle, parabola, ellipse, or hyperbola (although it may degenerate). To find the curve given its equation $Ax^2 + Bxy + Cy^2 + Dx + Ey + F = 0$,

(1) First rotate axes, to force $B = 0$, through an angle α satisfying $\cot 2\alpha =$ ______ .

(2) Next, ____________ axes by completing the squares (if necessary) to reduce the equation to a standard form.

OBJECTIVE: Given an equation of the form $Ax^2 + Bxy + Cy^2 + F = 0$, transform the equation by a rotation of axes into an equation that has no cross-product term.

34. Consider the equation $5x^2 - 2\sqrt{3}\,xy + 7y^2 = 6$. The required angle of rotation to eliminate the xy term satisfies

$\cot 2\alpha = \dfrac{5-7}{\_\_\_\_} =$ ____ ; hence $2\alpha =$ ___ radians, or $\alpha =$ ___ .

Thus, $\sin\alpha =$ ___ and $\cos\alpha =$ ___ . The equations for the rotation of axes are

$x = \frac{\sqrt{3}}{2}x' - \frac{1}{2}y'$ and ____________________ .

Substitution for x and y into the given second-degree equation gives

$$5\left(\tfrac{3}{4}x'^2 - \_\_\_\_\_\_ + \tfrac{1}{4}y'^2\right) - 2\sqrt{3}\left(\tfrac{\sqrt{3}}{4}x'^2 + \_\_\_\_\_\_ - \tfrac{\sqrt{3}}{4}y'^2\right)$$

$$+\ 7\left(\tfrac{1}{4}x'^2 + \_\_\_\_\_\_ + \tfrac{3}{4}y'^2\right) = 6$$

Simplifying algebraically,

$$\left(\tfrac{15}{4} - \_\_\_ + \tfrac{7}{4}\right)x'^2 + \left(\_\_\_ + \tfrac{3}{2} + \tfrac{21}{4}\right)y'^2 = 6\text{, or}$$

____________________ . This is an equation of ______________ .

---

33. $\frac{A-C}{B}$, translate

34. $-2\sqrt{3}$, $\frac{1}{\sqrt{3}}$, $\frac{\pi}{3}$, $\frac{\pi}{6}$, $\frac{1}{2}$, $\frac{\sqrt{3}}{2}$, $y = \frac{1}{2}x' + \frac{\sqrt{3}}{2}y'$, $\frac{\sqrt{3}}{2}x'y'$, $\frac{1}{2}x'y'$, $\frac{\sqrt{3}}{2}x'y'$, $\frac{3}{2}$, $\frac{5}{4}$, $4x'^2 + 8y'^2 = 6$, an ellipse

35. Consider the equation $3x^2 + 2xy + y^2 = 8$. The angle of rotation for elimination of the cross-product term satisfies

$$\cot 2\alpha = \frac{\_\_\_}{-2} = \_\_\_ ; \quad \text{thus,} \quad 2\alpha = \_\_\_ \text{ radians.}$$

It follows that $\cos 2\alpha =$ ____ . Thus, using the trigonometric half-angle formulas we find

$$\sin \alpha = \sqrt{\frac{1 - \cos 2\alpha}{2}} = \_\_\_\_\_\_\_\_ , \quad \text{and}$$

$$\cos \alpha = \sqrt{\frac{1 + \cos 2\alpha}{2}} = \_\_\_\_\_\_\_\_ .$$

Therefore, the rotational equations are

$$x = \frac{1}{2}\sqrt{2 - \sqrt{2}}\, x' - \frac{1}{2}\sqrt{2 + \sqrt{2}}\, y' \quad \text{and}$$

________________________ .

Substitution for x and y into the original second degree equation gives,

____________________ .

This equation represents __________ .

8-9 Invariants and the Discriminant.

OBJECTIVE A: Given a second degree equation of the form
$Ax^2 + Bxy + Cy^2 + Dx + Ey + F = 0$,
use the discriminant to classify it as representing a circle, an ellipse, a parabola, or a hyperbola.

36. The discriminant is the invariant expression ________ . Both the discriminant and the expression _____ are invariant under rotations of axes.

37. If the discriminant is positive, the equation represents __________ .

---

35. $3 - 1$, -1, $\frac{3\pi}{4}$, $-\frac{\sqrt{2}}{2}$, $\frac{1}{2}\sqrt{2 + \sqrt{2}}$, $\frac{1}{2}\sqrt{2 - \sqrt{2}}$,

$y = \frac{1}{2}\sqrt{2 + \sqrt{2}}\, x' + \frac{1}{2}\sqrt{2 - \sqrt{2}}\, y'$, $\left(2 - \sqrt{2}\right)x'^2 + \left(2 + \sqrt{2}\right)y'^2 = 8$,

an ellipse   36. $B^2 - 4AC$, $A + C$ 37. a hyperbola

38. If the discriminant is zero, the equation represents ________________ .

39. If the discriminant is negative, the equation represents ________________ .

40. In order that the equation represent a circle, it is necessary that the discriminant be ___________ and that ________ .

41. Consider the equation given by $2x^2 - 4xy - y^2 + 20x - 2y + 17 = 0$. The discriminant is $B^2 - 4AC = 16 - (\_\_\_) = \_\_\_$. Thus, the equation represents ________________ .

42. For $4x^2 - 12xy + 9y^2 - 52x + 26y + 81 = 0$ the discriminant is $B^2 - 4AC = \_\_\_\_ - 4(36) = \_\_\_$. Thus, the equation represents ________________ .

OBJECTIVE B: Use the invariants $B^2 - 4AC$ and $A + C$ to determine an equation to which $Ax^2 + Bxy + Cy^2 + F = 0$ reduces when the axes are rotated to eliminate the cross-product term.

43. Consider again the equation $3x^2 + 2xy + y^2 = 8$, as in Problem 35 above. Now, $B^2 - 4AC = \_\_\_$ and $A + C = \_\_\_$. Thus, $A' + C' = \_\_\_$ and $-4A'C' = \_\_\_$ or $C' = \_\_\_\_$. Substitution into the first equation $A' + C' = 4$ gives $4 = A' + C' = A' + \_\_\_\_$ or $A'^2 - 4A' + 2 = 0$. Solving this quadratic equation for A' gives $A' = \_\_\_\_\_$ or $A' = \_\_\_\_\_$; thus, $C' = 4 - A'$ is given by $C' = \_\_\_\_\_\_$ or $C' = \_\_\_\_\_\_$.

Thus, we find an equation ________________________________ or

________________________________ with no cross-product term.

8-10 Sections of a Cone.

44. The circle, parabola, ellipse, and hyperbola are known as ________________ because each may be obtained by cutting a ________ by a _________ .

---

38. a parabola 39. an ellipse 40. negative, $A = C$

41. -8, 24, a hyperbola 42. 144, 0, a parabola

43. -8, 4, 4, -8, $\frac{2}{A'}$, $\frac{2}{A'}$, $2 + \sqrt{2}$, $2 - \sqrt{2}$, $2 - \sqrt{2}$, $2 + \sqrt{2}$,
$\left(2 + \sqrt{2}\right)x'^2 + \left(2 - \sqrt{2}\right)y'^2 = 8$ or $\left(2 - \sqrt{2}\right)x'^2 + \left(2 + \sqrt{2}\right)y'^2 = 8$

44. conic sections, cone, plane

45. The type of section is determined by the relationship between the angle α the cutting plane makes with the ________________ and the angle β which __________ the cone.

46. The section is

________________ , if $\alpha = \beta$;

________________ , if $0 \le \alpha < \beta$;

________________ , if $\alpha = 90°$

________________ , if $\beta < \alpha < 90°$.

CHAPTER 8 OBJECTIVE - PROBLEM KEY

Objective	Problems in Thomas/Finney Text	Objective	Problems in Thomas/Finney Text
8-1	p. 394, 1-10	8-5 A	p. 410, 1-6
8-2 A	p. 398, 1-4	B	p. 410, 7-12
B	p. 399, 13,17,20,21	C	p. 410, 13-22
8-3	p. 400, 1-8	8-6 A	p. 417, 7,8
8-4 A	p. 404, 1-4	B	p. 417, 1-5
B	p. 404, 7-10	8-7	p. 425, 1-8
C	p. 404, 17	8-8	p. 429, 1-6, 11
D	p. 404, 13-15, 21	8-9 A	p. 432, 1-9

CHAPTER 8 SELF-TEST

1. Analyze the equation $x^{2/3} + y^{2/3} = 1$ to investigate symmetry, extent, intercepts, asymptotes, and slope at the intercepts. Sketch the graph.

2. Find equations for the tangent line and the normal line to the curve $9y^2 = x^3 + 3x^2$ at the point $\left(1, \frac{2}{3}\right)$.

3. Find the angle between the circle $x^2 + y^2 = 2$ and the hyperbola $x^2 - y^2 = 1$ in the first quadrant.

4. Find an equation for the points $P(x,y)$ whose distances from the point $(-4,2)$ are always twice their distance from the point $(6,-2)$.

5. Write an equation for the circle with center $C(4,6)$ and radius $r = 3$.

6. Find an equation of the circle with center at $C(1,2)$ and tangent to the line $x + 2y = 10$.

---

45. axis of the cone, generates 46. a parabola, a hyperbola, a circle, an ellipse

7. Find an equation of the circle passing through the three points $A(4,2)$, $B(6,0)$, and $C(0,-6)$.

8. Find an equation of the circle inscribed in the triangle formed by the lines $y = 0$, $x = 0$, and $x + y = 8$.

9. Find an equation of the parabola with vertex $V(2,0)$ and focus the origin. Sketch the graph.

10. Find an equation of the parabola whose directrix is the line $y = -5$ with vertex $V(4,-3)$. Sketch.

11. Find the vertex, the focus, and the directrix of the parabola $x^2 - 4x + 6y + 34 = 0$, and sketch the graph.

12. Find the center, vertices, foci, and eccentricity of the ellipse $9x^2 + 16y^2 - 54x + 128y + 193 = 0$, and sketch the graph.

13. Write an equation of the ellipse with center $C(2,5)$, focus $F(-1,5)$, and semiminor axis $= 4$.

14. Find the center, vertices, foci, and asymptotes of the hyperbola $9x^2 - 4y^2 - 18x + 32y - 91 = 0$, and sketch the graph.

15. Consider the equation $x^2 - 2xy - y^2 = 12$.
    (a) Use the discriminant to classify it.
    (b) Transform the equation by a rotation of axes into an equation with no cross-product term.

16. Use the invariants $B^2 - 4AC$ and $A + C$ to determine an equation to which $x^2 + xy + y^2 = 10$ reduces when the axes are rotated to eliminate the cross-product term.

SOLUTIONS TO CHAPTER 8 SELF-TEST

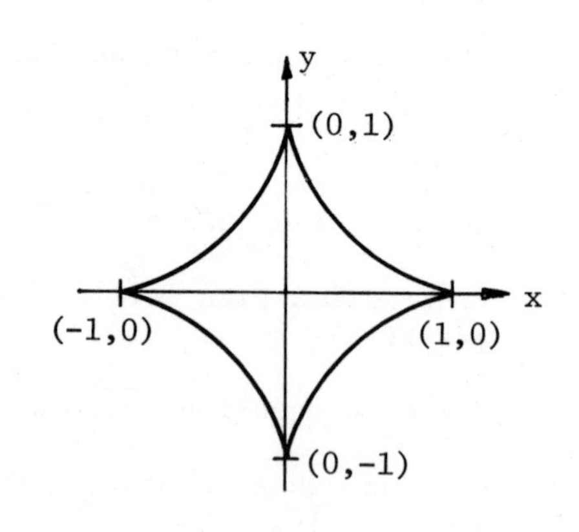

1. Since $x^{2/3} + y^{2/3} = 1$ can be written $\left(x^{1/3}\right)^2 + \left(y^{1/3}\right)^2 = 1$, the curve is symmetric with respect to both axes and the origin. Also, $1 - x^{2/3}$ and $1 - y^{2/3}$ must both be nonnegative so $-1 \le x \le 1$ and $-1 \le y \le 1$ defines the extent of the curve in the x and y directions. The points $(-1,0)$, $(1,0)$, $(0,-1)$, and $(0,1)$ are the intercepts. Differentiating the equation implicitly, we find

$$\frac{2}{3}x^{-1/3} + \frac{2}{3}y^{-1/3}y' = 0 \quad \text{or} \quad \frac{dy}{dx} = \left(\frac{-y}{x}\right)^{1/3}, \quad x \neq 0 .$$

Hence, when $x = \pm 1$ and $y = 0$ the curve is tangent to the x-axis. Similarly, since

$$\frac{dx}{dy} = \left(-\frac{x}{y}\right)^{1/3}, \quad y \neq 0,$$

the curve is tangent to the y-axis at $(0,1)$ and $(0,-1)$. The graph is sketched above.

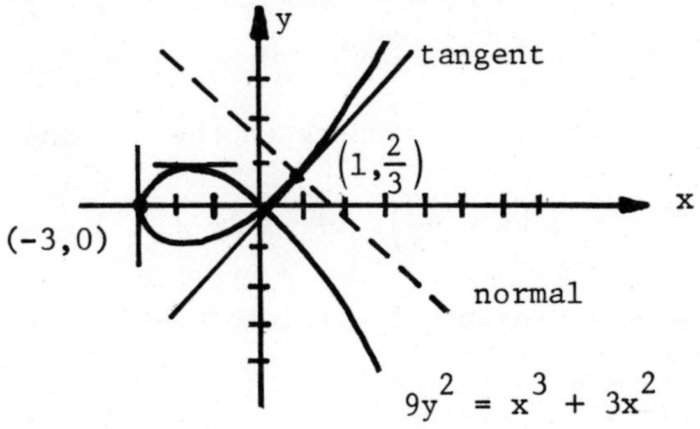

2. Differentiating implicitly,

$$18yy' = 3x^2 + 6x, \quad \text{or}$$

$$\frac{dy}{dx} = \frac{x(x+2)}{6y}, \quad y \neq 0.$$

Thus, at $\left(1, \frac{2}{3}\right)$, $y' = \frac{3}{4}$ is the slope of the tangent line. Hence, equations for the tangent and normal are,

tangent
$(1,\frac{2}{3})$
x
y
(-3,0)
normal
$9y^2 = x^3 + 3x^2$

$$y - \frac{2}{3} = \frac{3}{4}(x-1), \text{ or } \quad 12y - 9x = -1, \quad \text{tangent;}$$

$$y - \frac{2}{3} = -\frac{4}{3}(x-1), \text{ or } \quad 12y + 9x = 17, \quad \text{normal.}$$

A sketch of the curve is shown above.

Notice that when $x = -2$, $y = \pm\frac{2}{3}$ and the curve has slope zero. On the other hand, at the point $(-3,0)$ $dx/dy = 0$, and the curve has a vertical tangent.

3. First we find the point of intersection of the two curves: summing their equations gives $2x^2 = 3$ or, since $x \geq 0$, $x = \sqrt{3}/\sqrt{2}$. Then $y^2 = 2 - x^2$ and $y = 1/\sqrt{2}$. Hence, $P\left(\frac{\sqrt{3}}{\sqrt{2}}, \frac{1}{\sqrt{2}}\right)$ is the point of intersection in the first quadrant.

By implicit differentiation, the slope of the circle at P is

$$m_1 = -\frac{x}{y}\bigg|_P = -\sqrt{3}.$$

Similarly, the slope of the hyperbola at P is

$$m_2 = \frac{x}{y}\bigg|_P = \sqrt{3}.$$

Then the angle ϕ between the curves at P satisfies,

$$\tan\phi = \frac{m_2 - m_1}{1 + m_2 m_1} = \frac{\sqrt{3} + \sqrt{3}}{1 - 3} = -\sqrt{3}.$$

Therefore, $\phi = -\pi/3$ radians. The negative sign means the angle from the circle to the hyperbola is $60°$ in the clockwise direction.

4. Let d_1 denote the distance from P to (-4,2) and d_2 the distance from P to (6,-2). Then $d_1 = 2d_2$, or $d_1^2 = 4d_2^2$. Therefore,

$$(x+4)^2 + (y-2)^2 = 4\left[(x-6)^2 + (y+2)^2\right] ,$$

or expanding algebraically and collecting like terms,

$$3x^2 - 56x + 3y^2 + 20y + 140 = 0 .$$

Completing the squares in the x and y terms,

$$3\left(x - \frac{28}{3}\right)^2 + 3\left(y + \frac{10}{3}\right)^2 = \frac{16 \cdot 29}{3} .$$

This represents a circle of radius $r = \frac{4}{3}\sqrt{29}$ with center $C\left(\frac{28}{3}, \frac{-10}{3}\right)$.

5. $(x-4)^2 + (y-6)^2 = 9$, or $x^2 + y^2 - 8x - 12y + 43 = 0$.

6. The distance between the point (h,k) and the line $ax + by = c$ is

$$\frac{|ah + bk - c|}{\sqrt{a^2 + b^2}}$$

(see Problem 21 on page 404 of the text). Now, the radius of the circle is the distance from its center to the tangent line, and we find this to be

$$r = \frac{|1 \cdot 1 + 2 \cdot 2 - 10|}{\sqrt{1+4}} = \sqrt{5} .$$

Thus, an equation of the circle is $(x-1)^2 + (y-2)^2 = 5$.

7. Using the equation $C_1x + C_2y + C_3 = -(x^2 + y^2)$ for a circle, by substitutions we obtain the equations,

$$4C_1 + 2C_2 + C_3 = -20 \quad \text{from point A},$$

$$6C_1 + C_3 = -36 \quad \text{from point B},$$

$$-6C_2 + C_3 = -36 \quad \text{from point C}.$$

We solve these by elimination: $-\frac{6}{4}$ times the first equation added to the second equation gives

$$-3C_2 - \frac{1}{2}C_3 = -6 , \quad \text{or} \quad -6C_2 - C_3 = -12 .$$

Addition of this last equation and the third equation results in $-12C_2 = -48$ or $C_2 = 4$.

From the third equation, $C_3 = -36 + 6C_2$, and substitution of the value $C_2 = 4$ then gives $C_3 = -12$.

From the second equation $6C_1 = -36 - C_3$, and substitution of $C_3 = -12$ gives $6C_1 = -24$, or $C_1 = -4$.

Thus, an equation for the circle is $x^2 + y^2 - 4x + 4y - 12 = 0$. Completing the squares, $(x-2)^2 + (y+2)^2 = 20$.

8. Since the circle is tangent to the line $x + y = 8$ with slope $m = -1$, the center of the circle must lie on the line $y = x$. Let $C(h,h)$ denote the center. Then the radius of the circle is $r = h$ because the circle is tangent to each coordinate axis. However, r is also equal to the distance from the center to the line $x + y = 8$. Thus,

$$h = \frac{|1 \cdot h + 1 \cdot h - 8|}{\sqrt{1+1}} = \frac{|2h - 8|}{\sqrt{2}} .$$

Squaring both sides and simplifying we obtain,

$$h^2 - 16h + 32 = 0$$

Solution of this quadratic equation for $0 \le h \le 8$ gives, $h = 8 - 4\sqrt{2} \approx 2.343$ units. An equation for the circle is (approximately) given by,

$$(x - 2.343)^2 + (y - 2.343)^2 = 5.49 .$$

9. The parabola opens to the left and hence has an equation of the form

$$(y - k)^2 = -4p(x - h) .$$

The point (h,k) is the vertex $V(2,0)$. The value of p is the distance from the vertex to the focus, so $p = 2$. Thus an equation of the parabola is

$$y^2 = -8(x - 2) .$$

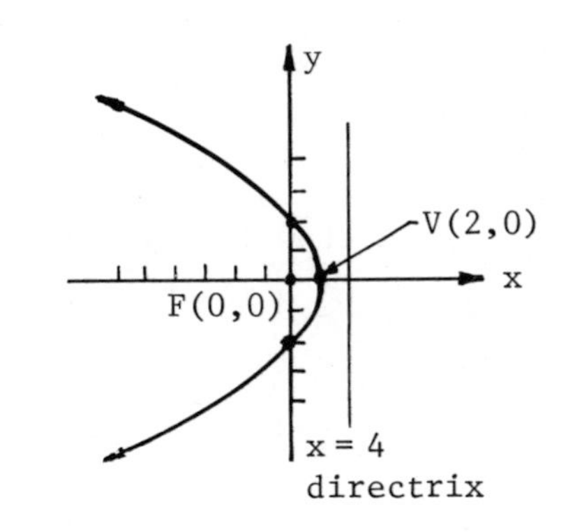

The directrix is the line $x = 4$.

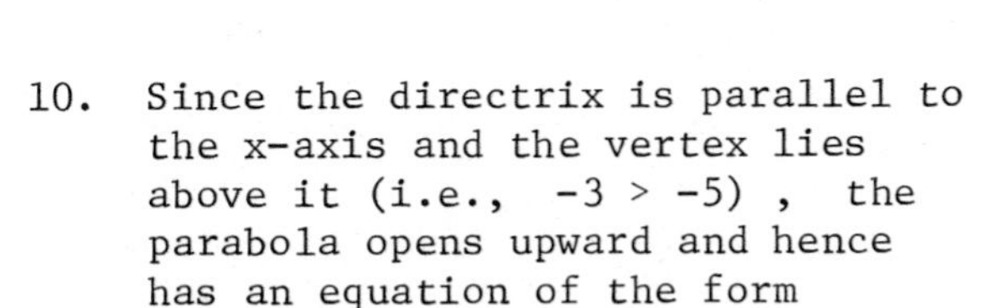

10. Since the directrix is parallel to the x-axis and the vertex lies above it (i.e., $-3 > -5$), the parabola opens upward and hence has an equation of the form

$$(x - h)^2 = 4p(y - k) .$$

The value of p is the distance from the vertex to the directrix, so $p = 2$. Thus an equation of the parabola is

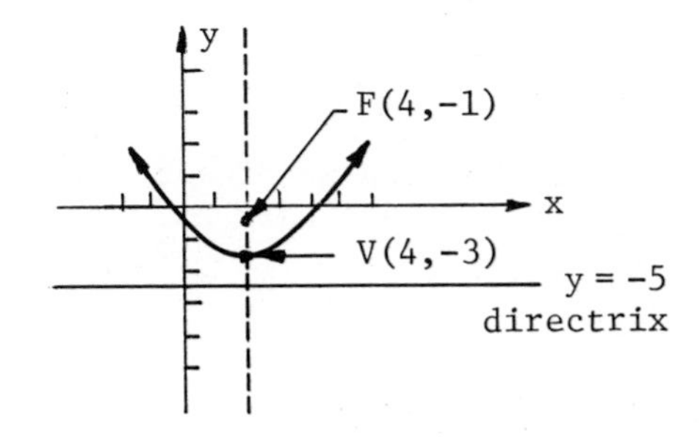

$(x - 4)^2 = 8(y + 3)$. The focus is $F(4,-1)$.

11. Completing the square in x,

$$(x-2)^2 = -6(y+5) .$$

The parabola opens downward with $4p = 6$ or $p = \frac{3}{2}$. The vertex is located at $V(2,-5)$. The focus is given by $F(2, -5-p) = F\left(2, \frac{-13}{2}\right)$. The directrix is the line $y = -5+p$ or $y = -\frac{7}{2}$. The parabola is sketched at the right.

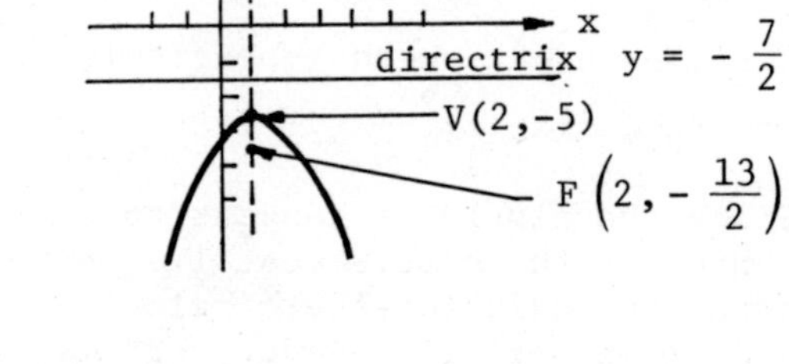

12. Completing the squares in the x and y terms,

$$9(x^2-6x+9) + 16(y^2+8y+16) = 81+256-193 ,$$

which may be written as

$$9(x-3)^2 + 16(y+4)^2 = 144 ,$$

or

$$\frac{(x-3)^2}{16} + \frac{(y+4)^2}{9} = 1 .$$

The center is $C(3,-4)$.
The semimajor axis is $a = 4$, and the semiminor axis is $b = 3$.
The foci lie on the line $y = -4$.
Now, $c^2 = a^2 - b^2 = 16-9 = 7$.
Thus, the foci are located at

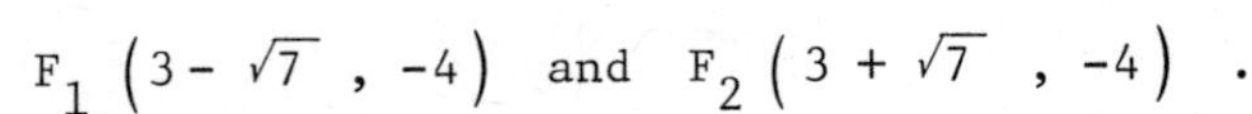

$$F_1\left(3-\sqrt{7}, -4\right) \text{ and } F_2\left(3+\sqrt{7}, -4\right) .$$

The vertices are $V_1(-1,-4)$, $V_2(7,-4)$, $V_3(3,-1)$, and $V_4(3,-7)$. The graph is shown at the right. The eccentricity is $e = c/a = \sqrt{7}/4 \approx 0.661$.

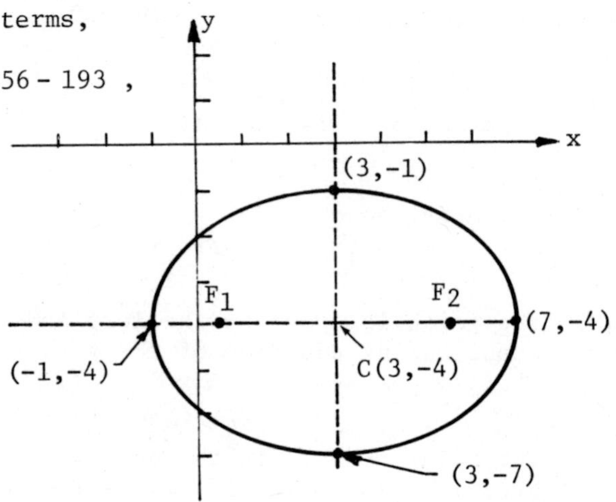

13. The focus lies on the major axis at a distance $c = 2-(-1) = 3$ units from the center $C(2,5)$. The major axis is along the line $y = 5$ which contains the center and focus. Now, $a^2 = b^2 + c^2 = 16+9 = 25$. Hence, an equation of the ellipse is

$$\frac{(x-2)^2}{25} + \frac{(y-5)^2}{16} = 1 \quad \text{or} \quad 16(x-2)^2 + 25(y-5)^2 = 400 .$$

14. Completing the squares in the x and y terms,

$$9(x^2 - 2x + 1) - 4(y^2 - 8y + 16) = 36 ,$$

or

$$\frac{(x-1)^2}{4} - \frac{(y-4)^2}{9} = 1 .$$

The center is $C(1,4)$. Now, $c^2 = a^2 + b^2 = 4 + 9 = 13$.

Since $b > a$, the hyperbola opens to the left and to the right. Thus, the foci are $F_1\left(1 - \sqrt{13}\ , 4\right)$ and $F_2\left(1 + \sqrt{13}\ , 4\right)$. The vertices are found by setting $y = 4$; whence $x - 1 = \pm 2$. We find $V_1(-1,4)$ and $V_2(3,4)$. Finally, the asymptotes are obtained by setting

$$\frac{(x-1)^2}{4} + \frac{(y-4)^2}{9} \quad \text{equal to zero,}$$

so $y - 4 = \pm \frac{3}{2}(x-1)$. The graph is sketched at the right.

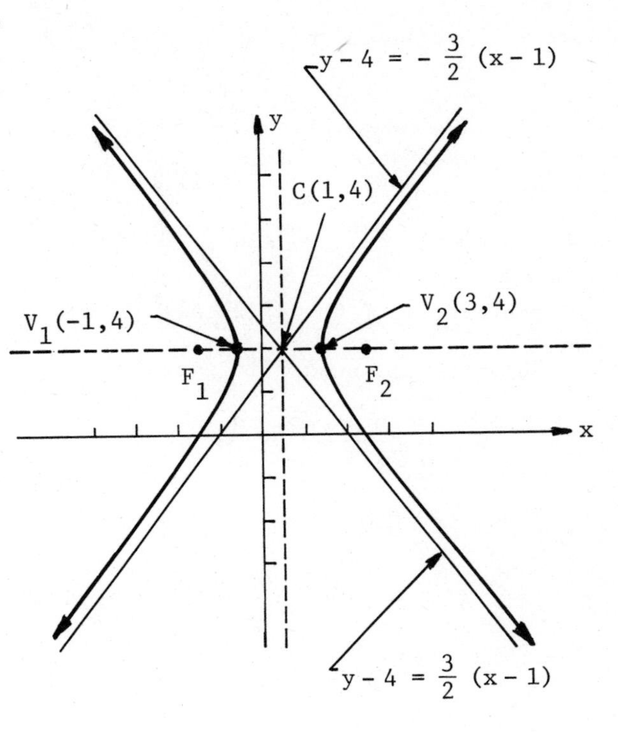

15. (a) The discriminant is $B^2 - 4AC = (-2)^2 - 4(1)(-1) = 8 > 0$. Therefore, the equation represents a hyperbola.

(b) $\cot 2\alpha = \frac{A-C}{B} = \frac{1+1}{-2} = -1$; thus $2\alpha = \frac{3\pi}{4}$ radians.

Hence, $\cos 2\alpha = -1/\sqrt{2}$. Using the half-angle formulas,

$$\sin\alpha = \sqrt{\frac{1 - \cos 2\alpha}{2}} = \sqrt{\frac{1 + \sqrt{2}}{2\sqrt{2}}} = \frac{1}{2}\sqrt{2 + \sqrt{2}} \quad , \text{ and}$$

$$\cos\alpha = \sqrt{\frac{1 + \cos 2\alpha}{2}} = \sqrt{\frac{1 - \sqrt{2}}{2\sqrt{2}}} = \frac{1}{2}\sqrt{2 - \sqrt{2}} \quad .$$

Now,

$$A' = A\cos^2\alpha + B\cos\alpha\sin\alpha + C\sin^2\alpha$$

$$= \frac{1}{4}\left(2 - \sqrt{2}\right) - \frac{2}{4}\left(\sqrt{-2+4}\right) - \frac{1}{4}\left(2 + \sqrt{2}\right) = -\sqrt{2}$$

$$C' = A\sin^2\alpha - B\cos\alpha\sin\alpha + C\cos^2\alpha$$

$$= \frac{1}{4}\left(2 + \sqrt{2}\right) + \frac{2}{4}\left(\sqrt{-2+4}\right) - \frac{1}{4}\left(2 - \sqrt{2}\right) = \sqrt{2} \quad .$$

Thus, since $B' = 0$ because of the choice $\alpha = \frac{3\pi}{8}$, the original equation is reduced to $A'x'^2 + C'y'^2 = F$, or

$$-\sqrt{2}\, x'^2 + \sqrt{2}\, y'^2 = 12 , \quad \text{or} \quad y'^2 - x'^2 = 6\sqrt{2} .$$

16. Now, $B^2 - 4AC = 1 - 4(1)(1) = -3 = -4A'C'$; thus

$A'C' = \frac{3}{4}$. Also, $A + C = 2 = A' + C'$.

Therefore, $\frac{3}{4} = A'(2 - A') = 2A' - A'^2$, or $4A'^2 - 8A' + 3 = 0$.

Solving this quadratic equation,

$$A' = \frac{8 \pm \sqrt{64 - 48}}{8} .$$

Thus, $A' = \frac{3}{2}$ or $A' = \frac{1}{2}$.

For $A' = \frac{3}{2}$, $C' = \frac{1}{2}$ and the original equation reduces to

$$\frac{3}{2} x'^2 + \frac{1}{2} y'^2 = 10 \quad \text{or} \quad 3x'^2 + y'^2 = 20 .$$

On the other hand, if $A' = \frac{1}{2}$, $C' = \frac{3}{2}$ and the original equation reduces to

$$x'^2 + 3y'^2 = 20 .$$

We recognize either of these equations as representing an ellipse.

CHAPTER 9 HYPERBOLIC FUNCTIONS

9-1 Introduction.

In this chapter you will study the hyperbolic functions and see that the identities and differentiation formulas associated with them bear a striking resemblance to those associated with the trigonometric functions. Also there is a direct connection between the inverse hyperbolic functions and the natural logarithm. These functions are useful in applied mathematics, in certain integration methods, and in solving differential equations. They arise in the study of falling bodies, hanging cables, ocean waves, and other phenomena studied in science and engineering.

9-2 Definitions and Identities.

OBJECTIVE A: Define the six hyperbolic functions, and graph them.

1. cosh x = _______________ .

   Sketch the graph at the right.

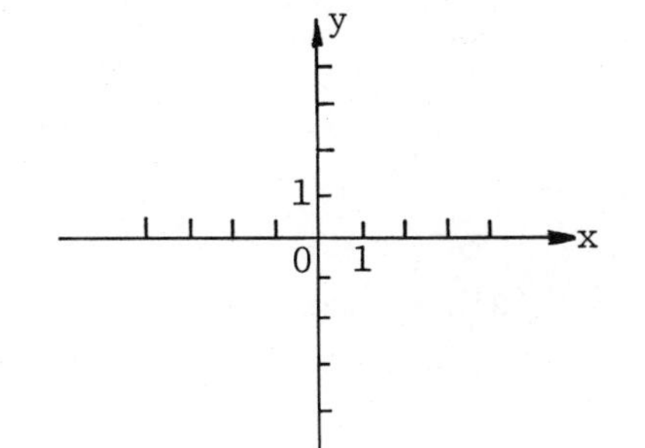

2. sinh x = _______________ .

   Sketch the graph at the right on the same coordinate system as the cosh x .

3. tanh x = ___________ .

4. coth x = ___________ .

   Sketch the graphs at the right.

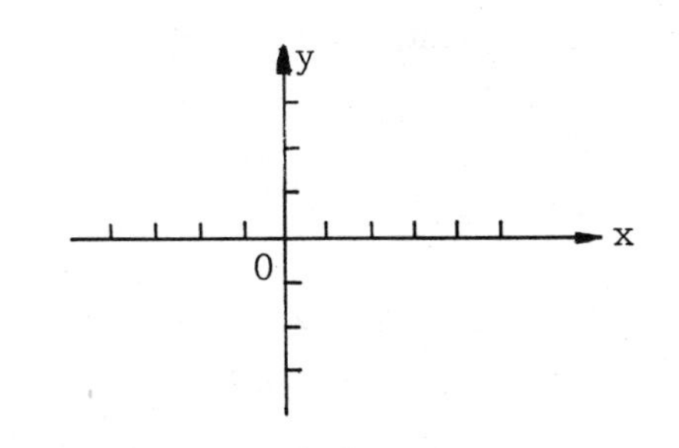

---

1. $\frac{1}{2}(e^{x} + e^{-x})$ 2. $\frac{1}{2}(e^{x} - e^{-x})$ 3. $\frac{\sinh x}{\cosh x}$ 4. $\frac{\cosh x}{\sinh x}$

y = cosh x
y = sinh x
y
0
x

y = coth x
y=1
y=-1
y = coth x
y = tanh x
y
x

5. $\operatorname{sech} x$ = __________ .

Sketch the graph at the right.

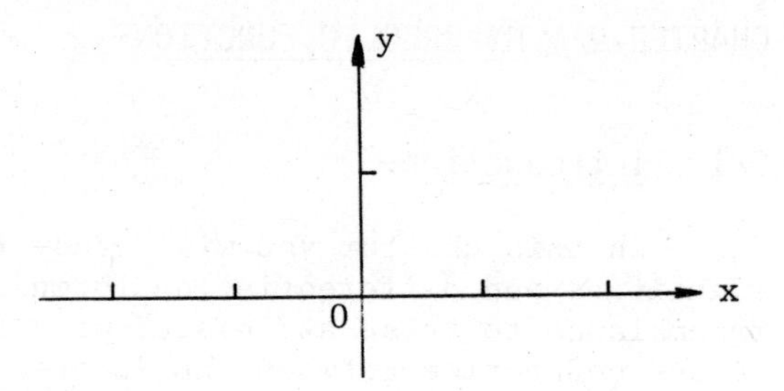

6. $\operatorname{csch} x$ = __________ .

OBJECTIVE B: Given the value for one of the six hyperbolic functions at a point, determine the values of the remaining five at that point.

7. Suppose $\tanh x = -\frac{\sqrt{3}}{2}$. Then,

$\operatorname{sech}^2 x = 1 -$ __________ = _____ or $\operatorname{sech} x$ = _____ .

Thus, $\cosh x = \frac{1}{\operatorname{sech} x}$ = ___ .

8. Continuing Problem 7, $\sinh^2 x = \cosh^2 x -$ ___ = ___ .
Since $\tanh x$ is negative and $\cosh x$ is positive, it follows that $\sinh x$ = _______ . Then, $\operatorname{csch} x$ = _______ and $\coth x$ = _______ .

9. $\cosh(-x)$ = __________ .

10. $\sinh(-x)$ = __________ .

11. $\sinh(x+y)$ = ______________________________ .

12. $\cosh(x+y)$ = ______________________________ .

13. $\sinh 2x$ = ________________ .

14. $\cosh 2x$ = ________________ .

15. $\cosh 2x - 1$ = ____________ .

---

5. $\frac{1}{\cosh x}$

6. $\frac{1}{\sinh x}$

y = 1

y = sech x

7. $\tanh^2 x$, $\frac{1}{4}$, $\frac{1}{2}$, 2

8. 1, 3, $-\sqrt{3}$, $-\frac{1}{\sqrt{3}}$, $-\frac{2}{\sqrt{3}}$

9. $\cosh x$

10. $-\sinh x$

11. $\sinh x \cosh y + \cosh x \sinh y$

12. $\cosh x \cosh y + \sinh x \sinh y$

13. $2 \sinh x \cosh x$

14. $\cosh^2 x + \sinh^2 x$

15. $2 \sinh^2 x$

9-3 Derivatives and Integrals.

OBJECTIVE A: Calculate the derivatives of functions expressed in terms of hyperbolic functions.

16. $\frac{d}{dx} \cosh u = $ ________ .

17. $\frac{d}{dx} \sinh u = $ ________ .

18. $\frac{d}{dx} \tanh u = $ ________ .

19. $y = \sinh^3 (3 - 2x^2)$

$\frac{dy}{dx} = 3 \sinh^2 (3 - 2x^2) \cdot \frac{d}{dx}$ ________

$= 3 \sinh^2 (3 - 2x^2) \cosh (3 - 2x^2) \frac{d}{dx}$ ________

$=$ ________ .

20. $y = e^x \tanh 2x$

$\frac{dy}{dx} = e^x \frac{d}{dx}$ (________) $+ e^x \tanh 2x$

$= e^x$ ________ $\frac{d}{dx}$ (____) $+ e^x \tanh 2x$

$=$ ________ .

21. $y = x^{\sinh x}, \quad x > 0$

$\frac{dy}{dx} = \frac{d}{dx} \left(e^{\sinh x \cdot \ln x} \right) = e^{\sinhx \cdot \ln x} \frac{d}{dx}$ (________)

$= e^{\sinh x \cdot \ln x} \left(\sinh x \cdot \frac{d}{dx} \ln x + \ln x \cdot \frac{d}{dx} \right.$ ________ $\left. \right)$

$= e^{\sinh x \cdot \ln x}$ (________)

$= x^{\sinh x - 1}$ (________) .

---

16. $\sinh u \frac{du}{dx}$ 17. $\cosh u \frac{du}{dx}$ 18. $\operatorname{sech}^2 u \frac{du}{dx}$

19. $\sinh (3 - 2x^2)$, $3 - 2x^2$, $-12x \sinh^2 (3 - 2x^2) \cosh (3 - 2x^2)$

20. $\tanh 2x$, $\operatorname{sech}^2 2x$, $2x$, $e^x\left(2 \operatorname{sech}^2 2x + \tanh 2x\right)$

21. $\sinh x \cdot \ln x$, $\sinh x$, $\frac{\sinh x}{x} + \cosh x \cdot \ln x$, $\sinh x + x \cosh x \ln x$

22. $e^y = \text{sech } x$

Differentiating implicitly, $\frac{d}{dx}(e^y) = \frac{d}{dx} \text{sech } x$, or

_______ $= - \text{sech } x \tanh x$. Thus,

$\frac{dy}{dx} = -e^{-y}$ ______________ $=$ __________ .

OBJECTIVE B: Integrate functions whose expressions involve hyperbolic functions.

23. $\int x \cosh (x^2 + 3)\, dx$

Let $u = x^2 + 3$, then $du =$ _______, and the integral becomes

$\int x \cosh (x^2 + 3)\, dx = \int$ __________ $du =$ __________ $+ C$

$=$ ____________________ .

24. $\int \sinh^2 x\, dx$

From the identities $\cosh 2x = \sinh^2 x + \cosh^2 x$ and $\cosh^2 x - \sinh^2 x = 1$, we have

$\cosh 2x =$ ______________ or $\sinh^2 x = \frac{1}{2}($ ____________ $)$.

Thus, $\int \sinh^2 x\, dx =$ __________________ $+ C$.

25. $\int \tanh x \ln (\cosh x)\, dx$

Let $u = \ln (\cosh x)$, so $du =$ ____________________ and the integral becomes

$\int \ln (\cosh x) \tanh x\, dx = \int$ ___ $du =$ ______ $+ C$

$=$ ____________________ .

---

22. $e^y \frac{dy}{dx}$, $\text{sech } x \tanh x$, $- \tanh x$

23. $2x\, dx$, $\frac{1}{2} \cosh u$, $\frac{1}{2} \sinh u$, $\frac{1}{2} \sinh (x^2 + 3) + C$

24. $2 \sinh^2 x + 1$, $\cosh 2x - 1$, $\frac{1}{4} \sinh 2x - \frac{1}{2} x$

25. $\frac{1}{\cosh x} \cdot \sinh x\, dx$, u, $\frac{1}{2} u^2$, $\frac{1}{2} \ln^2 (\cosh x) + C$

9-4 Geometric Significance of the Hyperbolic Radian.

26. Consider the variable u in the equations

$$x = \cosh u \quad , \quad y = \sinh u$$

as they relate to the point $P(x,y)$ on the hyperbola

$$x^2 - y^2 = 1 .$$

In reference to the accompanying figure at the right,

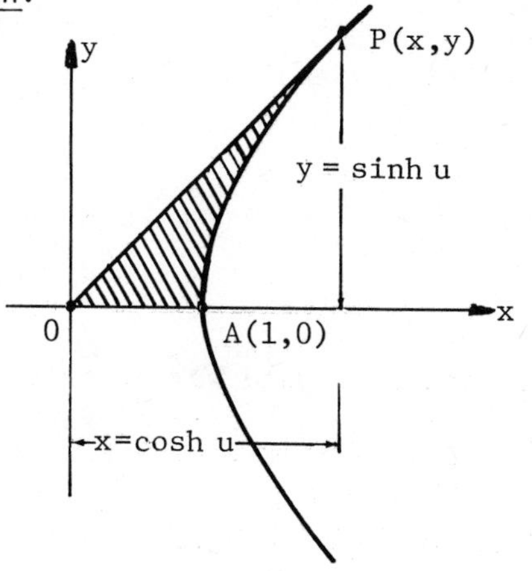

u = ______________________________ .

27. The variable u in Problem 26 simply refers to a dimensionless __________ __________ , sometimes called the __________ radian.

9-5 The Inverse Hyperbolic Functions.

28. $y = \sinh^{-1} x$ means __________ . Thus $x = \dfrac{e^y - \_\_\_\_}{2}$ or

$2x\,e^y$ = __________ or $e^{2y} - \_\_\_\_\_\_ - 1 = 0$.

Solution of this quadratic equation by the quadratic formula gives,

$e^y = \dfrac{2x \pm \sqrt{\_\_\_\_\_\_}}{2}$. Since $e^y > 0$ we must have e^y = __________

or $\sinh^{-1} x = y$ = __________ .

29. We can use the formula found in Problem 28 to calculate dy/dx for $y = \sinh^{-1} x$:

$$\frac{d}{dx} \ln\left(x + \sqrt{x^2+1}\right) = \frac{1}{x + \sqrt{x^2+1}} \cdot \frac{d}{dx}\left(\_\_\_\_\_\_\_\_\right)$$

---

26. twice the area of the sector AOP

27. real number, hyperbolic

28. $x = \sinh y$, e^{-y}, $e^{2y} - 1$, $2x\,e^y$, $4x^2 + 4$, $x + \sqrt{x^2+1}$, $\ln\left(x + \sqrt{x^2+1}\right)$

$$= \frac{1}{x + \sqrt{x^2+1}} \cdot \left(\underline{\hspace{3cm}} \right)$$

$$= \frac{1}{x + \sqrt{x^2+1}} \cdot \left(\frac{\underline{\hspace{2cm}}}{\sqrt{x^2+1}} \right) = \underline{\hspace{2cm}} .$$

30. An alternate way to calculate the derivative of $y = \sinh^{-1} x$ is as follows:

Differentiate $x = \sinh y$ implicitly:

$\frac{d}{dx} x = \frac{d}{dx} \sinh y$ or $1 = \underline{\hspace{3cm}}$. Thus,

$$\frac{dy}{dx} = \frac{1}{\underline{\hspace{1.5cm}}} = \frac{1}{\sqrt{1 + \underline{\hspace{1.5cm}}}} = \underline{\hspace{2cm}} .$$

The positive square root is taken in the penultimate step because $\cosh y$ is always ____________ .

OBJECTIVE A: Calculate the derivatives of functions expressed in terms of inverse hyperbolic functions.

31. $\frac{d}{dx} \tanh^{-1} e^x = \frac{1}{\underline{\hspace{1.5cm}}} \cdot \frac{d}{dx} e^x = \underline{\hspace{2cm}} .$

32. $\frac{d}{dx} \ln (\sinh^{-1} x) = \frac{1}{\sinh^{-1} x} \cdot \frac{d}{dx} \underline{\hspace{2cm}} = \underline{\hspace{3cm}} .$

---

29. $x + \sqrt{x^2+1}$, $1 + \frac{2x}{2\sqrt{x^2+1}}$, $\sqrt{x^2+1} + x$, $\frac{1}{\sqrt{x^2+1}}$

30. $\cosh y \cdot \frac{dy}{dx}$, $\cosh y$, $\sinh^2 y$, $\frac{1}{\sqrt{1+x^2}}$, positive

31. $1 - e^{2x}$, $\frac{e^x}{1-e^{2x}}$

32. $\sinh^{-1} x$, $\frac{1}{\sinh^{-1} x \cdot \sqrt{1+x^2}}$

33. $\frac{d}{dx}\sqrt{\coth^{-1} x} = \frac{1}{2}\left(\coth^{-1} x\right)^{-1/2} \cdot \frac{d}{dx}$ ________

$= $ ________________ .

34. $\frac{d}{dx}\cosh^{-1}\frac{3}{x^2} = $ ________ $\cdot \frac{-6}{x^3} = $ ________ .

OBJECTIVE B: Evaluate integrals using integration formulas for inverse hyperbolic functions.

35. $\int_{-3}^{-2} \frac{dx}{\sqrt{x^2+1}}$

Since $\int \frac{dx}{\sqrt{x^2+1}} = $ ________ $+ C = \ln\left(\text{\_\_\_\_\_\_\_\_}\right) + C$,

$\int_{-3}^{-2} \frac{dx}{\sqrt{x^2+1}} = $ ________________ $\Big]_{-3}^{-2}$

$= \ln\left(\sqrt{5} - 2\right) - $ ________ ≈ 0.375 .

36. $\int_{0.5}^{0.9} \frac{dx}{x\sqrt{1-x^2}}$

For $0 < x < 1$, $\int \frac{dx}{x\sqrt{1-x^2}} = $ ________ $+ C$.

Now, $\operatorname{sech}^{-1} x = \cosh^{-1}$ ____ $= \ln\left(\text{\_\_\_\_\_\_\_\_}\right) = \ln\left(\frac{\text{\_\_\_\_\_\_}}{x}\right)$.

---

33. $\coth^{-1} x$, $\frac{1}{2(1-x^2)\sqrt{\coth^{-1} x}}$

34. $\frac{1}{\sqrt{\frac{9}{x^4} - 1}}$, $\frac{-6}{x\sqrt{9 - x^4}}$

35. $\sinh^{-1} x$, $x + \sqrt{x^2+1}$, $\ln\left(x + \sqrt{x^2+1}\right)$, $\ln\left(\sqrt{10} - 1\right)$

36. $-\operatorname{sech}^{-1} x$, $\frac{1}{x}$, $\frac{1}{x} + \sqrt{\frac{1}{x^2} - 1}$, $1 + \sqrt{1-x^2}$, $-\ln\left(\frac{1+\sqrt{1-x^2}}{x}\right)$,
$\ln\left(\frac{1+\sqrt{1-.25}}{.5}\right)$, $\ln\left(2+\sqrt{3}\right)$

Thus,

$$\int_{0.5}^{0.9} \frac{dx}{x\sqrt{1-x^2}} = \underline{\hspace{4cm}}\Big]_{0.5}^{0.9}$$

$$= -\ln\left(\frac{1 + \sqrt{1-.81}}{.9}\right) + \underline{\hspace{4cm}}$$

$$= -\ln\left(\frac{10+\sqrt{19}}{9}\right) + \underline{\hspace{3cm}}$$

$$\approx 0.850 \ .$$

37. $\int \frac{dx}{16 - x^2} = \frac{1}{16}\int \underline{\hspace{3cm}}$

Let $u = \frac{x}{4}$, $du =$ ______ , and the integral becomes

$$\int \frac{dx}{16-x^2} = \int \frac{4\,du}{\underline{\hspace{2.5cm}}} = \frac{1}{4}\left(\underline{\hspace{3cm}}\right) + C$$

$$= \underline{\hspace{3cm}} + C \ .$$

9-6 The Hanging Cable.

38. To find an equation of a 50 foot long cable, weighing 5 lb/ft, and hanging under its own weight between two supports 30 feet apart, first choose the origin so that the y axis goes through the lowest point of the cable. Then an equation of the cable has the form

$$y = \underline{\hspace{3cm}} \quad \text{and} \quad \frac{dy}{dx} = \underline{\hspace{3cm}} \ .$$

---

37. $\frac{dx}{1 - (x^2/16)}$, $\frac{1}{4}\,dx$, $16(1-u^2)$, $\frac{1}{2}\ln\left|\frac{1+u}{1-u}\right|$, $\frac{1}{8}\ln\left|\frac{4+x}{4-x}\right|$

38. $\frac{H}{w}\cosh\left(\frac{w}{H}x\right)$, $\sinh\left(\frac{w}{H}x\right)$, $1+\sinh^2\left(\frac{w}{H}x\right)$, $\cosh\left(\frac{w}{H}x\right)$,

$\frac{H}{w}\sinh\frac{wL}{H}$, $\frac{H}{5}\sinh\frac{5}{H}15$, $8.16\cosh\frac{x}{8.16}$

The length of the cable from $x = 0$ to $x = L$ is

$$s = \int_0^L \sqrt{1 + \left(\frac{dy}{dx}\right)^2}\, dx = \int_0^L \sqrt{\underline{\qquad\qquad\qquad}}\, dx$$

$$= \int_0^L \underline{\qquad\qquad} = \frac{H}{w} \sinh\left(\frac{w}{H} x\right)\Big]_0^L = \underline{\qquad\qquad} .$$

Substituting the conditions $w = 5$, $L = 15$, $s = 25$ gives

$$25 = \underline{\qquad\qquad} \quad \text{or} \quad \frac{125}{H} = \sinh \frac{75}{H} .$$

Using a table of values of $\sinh z$, the approximate value of H which will satisfy this last equation is $H = 40.8$ lb. Thus, an equation for the hanging cable is

$$y = \underline{\qquad\qquad\qquad} .$$

39. The lowest point on the cable in Problem 38 is the point ________ . The highest point occurs when $x = 15$ ft. , so $y =$ ________ , or

$y \approx (8.16)(3.22) = 26.28$ ft. Therefore, the amount of sag in the cable is

$$\text{sag} = 26.28 - \underline{\qquad} = \underline{\qquad} \text{ feet} .$$

CHAPTER 9 OBJECTIVE - PROBLEM KEY

Objective	Problems in Thomas/Finney Text	Objective	Problems in Thomas/Finney Text
9-2 A	See Figure 9-2(a)-(d)	9-3 B	p. 447, 10-18
B	p. 443, 6	9-5 A	p. 454, 3-10
9-3 A	p. 447, 1-9	9-5 B	p. 454, 11-15

CHAPTER 9 SELF-TEST

1. Define the hyperbolic function $y = \operatorname{csch} x$, and sketch its graph.

2. Given that $\cosh x = 2$, $x < 0$, find the values of the remaining hyperbolic functions at x .

---

39. $(0, 8.16)$, $8.16 \cosh \frac{15}{8.16}$, 8.16, 18.12

Find dy/dx in each of the following:

3. $y = \tanh(\sin x)$

4. $y = \coth^{-1}(\ln x)$

5. $y = \sqrt{\cosh^{-1} x^2}$

6. $y = \ln(\sinh x^3)$

7. $y = x^{-1} \tanh^{-1} x^2$

8. $y = \sinh^{-1}(\tan x)$

Integrate each of the following:

9. $\displaystyle\int \frac{4\,dx}{\left(e^x - e^{-x}\right)^2}$

10. $\displaystyle\int \frac{\sinh(\ln x)\,dx}{x}$

11. $\displaystyle\int \sqrt{1 + \cosh x}\;dx$

12. $\displaystyle\int_3^7 \frac{dx}{\sqrt{x^2 - 1}}$

13. $\displaystyle\int_0^{1/2} \frac{\cosh x\,dx}{1 - \sinh^2 x}$

14. $\displaystyle\int_0^1 \frac{dx}{\sqrt{e^{2x} + 1}}$

15. Find the length of the catenary $y = 3 \cosh \frac{x}{3}$ from $x = 0$ to $x = 3$.

SOLUTIONS TO CHAPTER 9 SELF-TEST

1. $y = \operatorname{csch} x = \dfrac{1}{\sinh x}$,

where $\sinh x = \frac{1}{2}(e^x - e^{-x})$.

The graph is sketched at the right.

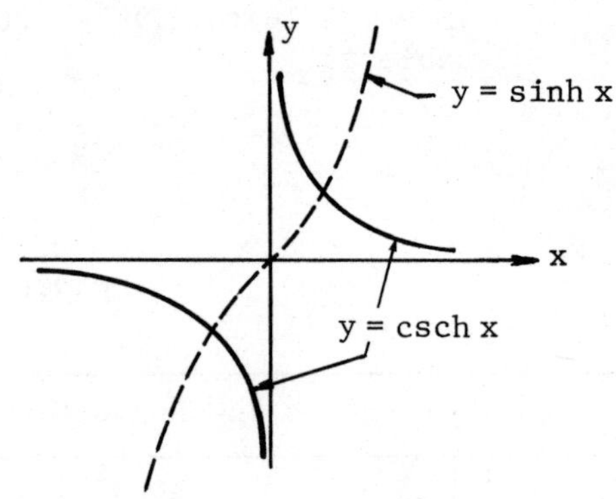

2. $\operatorname{sech} x = \dfrac{1}{\cosh x} = \dfrac{1}{2}$; $\tanh^2 x = 1 - \operatorname{sech}^2 x = \dfrac{3}{4}$ so that $\tanh x = \dfrac{-\sqrt{3}}{2}$

since $x < 0$; $\sinh x = \cosh x \tanh x = -\sqrt{3}$;

$\coth x = \dfrac{1}{\tanh x} = -\dfrac{2}{\sqrt{3}}$; and $\operatorname{csch} x = \dfrac{1}{\sinh x} = -\dfrac{1}{\sqrt{3}}$.

3. $\dfrac{dy}{dx} = \operatorname{sech}^2(\sin x) \cdot \cos x$

4. $\dfrac{dy}{dx} = \dfrac{1}{\left[1 - (\ln x)^2\right]} \cdot \dfrac{1}{x}$, $\ln x > 1$

5. $\frac{dy}{dx} = \frac{1}{2}\left(\cosh^{-1} x^2\right)^{-1/2} \cdot \frac{d}{dx}\left(\cosh^{-1} x^2\right)$

$= \frac{1}{2}\left(\cosh^{-1} x^2\right)^{-1/2} \cdot \frac{1}{\sqrt{x^4 - 1}} \cdot \frac{d}{dx}(x^2)$

$= \frac{x}{\sqrt{(x^4 - 1)\cosh^{-1} x^2}}, \quad x > 1$

6. $\frac{dy}{dx} = \frac{1}{\sinh x^3} \cdot \frac{d}{dx}(\sinh x^3) = \frac{1}{\sinh x^3} \cdot \cosh x^3 \cdot \frac{d}{dx}(x^3) = 3x^2 \coth x^3$

7. $\frac{dy}{dx} = -\frac{1}{x^2}\tanh^{-1} x^2 + x^{-1} \cdot \frac{1}{1 - x^4} \cdot 2x$

$= -x^{-2}\tanh^{-1} x^2 + \frac{2}{1 - x^4}, \quad x^4 < 1$

8. $\frac{dy}{dx} = \frac{1}{\sqrt{\tan^2 x + 1}} \cdot \sec^2 x = \frac{\sec^2 x}{\sqrt{\sec^2 x}} = \sec x, \quad \text{if } -\frac{\pi}{2} < x < \frac{\pi}{2}$

9. $\int \frac{4\,dx}{\left(e^x - e^{-x}\right)^2} = \int \operatorname{csch}^2 x\,dx = -\coth x + C$

10. Let $u = \ln x$, $du = \frac{1}{x}\,dx$, and the integral becomes

$\int \frac{\sinh(\ln x)\,dx}{x} = \int \sinh u\,du = \cosh u + C = \cosh(\ln x) + C$

11. From the identity $2\cosh^2 x = \cosh 2x + 1$, we find that

$\sqrt{2}\cosh\frac{x}{2} = \sqrt{\cosh x + 1}$. Thus,

$\int \sqrt{\cosh x + 1}\,dx = \sqrt{2}\int \cosh\frac{x}{2}\,dx = 2\sqrt{2}\sinh\frac{x}{2} + C$

12. $\int_3^7 \frac{dx}{\sqrt{x^2 - 1}} = \cosh^{-1} x \Big]_3^7 = \ln\left(x + \sqrt{x^2 - 1}\right)\Big]_3^7$

$= \ln\left(7 + \sqrt{48}\right) - \ln\left(3 + \sqrt{8}\right)$

$= \ln\left(\frac{7 + 4\sqrt{3}}{3 + 2\sqrt{2}}\right) \approx 2.39 .$

13. Let $u = \sinh x$, $du = \cosh x\,dx$, so that

$$\int_0^{1/2} \frac{\cosh x\,dx}{1 - \sinh^2 x} = \tanh^{-1}(\sinh x)\Big]_0^{1/2} = \frac{1}{2}\ln\left[\frac{1 + \sinh x}{1 - \sinh x}\right]_0^{1/2}$$

$$= \frac{1}{2}\ln\left[\frac{1 + 0.5211}{1 - 0.5211}\right] \approx 0.5778 \text{ by Tables.}$$

14. $$\int_0^1 \frac{dx}{\sqrt{e^{2x}+1}} = \int_0^1 \frac{e^x\,dx}{e^x\sqrt{e^{2x}+1}}, \text{ which is of the form}$$

$$\int \frac{du}{|u|\sqrt{u^2+1}} = -\operatorname{csch}^{-1} u + C, \text{ for } u = e^x \neq 0.$$

Thus, $$\int_0^1 \frac{dx}{\sqrt{e^{2x}+1}} = -\operatorname{csch}^{-1} e^x\Big]_0^1 = -\ln\left(\frac{1 + \sqrt{1+e^{2x}}}{e^x}\right)\Big]_0^1$$

$$= -\ln\left(\frac{1 + \sqrt{1+e^2}}{e}\right) + \ln\left(1 + \sqrt{2}\right)$$

$$\approx 0.52.$$

15. $$s = \int_0^3 \sqrt{1 + \left(\frac{dy}{dx}\right)^2}\,dx = \int_0^3 \sqrt{1 + \sinh^2 \frac{x}{3}}\,dx$$

$$= \int_0^3 \cosh\frac{x}{3}\,dx = 3\sinh\frac{x}{3}\Big]_0^3 = 3\sinh 1 \approx 3.53.$$

CHAPTER 10 POLAR COORDINATES

10-1 The Polar Coordinate System.

OBJECTIVE A: Given a point P in polar coordinates (r,θ), give the cartesian coordinates (x,y) of P.

1. The polar and cartesian coordinates are related by the equations

   $x =$ ________ and $y =$ ________ .

2. If P is the point $\left(-2, \frac{\pi}{6}\right)$ in polar coordinates, then

   $x =$ ________ and $y =$ ________ so that P can be expressed in cartesian coordinates by (____ , ____) .

3. For $P = \left(-2, -\frac{\pi}{6}\right)$ in polar coordinates, $x =$ ________ and $y =$ ________ so that $P =$ (____ , ____) in cartesian coordinates.

OBJECTIVE B: Given a simple equation in polar coordinates, write an equivalent equation in cartesian coordinates and sketch the graph.

4. Consider the equation $r = -3 \sec \theta$. Then, $-3 = r$ ________ , or since $x =$ ________ the equation is equivalent to ________ . This is an equation of a ________ line 3 units to the left of the ___ axis .

5. Consider the equation $r \sin\left(\theta - \frac{\pi}{3}\right) = \frac{1}{2}$. By the trigonometric summation identities, $\sin\left(\theta - \frac{\pi}{3}\right) = \sin\theta \cos\frac{\pi}{3} -$ ________

   $=$ ________ .

   Therefore, the polar equation can be written

   $\frac{1}{2} = \frac{1}{2} r \sin\theta -$ ________ $= \frac{1}{2} y -$ ____ .

   Simplifying algebraically, $y =$ ________ . This is an equation of a line with slope $m =$ ____ and y-intercept $b =$ ____ .

---

1. $r \cos\theta$, $r \sin\theta$
2. $-2 \cos\frac{\pi}{6}$, $-2 \sin\frac{\pi}{6}$, $\left(-\sqrt{3}, -1\right)$
3. $-2 \cos\left(-\frac{\pi}{6}\right)$, $-2 \sin\left(-\frac{\pi}{6}\right)$, $\left(-\sqrt{3}, 1\right)$
4. $\cos\theta$, $r \cos\theta$, $x = -3$, vertical, y
5. $\cos\theta \sin\frac{\pi}{3}$, $\frac{1}{2}\sin\theta - \frac{\sqrt{3}}{2}\cos\theta$, $\frac{\sqrt{3}}{2} r \cos\theta$, $\frac{\sqrt{3}}{2} x$, $\sqrt{3}x + 1$, $\sqrt{3}$, 1

6. Suppose $r = \tan\theta \sec\theta$. Then, $r = \sin\theta \cdot$ ______ or $r = \frac{\sin\theta}{\_\_\_\_\_\_}$.

Equivalently, $\sin\theta =$ __________ or $r \sin\theta =$ __________ . In terms of cartesian coordinates the equation becomes ______ , which is readily recognized as an equation of ____________ .

OBJECTIVE C: Graph the points $P(r,\theta)$ whose polar coordinates satisfy a given equation, inequality or inequalities.

7. $\theta = -\frac{\pi}{4}$, $-2 \le r$

Sketch the graph at the right.

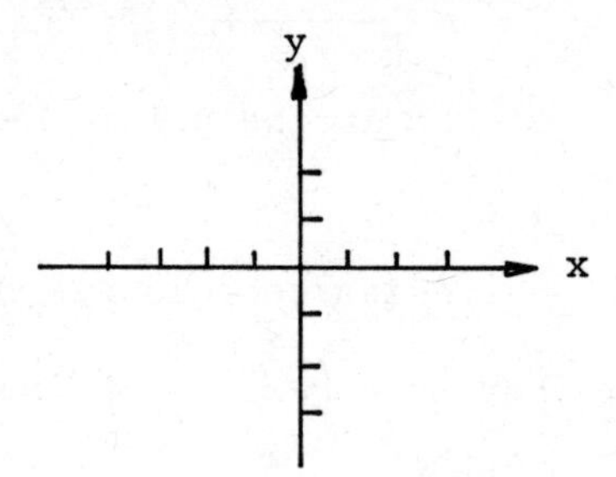

8. $r = -2$, $-\frac{3\pi}{4} < \theta \le \frac{\pi}{6}$

Sketch the graph at the right.

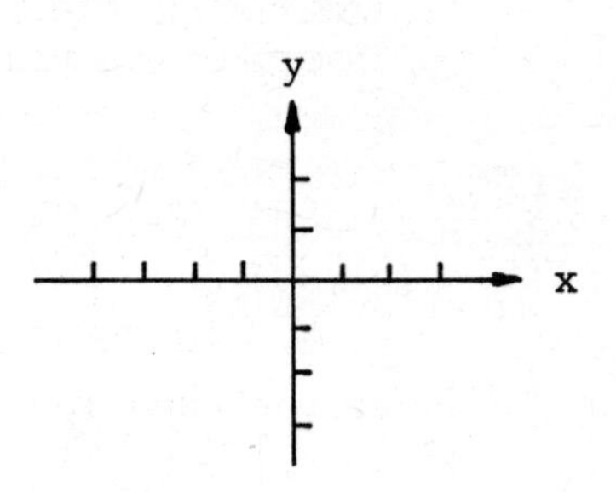

---

6. $\sec^2\theta$, $\cos^2\theta$, $r\cos^2\theta$, $r^2\cos^2\theta$, $y = x^2$, a parabola

7.

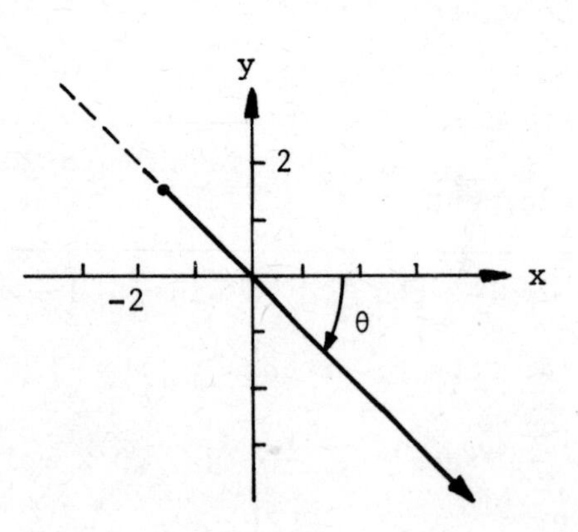

8.

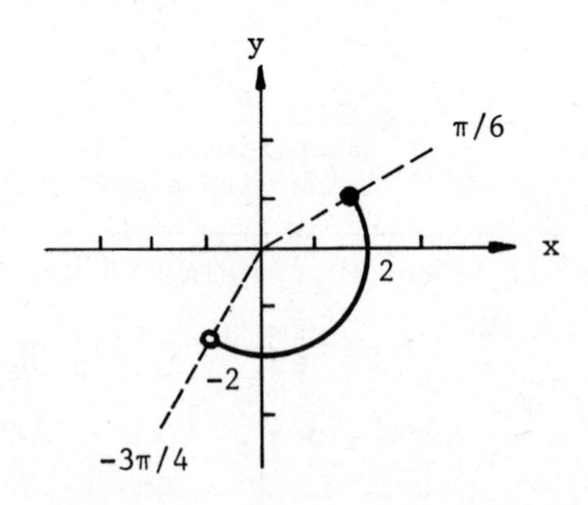

10-2 Graphs of Polar Equations.

OBJECTIVE: Given an equation $F(r,\theta) = 0$ in polar coordinates, analyze and sketch its graph.

9. The graph of $F(r,\theta) = 0$ is symmetric about the x-axis if the equation is unchanged when ___ is replaced by ___ .

10. The graph of $F(r,\theta) = 0$ is symmetric about the origin if the equation is unchanged when ___ is replaced by ___ .

11. The graph of $F(r,\theta) = 0$ is symmetric about the y-axis if the equation is unchanged when ___ is replaced by ______ .

12. Consider the curve given by

$r = 1 - 2 \cos \theta$.

Since $\cos(-\theta) = \cos\theta$, the curve is symmetric about the ________ .

Next, $\frac{dr}{d\theta}$ = __________ . Thus, as θ varies from 0 to $\frac{\pi}{3}$,

r increases from r = ___ to r = ___ ; and as θ varies from $\frac{\pi}{2}$ to π , r increases from r = ___ to r = ___ .

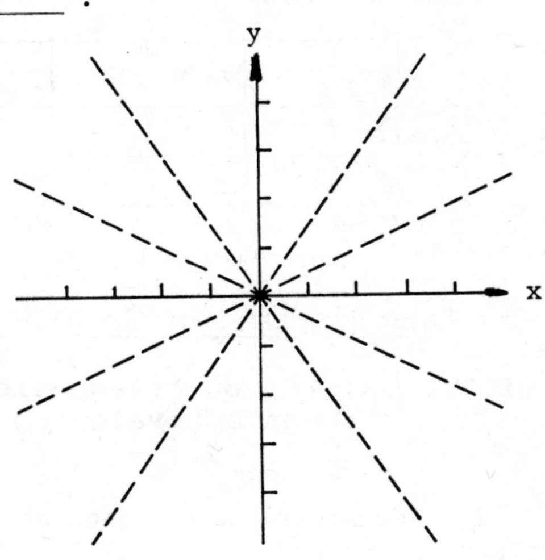

Complete the following table of values for the curve, and sketch its graph using its symmetries.

θ	0	$\pi/6$	$\pi/3$	$\pi/2$	$2\pi/3$	$5\pi/6$	π
r							

---

9. θ, $-\theta$

10. r, $-r$

11. θ, $\pi - \theta$

12. x-axis, $2 \sin\theta$, -1, 0, 1, 3

θ	0	$\pi/6$	$\pi/3$	$\pi/2$	$2\pi/3$	$5\pi/6$	π
r	-1	$1-\sqrt{3}$	0	1	2	$1+\sqrt{3}$	3

The dashed portion of the curve is the rest of it due to its symmetry about the x-axis.

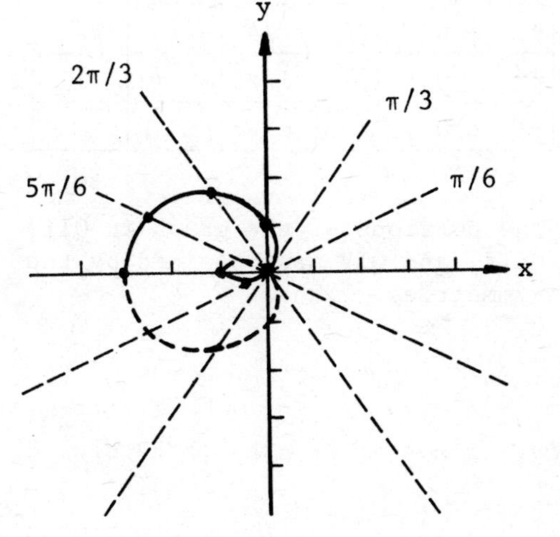

13. Consider the curve given by $r^2 = \sin\theta$. Since the $\sin\theta$ must be nonnegative in order to equal the square of a real number, we must restrict θ to the interval ________. Since the equation remains unchanged when r is replaced by $-r$, the curve is symmetric about the ________. Also, since $\sin(\pi - \theta) = \sin\theta$, the curve is symmetric about the ________. Complete the following table and sketch the graph using these symmetries:

θ	0	$\pi/6$	$\pi/4$	$\pi/3$	$\pi/2$
r^2					
r					

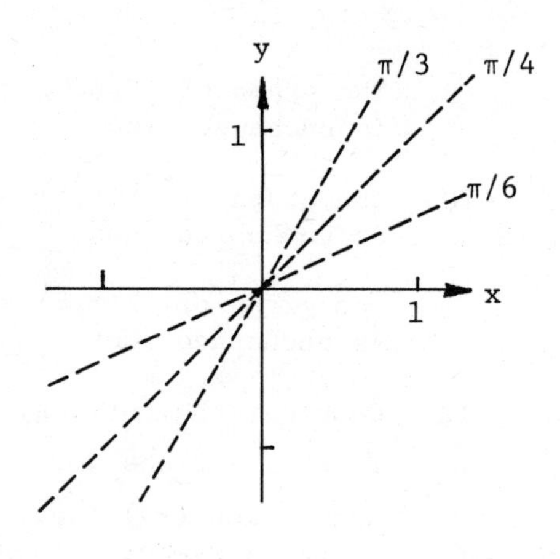

10-3 Polar Equations of the Conic Sections and Other Curves.

OBJECTIVE A: Given an equation in cartesian coordinates, write an equivalent equation in polar coordinates.

14. Consider the equation of the ellipse $9x^2 + (y-2)^2 = 4$. Expanding algebraically, and rearranging terms, we find $9x^2 + y^2 =$ ____. To transform the equation into polar coordinates, substitute $x =$ ________ and $y =$ ________ to obtain:

$$9r^2 \cos^2\theta + \_\_\_\_\_\_\_\_ = 4r\sin\theta, \text{ or}$$

$$9r^2(1 - \sin^2\theta) + \_\_\_\_\_\_\_\_ = 4r\sin\theta .$$

---

13. $0 \le \theta \le \pi$, origin, y-axis

θ	0	$\pi/6$	$\pi/4$	$\pi/3$	$\pi/2$
r^2	0	1/2	$1/\sqrt{2}$	$\sqrt{3}/2$	1
r	0	±.71	±.84	±.93	±1

The portions of the graph in QII, QIII, and QIV are obtained by the symmetries.

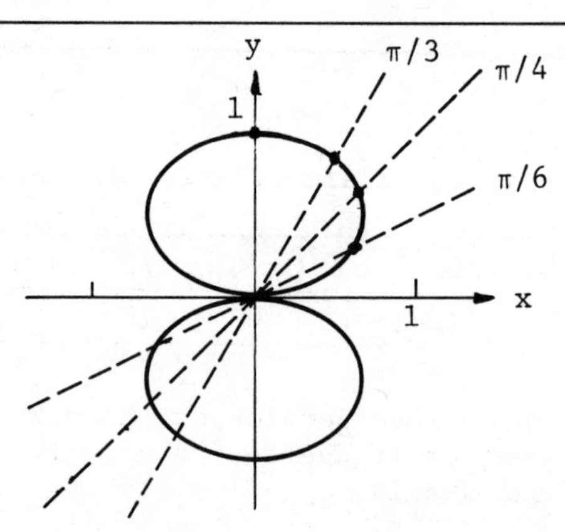

14. $4y$, $r\cos\theta$, $r\sin\theta$, $r^2\sin^2\theta$, $r^2\sin^2\theta$, $9 - 8\sin^2\theta$, $\dfrac{4\sin\theta}{9 - 8\sin^2\theta}$

Simplifying algebraically,

$r^2($ ______________ $) = 4r \sin\theta$.

Either $r = 0$, or $r =$ ________________ . However, the latter equation includes the origin among its points and therefore represents the entire ellipse.

OBJECTIVE B: Given an equation in polar coordinates, find an equivalent cartesian equation and sketch the graph.

15. Given the equation $r = \dfrac{1}{3\cos\theta + 2\sin\theta}$, clear fractions and obtain $3r\cos\theta +$ __________ $= 1$. Next, substitute $x = r\cos\theta$ and $y =$ __________ to obtain $3x +$ ____ $= 1$. This is an equation of a straight line with slope $m =$ ____ and y-intercept $b =$ ____ .

16. The equation $r^2 - 5r + 4 = 0$ factors into $(r-4)($ ______ $) = 0$. Thus, $r = 4$ or $r =$ ___ . The graph is two concentric circles, one of radius 4 and the other of radius 1 , centered at the origin.

17. Consider $r^2 = 2\csc 2\theta$. Then, $r^2 \sin 2\theta =$ ___ . Now, $\sin 2\theta = 2$ ______________ , so the equation becomes $2r\sin\theta \cdot$ __________ $= 2$ or ______ $= 2$. That is, $xy = 1$ which is an equation of ______________ with center __________ .

18. Given the equation $r = \dfrac{3}{1 - 2\cos\theta}$, clear fractions to obtain $r -$ __________ $= 3$; or substituting $x = r\cos\theta$, $r =$ _________ . Hence, $r^2 =$ ____________ $= 9 + 12x +$ _____ . Since $r^2 = x^2 + y^2$ this last equation simplifies to $y^2 = 9 + 12x +$ ____ or $y^2 = 3(x+2)^2 + ($ ___ $)$. Therefore, $(x+2)^2 - \dfrac{\quad}{\quad} = 1$. This is an equation of ______________ with center _________ .

19. For $r = 2\sin\left(\theta + \frac{\pi}{4}\right)$ we can expand the right side by the summation formula for the sine: $r = 2\sin\theta\cos\frac{\pi}{4} +$ _______________ or $r =$ ________________________ .

---

15. $2r\sin\theta$, $r\sin\theta$, $2y$, $-\frac{3}{2}$, $\frac{1}{2}$

16. $r - 1$, 1

17. 2, $\sin\theta\cos\theta$, $r\cos\theta$, $2xy$, a hyperbola, $C(0,0)$

18. $2r\cos\theta$, $3 + 2x$, $(3+2x)^2$, $4x^2$, $3x^2$, -3, $\frac{y^2}{3}$, a hyperbola, $C(-2,0)$

19. $2\cos\theta\sin\frac{\pi}{4}$, $\sqrt{2}\sin\theta + \sqrt{2}\cos\theta$, $\sqrt{2}\, r\cos\theta$, $\left(y - \frac{\sqrt{2}}{2}\right)^2$, a circle, $C\left(\frac{\sqrt{2}}{2}, \frac{\sqrt{2}}{2}\right)$

Hence, $r^2 = \sqrt{2}\ r \sin\theta +$ ____________ . Since $x^2 + y^2 = r^2$, $x = r\cos\theta$, and $y = r\sin\theta$, substitution and algebraic simplification yields

$\left(x - \frac{\sqrt{2}}{2}\right)^2 +$ ____________ $= 1$. This equation represents ____________

with center ____________ and $r = 1$.

10-4 The Angle Between the Radius Vector and the Tangent Line.

OBJECTIVE A: Given a polar curve $r = f(\theta)$, calculate its arc length as θ varies from $\theta = a$ to $\theta = b$.

20. The differential element of arc length ds for the polar curve $r = f(\theta)$ satisfies the equation

$$ds^2 = \_\_\_\_\_\_\_\_\_\_\_\_ .$$

Thus the length of arc traced out by the curve as θ varies from $\theta = a$ to $\theta = b$ is given by

$$s = \_\_\_\_\_\_\_\_\_\_\_\_ .$$

21. To determine the length of the curve $r = 3\sec\theta$ as θ varies from $\theta = 0$ to $\theta = \frac{\pi}{4}$, we find $\frac{dr}{d\theta} =$ ____________ . Then the arc length is,

$$s = \int_0^{\pi/4} \sqrt{\_\_\_\_\_\_\_\_\_\_\_\_}\ d\theta$$

$$= 3\int_0^{\pi/4} \sec\theta \sqrt{\_\_\_\_\_\_\_\_\_\_\_\_}\ d\theta$$

$$= 3\int_0^{\pi/4} \_\_\_\_\_\_\_\_\ d\theta = \_\_\_\_\_\_\_\_ \Big]_0^{\pi/4} = \_\_\_ .$$

---

20. $r^2\, d\theta^2 + dr^2$, $\int_a^b \sqrt{r^2 + \left(\frac{dr}{d\theta}\right)^2}\ d\theta$

21. $3\sec\theta\tan\theta$, $9\sec^2\theta + 9\sec^2\theta\tan^2\theta$, $1 + \tan^2\theta$, $\sec^2\theta$, $3\tan\theta$, 3

22. Consider the polar curve $r = \cos^4 \frac{\theta}{4}$.
As θ varies from $\theta = 0$ to $\theta = 2\pi$, the equation describes the path shown in the figure at the right. As θ varies from $\theta = 2\pi$ to $\theta = 4\pi$ the curve shown is reflected across the x-axis. Thus, the total arc length is given by

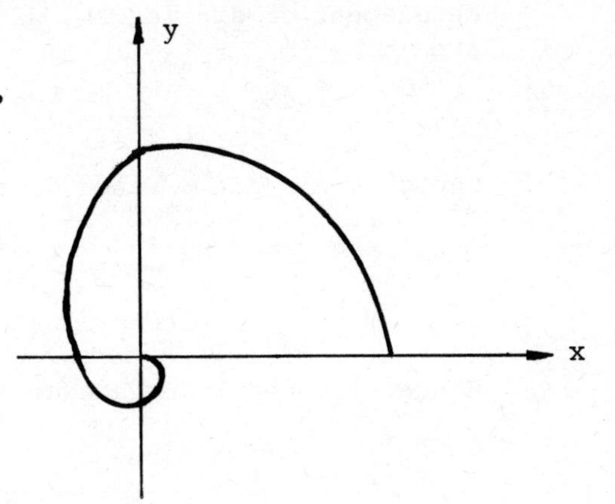

$$s = 2 \int_0^{2\pi} \sqrt{r^2 + \underline{\qquad}}\ d\theta .$$

Now, $\frac{dr}{d\theta} = \underline{\qquad\qquad}$ so that

$$r^2 + \left(\frac{dr}{d\theta}\right)^2 = \cos^8 \frac{\theta}{4} + \underline{\qquad\qquad} = \cos^6 \frac{\theta}{4} .$$

Thus, $s = 2 \int_0^{2\pi} \underline{\qquad\qquad}\ d\theta$.

Since $\cos \frac{\theta}{4} \geq 0$ for $0 \leq \theta \leq 2\pi$, the integral becomes

$$s = 2 \int_0^{2\pi} \left(1 - \sin^2 \frac{\theta}{4}\right) \underline{\qquad}\ d\theta$$

$$= 2 \int_{\_}^{\_} \underline{\qquad\qquad}\ du , \quad \text{where } u = \sin \frac{\theta}{4}$$

$$= \underline{\quad} \left(\underline{\qquad\qquad} \right]_{\_}^{\_} = \underline{\quad} .$$

OBJECTIVE B: Find the area of the surface generated when a polar graph is revolved about the x-axis or the y-axis.

23. The graph of the polar equation $r = 5 \cos \theta$ is the circle shown at the right. If the graph is rotated about the x-axis, the total surface area is generated by that portion of the graph as θ varies from $\theta = 0$ to $\theta = \underline{\quad}$ because of the symmetry of the graph across the x-axis.

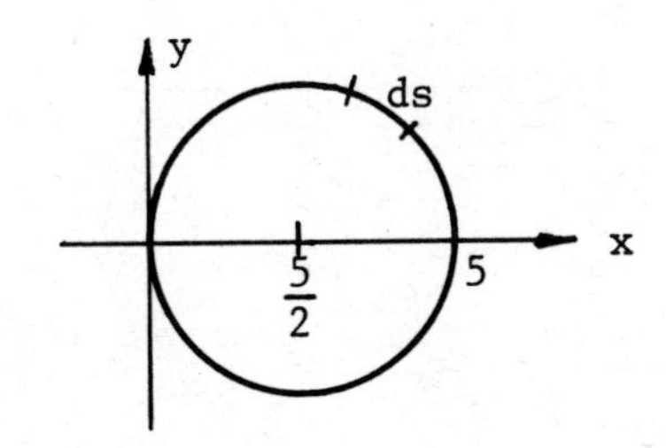

---

22. $\left(\frac{dr}{d\theta}\right)^2$, $-\cos^3 \frac{\theta}{4} \sin \frac{\theta}{4}$, $\cos^6 \frac{\theta}{4} \sin^2 \frac{\theta}{4}$, $\left|\cos^3 \frac{\theta}{4}\right|$, $\cos \frac{\theta}{4}$
$\int_0^1 4(1 - u^2)\ du$, $8 \left(u - \frac{1}{3} u^3\right)\Big]_0^1$, $\frac{16}{3}$

An element of arc length ds (see the figure) generates a portion of surface area

$$dS = \_\_\_\_\_\_ ds ,$$

where $y = r \sin \theta$ and $ds = \_\_\_\_\_\_\_\_\_\_\_\_$.

Thus, $dS = 2\pi \cdot \_\_\_\_\_\_\_\_\_\_\_\_\_\_\_\_\_\_\_\_ d\theta$

$$= 2\pi(5 \cos \theta) \_\_\_\_\_\_\_\_\_ d\theta .$$

Hence, the total surface area is given by

$$S = 50\pi \int_{\_\_}^{\_\_} \_\_\_\_\_\_\_\_\_\_ d\theta = \_\_\_\_\_\_\_\_\_\_ \Big]_{\_\_}^{\_\_}$$

$$= \_\_\_\_\_ \approx 78.54 .$$

24. If the graph in Problem 23 is rotated about the axis $\theta = \pi/2$, the total surface area is generated by the graph as θ varies from $\theta = 0$ to $\theta = \_\_\_$. An element of arc length ds now generates a portion of the surface area

$$dS = \_\_\_\_\_\_\_\_ = 2\pi(\_\_\_\_\_\_\_\_) \cdot 5 \, d\theta .$$

Thus, the total surface area is given by

$$S = 50\pi \int_{\_\_}^{\_\_} \_\_\_\_\_\_\_\_\_\_ d\theta = 50\pi \Big[\_\_\_\_\_\_\_\_\_\_ \Big]_{\_\_}^{\_\_}$$

$$= \_\_\_\_\_ \approx 246.74 .$$

OBJECTIVE C: Find the angle ψ from the radius vector to the tangent line for a given polar curve $r = f(\theta)$.

25. The angle ψ satisfies the equation $\tan \psi = \_\_\_\_\_$.

---

23. $\frac{\pi}{2}$, $2\pi y$, $\sqrt{dr^2 + r^2 d\theta^2}$, $r \sin \theta \sqrt{25 \sin^2 \theta + 25 \cos^2 \theta}$, $5 \sin \theta$,

$\int_0^{\pi/2} \sin \theta \cos \theta \, d\theta$, $50\pi \cdot \frac{1}{2} \sin^2 \theta \Big]_0^{\pi/2}$, 25π

24. π, $2\pi x \, ds$, $r \cos \theta$, $\int_0^{\pi} \cos^2 \theta \, d\theta$, $\frac{1}{2} \theta + \frac{1}{4} \sin 2\theta \Big]_0^{\pi}$, $25\pi^2$

25. $\frac{r \, d\theta}{dr}$

26. To find $\tan \psi$ for the curve $r = \sin \theta$ at the point $\left(\frac{1}{2}, \frac{\pi}{6}\right)$, calculate $\frac{dr}{d\theta}$ = ________ . When $(r,\theta) = \left(\frac{1}{2}, \frac{\pi}{6}\right)$, then $\frac{dr}{d\theta}$ = ______ so that $\tan \psi = \frac{r}{dr/d\theta} = \frac{1/2}{\_\_\_\_} =$ ____ . It follows that ψ = ____ . Therefore the slope of the tangent line to the polar graph at $\left(\frac{1}{2}, \frac{\pi}{6}\right)$ is given by $\tan(\theta + \psi)$ = tan (____) = ____ .

27. To calculate an equation of the tangent line to the curve $r = \sin \theta$ at $\left(\frac{1}{2}, \frac{\pi}{6}\right)$ in Problem 26, one can first express the polar point $\left(\frac{1}{2}, \frac{\pi}{6}\right)$ in cartesian form: $x = \frac{1}{2} \cos$ ____ = ____ and y = ____ $\sin \frac{\pi}{6}$ = ____ . Using the slope m = ____ calculated in the previous problem, a cartesian equation for the tangent line is given by $y - \frac{1}{4}$ = __________ , or simplifying algebraically, $2\left(\sqrt{3}\,x - y\right)$ = ___ .

28. An equation of the tangent line in polar coordinates may be obtained by substituting $x = r \cos \theta$ and $y = r \sin \theta$ into the final equation in Problem 27 and then solving for r :

$2\left(\sqrt{3}\, r \cos \theta - r \sin \theta\right)$ = ____ , or r = ____________________ .

10-5 Plane Areas in Polar Coordinates.

OBJECTIVE: Find the total plane area enclosed by a polar graph $r = f(\theta)$ and the rays $\theta = \alpha$, $\theta = \beta$.

29. The area bounded by the polar curve $r = f(\theta)$ and the rays $\theta = \alpha$, $\theta = \beta$ is given by the integral

A = ____________________ .

30. Find the area inside the larger loop and outside the smaller loop of the polar graph $r = 1 - 2 \cos \theta$ given in Problem 12.

Solution: The graph of the curve is sketched in the figure on the next page. That part of the curve traced out as θ varies from $\theta = 0$ to $\theta = \pi$ is drawn in with a broader ink stroke.

---

26. $\cos \theta$, $\frac{\sqrt{3}}{2}$, $\frac{\sqrt{3}}{2}$, $\frac{\sqrt{3}}{3}$, $\frac{\pi}{6}$, $\frac{\pi}{3}$, $\sqrt{3}$

27. $\frac{\pi}{6}$, $\frac{\sqrt{3}}{4}$, $\frac{1}{2}$, $\frac{1}{4}$, $\sqrt{3}$, $\sqrt{3}\left(x - \frac{\sqrt{3}}{4}\right)$, 1

28. 1, $\frac{1}{2\left(\sqrt{3}\cos \theta - \sin \theta\right)}$

29. $\int_{\alpha}^{\beta} \frac{1}{2}\left[f(\theta)\right]^2 d\theta$

Now, as θ varies from $\frac{\pi}{3}$ to π, the radius vector r sweeps out the larger loop of the curve including that portion of the smaller loop lying above the x-axis. By symmetry, the area of that smaller half-loop is the same as the area of the half-loop swept out by r as θ varies from 0 to $\frac{\pi}{3}$. Thus, the total area inside the larger loop and outside the smaller loop is

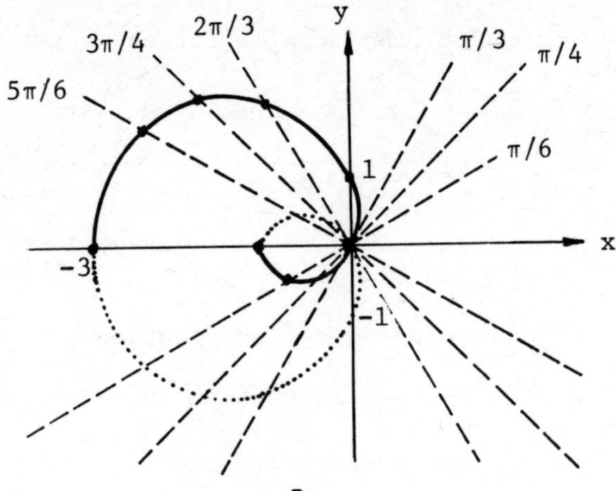

$$A = 2\left[\int_{\pi/3}^{\pi} \frac{1}{2} f^2(\theta)\, d\theta - \underline{\qquad\qquad\qquad\qquad}\right]$$

$$= \int_{\pi/3}^{\pi} \left(1 - 4\cos\theta + 4\cos^2\theta\right) d\theta - \underline{\qquad\qquad\qquad\qquad}$$

$$= \left[\theta - 4\sin\theta + 4\left(\frac{\theta}{2} + \frac{1}{4}\sin 2\theta\right)\right]_{\pi/3}^{\pi}$$

$$- \underline{\qquad\qquad\qquad\qquad}$$

$$= (\pi + 2\pi) - \left[\frac{\pi}{3} - 4\left(\frac{\sqrt{3}}{2}\right) + 4\left(\frac{\pi}{6} + \frac{1}{4}\cdot\frac{\sqrt{3}}{2}\right)\right]$$

$$- \underline{\qquad\qquad\qquad\qquad}$$

$$= \underline{\qquad\qquad} \approx 8.338 .$$

---

30. $\int_0^{\pi/3} \frac{1}{2} f^2(\theta)\, d\theta$, $\int_0^{\pi/3} \left(1 - 4\cos\theta + 4\cos^2\theta\right) d\theta$,

$\left[\theta - 4\sin\theta + 4\left(\frac{\theta}{2} + \frac{1}{4}\sin 2\theta\right)\right]_0^{\pi/3}$, $\left[\frac{\pi}{3} - 4\left(\frac{\sqrt{3}}{2}\right) + 4\left(\frac{\pi}{6} + \frac{1}{4}\cdot\frac{\sqrt{3}}{2}\right)\right]$,

$3\sqrt{3} + \pi$

CHAPTER 10 OBJECTIVE - PROBLEM KEY

Objective	Problems in Thomas/Finney Text	Objective	Problems in Thomas/Finney Text
10-1 A	p. 463, 1,10	10-3 B	p. 472, 8-14
B	p. 463, 2, 11-18	10-4 A	p. 477, 5,8,9,10
	p. 467, 2	B	p. 477, 7,11
C	p. 463, 3-9	C	p. 477, 1,4
10-2	p. 467, 3-10, 12	10-5	p. 479, 1-7
10-3 A	p. 471, 3-7		

CHAPTER 10 SELF-TEST

1. Convert the following from polar coordinates to cartesian coordinates.

   (a) $\left(-6, \frac{\pi}{4}\right)$ (b) $\left(1, -\frac{5\pi}{6}\right)$ (c) $\left(-2, -\frac{7\pi}{12}\right)$

2. Write the following simple polar equations in cartesian form.

   (a) $r = 5$ (b) $\theta = \frac{3\pi}{4}$ (c) $r = -5 \csc \theta$

In Problems 3 and 4, graph the polar equation.

3. $r = 2 \cos 4\theta$ 4. $r^2 = -\sin 2\theta$

5. Determine a cartesian equation, and sketch the curve, for $r = 5 \cos \theta + 5 \sin \theta$.

6. Find a polar equation of the line with slope $m = -2$ passing through the cartesian point $(1,-3)$.

7. Find a polar equation of the circle centered at the cartesian point $\left(\frac{1}{3}, 0\right)$ and passing through the origin.

8. Find the tangent of the angle ψ between the radius vector to the polar curve $r = 1 + 2 \sin \theta$ and the tangent line at the point $(2, \pi/6)$. Use this result to calculate the slope of the tangent line at that point.

9. Find the length of the polar curve $r = a \cos(\theta + b)$ from $\theta = 0$ to $\theta = \pi$, where a and b are constants.

10. Find the area of the region bounded on the outside by the graph of $r = 2 + 2 \sin \theta$ for $\theta = 0$ to $\theta = \pi$, and on the inside by the graph of $r = 2 \sin \theta$.

11. Write an integral expressing the surface area generated by rotating the portion of the polar curve $r = 1 + \cos \theta$ in the first quadrant about $\theta = \pi/2$.

SOLUTIONS TO CHAPTER 10 SELF-TEST

1. (a) $x = -6\cos\frac{\pi}{4} = -6\cdot\frac{\sqrt{2}}{2} = -3\sqrt{2}$; $y = -6\sin\frac{\pi}{4} = -3\sqrt{2}$

(b) $x = 1\cos\left(-\frac{5\pi}{6}\right) = \cos\frac{5\pi}{6} = -\frac{\sqrt{3}}{2}$; $y = 1\sin\left(-\frac{5\pi}{6}\right) = -\sin\frac{5\pi}{6} = -\frac{1}{2}$

(c) $x = -2\cos\left(-\frac{7\pi}{12}\right) = -2\cos\frac{7\pi}{12} = \frac{\sqrt{2}}{2}\left(\sqrt{3}-1\right) \approx 0.518$

$y = -2\sin\left(-\frac{7\pi}{12}\right) = 2\sin\frac{7\pi}{12} = \frac{\sqrt{2}}{2}\left(\sqrt{3}+1\right) \approx 1.932$

2. (a) $\pm\sqrt{x^2+y^2} = 5$ or $x^2+y^2 = 25$

(b) $y = -x$

(c) Equivalently, $r\sin\theta = -5$, or $y = -5$

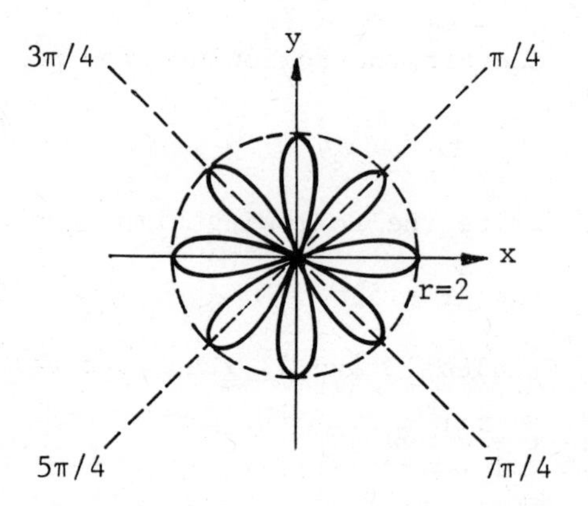

3. $r = 2\cos 4\theta$ is symmetric about the x-axis, the y-axis, and the origin.

The graph is the eight-leafed rose sketched at the right.

4. $r^2 = -\sin 2\theta$ is symmetric about the origin since the equation is unchanged when r is replaced by $-r$. Notice that $-\sin 2\theta$ must be nonnegative. If θ is restricted to the interval $[0, 2\pi]$, then $-\sin 2\theta \geq 0$ if and only if θ is in $\left[\frac{\pi}{2}, \pi\right]$ or $\left[\frac{3\pi}{2}, 2\pi\right]$. Using the following table and symmetry we obtain the graph skecthed at the right:

θ	$\pi/2$	$7\pi/12$	$3\pi/4$	π	$3\pi/2$
r	0	$\sqrt{2}/2$	1	0	0

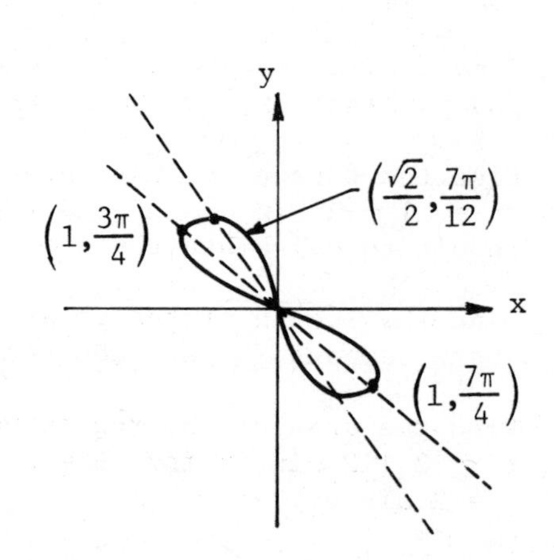

The curve is a lemniscate.

5. Since $r = 5\cos\theta + 5\sin\theta$, for $r \neq 0$ ($r = 0$ is on the graph at $\theta = \frac{3\pi}{4}$),

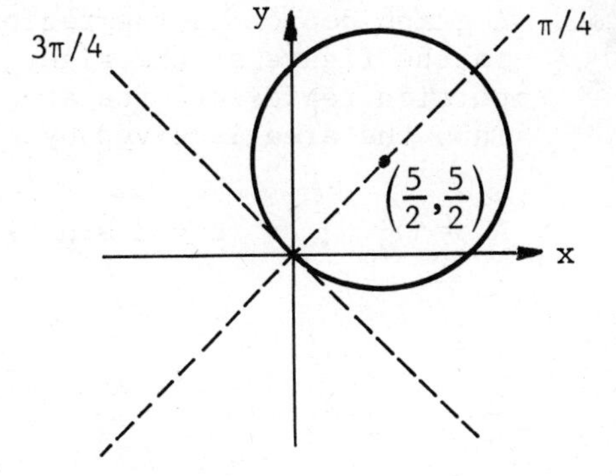

$$r^2 = 5r\cos\theta + 5r\sin\theta, \quad \text{or}$$

$$x^2 + y^2 = 5x + 5y.$$

Then, completing the squares in the x and y terms gives

$$\left(x - \frac{5}{2}\right)^2 + \left(y - \frac{5}{2}\right)^2 = \frac{25}{2}.$$

This is a circle with center $\left(\frac{5}{2}, \frac{5}{2}\right)$ and radius $\frac{5}{\sqrt{2}}$.

6. A cartesian equation of the line is given by $y + 3 = -2(x-1)$ or $2x + y = -1$. Thus, $2r\cos\theta + r\sin\theta = -1$, or solving for r,

$$r = \frac{-1}{2\cos\theta + \sin\theta}.$$

7. A cartesian equation of the circle is given by $\left(x - \frac{1}{3}\right)^2 + y^2 = \frac{1}{9}$ or $x^2 + y^2 = \frac{2}{3}x$. Thus, $3r^2 = 2r\cos\theta$, or since $r = 0$ lies on the graph $3r = 2\cos\theta$ when $\theta = \frac{\pi}{2}$, the latter gives a polar equation of the circle.

8. For $r = 1 + 2\sin\theta$, $\frac{dr}{d\theta} = 2\cos\theta = \sqrt{3}$ when $\theta = \pi/6$. Hence, $\tan\psi = \frac{r\,d\theta}{dr} = \frac{2}{\sqrt{3}}$. The slope of the tangent line is then given by

$$m = \tan(\theta + \psi) = \tan\left(\frac{\pi}{6} + \psi\right) = \frac{\tan(\pi/6) + \tan\psi}{1 - \tan(\pi/6)\tan\psi}$$

$$= \frac{(1/\sqrt{3}) + (2/\sqrt{3})}{1 - (1/\sqrt{3})(2/\sqrt{3})} = 3\sqrt{3}.$$

9. For $r = a\cos(\theta + b)$, $\frac{dr}{d\theta} = -a\sin(\theta + b)$ so that

$$r^2 + \left(\frac{dr}{d\theta}\right)^2 = a^2\cos^2(\theta + b) + a^2\sin^2(\theta + b) = a^2.$$

Therefore, the arc length is given by the integral

$$s = \int_0^{\pi} \sqrt{a^2}\, d\theta = |a|\,\pi.$$

10. A graph depicting the region is shown in the figure at the right (the shaded portion represents the area we seek). Thus the area is given by

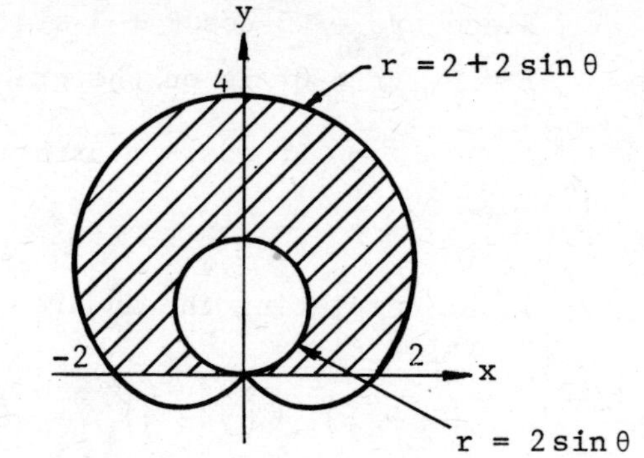

$$A = \frac{1}{2}\int_0^{\pi} (2 + 2\sin\theta)^2\, d\theta$$

$$- \frac{1}{2}\int_0^{\pi} (2\sin\theta)^2\, d\theta$$

$$= \frac{1}{2}\int_0^{\pi} (4 + 8\sin\theta)\, d\theta = 2(\theta - 2\cos\theta)\Big]_0^{\pi} = 2\pi + 8 \approx 14.28 .$$

11. A sketch of the surface is shown in the figure at the right. An element of arc length ds generates a portion of surface area

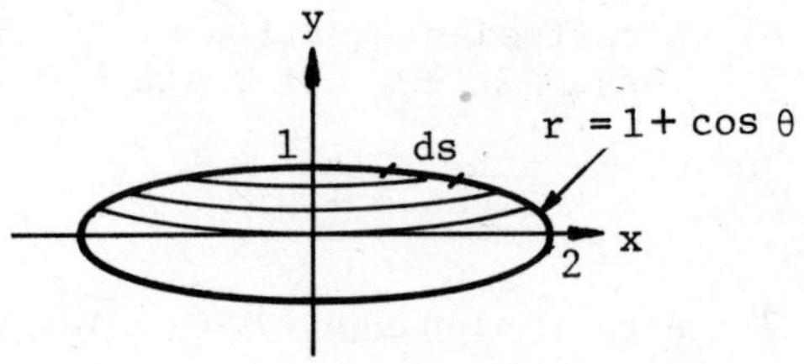

$$dS = 2\pi x\, ds .$$

Now, $$\frac{ds}{d\theta} = \sqrt{r^2 + \left(\frac{dr}{d\theta}\right)^2}$$

$$= \sqrt{(1 + 2\cos\theta + \cos^2\theta) + \sin^2\theta} = \sqrt{2}\,\sqrt{1 + \cos\theta} .$$

Hence, $dS = 2\pi x\, ds = 2\pi r\cos\theta\, ds = 2\sqrt{2}\,\pi\,(1 + \cos\theta)^{3/2}\cos\theta\, d\theta$.

Therefore, the total surface area generated is given by the integral

$$S = 2\sqrt{2}\,\pi\int_0^{\pi/2} (1 + \cos\theta)^{3/2}\cos\theta\, d\theta .$$

(The definite integral can be evaluated by using the identity

$\cos^2\frac{\theta}{2} = 1 + \cos\theta$, but it is tedious to carry out the calculations.

Using Simpson's rule with $n = 12$, an approximate value to the integral is 39.31 . This calculation was made using a calculator.)

CHAPTER 11 VECTORS AND PARAMETRIC EQUATIONS

11-1 Vector Components and the Unit Vectors **i** and **j** .

OBJECTIVE A: Given the points P_1 and P_2 in the plane, express the vector $\overrightarrow{P_1P_2}$ in the form $a\mathbf{i} + b\mathbf{j}$.

1. Consider the points $P_1(x_1,y_1)$ and $P_2(x_2,y_2)$. The directed change in the x-direction from P_1 to P_2 is a = ________ , and in the y-direction it is b = ________ . Hence, the vector $\overrightarrow{P_1P_2}$ can be expressed as $\overrightarrow{P_1P_2}$ = ____________________ .

2. The vector from the point $P_1(-7,4)$ to the point $P_2(3,-5)$ is $\overrightarrow{P_1P_2} = (3 -$ ___ $)\mathbf{i} + ($ ___ $- 4)\mathbf{j}$ or $\overrightarrow{P_1P_2}$ = ________ .

3. The vector from $P(2,-9)$ to the origin is $\overrightarrow{PO}$ = __________ .

OBJECTIVE B: Express the sum and difference of two given vectors, and multiples of given vectors by scalars, in the form $a\mathbf{i} + b\mathbf{j}$.

4. The sum of $\mathbf{v}_1 = a_1\mathbf{i} + b_1\mathbf{j}$ and $\mathbf{v}_2 = a_2\mathbf{i} + b_2\mathbf{j}$ is

   $\mathbf{v}_1 + \mathbf{v}_2$ = ____________________ .

   The difference $\mathbf{v}_2 - \mathbf{v}_1$ is given by

   $\mathbf{v}_2 - \mathbf{v}_1$ = ____________________ .

5. If $\mathbf{v}_1 = 3\mathbf{i} - 4\mathbf{j}$ and $\mathbf{v}_2 = -5\mathbf{i} + 2\mathbf{j}$, then

   $\mathbf{v}_1 + \mathbf{v}_2$ = ________ and $\mathbf{v}_2 - \mathbf{v}_1$ = ________ .

6. If $\mathbf{v} = -3\mathbf{i} + 7\mathbf{j}$, then

   $-\mathbf{v}$ = ________ , $2\mathbf{v}$ = ________ ,

   $-5\mathbf{v}$ = ________ , and $\frac{1}{4}\mathbf{v}$ = ________ .

---

1. $x_2 - x_1$, $y_2 - y_1$, $(x_2 - x_1)\mathbf{i} + (y_2 - y_1)\mathbf{j}$
2. -7, -5, $10\mathbf{i} - 9\mathbf{j}$
3. $-2\mathbf{i} + 9\mathbf{j}$
4. $(a_1 + a_2)\mathbf{i} + (b_1 + b_2)\mathbf{j}$, $(a_2 - a_1)\mathbf{i} + (b_2 - b_1)\mathbf{j}$
5. $-2\mathbf{i} - 2\mathbf{j}$, $-8\mathbf{i} + 6\mathbf{j}$
6. $3\mathbf{i} - 7\mathbf{j}$, $-6\mathbf{i} + 14\mathbf{j}$, $15\mathbf{i} - 35\mathbf{j}$, $-\frac{3}{4}\mathbf{i} + \frac{7}{4}\mathbf{j}$

OBJECTIVE C: Given a vector $\mathbf{v} = a\mathbf{i} + b\mathbf{j}$, calculate its length or magnitude, and the angle it makes with the positive x-axis.

7. The length or magnitude of the vector $\mathbf{v} = a\mathbf{i} + b\mathbf{j}$ is given by

$|\mathbf{v}|$ = ____________ .

8. The angle θ that $\mathbf{v} = a\mathbf{i} + b\mathbf{j}$ makes with the positive x-axis satisfies

$\cos\theta$ = ____________ and $\sin\theta$ = ____________ .

9. The length of $\mathbf{v} = -\sqrt{2}\,\mathbf{i} + 7\mathbf{j}$ is $|\mathbf{v}|$ = ____________ .

10. Consider the vector $\mathbf{v} = -\mathbf{i} + \sqrt{3}\,\mathbf{j}$. The length of $\mathbf{v}$ is $|\mathbf{v}|$ = ________ so a unit vector $\mathbf{u}$ having the same direction as $\mathbf{v}$ is given by $\mathbf{u} = \frac{\mathbf{v}}{|\mathbf{v}|}$ = ______________ . Thus, if θ is the angle $\mathbf{v}$ (or $\mathbf{u}$) makes with the positive x-axis, then

$\cos\theta$ = ____ and $\sin\theta$ = ____ .

It follows that $\theta = \frac{\pi}{2} +$ ____ = ____ radians.

OBJECTIVE D: Find unit vectors tangent and normal to a given curve $y = f(x)$ at a specified point $P(a,b)$.

11. Consider the curve $y = 1 - 2e^x$. Observe that the point $P(0,-1)$ lies on the curve. The slope of the line tangent to the curve at P is

$y' =$ ______$\Big|_{x=0}$ = ___ . A vector $\mathbf{v}$ having this slope is given by

$\mathbf{v}$ = ___ $\mathbf{i}$ + ____ $\mathbf{j}$. (Any nonzero multiple of $\mathbf{v}$ will also have the slope -2 , so $\mathbf{v}$ is not unique.) A unit vector parallel to $\mathbf{v}$ is given by

$\mathbf{u} = \frac{\mathbf{v}}{|\mathbf{v}|}$ = ______________ .

The vector $\mathbf{u}$ is a unit vector tangent to the curve at $P(0,-1)$; the vector $-\mathbf{u}$ is also a unit vector tangent to the curve at $P(0,-1)$.

---

7. $\sqrt{a^2+b^2}$ 8. $\frac{a}{\sqrt{a^2+b^2}}$, $\frac{b}{\sqrt{a^2+b^2}}$ 9. $\sqrt{2+49} = \sqrt{51}$

10. $\sqrt{1+3} = 2$, $-\frac{1}{2}\mathbf{i} + \frac{\sqrt{3}}{2}\mathbf{j}$, $-\frac{1}{2}$, $\frac{\sqrt{3}}{2}$, $\frac{\pi}{6}$, $\frac{2\pi}{3}$

11. $-2e^x$, -2, 1, -2, $\frac{1}{\sqrt{5}}\mathbf{i} - \frac{2}{\sqrt{5}}\mathbf{j}$

12. One unit vector normal to the curve at P for Problem 11 above is

n = ________________ . A second vector is **-n** = ________________ .

Either vector will do as a unit vector normal to $y = 1 - 2e^x$ at $P(0,-1)$.

11-2 Parametric Equations in Kinematics.

OBJECTIVE A: Use the parametric equations describing the motion of a projectile fired with initial velocity v_0 at an angle of elevation α to answer questions concerning the motion of the projectile. Assume that air resistance is neglected and gravity is the only force acting on the projectile.

13. For a projectile fired with an initial velocity v_0 ft/sec at an angle of elevation α , the parametric equations for its motion are $x = c_1 t +$ ___ ft and $y =$ ___ $+ c_3 t + c_4$ ft, where c_1, c_2, c_3, and c_4 are constants determined from the initial conditions. Here t is measured in __________ .

14. The initial conditions give the ____________________ $x(t_0)$ and $y(t_0)$, and the ____________________ $v_x(t_0)$ and $v_y(t_0)$ at some given instant $t = t_0$.

15. Suppose a boy throws a ball at an angle of 45° with the horizontal and with an initial speed of 30 ft/sec toward a building 35 ft away . If the ball leaves the boy's hand from a height of 3 ft above the ground, will it reach the building?

Solution. For this problem, the initial conditions are $t = 0$ sec , $x =$ ___ ft , $y =$ ___ ft , $v_x(0) = 30 \cos$ ___ ≈ 21.21 ft/sec , and $v_y(0) =$ ___ $\sin 45° \approx$ ______ ft/sec . The parametric equations for the motion are $x = c_1 t + c_2$ and $y = -\frac{1}{2}gt^2 + c_3 t + c_4$. From the initial conditions, we find $c_2 = 0$, $c_1 =$ ______ , $c_4 =$ ___ , and $c_3 = 21.21$.

To find the horizontal range R , we set $y = -\frac{1}{2}gt^2 + 21.21t + 3$ equal to _______ and solve for t : using $g = 32$ ft/sec^2 ,

---

12. $\frac{2}{\sqrt{5}}\,\mathbf{i} + \frac{1}{\sqrt{5}}\,\mathbf{j}$, $-\frac{2}{\sqrt{5}}\,\mathbf{i} - \frac{1}{\sqrt{5}}\,\mathbf{j}$ 13. c_2, $-\frac{1}{2}gt^2$, seconds

14. position coordinates, velocity components

15. 0, 3, 45°, 30, 21.21, 21.21, 3, zero, 192, 21.21, x, 30.75, does not

$$t \approx \frac{-21.21 \pm \sqrt{(21.21)^2 + \_\_\_\_\_\_}}{-32} \approx 1.45 \text{ sec.}$$

(since t must be nonnegative we take the negative square root) . Then, $R = \_\_\_\_\_\_ \cdot (1.45)$, the value of ___ when $t = 1.45$. Hence, $R \approx$ ________ ft . Since $R < 35$ ft , the ball ___________ reach the building.

OBJECTIVE B: Find parametric equations for the curve described by the point $P(x,y)$ for $t \geq 0$ if its coordinates satisfy given elementary differential equations of the type $dx/dt = f(x)$ and $dy/dt = g(x)$, or $dx/dt = f(y)$ and $dy/dt = g(y)$, together with initial conditions.

16. Consider the equations $dx/dt = 1 + x^2$, $dy/dt = x$, with initial conditions $t = 0$, $x = 0$, and $y = 2$. Then,

$$\frac{dx}{1+x^2} = dt \quad \text{or} \quad \_\_\_\_\_\_\_\_ = t + c_1 .$$

Since $x = 0$ when $t = 0$, it follows that $c_1 =$ ____ .
Thus, $x =$ ________ .
Next,

$$\frac{dy}{dt} = x = \_\_\_\_\_\_\_\_ \quad \text{or} \quad dy = \tan t \, dt .$$

Hence,

$$y = \_\_\_\_\_\_\_\_\_\_\_\_ + c_2 .$$

Then, $y = 2$ when $t = 0$ implies $c_2 =$ ____ .

Therefore, $y =$ ______________ .

11-3 Parametric Equations in Analytic Geometry.

OBJECTIVE A: Sketch the graph of a curve given in parametric form $x = f(t)$ and $y = g(t)$ as the parameter t varies over a given domain. Also, find a cartesian equation for the curve.

17. Consider the curve given by the parametric equations $x = t - 2$, $y = 2t + 3$, $-\infty < t < \infty$. Complete the following table providing some of the points $P(x,y)$ on the curve:

---

16. $\tan^{-1} x$, 0, $\tan t$, $\tan t$, $-\ln |\cos t|$, 2, $2 - \ln |\cos t|$

t	-2	-1	0	1	2	3
x	-4					
y	-1					

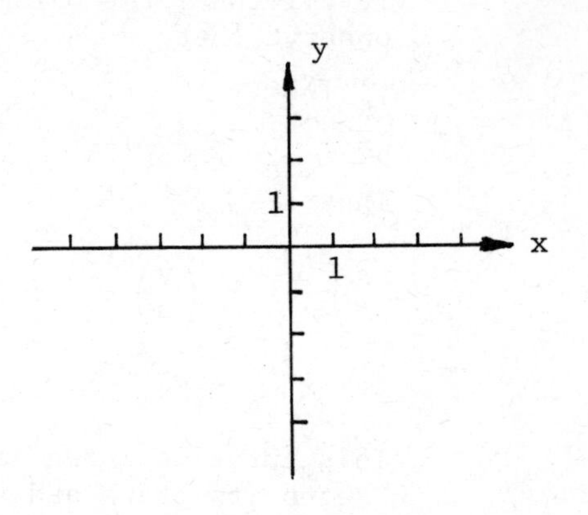

To eliminate the parameter t, note that $t = x+2$. Substitution for t in the parametric equation for y gives y = ________ . This is a cartesian equation for a ________ with slope m = ___ and y-intercept b = ___ . Sketch the curve in the coordinate system at the right.

18. Consider the curve $x = a\cos^3 t$, $y = a\sin^3 t$, $-\infty < t < \infty$. Complete the following table providing some of the points $P(x,y)$ on the curve:

t	0	$\frac{\pi}{6}$	$\frac{\pi}{4}$	$\frac{\pi}{3}$	$\frac{\pi}{2}$	$\frac{3\pi}{4}$	π	$\frac{3\pi}{2}$	2π
x	a								
y	0								

17.

t	-2	-1	0	1	2	3
x	-4	-3	-2	-1	0	1
y	-1	1	3	5	7	9

$2x+7$, line, 2, 7

$x = t - 2$
$y = 2t + 3$
(-3,1)
(-4,-1)

18.

t	0	$\frac{\pi}{6}$	$\frac{\pi}{4}$	$\frac{\pi}{3}$	$\frac{\pi}{2}$	$\frac{3\pi}{4}$	π	$\frac{3\pi}{2}$	2π
x	a	$\approx .6a$	$\frac{a}{2\sqrt{2}}$	$\frac{a}{8}$	0	$-\frac{a}{2\sqrt{2}}$	-a	0	a
y	0	$\frac{a}{8}$	$\frac{a}{2\sqrt{2}}$	$\approx .6a$	a	$\frac{a}{2\sqrt{2}}$	0	-a	0

$\cos t$, $\sin t$, $\cos^2 t + \sin^2 t$, 1, $a^{2/3}$

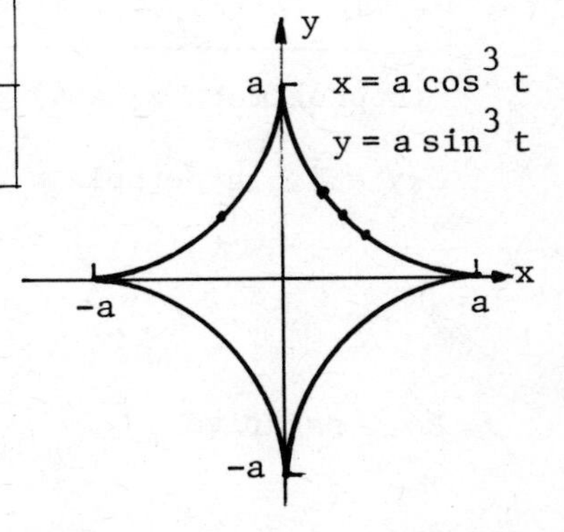

To eliminate the parameter t, observe that

$$\left(\frac{x}{a}\right)^{1/3} = \_\_\_\_\_\_\_\_ \quad \text{and} \quad \left(\frac{y}{a}\right)^{1/3} = \_\_\_\_\_\_\_\_ .$$

Thus,

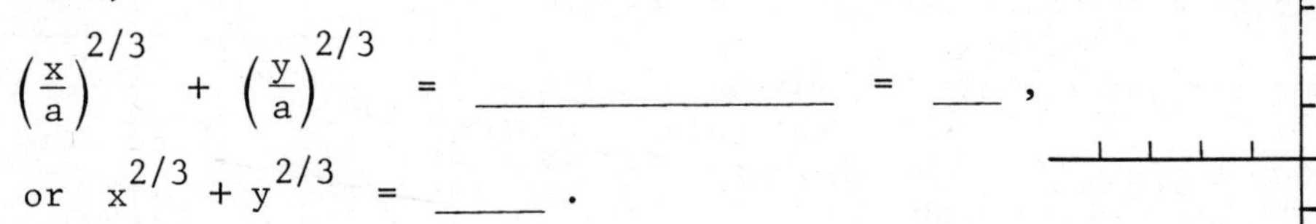

$$\left(\frac{x}{a}\right)^{2/3} + \left(\frac{y}{a}\right)^{2/3} = \_\_\_\_\_\_\_\_\_\_\_\_\_\_\_\_ = \_\_\_ ,$$

or $x^{2/3} + y^{2/3} = \_\_\_\_\_$.

This curve is known as a hypocycloid. Sketch its graph and notice its symmetries.

19. For the curve given by the parametric

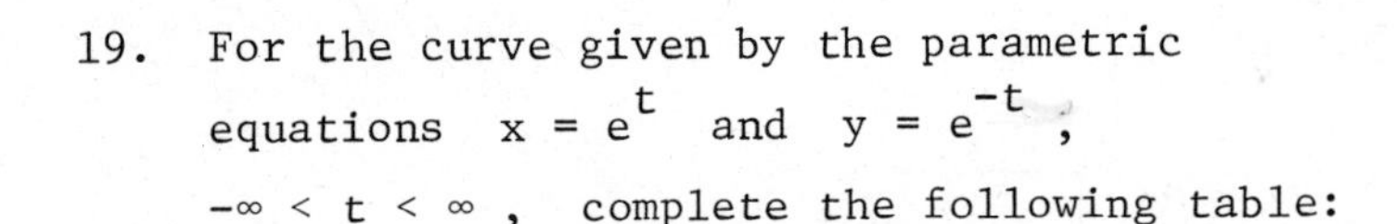

equations $x = e^t$ and $y = e^{-t}$, $-\infty < t < \infty$, complete the following table:

t	-2	-1	0	1	2	3
x						
y						

To eliminate the parameter t, notice that $xy = \_\_\_$. This equation describes a ____________ . Sketch the graph in the coordinate system at the right.

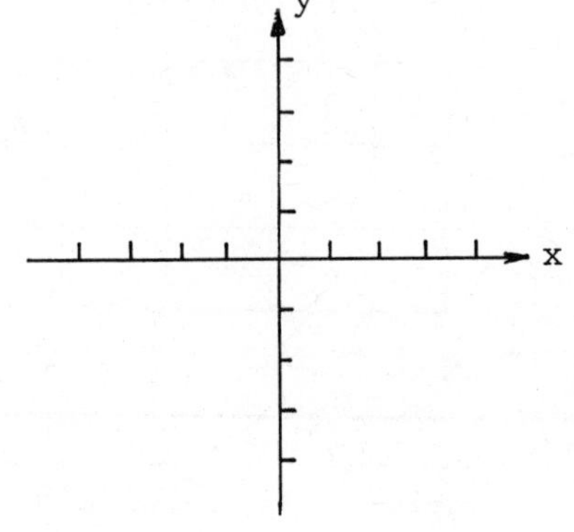

20. For the curve given parametrically in Problem 19, notice that x and y are always positive. Are the parametric equations and the cartesian equation coextensive? _____ , because x and y can both be __________ in the cartesian equation $xy = 1$.

19.

t	-2	-1	0	1	2	3
x	.14	.37	1	2.7	7.4	20
y	7.4	2.7	1	.37	.14	.05

(approximate values)

$xy = 1$, hyperbola

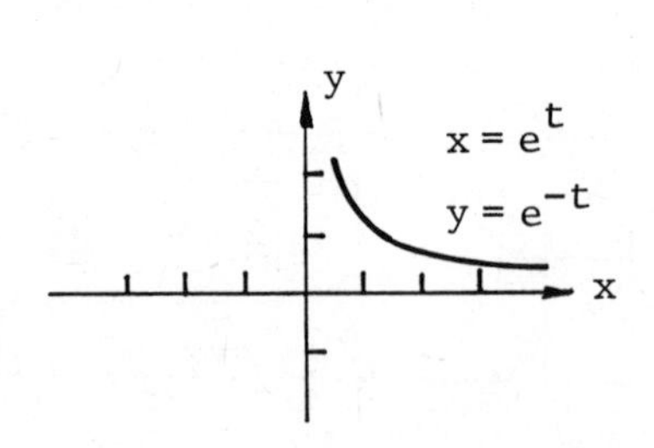

20. No, negative

OBJECTIVE B: Find parametric equations for a curve described geometrically, or by an equation, in terms of some specified or arbitrary parameter.

21. Find parametric equations for the circle with center $C(-2,3)$ and radius $r = \sqrt{2}$.

Solution. An equation of the circle is $(x+2)^2 +$ ______ = ___ , or ____________ = 1 . This suggests the substitutions

$\frac{x+2}{\sqrt{2}} = \sin\theta$ and $\frac{y-3}{\sqrt{2}} =$ ______ . Hence, parametric equations for the circle are $x =$ _________ and $y =$ _________ , $0 \le \theta \le 2\pi$.

22. Find parametric equations for the line in the plane through the point (a,b) with slope m , where the parameter t is the change $x-a$.

Solution. For any point $P(x,y)$ on the line, $y-b = m($ ____ $)$ = ____ . Thus, $x =$ ______ and $y =$ ______ give parametric equations of the line in terms of the specified parameter t .

11-4 Space Coordinates.

OBJECTIVE A: Describe the set of points in space whose cartesian, cylindrical, or spherical coordinates satisfy given pairs of simultaneous equations.

23. Consider the pair of cartesian equations $y^2 - z^2 = 1$, $x = 2$. The equation $y^2 - z^2 = 1$ represents a _________ . This curve lies in the _______ $x = 2$ that is _________ to the yz-plane and 2 units from it. The set of points representing the curve is sketched at the right.

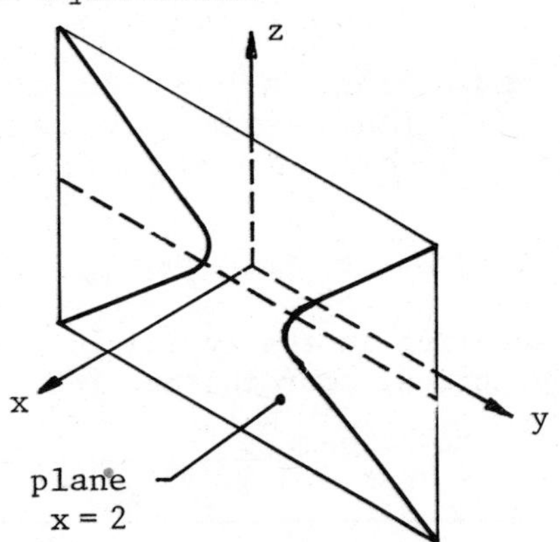

---

21. $(y-3)^2$, 2, $\left(\frac{x+2}{\sqrt{2}}\right)^2 + \left(\frac{y-3}{\sqrt{2}}\right)^2$, $\cos\theta$, $\sqrt{2}\sin\theta - 2$, $3 + \sqrt{2}\cos\theta$

22. $x-a$, mt, $a+t$, $b+mt$

23. hyperbola, plane, parallel

24. Consider the pair of cylindrical equations $z = 5$, $r = \theta$. The equation $z = 5$ represents a ________ that is parallel to the __________ and 5 units above it. The equation $r = \theta$ represents a __________ on that plane. It is sketched at the right.

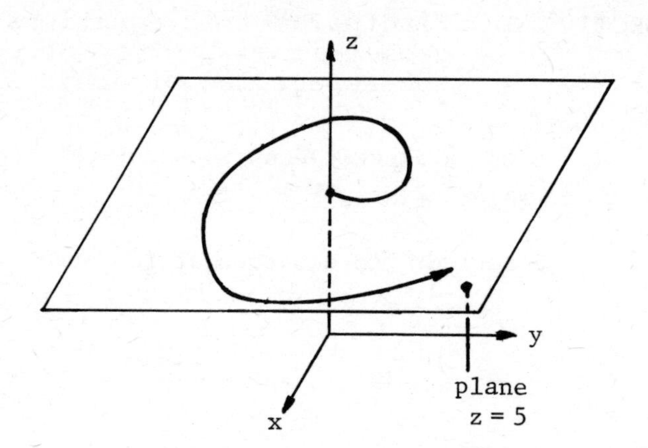

25. Consider the pair of equations $\theta = \pi/4$, $\rho = 3 \sec \phi$ representing a set of points in spherical coordinates. The equation $\rho = 3 \sec \phi$ can be written $3 =$ __________, or $3 =$ ___. Thus it represents a ________ parallel to the __________ and 3 units above it. The equation $\theta = \pi/4$ represents a ____________ so the set of points described by the pair of equations is a half-line or ray at an angle $\pi/4$ with the __________. It is sketched at the right.

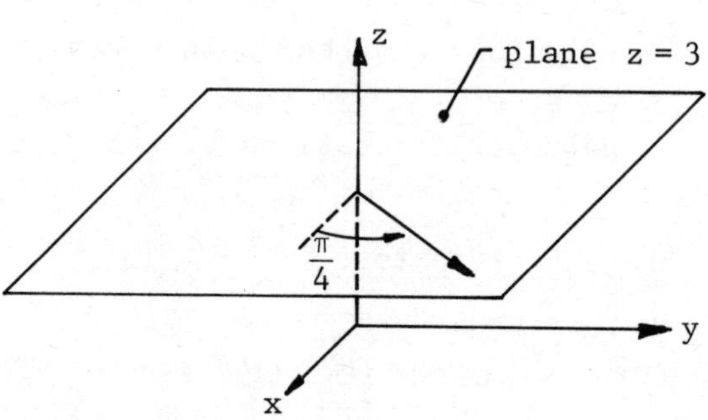

OBJECTIVE B: Translate an equation from a given coordinate system (cartesian, cylindrical, or spherical) into forms that are appropriate to the other two systems.

26. To translate the equation $z = x^2 + y^2$ into cylindrical coordinates, let $x =$ __________ and $y =$ __________. Then, $z =$ ____ is an appropriate equation in the cylindrical coordinate system. To translate this into spherical coordinates, let $z =$ __________ and $r =$ __________ yielding the equation ______________________ or $\cos \phi =$ __________.

---

24. plane, xy-plane, spiral

25. $\rho \cos \phi$, z, plane, xy-plane, half-plane, xz-plane, $z = 3$

26. $r \cos \theta$, $r \sin \theta$, r^2, $\rho \cos \phi$, $\rho \sin \phi$, $\rho \cos \phi = \rho^2 \sin^2 \phi$, $\rho \sin^2 \phi$

27. To translate the equation $\phi = \pi/4$ into cylindrical coordinates, let $r = $ __________ $=$ _______ . Then $2r^2 = $ ___ $= z^2 +$ ___ , or $r^2 = $ ___ . To translate this into cartesian equations, note that $r^2 = $ __________ so the equation becomes ______________ . The equation $\phi = \pi/4$ describes a ________ with vertex at the origin 0 , axis Oz , and generating angle $\pi/4$.

11-5 Vectors in Space.

OBJECTIVE A: Given two points P_1 and P_2 in space, express the vector $\overrightarrow{P_1P_2}$ from P_1 to P_2 in the form $a\mathbf{i} + b\mathbf{j} + c\mathbf{k}$. Also, calculate the distance between P_1 and P_2 .

28. The vector from the point $(-1,3,0)$ to the point $(4,-2,1)$ is given by $(4+1)\mathbf{i}$ + ________ $\mathbf{j}$ + ________ $\mathbf{k}$ or ______________ .

29. The distance between $(-1,3,0)$ and $(4,-2,1)$ is the length of the vector found in Problem 28. The distance is given by

$$\sqrt{(4+1)^2 + (\_\_\_\_\_\_) + (\_\_\_\_\_\_)} = \sqrt{\_\_\_\_} \approx 7.14 .$$

OBJECTIVE B: Find the length of any space vector.

30. The length of the vector $\mathbf{v} = -2\mathbf{i} + 6\mathbf{j} - 5\mathbf{k}$ is

$$|\mathbf{v}| = \sqrt{(-2)^2 + (\_\_\_\_) + (\_\_\_\_)} = \sqrt{\_\_\_} \approx 8.06 .$$

OBJECTIVE C: Given a cartesian equation of a sphere in space, find the coordinates of its center and the radius.

31. Consider the equation of the sphere given by $x^2 + y^2 + z^2 + 2x - 6y + 4z + 9 = 0$. Complete the squares in the given equation to obtain,

$(x^2 + 2x + 1) + (y^2 - 6y + 9) + ($ ____________ $) = $ _______ , or

$(x+1)^2 + (y-3)^2 + (z+2)^2 = 5$.

Thus, the center is ______________ and the radius is ____ .

---

27. $\rho \sin \phi$, $\frac{\sqrt{2}}{2}\rho$, ρ^2, r^2, z^2, $x^2 + y^2$, $z^2 = x^2 + y^2$, cone

28. $(-2-3)$, $(1-0)$, $5\mathbf{i} - 5\mathbf{j} + \mathbf{k}$

29. $(-2-3)^2$, $(1-0)^2$, 51

30. 6^2, $(-5)^2$, 65

31. $z^2 + 4z + 4$, $1 + 4$, $(z+2)^2$, 5, $(-1,3,-2)$, $\sqrt{5}$

OBJECTIVE D: Given a nonzero space vector find its direction.

32. If $\mathbf{A}$ is a nonzero vector, <u>the direction of</u> $\mathbf{A}$ is a _______ vector obtained from $\mathbf{A}$ by dividing $\mathbf{A}$ by _______ .

33. To compute the direction of the vector $\mathbf{A} = -2\mathbf{i} + 6\mathbf{j} - 5\mathbf{k}$, we divide $\mathbf{A}$ by $|\mathbf{A}| =$ _____ found in Problem 30. This gives the unit vector

$\mathbf{u} = \dfrac{\mathbf{A}}{|\mathbf{A}|} =$ ______________________________ as the direction of $\mathbf{A}$.

11-6 The Scalar Product of Two Vectors.

OBJECTIVE A: Find the scalar product of two vectors in space and the cosine of the angle between them.

34. The scalar or dot product of the vectors $\mathbf{A} = 2\mathbf{i} + \mathbf{j} - 3\mathbf{k}$ and $\mathbf{B} = 3\mathbf{i} - 6\mathbf{j} + 2\mathbf{k}$ is $\mathbf{A} \cdot \mathbf{B} = (2)(3) + (1)(\_\_) + (\_\_)(\_\_) = \_\_\_$.

35. For the vectors in Problem 34,

$|\mathbf{A}| = \sqrt{4+1+\_\_\_} = \_\_\_\_\_$ and $|\mathbf{B}| = \sqrt{9+36+\_\_\_} = \_\_\_\_\_$.

Thus, the cosine of the angle θ between them is

$\cos\theta = \dfrac{\_\_\_\_\_}{|\mathbf{A}||\mathbf{B}|} = \dfrac{-6}{\_\_\_\_\_} \approx -.229$, or $\theta \approx 103°$.

OBJECTIVE B: Given two space vectors $\mathbf{A}$ and $\mathbf{B}$ find the projection vector $\text{proj}_{\mathbf{A}}\,\mathbf{B}$ and the component of $\mathbf{B}$ in the direction $\mathbf{A}$.

36. The projection of $\mathbf{B}$ onto $\mathbf{A}$ is the vector

$\text{proj}_{\mathbf{A}}\,\mathbf{B} =$ ____________ .

37. The component of $\mathbf{B}$ in the direction $\mathbf{A}$ is the scalar __________ .

Since $\mathbf{A} \cdot \mathbf{A} = |\mathbf{A}|^2$, the magnitude of the vector $\text{proj}_{\mathbf{A}}\,\mathbf{B}$ is given by

$|\text{proj}_{\mathbf{A}}\,\mathbf{B}| = \dfrac{|\mathbf{B}\cdot\mathbf{A}|}{|\mathbf{A}|^2}\,|\mathbf{A}| =$ __________ .

---

32. unit, $|\mathbf{A}|$

33. $\sqrt{65}$, $-\dfrac{2}{\sqrt{65}}\mathbf{i} + \dfrac{6}{\sqrt{65}}\mathbf{j} - \dfrac{5}{\sqrt{65}}\mathbf{k}$

34. -6, -3, 2, -6

35. 9, $\sqrt{14}$, 4, 7, $\mathbf{A}\cdot\mathbf{B}$, $7\sqrt{14}$

36. $\left(\dfrac{\mathbf{B}\cdot\mathbf{A}}{\mathbf{A}\cdot\mathbf{A}}\right)\mathbf{A}$

37. $\dfrac{\mathbf{B}\cdot\mathbf{A}}{|\mathbf{A}|}$, $|\mathbf{A}|^2$, $\dfrac{|\mathbf{B}\cdot\mathbf{A}|}{|\mathbf{A}|}$, $\text{proj}_{\mathbf{A}}\,\mathbf{B}$

Thus, the absolute value of the scalar component of B in the direction A is simply the magnitude of __________ .

38. Consider the vectors $A = -2i + j + 2k$ and $B = 2i - 6j - 3k$. Then,

$\text{proj}_A B = \left(\frac{B \cdot A}{A \cdot A}\right) A = \frac{\quad}{4+1+4} A = \underline{\quad} A = $ __________ .

Then the component of B in the direction A is the scalar

$\frac{B \cdot A}{|A|} = $ ____ . The fact that this component is negative means that the vector $\text{proj}_A B$ points in the __________ direction of A .

39. For the vectors in Problem 38 ,

$\text{proj}_B A = \left(\frac{A \cdot B}{B \cdot B}\right) \underline{\quad} = \frac{-16}{\underline{\quad}} B = $ __________ .

The component of A in the direction of B is $\frac{A \cdot B}{\underline{\quad}} = $ ____ .

OBJECTIVE C: Use vector methods to calculate the distance between a given point P(x,y) and a given line L in the xy-plane.

40. Find the distance between the point P(1,-7) and the line L: $5y - 7x + 14 = 0$.

Solution. If we write the equation of the line L in the form $7x - 5y = 14$, we find a normal vector to the line is $N =$ __________ . The point B(0, ____) lies on the line. Thus, the vector $\overrightarrow{BP} =$ __________ is a vector from the line L to P . The magnitude of the projection of $\overrightarrow{BP}$ onto the normal N gives the distance between P and L :

$d = |\text{proj}_N \overrightarrow{BP}| = \frac{|\underline{\quad} \cdot N|}{|N|} = \frac{|\underline{\quad}|}{\sqrt{74}} = $ __________ .

Notice that since $|\text{proj}_N \overrightarrow{BP}|$ is the <u>absolute value</u> of the component of $\overrightarrow{BP}$ in the direction N we need not be concerned with whether $\overrightarrow{BP}$ points in the same or opposite direction to N : either way the procedure correctly computes a positive distance between P and L .

---

38. $-4-6-6$, $-\frac{16}{9}$, $\frac{32}{9} i - \frac{16}{9} j - \frac{32}{9} k$, $-\frac{16}{3}$, opposite

39. B, 49, $-\frac{32}{49} i + \frac{96}{49} j + \frac{48}{49} k$, $|B|$, $-\frac{16}{7}$

40. $7i - 5j$, $-\frac{14}{5}$, $i - \frac{21}{5} j$, $\overrightarrow{BP}$, $7+21$, $\frac{14\sqrt{74}}{37}$

OBJECTIVE D: Given a vector $\mathbf{B}$ and a vector $\mathbf{A}$, write $\mathbf{B}$ as the sum of a vector $\mathbf{B}_1$ parallel to $\mathbf{A}$ and a vector $\mathbf{B}_2$ perpendicular to $\mathbf{A}$.

41. Let $\mathbf{B} = -2\mathbf{i} + 3\mathbf{j} - \frac{1}{2}\mathbf{k}$ and $\mathbf{A} = \mathbf{i} + 4\mathbf{j} - 8\mathbf{k}$. The vector $\mathbf{B}_1 = \text{proj}_{\mathbf{A}}\mathbf{B}$ is a vector parallel to $\mathbf{A}$:

$$\text{proj}_{\mathbf{A}}\mathbf{B} = \left(\_\_\_\_\_\right)\mathbf{A} = \frac{\_\_\_}{81}\mathbf{A} = \_\_\_\_\_\_\_\_\_\_\_\_\_\_\_\_ .$$

The vector $\mathbf{B}_2 = \mathbf{B} - \mathbf{B}_1$ is perpendicular to $\mathbf{A}$:

$$\mathbf{B}_2 = \left(-2 - \frac{14}{81}\right)\mathbf{i} + \left(3 - \_\_\_\right)\mathbf{j} + \left(-\frac{1}{2} - \_\_\_\right)\mathbf{k}$$

$$= \_\_\_\_\_\_\_\_\_\_\_\_\_\_\_\_\_\_\_\_ .$$

OBJECTIVE E: Know the following five properties of the scalar product.

42. $\mathbf{B}\cdot\mathbf{A} = \_\_\_\_\_\_$.

43. $\mathbf{A}\cdot(\mathbf{B}+\mathbf{C}) = \_\_\_\_\_\_\_\_\_\_$.

44. $\mathbf{A}\cdot\mathbf{A} = 0$ if and only if $\_\_\_\_\_$.

45. $\mathbf{A}\cdot(c\mathbf{B}) = (c\mathbf{A})\cdot\mathbf{B} = \_\_\_\_\_\_\_$.

46. $\mathbf{A}\cdot\mathbf{A}$ is ________ for every nonzero vector $\mathbf{A}$.

11-7 The Vector Product of Two Vectors in Space.

OBJECTIVE A: Define the cross product of two vectors in space, and give at least five properties of the cross product.

47. $\mathbf{A}\times\mathbf{B} = (\_\_\_\_\_\_\_\_\_\_\_\_)\mathbf{n}$, where $\mathbf{n}$ is a ______ vector ___________ to the plane of $\mathbf{A}$ and $\mathbf{B}$, and pointing in the direction a right-threaded screw advances when its head is rotated from ___ to ___ through the angle θ from $\mathbf{A}$ to $\mathbf{B}$.

48. $\mathbf{B}\times\mathbf{A} = \_\_\_\_\_\_\_$.

49. $\mathbf{A}\times(\mathbf{B}+\mathbf{C}) = \_\_\_\_\_\_\_\_\_\_\_$.

---

41. $\frac{\mathbf{B}\cdot\mathbf{A}}{\mathbf{A}\cdot\mathbf{A}}$, 14, $\frac{14}{81}\mathbf{i} + \frac{56}{81}\mathbf{j} - \frac{112}{81}\mathbf{k}$, $\frac{56}{81}$, $-\frac{112}{81}$, $-\frac{176}{81}\mathbf{i} + \frac{187}{81}\mathbf{j} + \frac{143}{162}\mathbf{k}$

42. $\mathbf{A}\cdot\mathbf{B}$

43. $\mathbf{A}\cdot\mathbf{B} + \mathbf{A}\cdot\mathbf{C}$

44. $\mathbf{A} = \vec{0}$

45. $c(\mathbf{A}\cdot\mathbf{B})$

46. positive

47. $|\mathbf{A}||\mathbf{B}|\sin\theta$, unit, perpendicular, $\mathbf{A}$, $\mathbf{B}$

48. $-\mathbf{A}\times\mathbf{B}$

49. $(\mathbf{A}\times\mathbf{B}) + (\mathbf{A}\times\mathbf{C})$

50. $(\mathbf{B}+\mathbf{C}) \times \mathbf{A}$ = ________________ . 51. $(r\mathbf{A}) \times (s\mathbf{B})$ = ___________ .

52. If $\mathbf{A}\times\mathbf{B} = \vec{0}$, then $\mathbf{A} = \vec{0}$ or $\mathbf{B} = \vec{0}$, or $\mathbf{A}$ and $\mathbf{B}$ are ____________ vectors.

53. $\mathbf{i}\times\mathbf{j}$ = ___ , $\mathbf{j}\times\mathbf{k}$ = ___ , and $\mathbf{k}\times\mathbf{i}$ = ___ .

These three equations are easily remembered by writing the vectors in cyclic order $\mathbf{i}, \mathbf{j}, \mathbf{k}, \mathbf{i}, \mathbf{j}$ and realizing that the cross product of any two results in the next one in line.

OBJECTIVE B: Calculate the cross product of any two vectors in space whose $\mathbf{i}$- , $\mathbf{j}$- , and $\mathbf{k}$- components are given.

54. Let $\mathbf{A} = \mathbf{i}+\mathbf{j}+\mathbf{k}$ and $\mathbf{B} = -5\mathbf{i}+3\mathbf{j}-2\mathbf{k}$. Then

$$\mathbf{A}\times\mathbf{B} = \begin{vmatrix} \mathbf{i} & \mathbf{j} & \mathbf{k} \\ \_\_ & \_\_ & \_\_ \\ \_\_ & \_\_ & \_\_ \end{vmatrix} = (-2-3)\mathbf{i} + (-5+\_\_)\mathbf{j} + (\_\_\_\_)\mathbf{k}$$

$$= \_\_\_\_\_\_\_\_\_\_\_\_ .$$

55. For the vectors in Problem 54, the magnitude $|\mathbf{A}\times\mathbf{B}|$ = ______ represents the area of the ______________ with sides ___ and $\mathbf{B}$.

OBJECTIVE C: Find a vector that is perpendicular to two given vectors in space.

56. Let $\mathbf{A} = \mathbf{i}+\mathbf{k}$ and $\mathbf{B} = \mathbf{i}+2\mathbf{j}-\mathbf{k}$. Then a vector perpendicular to both $\mathbf{A}$ and $\mathbf{B}$ is $\mathbf{A}\times\mathbf{B}$ = ________________ .

57. The three points $A(1,0,2)$, $B(3,1,4)$, and $C(-1,5,1)$ determine a plane in space. The vectors $\overrightarrow{AB}$ = ______________ and $\overrightarrow{AC}$ = ____________ both lie in that plane. Thus, the cross product $\overrightarrow{AB}\times\overrightarrow{AC}$ = ________________ gives a vector that is perpendicular to the plane containing A, B, and C . Other vectors perpendicular to that plane are $\overrightarrow{CB}\times\overrightarrow{CA}$, $\overrightarrow{BA}\times\overrightarrow{BC}$, $\overrightarrow{AB}\times\overrightarrow{BC}$, etc.

---

50. $(\mathbf{B}\times\mathbf{A}) + (\mathbf{C}\times\mathbf{A})$

51. $rs(\mathbf{A}\times\mathbf{B})$

52. parallel

53. $\mathbf{k}, \mathbf{i}, \mathbf{j}$

54. $\begin{vmatrix} \mathbf{i} & \mathbf{j} & \mathbf{k} \\ 1 & 1 & 1 \\ -5 & 3 & -2 \end{vmatrix}$, 2, $3+5$, $-5\mathbf{i}-3\mathbf{j}+8\mathbf{k}$

55. $7\sqrt{2}$, parallelogram, $\mathbf{A}$

56. $-2\mathbf{i}+2\mathbf{j}+2\mathbf{k}$

57. $2\mathbf{i}+\mathbf{j}+2\mathbf{k}$, $-2\mathbf{i}+5\mathbf{j}-\mathbf{k}$, $-11\mathbf{i}-2\mathbf{j}+12\mathbf{k}$

OBJECTIVE D: Find the area of any triangle with specified vertices in space, and find the distance between the origin and the plane determined by that triangle.

58. The area of the triangle determined by the points in Problem 57 above is

$$\text{area} = \frac{1}{2}\ \underline{\hspace{3cm}} = \frac{1}{2}\sqrt{|2| + \underline{\hspace{1cm}} + \underline{\hspace{1cm}}} = \underline{\hspace{2cm}}\ .$$

59. To calculate the distance between the origin and the plane determined by the triangle in Problem 57, simply calculate the magnitude of the projection of any vector $\overrightarrow{OP}$ from the origin O to a point P on the plane onto a direction vector perpendicular to the plane. For instance, the point A(1,0,2) lies on the plane and $\overrightarrow{OA} = \underline{\hspace{2cm}}$. A vector orthogonal to the plane is $\mathbf{N} = \overrightarrow{AB} \times \overrightarrow{AC}$ which was calculated in Problem 57. Hence,

$$|\text{proj}_{\mathbf{N}}\ \overrightarrow{OA}| = \frac{|\overrightarrow{OA} \cdot \underline{\hspace{1cm}}|}{|\mathbf{N}|} = \frac{|\underline{\hspace{3cm}}|}{\sqrt{269}} = \underline{\hspace{2cm}}$$

gives the distance between the origin and the plane of the triangle.

11-8 Equations of Lines and Planes.

OBJECTIVE A: Write parametric and cartesian equations of a line in space given (a) two points on the line, or (b) a point on the line and a vector parallel to the line.

60. Consider the line L determined by the two points $P\left(\frac{1}{2}, 0, 1\right)$ and $Q\left(\frac{1}{2}, -1, 2\right)$. The vector $\overrightarrow{PQ} = \underline{\hspace{2cm}}$ lies in a direction parallel to the line. Then, using the point P on the line,

$$x = \frac{1}{2} + 0t\ ,\quad y = \underline{\hspace{2cm}}\ ,\quad \text{and}\quad z = \underline{\hspace{2cm}}$$

are parametric equations for the line. Using the point Q ,

$$x = \frac{1}{2} + 0s\ ,\quad y = \underline{\hspace{2cm}}\ ,\quad \text{and}\quad z = \underline{\hspace{2cm}}$$

is another set of parametric equations representing the same line. The point $\left(\frac{1}{2}, -2, 3\right)$ lies on the line when $t = 2$ and when $s = \underline{\hspace{1cm}}$. The two sets of equations describe the same set of points (a straight line), but any particular point on that line will be given by different values of the parameters t and s in each representation.

---

58. $|\overrightarrow{AB} \times \overrightarrow{AC}|$, 4, 144, $\frac{1}{2}\sqrt{269}$ 59. $\mathbf{i} + 2\mathbf{k}$, $\mathbf{N}$, $-11 + 24$, $\frac{13}{\sqrt{269}} \approx .79262$

60. $-\mathbf{j} + \mathbf{k}$, $0 - t$, $1 + t$, $-1 - s$, $2 + s$, 1

61. To obtain cartesian equations for the line in Problem 60, solve for the parameter in each equation (if possible) of a set of parametric equations for the line, and then equate the parameters: For instance, from the parametric equations

$$x = \frac{1}{2} + 0t, \quad y = -t, \quad \text{and} \quad z = 1+t$$

we find that $0t =$ ______ so $x =$ ___ must hold. From the second equation $t =$ ____, and from the third $t =$ _______ . Thus, $x = \frac{1}{2}$ and $y =$ _______ give cartesian equations for the line. However, these cartesian equations are not unique. Using the second set of parametric equations in Problem 60 gives $x = \frac{1}{2}$, $s = -y - 1$, and $s = z - 2$; or $x = \frac{1}{2}$ and $z - 2 = -y - 1$ as cartesian equations for the same line

62. A line through the point $(4,-2,2)$ in the direction parallel to the vector $\mathbf{A} = -2\mathbf{i} + \mathbf{j} + \mathbf{k}$ is given by $x =$ _______, $y =$ ________, and $z =$ ________ .

OBJECTIVE B: Write an equation of a plane in space given (a) a point on the plane and a vector normal to the plane, or (b) three noncollinear points on the plane.

63. An equation of the plane in space through the point $(1,-1,7)$ and perpendicular to the vector $\mathbf{v} = -2\mathbf{i} - 4\mathbf{j} + 5\mathbf{k}$ is given by $-2(x-1) + (-4)($ ______ $) +$ ___ $(z-7) = 0$, or simplifying algebraically, ______________ $= -37$.

64. To find an equation of the plane containing the three points $A(4,-2,3)$, $B(3,2,-1)$, and $C(-1,1,1)$ observe that the vectors $\overrightarrow{AB} =$ ______________ and $\overrightarrow{AC} =$ ________________ both lie in the plane. Hence $\overrightarrow{AB} \times \overrightarrow{AC} =$ ________________ is normal to the plane. Using the point A on the plane, an equation for it is $4(x-4) +$ ___ $(y+2) + 17($ ______ $) = 0$, or ___________________ $= 31$.

OBJECTIVE C: Find the distance between a given point and plane in space.

65. To find the distance between the point P and a plane, pick any point Q on the plane and calculate the magnitude of the projection vector of $\overrightarrow{PQ}$ onto a vector perpendicular to the plane. Let us find the distance between the point $P(5,-1,0)$ and the plane $2x + 4y - 5z = -37$.

---

61. $x - \frac{1}{2}$, $\frac{1}{2}$, $-y$, $z - 1$, $1 - z$

62. $4 - 2t$, $-2 + t$, $2 + t$

63. $y + 1$, 5, $2x + 4y - 5z$

64. $-\mathbf{i} + 4\mathbf{j} - 4\mathbf{k}$, $-5\mathbf{i} + 3\mathbf{j} - 2\mathbf{k}$, $4\mathbf{i} + 18\mathbf{j} + 17\mathbf{k}$, 18, $z - 3$, $4x + 18y + 17z$

The point $Q(0,0,\_\_\_)$ lies on the plane. A vector normal to the plane is $\mathbf{N} = 2\mathbf{i} + \_\_\_\_\_\_\_\_$, and the vector $\overrightarrow{PQ} = \_\_\_\_\_\_\_\_\_\_\_\_$. Thus,

$$|\text{proj}_{\mathbf{N}}\ \overrightarrow{PQ}| = \frac{|\overrightarrow{PQ} \cdot \mathbf{N}|}{\_\_\_\_} = \frac{|\_\_\_|}{\sqrt{45}} \approx 6.41$$

gives the distance between the point and the plane.

OBJECTIVE D: Find the distance between a given point and line in space.

66. Suppose it is desired to find the distance between the point $P(7,3,-1)$ and the line $L: x = 4 - 2t$, $y = -2 + t$, and $z = 2 + t$. The vector $\mathbf{v} = -2\mathbf{i} + \mathbf{j} + \mathbf{k}$ is ________ to L (see figure at the right). Choose any point on the line L, say $Q(4,-2,2)$. Let θ denote the angle between $\mathbf{v}$ and $\overrightarrow{QP}$. Then, from the above diagram,

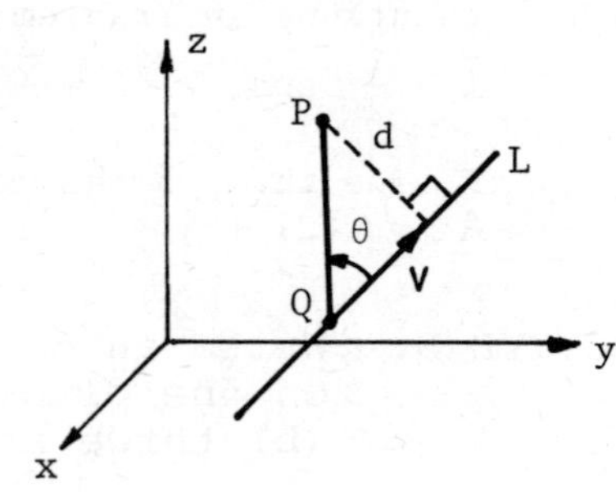

$d = |\overrightarrow{QP}| \sin\theta$ so that $|\mathbf{v}|\, d = |\mathbf{v}|\,|\overrightarrow{QP}| \sin\theta = \_\_\_\_\_\_\_\_$. Thus,

$d = \dfrac{|\mathbf{v} \times \overrightarrow{QP}|}{|\mathbf{v}|}$ gives the desired distance. Now, $\overrightarrow{QP} = \_\_\_\_\_\_\_\_$ and

$$\mathbf{v} \times \overrightarrow{QP} = \begin{vmatrix} \mathbf{i} & \mathbf{j} & \mathbf{k} \\ -2 & 1 & 1 \\ \_\_ & \_\_ & \_\_ \end{vmatrix} = \_\_\_\_\_\_\_\_\_\_ .$$

Hence, $d = \dfrac{\overline{\qquad\qquad}}{\sqrt{4+1+1}} = \_\_\_\_\_$.

11-9 Products of Three Vectors or More.

OBJECTIVE: Given any three vectors $\mathbf{A}$, $\mathbf{B}$, and $\mathbf{C}$ in space, calculate the triple scalar product $\mathbf{A} \cdot (\mathbf{B} \times \mathbf{C})$.

67. For the vectors $\mathbf{A} = 3\mathbf{i} + 5\mathbf{j} - 2\mathbf{k}$, $\mathbf{B} = -\mathbf{i} + 4\mathbf{j} - 4\mathbf{k}$, and $\mathbf{C} = 4\mathbf{j} - \mathbf{k}$, the triple scalar product $\mathbf{A} \cdot (\mathbf{B} \times \mathbf{C})$ is given by the determinant

---

65. $\frac{37}{5}$, $4\mathbf{j} - 5\mathbf{k}$, $-5\mathbf{i} + \mathbf{j} + \frac{37}{5}\mathbf{k}$, $|\mathbf{N}|$, -43

66. parallel, $|\mathbf{v} \times \overrightarrow{QP}|$, $3\mathbf{i} + 5\mathbf{j} - 3\mathbf{k}$, 3, 5, -3, $-8\mathbf{i} - 3\mathbf{j} - 13\mathbf{k}$,

$\sqrt{64+9+169}$, $\sqrt{\dfrac{121}{3}} = \dfrac{11}{3}\sqrt{3} \approx 6.35$

$$\mathbf{A} \cdot (\mathbf{B} \times \mathbf{C}) = \begin{vmatrix} \_\_ & \_\_ & \_\_ \\ \_\_ & \_\_ & \_\_ \\ \_\_ & \_\_ & \_\_ \end{vmatrix} = \_\_\_\_ .$$

68. $\mathbf{A} \cdot (\mathbf{B} \times \mathbf{C}) = (\mathbf{A} \times \_\_) \cdot \_\_$ so that the dot and the cross may be ______________ in the triple scalar product.

OBJECTIVE B: Find the volume of the box determined by three given vectors $\mathbf{A}$, $\mathbf{B}$, and $\mathbf{C}$.

69. For the three vectors given in Problem 67 above, the volume of the parallelepiped determined by them is the absolute value of the triple scalar product $\mathbf{A} \cdot (\_\_\_\_)$:

$$|\mathbf{A} \cdot (\_\_\_\_)| = \_\_\_ .$$

OBJECTIVE C: Given three vectors $\mathbf{A}$, $\mathbf{B}$, and $\mathbf{C}$ in space, find the triple vector product $(\mathbf{A} \times \mathbf{B}) \times \mathbf{C}$ by two methods.

70. $(\mathbf{A} \times \mathbf{B}) \times \mathbf{C} = (\mathbf{A} \cdot \mathbf{C})\mathbf{B} - \_\_\_\_ .$

71. For the vectors $\mathbf{A} = 2\mathbf{i} - 3\mathbf{j} + 4\mathbf{k}$, $\mathbf{B} = \frac{1}{2}\mathbf{i} + \mathbf{j} - \mathbf{k}$, and $\mathbf{C} = 3\mathbf{i} - 5\mathbf{j} + 2\mathbf{k}$,

$\mathbf{A} \times \mathbf{B} = \_\_\_\_\_\_\_\_$ so that $(\mathbf{A} \times \mathbf{B}) \times \mathbf{C} = \_\_\_\_\_\_\_\_ .$

72. Using the formula in Problem 70,

$\mathbf{A} \cdot \mathbf{C} = \_\_\_$ so $(\mathbf{A} \cdot \mathbf{C})\mathbf{B} = \_\_\_\_\_\_\_\_$, and

$\mathbf{B} \cdot \mathbf{C} = \_\_\_$ so $(\mathbf{B} \cdot \mathbf{C})\mathbf{A} = \_\_\_\_\_\_\_\_ .$

Hence, $(\mathbf{A} \cdot \mathbf{C})\mathbf{B} - (\mathbf{B} \cdot \mathbf{C})\mathbf{A} = \_\_\_\_\_\_\_\_$ in agreement with $\mathbf{A} \times (\mathbf{B} \times \mathbf{C})$ as found in Problem 71.

---

67. $\begin{vmatrix} 3 & 5 & -2 \\ -1 & 4 & -4 \\ 0 & 4 & -1 \end{vmatrix}$, 39

68. $\mathbf{B}$, $\mathbf{C}$, interchanged

69. $\mathbf{B} \times \mathbf{C}$, $\mathbf{B} \times \mathbf{C}$, 39

70. $(\mathbf{B} \cdot \mathbf{C})\mathbf{A}$

71. $-\mathbf{i} + 4\mathbf{j} + \frac{7}{2}\mathbf{k}$, $\frac{51}{2}\mathbf{i} + \frac{25}{2}\mathbf{j} - 7\mathbf{k}$

72. 29, $\frac{29}{2}\mathbf{i} + 29\mathbf{j} - 29\mathbf{k}$, $-\frac{11}{2}$, $-11\mathbf{i} + \frac{33}{2}\mathbf{j} - 22\mathbf{k}$, $\frac{51}{2}\mathbf{i} + \frac{25}{2}\mathbf{j} - 7\mathbf{k}$

OBJECTIVE D: Find the volume of a tetrahedron in space given its four vertices.

73. Consider the tetrahedron with vertices A, B, C, D shown in the figure at the right. The volume V is

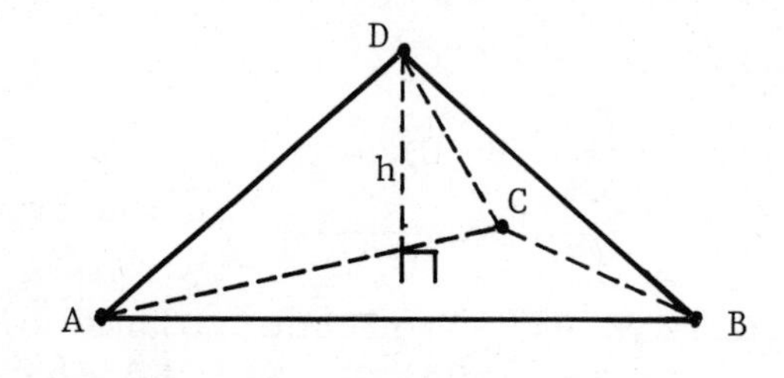

$V = \frac{1}{3}$ (area base) $\cdot$ height .

Now, the area of the base triangle ABC is given by

area $\Delta ABC = \frac{1}{2} \mid$ ________ $\mid$.

The vector $\mathbf{N} = \overrightarrow{AC} \times \overrightarrow{AB}$ is also ________ to the base. Hence, the height h of the tetrahedron is the magnitude of the projection of $\overrightarrow{AD}$ onto $\mathbf{N}$:

$h = |\text{proj}_{\mathbf{N}}\ \overrightarrow{AD}| = \dfrac{|\overrightarrow{AD} \cdot (\overrightarrow{AC} \times \overrightarrow{AB})|}{\_\_\_\_\_\_\_\_}$. Therefore,

$V = \frac{1}{3} \cdot \frac{1}{2} |\overrightarrow{AC} \times \overrightarrow{AB}| \cdot h =$ ____________ .

74. Given the tetrahedron with vertices A(1,2,3) , B(3,2,1) , C(1,0,1) , and D(1,5,7) ,

$\overrightarrow{AD} =$ ________ , $\overrightarrow{AC} =$ ________ and $\overrightarrow{AB} =$ ________ .

Thus, the volume is given by

$$V = \frac{1}{6} |\overrightarrow{AD} \cdot (\overrightarrow{AC} \times \overrightarrow{AB})| = \frac{1}{6} \begin{vmatrix} \_\_ & \_\_ & \_\_ \\ \_\_ & \_\_ & \_\_ \\ \_\_ & \_\_ & \_\_ \end{vmatrix} = \_\_ .$$

OBJECTIVE E: Find the area of the orthogonal projection of a given parallelogram onto a specified plane.

75. If the parallelogram is determined by the vectors $\mathbf{A}$ and $\mathbf{B}$, and if $\mathbf{n}$ is a unit normal to the specified plane, then the area of the orthogonal projection is the absolute value of ________ .

---

73. $\overrightarrow{AC} \times \overrightarrow{AB}$, normal, $|\overrightarrow{AC} \times \overrightarrow{AB}|$, $\frac{1}{6} |\overrightarrow{AD} \cdot (\overrightarrow{AC} \times \overrightarrow{AB})|$

74. $3\mathbf{j} + 4\mathbf{k}$, $-2\mathbf{j} - 2\mathbf{k}$, $2\mathbf{i} - 2\mathbf{k}$, $\begin{vmatrix} 0 & 3 & 4 \\ 0 & -2 & -2 \\ 2 & 0 & -2 \end{vmatrix}$, $\frac{2}{3}$

75. $(\mathbf{A} \times \mathbf{B}) \cdot \mathbf{n}$

76. The area of the projection of the parallelogram with sides $\mathbf{A} = 4\mathbf{i} + 6\mathbf{j} + 9\mathbf{k}$ and $\mathbf{B} = 2\mathbf{i} - 3\mathbf{j}$ onto the xy-plane is

| ____________ | = |(________________) · **k**| = ___ .

11-10 Cylinders.

OBJECTIVE: Discuss and sketch cylinders whose equations are given.

77. Consider the surface given by $y = \sin x$. It is a cylinder with elements parallel to the ___- axis. It extends indefinitely in both the positive and negative directions along the ___- axis. Sketch the graph in the coordinate system at the right.

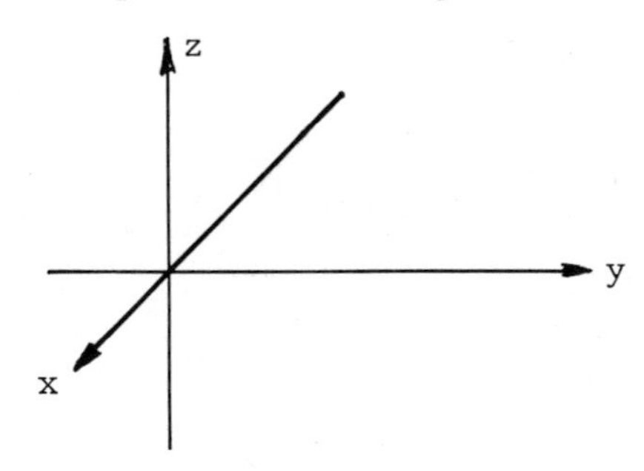

78. Consider the surface $z = x^3$. It is a cylinder with elements parallel to the ___- axis. It extends indefinitely in both the positive and negative directions along the y-axis, with parallel lines passing through the curve $z = x^3$ in the ___- plane.

Sketch the graph at the right.

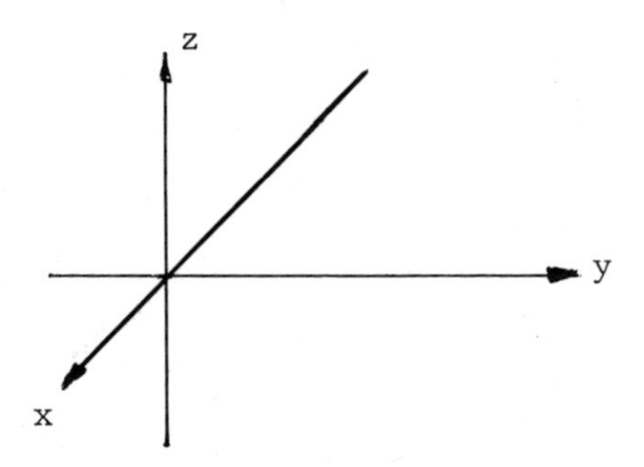

---

76. $(\mathbf{A} \times \mathbf{B}) \cdot \mathbf{k}$, $27\mathbf{i} + 18\mathbf{j} - 24\mathbf{k}$, 24

77. z, z

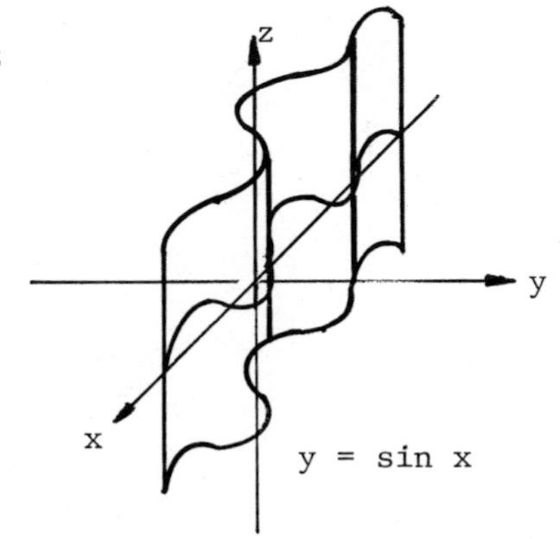

78. y, xz

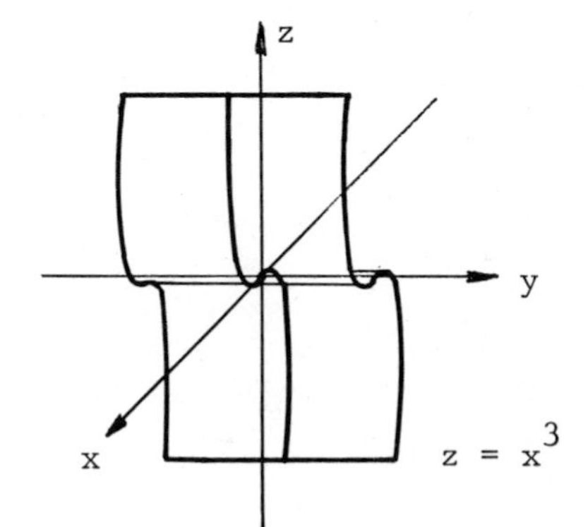

11-11 Quadric Surfaces.

OBJECTIVE: Discuss and sketch a given surface whose equation $F(x,y,z) = 0$ is a quadratic in the variables x, y, and z. The equation may also be given in cylindrical coordinates (r,θ,z).

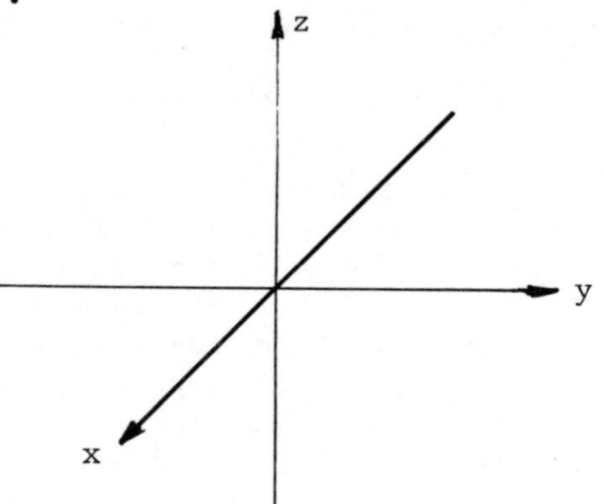

79. Consider the surface given by the equation
$x^2 - 2y^2 + z^2 - 4x - 12y = 20$.
Completing the squares we obtain
$(x-2)^2 +$ ________________ $= 6$
or, dividing by 6,

______________________________ .

This is a ______________ of ______ sheet. It is centered at the point ____________ . Cross sections parallel to the plane $y = 0$ are ____________ while those parallel to the other coordinate planes are ______________ . Sketch the surface.

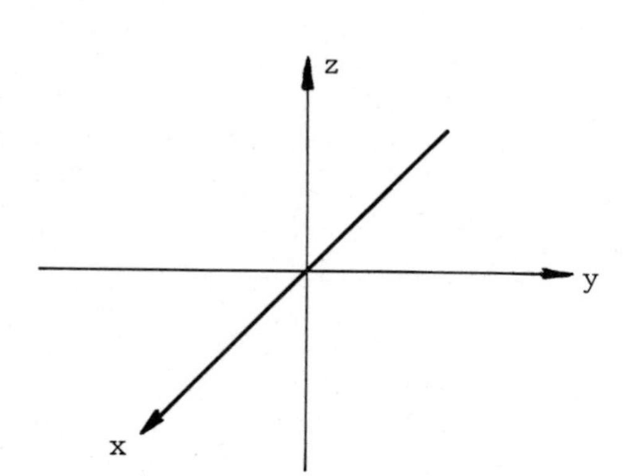

80. Consider the surface given in cylindrical coordinates by the equation
$6z = 12 + r^2 \cos 2\theta$.
Rewriting the equation in cartesian coordinates, we find

$6(z-2) = r^2(\cos^2\theta -$ __________ $)$

$=$ __________ $- r^2 \sin^2\theta$

$=$ __________ or,

$z - 2 = \dfrac{}{\text{\_\_\_\_\_\_\_\_\_\_}}$.

---

79. $-2(y+3)^2 + z^2$,

$\dfrac{(x-2)^2}{6} - \dfrac{(y+3)^2}{3} + \dfrac{z^6}{6} = 1$,

hyperboloid, one, (2,-3,0),

circles, hyperbolas

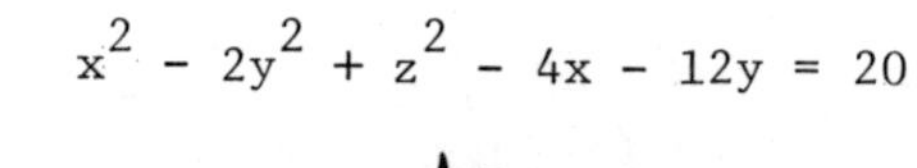

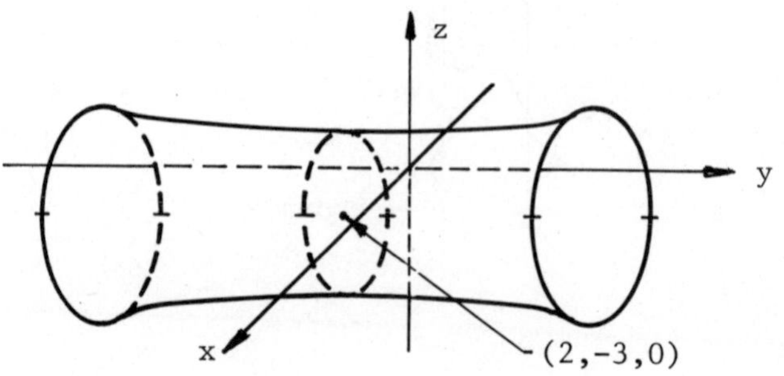

This equation represents a ______________________ . Cross sections parallel to the xy-plane are ____________ . If x or y is fixed we obtain ____________ . The surface is centered at the point __________ . Sketch the surface.

CHAPTER 11 OBJECTIVE - PROBLEM KEY

Objective		Problems in Thomas/Finney Text	Objective		Problems in Thomas/Finney Text
11-1	A	p. 487, 1-3	11-6	E	p. 507, 16,17
	B	p. 487, 4			pp. 503-505, Eqns (2),(7),(8)
	C	p. 487, 11-16	11-7	A	p. 512, 9, Eqns (2),(3)
	D	p. 487, 8-10			pp. 508-510 (4),(5),(7)
11-2	A	p. 490, 1-4		B	p. 512, 1
	B	p. 491, 5-7		C	p. 512, 2,5,6
11-3	A	p. 496, 1-10		D	p. 512, 3,4
	B	p. 496, 11-17, 21	11-8	A	p. 517, 2
11-4	A	p. 500, 1-12		B	p. 517, 5, 8-12, 20
	B	p. 500, 13-16		C	p. 517, 3
11-5	A	p. 502, 5,6		D	p. 517, 13
	B	p. 502, 7-10	11-9	A	p. 523, 7
	C	p. 502, 1-4		B	p. 523, 2
	D	p. 502, 11		C	p. 523, 1,3,4,8
11-6	A	p. 507, 3,6,7,8,10,11		D	p. 523, 6
	B	p. 507, 2,5		E	p. 523, 11
	C	p. 507, 14	11-10		p. 526, 1-8, 13-15
	D	p. 505, Example 2	11-11		p. 531, 1-24

CHAPTER 11 SELF-TEST

1. Find unit vectors tangent and normal to the curve $6y = x^3 - 6x^2 + 9x + 6$ at the point $P(0,1)$.

---

80. $\sin^2 \theta$, $r^2 \cos^2 \theta$,
$x^2 - y^2$, $\frac{x^2}{6} - \frac{y^2}{6}$,
hyperbolic paraboloid,
hyperbolas, parabolas,
(0,0,2)

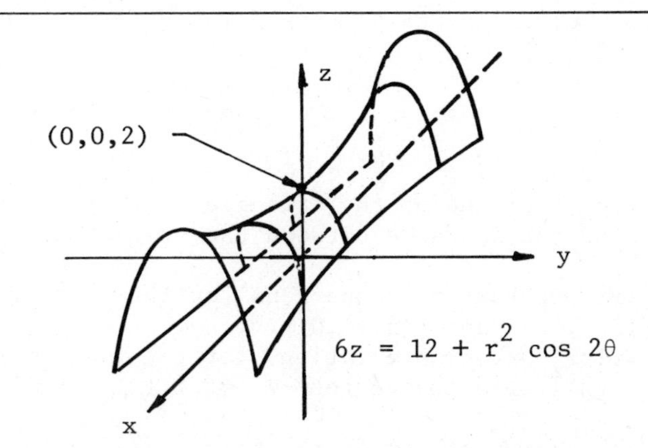

2. A cannon whose muzzle is tilted at an angle of elevation of $30°$ shoots a ball with an initial velocity of 1200 m/sec.
   (a) Find parametric equations describing the motion of the cannonball.
   (b) How far from the cannon does the ball land, and what is its velocity and speed at the time of impact with the earth?
   (c) How high does the ball get and what is the total time the ball spends in the air?
   Assume that the mouth of the cannon is at ground level and that gravity is the only force acting on the cannonball.

3. Find parametric equations for the curve described by the point $P(x,y)$ for $t \geq 0$ if its coordinates satisfy

$$\frac{dx}{dt} = \sqrt{y} \ , \quad \frac{dy}{dt} = 2y \ ; \qquad t = 0 \, , \ x = -5 \, , \ y = 4 \ .$$

4. Sketch the graph of the curve described by $x = t - 2$ and $y = t^2 - t + 1$ for $-\infty < t < \infty$. Also find a cartesian equation of the curve.

5. Find parametric equations for the circle $x^2 + y^2 = 2x$, using as parameter the arc length s measured counterclockwise from the point $(2,0)$ to the point (x,y) .

6. Describe the set of points whose cylindrical or spherical coordinates satisfy the given pairs of simultaneous equations. Sketch.

   (a) $\theta = \frac{\pi}{4}$, $z = r^2$ (b) $\phi = \frac{\pi}{3}$, $\rho = 2$

7. Translate the equation $r^2 = z(4 - z)$ from cylindrical into spherical and cartesian coordinates.

8. Given the vectors $\mathbf{A} = \mathbf{i} - \mathbf{j} + \mathbf{k}$ and $\mathbf{B} = -2\mathbf{j} + 3\mathbf{k}$:
   (a) Determine the cosine of the angle between $\mathbf{A}$ and $\mathbf{B}$.
   (b) Find $\mathbf{A} \times \mathbf{B}$.
   (c) Find the projection vector $\text{proj}_{\mathbf{A}} \mathbf{B}$, and the component of $\mathbf{B}$ in the direction $\mathbf{A}$.

9. Find the distance between the point $P(5,0)$ in the xy-plane and the line L : $2y = -3x + 7$.

10. (a) Find an equation of the plane in space containing the points $A\left(\frac{1}{2}, 0, 1\right)$, $B\left(-\frac{1}{2}, 1, -1\right)$, and $C\left(-\frac{3}{2}, 2, 1\right)$.
    (b) Find the area of the triangle with vertices A, B, C .
    (c) Find the distance between the origin and the plane determined in part (a).

11. (a) Write parametric equations for the line in space through the points $(7,1,-2)$ and $(5,-1,0)$.
    (b) Write cartesian equations for the line through the point $(12,0,-2)$ and parallel to the vector $\mathbf{v} = \mathbf{i} + \mathbf{j} - \mathbf{k}$.

12. Find the distance between the point $P(-3,1,0)$ and the line L : $\frac{x - 2}{1} = \frac{y + 3}{2} = \frac{z - 4}{2}$.

13. Find the volume of the parallelepiped ("box") determined by $\mathbf{u} = 2\mathbf{i} - \mathbf{j} + \mathbf{k}$, $\mathbf{v} = 3\mathbf{i} + 2\mathbf{j} - 2\mathbf{k}$, $\mathbf{w} = 3\mathbf{i} + 2\mathbf{j}$.

14. Find the volume of the tetrahedron with vertices $P(1,2,3)$, $Q(3,3,6)$, $R(4,2,-1)$, and $S(6,3,5)$.

15. Find the vector triple product $(\mathbf{A} \times \mathbf{B}) \times \mathbf{C}$, where $\mathbf{A} = \mathbf{i} - \mathbf{j} + \mathbf{k}$, $\mathbf{B} = -3\mathbf{i} - 6\mathbf{j} + 2\mathbf{k}$, and $\mathbf{C} = 4\mathbf{i} + \mathbf{j} - 3\mathbf{k}$ by two methods.

16. Find the area of the orthogonal projection of the parallelogram with sides determined by $\mathbf{A} = 3\mathbf{i} + 4\mathbf{j} + 5\mathbf{k}$ and $\mathbf{B} = -2\mathbf{i} + \mathbf{j} + 3\mathbf{k}$ onto the plane $x+y+z = 1$.

17. Sketch the cylinder $y = -\sin z$ in space.

18. Sketch the surface $z = \frac{1}{2}\sqrt{x^2+y^2}$.

19. Sketch the quadric surface $z = x^2+y^2-2x-2y+2$, and identify its type.

SOLUTIONS TO CHAPTER 11 SELF-TEST

1. Observe that the point P does lie on the curve since the coordinates of P satisfy the equation for the curve. Now, $y' = \frac{1}{2}(x^2-4x+3)$ so that $y'(0) = \frac{3}{2}$. One vector having this slope is $\mathbf{v} = 2\mathbf{i} + 3\mathbf{j}$, so a tangent vector having this slope is $\frac{\mathbf{v}}{|\mathbf{v}|} = \frac{2}{\sqrt{13}}\mathbf{i} + \frac{3}{\sqrt{13}}\mathbf{j}$. A unit normal vector is $\mathbf{n} = -\frac{3}{\sqrt{13}}\mathbf{i} + \frac{2}{\sqrt{13}}\mathbf{j}$.

2. (a) $\frac{d^2x}{dt^2} = 0 \text{ m/sec}^2$, $\frac{dx}{dt} = c_1 \text{ m/sec}$, $x = c_1t + c_2 \text{ m}$,

$\frac{d^2y}{dt^2} = -g \text{ m/sec}^2$, $\frac{dy}{dt} = -gt + c_3 \text{ m/sec}$, $y = -\frac{1}{2}gt^2 + c_3t + c_4 \text{ m}$.

From the initial conditions, $t = 0$ sec, $x = 0$ m , $y = 0$ m ,

$v_x(0) = v_0 \cos 30° = 600\sqrt{3}$ and $v_y(0) = v_0 \sin 30° = 600$,

we find $c_1 = 600\sqrt{3}$, $c_2 = 0$, $c_3 = 600$, $c_4 = 0$.

Thus,

$$x = 600\sqrt{3}\,t \quad \text{and} \quad y = -\frac{1}{2}gt^2 + 600t$$

are parametric equations for the motion. Here, $g \approx 9.81$.

(b) The maximum range occurs when $y = 0$ or $t = \frac{2 \cdot 600}{g} \approx 122.3$ sec .

The range is then $R = 600\sqrt{3}\left(\frac{1200}{g}\right) \approx 127{,}123 \text{ m} = 1{,}271.23 \text{ km}$.

The velocity at any time is the vector $\mathbf{v} = \mathbf{i}\frac{dx}{dt} + \mathbf{j}\frac{dy}{dt}$.

When $t = \dfrac{1200}{g}$, we find

$$\mathbf{v} = 600\sqrt{3}\,\mathbf{i} + \left(600 - g\cdot\frac{1200}{g}\right)\mathbf{j} = 600\left(\sqrt{3}\,\mathbf{i} - \mathbf{j}\right),$$

as the velocity of impact. Then, $|\mathbf{v}| = 600\sqrt{4} \approx 1200$ m/sec is the speed on impact with the earth.

(c) The maximum height of the cannonball occurs when $\dfrac{dy}{dt} = 0$ or $t_m = \dfrac{600}{g}$; then

$$h = -\frac{1}{2}g\left(\frac{600}{g}\right)^2 + 600\left(\frac{600}{g}\right) \approx 183.49 \text{ km}$$

is the maximum height attained. Finally, the ball remains in the air

$$2t_m = 2\left(\frac{600}{g}\right) \approx 122.3 \text{ sec} .$$

3. From $\dfrac{dy}{dt} = 2y$ we obtain $\dfrac{dy}{y} = 2\,dt$. Thus, $\ln y = 2t + \ln C_1$ or $y = C_1e^{2t}$. From the initial condition $t = 0$ and $y = 4$ we find $C_1 = 4$ so that $y = 4e^{2t}$. Next, $\dfrac{dx}{dt} = \sqrt{y}$ implies $dx = 2e^t\,dt$, or $x = 2e^t + C_2$. From the initial condition $t = 0$ and $x = -5$ we find $C_2 = -7$; hence $x = 2e^t - 7$.

4. We have the following table giving some of the points on the curve:

t	-1	0	1	2	3
x	-3	-2	-1	0	1
y	3	1	1	3	7

$x = t - 2$
$y = t^2 - t + 1$

Substitution of $t = x + 2$ into the parametric equation for y gives $y = (x+2)^2 - (x+2) + 1$, or simplifying algebraically, $y = x^2 + 3x + 3$. This is a cartesian equation of the parabola sketched in the figure above.

5. Completing the square gives the equation $(x-1)^2 + y^2 = 1$ which we recognize as an equation of a unit circle centered at $(1,0)$ (see the figure at the right). Hence,

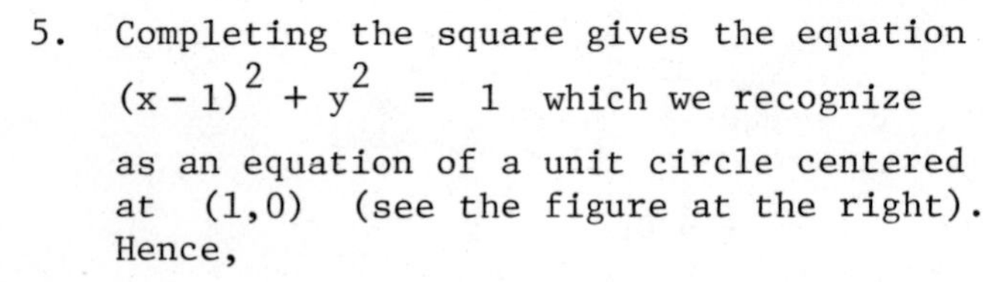

$$x - 1 = 1\cos s \quad\text{and}\quad y - 0 = 1\sin s$$

or,

$$x = 1 + \cos s \quad\text{and}\quad y = \sin s$$

are parametric equations for the circle in terms of the arc length parameter s .

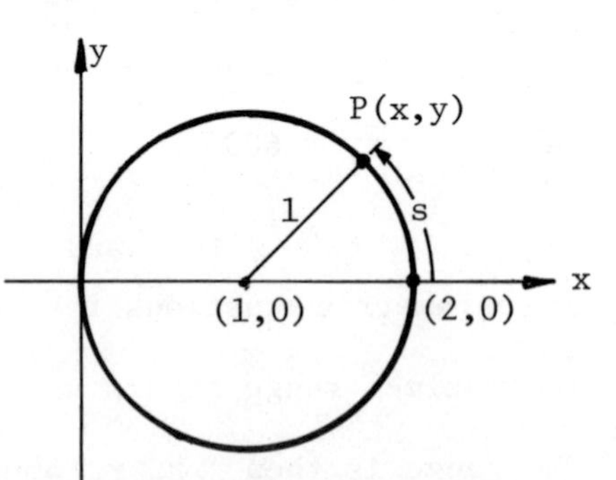

6. (a) The equation $z = r^2$ represents a parabola with vertical axis z and horizontal axis r. The parabola lies in the plane $\theta = \frac{\pi}{4}$ and is sketched at the right.

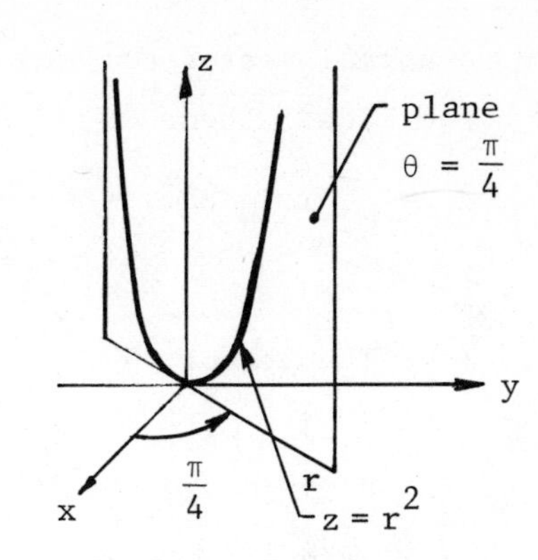

(b) The equation $\phi = \frac{\pi}{3}$ describes a cone in spherical coordinates. Also $z = \rho \cos \phi = 2 \cdot \frac{1}{2} = 1$ is constant. The intersection of the plane $z = 1$ and the cone $\phi = \frac{\pi}{3}$ results in a circle of radius $r = \rho \sin \phi = \sqrt{3}$ centered on the z-axis in the plane $z = 1$. It is sketched at the right.

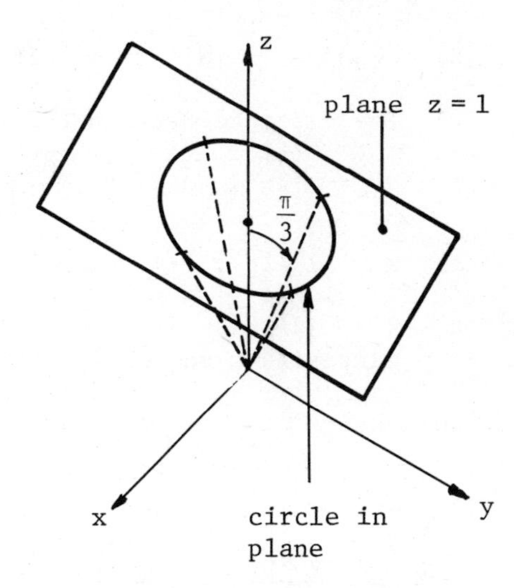

7. The equation can be written as $r^2 + z^2 = 4z$. In spherical coordinates $\rho^2 \sin^2 \phi + \rho^2 \cos^2 \phi = 4\rho \cos \phi$ or, $\rho = 4 \cos \phi$. (Note that $\rho = 0$ is included.)

In cartesian coordinates, $x^2 + y^2 + z^2 = 4z$ or, completing the square, $x^2 + y^2 + (z-2)^2 = 4$, which represents a sphere of radius 2 centered at the point $(0,0,2)$.

8. (a) $\cos \theta = \dfrac{\mathbf{A} \cdot \mathbf{B}}{|\mathbf{A}|\,|\mathbf{B}|} = \dfrac{5}{\sqrt{3}\,\sqrt{13}} = \dfrac{5}{\sqrt{39}} \approx 0.80064$.

(b) $\mathbf{A} \times \mathbf{B} = -\mathbf{i} - 3\mathbf{j} - 2\mathbf{k}$

(c) $\text{proj}_{\mathbf{A}}\, \mathbf{B} = \left(\dfrac{\mathbf{B} \cdot \mathbf{A}}{\mathbf{A} \cdot \mathbf{A}}\right)\mathbf{A} = \frac{5}{3}\,\mathbf{i} - \frac{5}{3}\,\mathbf{j} + \frac{5}{3}\,\mathbf{k}$, and the component of $\mathbf{B}$ in the direction $\mathbf{A}$ is $\dfrac{\mathbf{B} \cdot \mathbf{A}}{|\mathbf{A}|} = \dfrac{5}{\sqrt{3}}$.

9. A normal vector to $3x + 2y = 7$ is $\mathbf{N} = 3\mathbf{i} + 2\mathbf{j}$. The point $B(0, 7/2)$ lies on the line. Then $\overrightarrow{BP} = 5\mathbf{i} - \frac{7}{2}\mathbf{j}$, and the distance is given by

$$d = |\text{proj}_{\mathbf{N}} \overrightarrow{BP}| = \frac{|15 - 7|}{\sqrt{9 + 4}} = \frac{8}{\sqrt{13}} \approx 2.22 .$$

10. (a) The vector $\mathbf{N} = \overrightarrow{AC} \times \overrightarrow{AB} = \begin{vmatrix} \mathbf{i} & \mathbf{j} & \mathbf{k} \\ -2 & 2 & 0 \\ -1 & 1 & -2 \end{vmatrix} = -4\mathbf{i} - 4\mathbf{j}$ is normal to the plane. An equation of the plane is given by

$$-4\left(x - \frac{1}{2}\right) - 4(y - 0) + 0(z - 1) = 0 \quad \text{or,} \quad 4x + 4y = 2 .$$

(b) area $= \frac{1}{2}|\mathbf{N}| = \frac{1}{2}\sqrt{16 + 16} = 2\sqrt{2}$.

(c) The distance between the origin and the plane is given by

$$d = |\text{proj}_{\mathbf{N}} \overrightarrow{OA}| = \frac{|\overrightarrow{OA} \cdot \mathbf{N}|}{|\mathbf{N}|} = \frac{|-2 + 0 + 0|}{4\sqrt{2}} = \frac{1}{2\sqrt{2}} \approx 0.354 .$$

11. (a) $x = 7 - 2t$, $y = 1 - 2t$, $z = -2 + 2t$

(b) $x = 12 + t$, $y = t$, $z = -2 - t$ so that $y = x - 12 = -(z + 2)$ give cartesian equations of the line.

12. The vector $\mathbf{v} = \mathbf{i} + 2\mathbf{j} + 2\mathbf{k}$ is parallel to L, and the point $Q(2,-3,4)$ lies on L. The vector from Q to P is given by $\overrightarrow{QP} = -5\mathbf{i} + 4\mathbf{j} - 4\mathbf{k}$. The distance between P and the line L is then given by

$$d = \frac{|\mathbf{v} \times \overrightarrow{QP}|}{|\mathbf{v}|} = \frac{|-16\mathbf{i} - 6\mathbf{j} + 14\mathbf{k}|}{|\mathbf{i} + 2\mathbf{j} + 2\mathbf{k}|} = \frac{\sqrt{488}}{\sqrt{9}} \approx 7.36 .$$

13. The volume is the absolute value of the triple scalar product $\mathbf{u} \cdot (\mathbf{v} \times \mathbf{w})$. Thus,

$$\mathbf{u} \cdot (\mathbf{v} \times \mathbf{w}) = \begin{vmatrix} 2 & -1 & 1 \\ 3 & 2 & -2 \\ 3 & 2 & 0 \end{vmatrix} = 4 + (-1)(-6) + 1(6 - 6) = 10 ,$$

so the volume is 10 cubic units.

14. Let $\mathbf{u} = \overrightarrow{PQ} = 2\mathbf{i} + \mathbf{j} + 3\mathbf{k}$, $\mathbf{v} = \overrightarrow{PR} = 3\mathbf{i} - 4\mathbf{k}$, and $\mathbf{w} = \overrightarrow{PS} = 5\mathbf{i} + \mathbf{j} + 2\mathbf{k}$. Then the volume of the tetrahedron is given by the absolute value of $\frac{1}{6}\left[\mathbf{u} \cdot (\mathbf{v} \times \mathbf{w})\right]$. Thus,

$$\mathbf{u} \cdot (\mathbf{v} \times \mathbf{w}) = \begin{vmatrix} 2 & 1 & 3 \\ 3 & 0 & -4 \\ 5 & 1 & 2 \end{vmatrix} = 2(4) + 1(-26) + 3(3) = -9 .$$

Hence the volume is $V = \frac{1}{6}|-9| = \frac{3}{2}$.

15. Method 1. $A \times B = \begin{vmatrix} i & j & k \\ 1 & -1 & 1 \\ -3 & -6 & 2 \end{vmatrix} = 4i - 5j - 9k$ so that

$$(A \times B) \times C = \begin{vmatrix} i & j & k \\ 4 & -5 & -9 \\ 4 & 1 & -3 \end{vmatrix} = 24i - 24j + 24k .$$

Method 2. $(A \cdot C)B = (4-1-3)B = 0B = 0i + 0j + 0k$ and $(B \cdot C)A = (-12-6-6)A = -24A = -24i + 24j - 24k$. Hence,

$$(A \times B) \times C = (A \cdot C)B - (B \cdot C)A = 24i - 24j + 24k .$$

16. A unit normal to the plane is $n = \frac{1}{\sqrt{3}} i + \frac{1}{\sqrt{3}} j + \frac{1}{\sqrt{3}} k$,

$A \times B = \begin{vmatrix} i & j & k \\ 3 & 4 & 5 \\ -2 & 1 & 3 \end{vmatrix} = 7i - 19j + 11k$, and the area of the orthogonal projection is given by $|(A \times B) \cdot n| = \left| \frac{7}{\sqrt{3}} - \frac{19}{\sqrt{3}} + \frac{11}{\sqrt{3}} \right| = \frac{1}{\sqrt{3}} \approx 0.577$.

17.

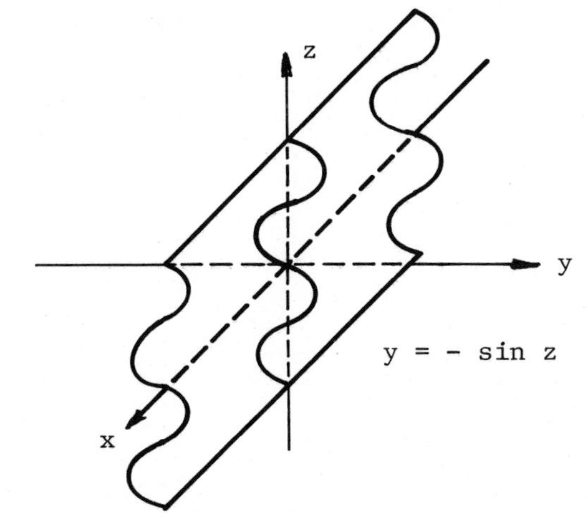

18.

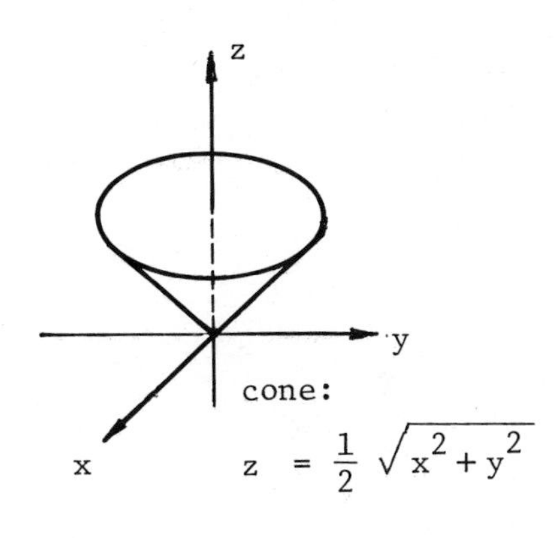

19. Completing the squares,

$$z = (x-1)^2 + (y-1)^2 ,$$

which represents the circular paraboloid sketched at the right.

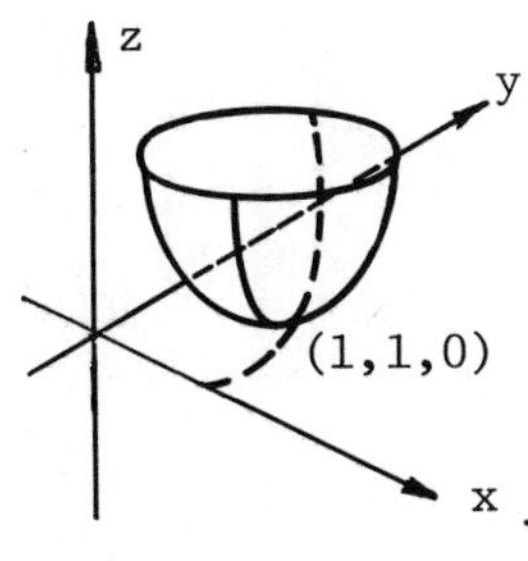

$z = x^2 + y^2 - 2x - 2y + 2$

12-1 Introduction.

1. A function $\mathbf{F}(t) = \mathbf{i}x(t) + \mathbf{j}y(t) + \mathbf{k}z(t)$ is called a __________ function of t.

2. The limit as $t \to a$ of $\mathbf{F}(t) = \mathbf{i}x(t) + \mathbf{j}y(t) + \mathbf{k}z(t)$ exists if and only if the limit as $t \to a$ of each __________ function $x(t)$, $y(t)$, and $z(t)$ exists.

3. If $\lim_{t \to a} x(t) = L_1$, $\lim_{t \to a} y(t) = L_2$, and $\lim_{t \to a} z(t) = L_3$, then

$$\lim_{t \to a} \mathbf{F}(t) = \underline{\qquad\qquad\qquad} .$$

4. The vector function $\mathbf{F}$ is continuous at $t = a$ if and only if each __________ __________ is continuous there.

OBJECTIVE: Find the derivative of a given vector function and give the domain of the derived function.

5. If $\mathbf{F}(t) = \mathbf{i}x(t) + \mathbf{j}y(t) + \mathbf{k}z(t)$ is a vector function and each component function is differentiable at $t = c$, then $\mathbf{F}$ is differentiable at $t = c$ and $\mathbf{F}'(c) = \underline{\qquad\qquad\qquad}$.

6. Consider the vector function

$$\mathbf{F}(t) = \mathbf{i}\sqrt{1-t^2} + \mathbf{j} \sinh t + \mathbf{k}\, 2^t .$$

The derivative is given by

$$\mathbf{F}'(t) = \underline{\qquad\qquad\qquad} .$$

The first component, $\sqrt{1-t^2}$, is differentiable for __________. The second component, $\sinh t$, is differentiable for __________; and the third component, 2^t, is differentiable for __________. Thus, the vector function $\mathbf{F}$ is differentiable for __________.

---

1. vector
2. component
3. $\mathbf{i}L_1 + \mathbf{j}L_2 + \mathbf{k}L_3$
4. component function
5. $\mathbf{i}x'(c) + \mathbf{j}y'(c) + \mathbf{k}z'(c)$
6. $-\mathbf{i}t\left(1-t^2\right)^{-1/2} + \mathbf{j} \cosh t + \mathbf{k}\, 2^t \ln 2$, $-1 < t < 1$, all t, all t, $-1 < t < 1$

12-2 Velocity and Acceleration.

OBJECTIVE A: Given the position vector of a moving point in the xy-plane at any time t, find the velocity and acceleration vectors. Evaluate these vectors and find the speed of the particle at a specified instant of time.

7. If the coordinates of a point $P(x,y)$ in the plane are given parametrically by $x = f(t)$ and $y = g(t)$ as functions of time, the vector from the origin to P is called the ____________ vector of P, and is given by $\mathbf{R}$ = ____________ or ____________________.

8. The derivative $\frac{d\mathbf{R}}{dt}$ of the position vector is called the ___________ vector. The slope of $\frac{d\mathbf{R}}{dt}$ is __________, so the vector $\frac{d\mathbf{R}}{dt}$ is ___________ to the curve. The magnitude of $\frac{d\mathbf{R}}{dt}$ is given by ____________________ = _____, which yields the instantaneous __________ of the curve.

9. The acceleration vector is the second derivative ______. Thus, for $\mathbf{R} = \mathbf{i}x + \mathbf{j}y$, $\mathbf{a}$ = ______________________.

10. If the position vector is $\mathbf{R} = \mathbf{i}\,e^t \cos t + \mathbf{j}\,e^t \sin t$, then the velocity and acceleration vectors are given by,

$$\mathbf{v} = \mathbf{i}\left(e^t \cos t - e^t \sin t\right) + \mathbf{j}\left(\underline{\hspace{6cm}}\right) \text{ and}$$

$$\mathbf{a} = \mathbf{i}\left(-2e^t \sin t\right) + \mathbf{j}\left(\underline{\hspace{3cm}}\right).$$

When $t = 0$, the position of the particle is the point ______. The velocity is $\mathbf{v}(0)$ = ________ and the acceleration is $\mathbf{a}(0)$ = ______. The speed of the particle at $t = 0$ is the magnitude of the velocity: thus, $|\mathbf{v}(0)|$ = _____.

---

7. position, $\mathbf{i}x + \mathbf{j}y$ or $\mathbf{i}\,f(t) + \mathbf{j}\,g(t)$

8. velocity, dy/dx, tangent, $\sqrt{\left(\frac{dx}{dt}\right)^2 + \left(\frac{dy}{dt}\right)^2}$ or $\left|\frac{ds}{dt}\right|$, speed

9. $\frac{d^2\mathbf{R}}{dt^2}$, $\mathbf{i}\frac{d^2x}{dt^2} + \mathbf{j}\frac{d^2y}{dt^2}$

10. $e^t \sin t + e^t \cos t$, $2e^t \cos t$, $(1,0)$, $\mathbf{i} + \mathbf{j}$, $2\mathbf{j}$, $\sqrt{2}$

OBJECTIVE B: Obtain the path of motion of a particle moving in the xy-plane when the force **F** acting on the particle is given as a function of time, and the initial position and initial velocity of the particle are known.

11. Suppose the force acting on a particle of mass m is given as a function of t by

$$\mathbf{F} = \mathbf{i}e^{t} + \mathbf{j}e^{-t},$$

and that the initial position at $t = 0$ is $\mathbf{R}_0 = \mathbf{i} + \mathbf{j}$ and the initial velocity is $\mathbf{v}_0 = \mathbf{i} - \mathbf{j}$.

From Newton's Second Law of Motion, $\mathbf{F} = m\mathbf{a}$, we find

$$m\frac{d\mathbf{v}}{dt} = \underline{\qquad\qquad}$$ or, separating variables,

$$m\,d\mathbf{v} = \left(\underline{\qquad\qquad}\right) dt.$$

Integrating this gives,

$$m\mathbf{v} = \underline{\qquad\qquad} + \mathbf{C}_1,$$

where $\mathbf{C}_1$ is a constant $\underline{\qquad\qquad}$. Using the initial velocity,

$$m(\mathbf{i} - \mathbf{j}) = \underline{\qquad\qquad} + \mathbf{C}_1, \text{ or } \mathbf{C}_1 = \underline{\qquad\qquad\qquad}.$$

Thus,

$$m\frac{d\mathbf{R}}{dt} = \mathbf{i}\left(m - 1 + e^{t}\right) + \mathbf{j}\left(\underline{\qquad\qquad}\right).$$

Another integration gives

$$m\mathbf{R} = \mathbf{i}\left(mt - t + e^{t}\right) + \mathbf{j}\left(\underline{\qquad\qquad}\right) + \mathbf{C}_2.$$

The initial position $\mathbf{R}_0 = \mathbf{i} + \mathbf{j}$ allows us to evaluate the constant vector $\mathbf{C}_2$:

$$m(\mathbf{i} + \mathbf{j}) = \mathbf{i}(0 + 1) + \mathbf{j}(\underline{\qquad}) + \mathbf{C}_2 \text{ or } \mathbf{C}_2 = \underline{\qquad\qquad\qquad}.$$

Therefore, the position vector $\mathbf{R}$ is given by

$$\mathbf{R} = \frac{1}{m}\left[\mathbf{i}(m - 1 + mt - t + e^{t}) + \mathbf{j}(\underline{\qquad\qquad\qquad})\right]$$

$$= \mathbf{i}\left(1 + t - \frac{1 + t - e^{t}}{m}\right) + \mathbf{j}\left(\underline{\qquad\qquad}\right).$$

---

11. $\mathbf{i}e^{t} + \mathbf{j}e^{-t}$, $\mathbf{i}e^{t} + \mathbf{j}e^{-t}$, $\mathbf{i}e^{t} - \mathbf{j}e^{-t}$, vector, $\mathbf{i} - \mathbf{j}$, $\mathbf{i}(m-1) + \mathbf{j}(1-m)$, $1 - m - e^{-t}$, $t - mt + e^{-t}$, $0 + 1$, $\mathbf{i}(m-1) + \mathbf{j}(m-1)$, $m - 1 + t - mt + e^{-t}$, $1 - t + \frac{t - 1 + e^{-t}}{m}$, $1 + t - \frac{1}{m}(1 + t - e^{-t})$

Therefore, parametric equations for the path are

$x = \underline{\hspace{4cm}}$ and $y = 1 - t + \frac{1}{m}(t - 1 + e^{-t})$.

OBJECTIVE C: Given the position vector of a moving point in space at any time t, find the velocity vector $\mathbf{v}$ and acceleration vector $\mathbf{a}$ as functions of t. Also find the angle between $\mathbf{v}$ and $\mathbf{a}$ at any specified time.

12. Consider the curve described by the position vector $\mathbf{R}(t) = \mathbf{i}\, a \cos t + \mathbf{j}\, a \sin t + \mathbf{k}\, bt$, where a and b are constants. The curve describes a circular helix. The velocity and acceleration vectors are given by $\mathbf{v}$ = ______________ and $\mathbf{a}$ = ______________, respectively. When $t = \pi/2$, the velocity is $\mathbf{v}(\pi/2)$ = __________ and the acceleration is $\mathbf{a}(\pi/2)$ = _____ . The cosine of the angle θ between the velocity and acceleration vectors when $t = \pi/2$ is given by

$$\cos\theta = \frac{\mathbf{a}\cdot\mathbf{v}}{|\mathbf{a}||\mathbf{v}|} = \underline{\hspace{2cm}}.$$ In fact, $\mathbf{a}\cdot\mathbf{v}$ = ___ for all t, so that the velocity and acceleration vectors are always perpendicular or orthogonal for the circular helix. Note that the acceleration vector $\mathbf{a}$ always points in the direction from the point $P(x,y,z)$ towards the z-axis in this example, because $\mathbf{i}\, a \cos t + \mathbf{j}\, a \sin t$ describes a circular path of radius a about the z-axis.

12-3 Tangential Vectors.

OBJECTIVE A: Given the position vector for the motion of a particle in the xy-plane, find the unit tangent vector $\mathbf{T}$ to the curve at any point of the curve.

13. Suppose the position vector of the motion is given by $\mathbf{R} = \mathbf{i} \ln t + \mathbf{j}\, t^2$ for $t > 0$. Then

$$\mathbf{v} = \frac{d\mathbf{R}}{dt} = \underline{\hspace{3cm}} ; \quad v = \sqrt{\underline{\hspace{2cm}}} = \frac{1}{t}\sqrt{\underline{\hspace{2cm}}} ;$$

$$\mathbf{T} = \frac{\mathbf{v}}{|\mathbf{v}|} = \mathbf{i}\left(1 + 4t^4\right)^{-1/2} + \mathbf{j}\ \underline{\hspace{3cm}} .$$

---

12. $-\mathbf{i}\, a \sin t + \mathbf{j}\, a \cos t + \mathbf{k}\, b$, $-\mathbf{i}\, a \cos t - \mathbf{j}\, a \sin t$, $-\mathbf{i}a + \mathbf{k}b$, $-\mathbf{j}a$, zero, 0

13. $\mathbf{i}\,\frac{1}{t} + \mathbf{j}\, 2t$, $\frac{1}{t^2} + 4t^2$, $1 + 4t^4$, $2t^2\left(1 + 4t^4\right)^{-1/2}$

OBJECTIVE B: Given the position vector or coordinates for the motion of a particle in space, find the unit tangent vector **T** to the curve at any point of the curve.

14. Consider the motion in space described by the position vector $\mathbf{R} = \mathbf{i}\tan t + \mathbf{j} + \mathbf{k}\sec t$, for $-\pi/2 < t < \pi/2$. Then,

$$\mathbf{v} = \frac{d\mathbf{R}}{dt} = \underline{\qquad\qquad\qquad\qquad} ;$$

$$|\mathbf{v}| = \sqrt{\sec^4 t + \underline{\qquad\qquad}}$$

$$= \sec t \sqrt{\underline{\qquad\qquad}} = \sec t \sqrt{2\sec^2 t - 1} ;$$

$$\mathbf{T} = \frac{\mathbf{v}}{|\mathbf{v}|} = \underline{\qquad\qquad\qquad\qquad\qquad\qquad} .$$

OBJECTIVE C: Given the coordinates for a curve in space in terms of some parameter t, find the length of the curve for a specified interval $a \le t \le b$.

15. Suppose the curve is given by $x = \frac{1}{3}(1+t)^{3/2}$, $y = \frac{1}{3}(1-t)^{3/2}$, $z = \frac{1}{2}t$. To find the length of the curve for $-1 \le t \le 1$, calculate

$$\frac{dx}{dt} = \underline{\qquad\qquad} , \quad \frac{dy}{dt} = \underline{\qquad\qquad} , \quad \frac{dz}{dt} = \frac{1}{2} .$$

Then, $$\frac{ds}{dt} = \sqrt{\left(\frac{dx}{dt}\right)^2 + \left(\frac{dy}{dt}\right)^2 + \left(\frac{dz}{dt}\right)^2}$$

$$= \sqrt{\underline{\qquad\qquad\qquad\qquad}} = \underline{\qquad} .$$

The length of the given curve is therefore

$$s = \int_{-1}^{1} \left(\frac{ds}{dt}\right) dt = \frac{1}{2}\sqrt{3} \int_{-1}^{1} \underline{\quad} \, dt = \underline{\qquad} .$$

---

14. $\mathbf{i}\sec^2 t + \mathbf{k}\sec t \tan t$, $\sec^2 t \tan^2 t$, $\sec^2 t + \tan^2 t$, $\mathbf{i}\sec t\left(2\sec^2 t - 1\right)^{-1/2} + \mathbf{k}\tan t\left(2\sec^2 t - 1\right)^{-1/2}$

15. $\frac{1}{2}(1+t)^{1/2}$, $-\frac{1}{2}(1-t)^{1/2}$, $\frac{1}{4}(1+t) + \frac{1}{4}(1-t) + \frac{1}{4}$, $\frac{1}{2}\sqrt{3}$, 1, $\sqrt{3}$

12-4 Curvature and Normal Vectors.

OBJECTIVE A: Given a curve in the xy-plane, find the curvature at any point on the curve.

16. Consider the curve $y = 1/x$, $x \neq 0$. To find the curvature, calculate

$\frac{dy}{dx} =$ ______ and $\frac{d^2y}{dx^2} =$ ______ . Then the curvature is given by

$$\kappa = \frac{|d^2y/dx^2|}{[\_\_\_\_]^{3/2}} = \frac{|2/x^3|}{[\_\_\_\_]^{3/2}} = \_\_\_\_\_\_ .$$

17. Suppose a curve is given parametrically by $x = t$ and $y = \frac{1}{3}t^3$. Then,

$\dot{x} =$ ___ , $\dot{y} =$ ___ , $\ddot{x} =$ ___ , $\ddot{y} =$ ___ , so the curvature is

given by

$$\kappa = \frac{\_\_\_\_\_\_}{\left[\dot{x}^2 + \dot{y}^2\right]^{3/2}} = \frac{|2t|}{\_\_\_\_\_\_} .$$

OBJECTIVE B: Given the coordinates for a curve in space in terms of some parameter t , find the unit tangent vector **T** , the principal normal vector **N** , the curvature κ , and the unit binormal vector **B** .

18. Suppose the position vector of the curve is given by

$\mathbf{R} = \mathbf{i}2t + \mathbf{j}t^2 + \mathbf{k}\frac{1}{3}t^3$. Then $\mathbf{v} = \frac{d\mathbf{R}}{dt} = \mathbf{i}2 + \mathbf{j}2t + \mathbf{k}$ ___ and

$$|\mathbf{v}| = \sqrt{\_\_\_\_\_\_} = \sqrt{(2+t^2)^2} = 2+t^2 ,$$

$$\mathbf{T} = \frac{\mathbf{v}}{|\mathbf{v}|} = \mathbf{i}\frac{2}{2+t^2} + \mathbf{j}\,\_\_\_\_\_ + \mathbf{k}\frac{t^2}{2+t^2} .$$

---

16. $-\frac{1}{x^2}$, $\frac{2}{x^3}$, $1 + \left(\frac{dy}{dx}\right)^2$, $1 + \frac{1}{x^4}$, $\frac{|2x^3|}{\left(1+x^4\right)^{3/2}}$

17. 1, t^2, 0, $2t$, $|\dot{x}\ddot{y} - \ddot{x}\dot{y}|$, $\left(1+t^4\right)^{3/2}$

18. t^2, $4+4t^2+t^4$, $\frac{2t}{2+t^2}$, $\frac{2(2-t^2)}{(2+t^2)^2}$, $-\mathbf{i}2t + \mathbf{j}(2-t^2) + \mathbf{k}2t$, $\frac{2}{(2+t^2)^3}$

$4t^2 + (2-t^2)^2 + 4t^2$, $\frac{2-t^2}{2+t^2}$ $\frac{2t}{2+t^2}$ $\frac{-2t}{2+t^2}$

Differentiation of T with respect to t gives

$$\frac{dT}{dt} = -\mathbf{i}\frac{4t}{\left(2+t^2\right)^2} + \mathbf{j}\,\underline{\qquad\qquad} + \mathbf{k}\frac{4t}{\left(2+t^2\right)^2}$$

$$= \frac{2}{\left(2+t^2\right)^2}\left[\underline{\qquad\qquad\qquad\qquad}\right].$$

By the rule $\frac{ds}{dt} = |\mathbf{v}|$ and the chain rule,

$$\frac{dT}{ds} = \frac{dT/dt}{ds/dt} = \underline{\qquad\qquad}\left[-\mathbf{i}2t + \mathbf{j}(2-t^2) + \mathbf{k}2t\right];$$

$$\kappa = \left|\frac{dT}{ds}\right| = \frac{2}{\left(2+t^2\right)^3}\sqrt{\underline{\qquad\qquad\qquad\qquad}} = \frac{2}{\left(2+t^2\right)^2}.$$

Then, $\mathbf{N} = \frac{dT/ds}{|dT/ds|} = -\mathbf{i}\frac{2t}{2+t^2} + \mathbf{j}\,\underline{\qquad\qquad} + \mathbf{k}\,\underline{\qquad\qquad}.$

Finally, $\mathbf{B} = \mathbf{T}\times\mathbf{N}$ gives

$$\mathbf{B} = \mathbf{i}\frac{t^2}{2+t^2} + \mathbf{j}\,\underline{\qquad\qquad} + \mathbf{k}\frac{2}{2+t^2}.$$

OBJECTIVE C: Given a curve in the xy-plane, find an equation of the osculating circle at a specified point.

19. Let us find an equation of the osculating circle to $y = \sin x$ at the point $(\pi/2, 1)$. First we calculate the curvature:

$y' = \underline{\qquad\qquad}$ and $y'' = \underline{\qquad\qquad}$ so that at $x = \pi/2$,

$y' = \underline{\qquad}$ and $y'' = \underline{\qquad}$. Thus, the curvature at $(\pi/2, 1)$ is given by

$$\kappa = \frac{|d^2y/dx^2|}{\left[1 + \left(\frac{dy}{dx}\right)^2\right]^{3/2}} = \underline{\qquad}.$$ The radius of curvature is $\rho = \frac{1}{\kappa} = \underline{\qquad}$

which is the radius of the osculating circle. The tangent to $y = \sin x$ when $x = \pi/2$ is horizontal, so the normal is in the $\underline{\qquad\qquad}$ direction.

Therefore, the center of the osculating circle is the point $\underline{\qquad\qquad}$.
An equation of the circle is given by

$\underline{\qquad\qquad\qquad\qquad\qquad\qquad}$.

---

19. $\cos x$, $-\sin x$, 0, -1, 1, 1, vertical, $\left(\frac{\pi}{2}, 0\right)$, $\left(x - \frac{\pi}{2}\right)^2 + y^2 = 1$

12-5 Differentiation of Products of Vectors.

OBJECTIVE A: Using the appropriate rules, calculate the derivatives of vector expressions involving the dot or cross products.

20. $\frac{d}{dt}(\mathbf{U} \cdot \mathbf{V}) =$ ______________ . 21. $\frac{d}{dt}(\mathbf{U} \times \mathbf{V}) =$ ______________ .

22. Suppose that $\mathbf{U}$ is a twice differentiable vector function of t . Then

$$\frac{d}{dt}\left(\mathbf{U} \times \frac{d\mathbf{U}}{dt}\right) = \frac{d\mathbf{U}}{dt} \times \frac{d\mathbf{U}}{dt} + \underline{\qquad\qquad} = \underline{\quad} \times \frac{d^2\mathbf{U}}{dt^2} .$$

Therefore, if $\mathbf{U}$ is parallel to $\frac{d^2\mathbf{U}}{dt^2}$ for all t , $\frac{d}{dt}\left(\mathbf{U} \times \frac{d\mathbf{U}}{dt}\right)$ is the ________ vector. It follows that $\mathbf{U} \times \frac{d\mathbf{U}}{dt}$ is a ____________ vector since the derivative of each component is zero (so that each component is a constant scalar function).

23. Suppose the tangent vector to a curve in the xy-plane, having position vector $\mathbf{R}(t)$, is everywhere perpendicular to some nonzero constant vector $\mathbf{A}$. Then,

$$\frac{d\mathbf{R}}{dt} = f(t)\mathbf{U} ,$$

where $\mathbf{U}$ is a unit vector perpendicular to $\mathbf{A}$. Let $g(t)$ be an indefinite integral of f , so that $\frac{d}{dt} g(t) =$ ______ . Then,

$$\frac{d}{dt}[\mathbf{R}(t) - g(t)\mathbf{U}] = \underline{\qquad\qquad\qquad} = \vec{0} .$$

It follows that $\mathbf{R}(t) - g(t)\mathbf{U}$ is a constant vector $\mathbf{C}$, so that $\mathbf{R}(t) =$ ____________ . Therefore, the curve must lie along the line $\mathbf{C} + s\mathbf{U}$, where $s = g(t)$. Thus, if the tangent vector to a curve is everywhere perpendicular to a constant nonzero vector $\mathbf{A}$, then the curve runs along a straight line perpendicular to $\mathbf{A}$.

OBJECTIVE B: Given the position vector for the motion of a particle in the xy-plane, find the velocity and acceleration vectors, the speed ds/dt , and the tangential and normal components of acceleration.

24. Consider the path described by $\mathbf{R} = \mathbf{i}t + \mathbf{j}\sin t$. Then,

$\mathbf{v} = \frac{d\mathbf{R}}{dt} =$ ______________ , the velocity vector;

---

20. $\frac{d\mathbf{U}}{dt} \cdot \mathbf{V} + \mathbf{U} \cdot \frac{d\mathbf{V}}{dt}$

21. $\frac{d\mathbf{U}}{dt} \times \mathbf{V} + \mathbf{U} \times \frac{d\mathbf{V}}{dt}$

22. $\mathbf{U} \times \frac{d^2\mathbf{U}}{dt^2}$, $\mathbf{U}$, zero, constant

23. $f(t)$, $\frac{d\mathbf{R}}{dt} - f(t)\mathbf{U}$, $\mathbf{C} + g(t)\mathbf{U}$

$\mathbf{a} = \frac{d\mathbf{v}}{dt} = $ ______________ , the acceleration vector;

$\frac{ds}{dt} = |\mathbf{v}| = $ __________________ , the speed;

$\frac{d^2s}{dt^2} = a_T = \frac{\overline{\qquad\qquad}}{\sqrt{1 + \cos^2 t}}$, the tangential component of acceleration;

and

$a_N = \sqrt{|\mathbf{a}|^2 - \_\_\_} = \sqrt{\sin^2 t - \frac{}{\_\_\_\_\_\_\_\_\_\_\_\_\_\_}}$

$= \sqrt{\frac{\overline{\qquad\qquad}}{1 + \cos^2 t}}$,

or $a_N = \frac{|\sin t|}{\sqrt{1 + \cos^2 t}}$, the normal component of acceleration.

25. For the path given by $\mathbf{R} = \mathbf{i}t^2 + \mathbf{j}t^3$, $t \geq 0$,

$\mathbf{v} = \frac{d\mathbf{R}}{dt} = $ ______________ , the velocity vector;

$\mathbf{a} = \frac{d\mathbf{v}}{dt} = $ ______________ , the acceleration vector;

$\frac{ds}{dt} = |\mathbf{v}| = $ ________________ , the speed;

$\frac{d^2s}{dt^2} = \sqrt{4 + 9t^2} + \frac{\overline{\qquad}}{\sqrt{4 + 9t^2}} = \frac{\overline{\qquad\qquad}}{\sqrt{4 + 9t^2}}$, the tangential component of acceleration; and

$a_N = \sqrt{\_\_\_\_ - a_T^2} = \sqrt{4 + 36t^2 - \frac{(4 + 18t^2)^2}{4 + 9t^2}}$

$\frac{\sqrt{(4 + 36t^2)\ \_\_\_\_\_\_\_ - (4 + 18t^2)^2}}{\sqrt{4 + 9t^2}} = \frac{\_\_\_}{\sqrt{4 + 9t^2}}$, the normal component of acceleration.

Remark: If $t < 0$, then $|\mathbf{v}| = -t\sqrt{4 + 9t^2}$.

---

24. $\mathbf{i} + \mathbf{j}\cos t$, $-\mathbf{j}\sin t$, $\sqrt{1 + \cos^2 t}$, $-\sin t \cos t$, a_T^2, $\frac{\sin^2 t \cos^2 t}{1 + \cos^2 t}$, $\sin^2 t$

25. $\mathbf{i}2t + \mathbf{j}3t^2$, $\mathbf{i}2 + \mathbf{j}6t$, $t\sqrt{4 + 9t^2}$, $9t^2$, $4 + 18t^2$, $|\mathbf{a}|^2$, $4 + 9t^2$, $6t$

OBJECTIVE C: Given the coordinates of a curve in space as differentiable functions of t, use equation (16) of this article to find the curvature at any point.

26. Consider the space curve with position described by $\mathbf{R} = \mathbf{i}(t+1) - \mathbf{j}t^2 + \mathbf{k}(1-2t)$. To calculate the curvature κ at any point, we find $\mathbf{v}$ = ________ and $\mathbf{a}$ = ____ . Hence,

$$\mathbf{v} \times \mathbf{a} = \begin{vmatrix} \mathbf{i} & \mathbf{j} & \mathbf{k} \\ 1 & -2t & -2 \\ 0 & -2 & 0 \end{vmatrix} = \text{\_\_\_\_\_\_\_\_}, \text{ and}$$

$|\mathbf{v} \times \mathbf{a}| = \sqrt{\_\_\_\_\_\_} = \_\_\_\_$. Also, $|\mathbf{v}| = \_\_\_\_\_\_\_\_$.

Therefore, $\kappa = \dfrac{|\mathbf{v} \times \mathbf{a}|}{\_\_\_\_} = \_\_\_\_\_\_\_\_$

gives the curvature for any value of t . In particular, for $t = 0$, $\kappa = \_\_\_\_$.

12-6 Polar Coordinates.

OBJECTIVE: Given a curve $r = f(\theta)$ in the plane in polar coordinates and the rate $d\theta/dt$, or given the polar coordinates r and θ as functions of the parameter t , find the velocity and acceleration vectors in terms of $\mathbf{u}_r$ and $\mathbf{u}_\theta$.

27. In terms of the vectors $\mathbf{u}_r$ and $\mathbf{u}_\theta$, $\mathbf{v} = \mathbf{u}_r \_\_\_\_ + \mathbf{u}_\theta \_\_\_\_$ and

$$\mathbf{a} = \mathbf{u}_r \left[\_\_\_\_\_\_\_\_ \right] + \mathbf{u}_\theta \left[\_\_\_\_\_\_\_\_ \right]$$

give the velocity and acceleration vectors, respectively, of the moving particle in polar coordinates.

28. Suppose a particle moves on the cardioid $r = a(1 + \cos\theta)$ and that the radius vector turns counterclockwise at the constant rate $d\theta/dt = 5/a\sqrt{3}$ rad/sec. Let us find the polar form of the velocity and acceleration vectors when $\theta = \frac{\pi}{3}$.

---

26. $\mathbf{i} - \mathbf{j}2t - \mathbf{k}2$, $-\mathbf{j}2$, $-\mathbf{i}4 - \mathbf{k}2$, $16 + 4$, $2\sqrt{5}$, $\sqrt{5 + 4t^2}$, $|\mathbf{v}|^3$, $\dfrac{2\sqrt{5}}{\left(5 + 4t^2\right)^{3/2}}$, $\dfrac{2}{5}$

27. $\dfrac{dr}{dt}$, $r\dfrac{d\theta}{dt}$, $\dfrac{d^2r}{dt^2} - r\left(\dfrac{d\theta}{dt}\right)^2$, $r\dfrac{d^2\theta}{dt^2} + 2\dfrac{dr}{dt}\dfrac{d\theta}{dt}$

Now,

$$\frac{dr}{dt} = \frac{dr}{d\theta}\cdot\frac{d\theta}{dt} = \underline{\qquad\qquad}\cdot\frac{5}{a\sqrt{3}} = \underline{\qquad\qquad};$$

$$r\frac{d\theta}{dt} = \underline{\qquad\qquad}.$$

Evaluation of these components at $\theta = \frac{\pi}{3}$ gives $\mathbf{v} = \mathbf{u}_r \underline{\quad} + \mathbf{u}_\theta \underline{\quad}$, the velocity vector. Next,

$$\frac{d^2r}{dt^2} = \frac{d\dot{r}}{d\theta}\ \frac{d\theta}{dt} = \underline{\qquad\qquad}\cdot\frac{5}{a\sqrt{3}} = \underline{\qquad\qquad} \quad \text{so that}$$

$$\frac{d^2r}{dt^2} - r\left(\frac{d\theta}{dt}\right)^2 = -\frac{25}{6a} - \left(\underline{\quad}\right)\frac{25}{3a^2} = \underline{\qquad} \quad \text{when} \quad \theta = \frac{\pi}{3};$$

also for $\theta = \frac{\pi}{3}$

$$r\frac{d^2\theta}{dt^2} + 2\ \frac{dr}{dt}\ \frac{d\theta}{dt} = \frac{3a}{2}(0) + 2\left(\underline{\quad}\right)\frac{5}{a\sqrt{3}} = \underline{\qquad}.$$

Therefore,

$$\mathbf{a} = \mathbf{u}_r\left[\underline{\qquad}\right] + \mathbf{u}_\theta\left[\underline{\qquad}\right], \quad \text{the acceleration.}$$

29. Consider a particle moving along the curve defined parametrically by $r = e^t$ and $\theta = t^2$. Then,

$$\frac{dr}{dt} = \underline{\quad}, \quad \frac{d\theta}{dt} = \underline{\quad};$$

$$\frac{d^2r}{dt^2} = \underline{\quad}, \quad \frac{d^2\theta}{dt^2} = \underline{\quad}.$$

The velocity at any time t is given by

$$\mathbf{v} = \mathbf{u}_r \underline{\quad} + \mathbf{u}_\theta \underline{\qquad};$$

and the acceleration is given by

$$\mathbf{a} = \mathbf{u}_r\left[e^t - \underline{\qquad}\right] + \mathbf{u}_\theta\left[2e^t + \underline{\qquad}\right]$$

$$= \mathbf{u}_r \underline{\qquad\qquad} + \mathbf{u}_\theta \underline{\qquad\qquad}.$$

---

28. $-a\sin\theta$, $-\frac{5}{\sqrt{3}}\sin\theta$, $\frac{5}{\sqrt{3}}(1+\cos\theta)$, $-\frac{5}{2}$, $\frac{5\sqrt{3}}{2}$, $-\frac{5}{\sqrt{3}}\cos\theta$, $\frac{-25}{3a}\cos\theta$, $\frac{3a}{2}$, $\frac{-50}{3a}$, $\frac{-5}{2}$, $\frac{-25}{a\sqrt{3}}$, $\frac{-50}{3a}$, $\frac{-25}{a\sqrt{3}}$

29. e^t, $2t$, e^t, 2, e^t, $2te^t$, $4t^2e^t$, $4te^t$, $e^t\left(1-4t^2\right)$, $2e^t\left(1+2t\right)$

CHAPTER 12 OBJECTIVE - PROBLEM KEY

Objective	Problems in Thomas/Finney Text	Objective	Problems in Thomas/Finney Text
12-1	p. 542, 1-5	12-4 A	p. 557, 1-12
12-2 A	p. 546, 1-8	B	p. 557, 15,16,17
B	p. 546, 9,10	C	p. 557, 13
C	p. 546, 11,12,13	12-5 A	p. 563, 2,3
12-3 A	p. 551, 1-8	B	p. 564, 5-9
B	p. 551, 9-12	C	p. 564, 14
C	p. 551, 13-16	12-6	p. 566, 2-6

CHAPTER 12 SELF-TEST

1. Find the first and second derivatives $F'(t)$ and $F''(t)$ of the following vector functions.

   (a) $F(t) = \mathbf{i}(\ln t) - \mathbf{j}t^{-2} + \mathbf{k}e^{t^2}$, $t > 0$

   (b) $F(t) = \mathbf{i}(t \sin t) + \mathbf{j}(\tan^{-1} t) + \mathbf{k}e^{-t^2}$

2. Find the velocity and acceleration vectors, and the speed of a particle moving along a curve according to the position vector given as a function of time.

   (a) $R = \mathbf{i}(3t^2 - 5) + \mathbf{j}(\ln t)$, $t > 0$

   (b) $R = \mathbf{i}e^{-2t} + \mathbf{j}\,5\cos 3t + \mathbf{k}\,5\sin 3t$

3. Find the cosine of the angle between the velocity and acceleration vectors at any time t in Problem 2(b).

4. The force acting on a particle of mass m is given as a function of t by

   $$F = \mathbf{i}\,6t + \mathbf{j}\sin t \quad .$$

   If the particle starts at the point $(0,-1)$ with initial velocity $\mathbf{v}_0 = \mathbf{i} + \mathbf{j}$, find the path of the particle.

5. Find the unit tangent vector, the principal normal, the curvature, and the length of the curve over the interval $1 \leq t \leq 2$, for the plane curve

   $$R = \mathbf{i}\,2\ln t - \mathbf{j}\left(\frac{1}{t} + t\right) \quad .$$

6. Find the unit tangent vector T, the principal normal vector N, the curvature κ, and the unit binormal vector B for the space curve given by

   $$R = \mathbf{i}(t+1) - \mathbf{j}t^2 + \mathbf{k}(1-2t) \; .$$

7. Find an equation of the osculating circle when $t = 1$ for the curve in Problem 5.

8. Establish the result

$$\frac{d}{dt}|\mathbf{U}| = \frac{\mathbf{U}\cdot\mathbf{U}'}{|\mathbf{U}|}$$

where $\mathbf{U}$ is a differentiable vector function of t for which $|\mathbf{U}|$ is never zero. Here $\mathbf{U}'$ means $d\mathbf{U}/dt$.

9. Find the velocity and acceleration vectors, the speed ds/dt, and the tangential and normal components of acceleration for the motion described by

$$\mathbf{R} = \mathbf{i}t + \mathbf{j}\ln t, \quad t > 0.$$

10. Find the curvature of the path

$$\mathbf{R} = \mathbf{i}e^t \cos t + \mathbf{j}e^t \sin t + \mathbf{k}e^t$$

when $t = 0$.

11. A particle moves counterclockwise on the circle $r = b\cos\theta$, $b > 0$, with constant speed v_0. Find the acceleration vector in terms of $\mathbf{u}_r$ and $\mathbf{u}_\theta$.

SOLUTIONS TO CHAPTER 12 SELF-TEST

1. (a) $\mathbf{F}'(t) = \mathbf{i}t^{-1} + \mathbf{j}2t^{-3} + \mathbf{k}2te^{t^2}$

$$\mathbf{F}''(t) = -\mathbf{i}t^{-2} - \mathbf{j}6t^{-4} + \mathbf{k}2e^{t^2}(1+2t^2)$$

(b) $\mathbf{F}'(t) = \mathbf{i}(\sin t + t\cos t) + \mathbf{j}\dfrac{1}{1+t^2} - \mathbf{k}2te^{-t^2}$

$$\mathbf{F}''(t) = \mathbf{i}(2\cos t - t\sin t) + \mathbf{j}\frac{-2t}{\left(1+t^2\right)^2} + \mathbf{k}2e^{-t^2}(2t^2-1)$$

2. (a) $\mathbf{v} = \dfrac{d\mathbf{R}}{dt} = \mathbf{i}6t + \mathbf{j}t^{-1}$, velocity

$$\mathbf{a} = \frac{d\mathbf{v}}{dt} = \mathbf{i}6 - \mathbf{j}t^{-2}, \quad \text{acceleration}$$

$$\frac{ds}{dt} = |\mathbf{v}| = \sqrt{36t^2 + t^{-2}}, \quad \text{speed}$$

(b) $\mathbf{v} = -\mathbf{i}2e^{-2t} - \mathbf{j}15\sin 3t + \mathbf{k}15\cos 3t$, velocity

$$\mathbf{a} = \frac{d\mathbf{v}}{dt} = \mathbf{i}4e^{-2t} - \mathbf{j}45\cos 3t - \mathbf{k}45\sin 3t, \quad \text{acceleration}$$

$$\frac{ds}{dt} = |\mathbf{v}| = \sqrt{4e^{-4t} + 225}, \quad \text{speed}$$

3. $\cos\theta = \dfrac{\mathbf{a}\cdot\mathbf{v}}{|\mathbf{a}|\,|\mathbf{v}|}$; $\mathbf{a}\cdot\mathbf{v} = -8e^{-4t}$ and $|\mathbf{a}| = \sqrt{16e^{-4t} + 2025}$.

Therefore, $\cos\theta = \dfrac{-8e^{-4t}}{\sqrt{4e^{-4t} + 225}\cdot\sqrt{16e^{-4t} + 2025}}$.

4. $\mathbf{F} = m\mathbf{a} = \mathbf{i}\,6t + \mathbf{j}\sin t$, and integration gives $m\mathbf{v} = \mathbf{i}\,3t^2 - \mathbf{j}\cos t + \mathbf{C}_1$.

At $t = 0$, $\mathbf{v} = \mathbf{i} + \mathbf{j}$ so that

$$m(\mathbf{i}+\mathbf{j}) = -\mathbf{j} + \mathbf{C}_1 \quad \text{or} \quad \mathbf{C}_1 = \mathbf{i}\,m + \mathbf{j}\,(m+1) , \quad \text{and}$$

$$m\mathbf{v} = \mathbf{i}\,(3t^2+m) + \mathbf{j}\,(m+1 - \cos t) .$$

Another integration gives $m\mathbf{R} = \mathbf{i}\,(t^3+mt) + \mathbf{j}\,(mt + t - \sin t) + \mathbf{C}_2$.

At $t = 0$, $\mathbf{R} = -\mathbf{j}$ so that

$$m(-\mathbf{j}) = \mathbf{C}_2 , \quad \text{and we find}$$

$$m\mathbf{R} = \mathbf{i}\,(t^3+mt) + \mathbf{j}\,(mt+t-m - \sin t) .$$

Therefore, $x = t + \frac{1}{m}t^3$ and $y = t - 1 + \frac{1}{m}(t - \sin t)$

give parametric equations of the path of motion.

5. $\mathbf{v} = \dfrac{d\mathbf{R}}{dt} = \mathbf{i}\,2t^{-1} + \mathbf{j}\,\dfrac{1-t^2}{t^2}$ so that

$|\mathbf{v}| = \sqrt{\dfrac{4}{t^2} + \dfrac{1-2t^2+t^4}{t^4}} = \dfrac{t^2+1}{t^2}$; thus the unit tangent vector is

$\mathbf{T} = \dfrac{\mathbf{v}}{|\mathbf{v}|} = \mathbf{i}\,\dfrac{2t}{1+t^2} + \mathbf{j}\,\dfrac{1-t^2}{1+t^2}$ and the principal normal is

$$\mathbf{N} = -\mathbf{i}\,\frac{1-t^2}{1+t^2} + \mathbf{j}\,\frac{2t}{1+t^2} .$$

The length of the curve is given by the integral

$$s = \int_1^2 |\mathbf{v}|\,dt = \int_1^2 (1+t^{-2})\,dt = t - \frac{1}{t}\Big|_1^2 = \frac{3}{2} .$$

To calculate the curvature, $\dot{x} = \frac{2}{t}$, $\dot{y} = \frac{1}{t^2} - 1$, $\ddot{x} = -\frac{2}{t^2}$, $\ddot{y} = -\frac{2}{t^3}$ gives

$$\kappa = \frac{|\dot{x}\ddot{y} - \dot{y}\ddot{x}|}{|\mathbf{v}|^3} = \frac{|-4t^{-4} + 2t^{-4} - 2t^{-2}|}{\left(t^2+1\right)^3 t^{-6}} = \frac{2t^2}{\left(t^2+1\right)^2} .$$

6. $\mathbf{v} = \dfrac{d\mathbf{R}}{dt} = \mathbf{i} - \mathbf{j}\,2t - \mathbf{k}\,2$, $|\mathbf{v}| = \dfrac{ds}{dt} = \sqrt{5+4t^2}$. Thus

$$\mathbf{T} = \frac{\mathbf{v}}{|\mathbf{v}|} = \left(5+4t^2\right)^{-1/2} (\mathbf{i} - \mathbf{j}\,2t - \mathbf{k}\,2) .$$

$$\frac{dT}{dt} = -\frac{1}{2}\left(5+4t^2\right)^{-3/2} 8t\left[\mathbf{i} - \mathbf{j}2t - \mathbf{k}2\right] + \left(5+4t^2\right)^{-1/2} (-\mathbf{j}2)$$

$$= 2\left(5+4t^2\right)^{-3/2}\left[-\mathbf{i}2t - \mathbf{j}5 + \mathbf{k}4t\right] ;$$

$$\frac{dT}{ds} = \frac{dT/dt}{ds/dt} = 2\left(5+4t^2\right)^{-2}\left[-\mathbf{i}2t - \mathbf{j}5 + \mathbf{k}4t\right] ;$$

$$\kappa = \left|\frac{dT}{ds}\right| = 2\left(5+4t^2\right)^{-2}\sqrt{20t^2+25} = 2\sqrt{5}\left(5+4t^2\right)^{-3/2} ;$$

$$\mathbf{N} = \frac{1}{\kappa}\frac{dT}{ds} = \left(5+4t^2\right)^{-1/2}\left[-\mathbf{i}\frac{2t}{\sqrt{5}} + \mathbf{j}\sqrt{5} + \mathbf{k}\frac{4t}{\sqrt{5}}\right]$$

$$\mathbf{B} = \mathbf{T}\times\mathbf{N} = -\mathbf{i}\frac{2}{\sqrt{5}} - \mathbf{k}\frac{1}{\sqrt{5}} .$$

7. From our calculation in the solution to Problem 5, $\rho = \frac{1}{\kappa} = \frac{\left(t^2+1\right)^2}{2t^2}$,

so $\rho = 2$ is the radius of curvature when $t = 1$. Also,

$$\mathbf{N} = -\mathbf{i}\frac{1-t^2}{1+t^2} + \mathbf{j}\frac{2t}{1+t^2}$$ so that $\mathbf{N} = \mathbf{j}$ when $t = 1$, and $\mathbf{R} = -\mathbf{j}2$ is the

position of the particle at $t = 1$; i.e., the particle is located at the point $(0,-2)$. The center is located $\rho = 2$ units from the position point $(0,-2)$ in the direction $\mathbf{N} = \mathbf{j}$; thus, the center of the osculating circle is $(0,0)$. Therefore,

$$x^2 + y^2 = 4$$

is an equation of the osculating circle when $t = 1$.

8. Now, $|\mathbf{U}| = \sqrt{\mathbf{U}\cdot\mathbf{U}}$ so that

$$\frac{d}{dt}|\mathbf{U}| = \frac{d}{dt}\sqrt{\mathbf{U}\cdot\mathbf{U}} = \frac{1}{2\sqrt{\mathbf{U}\cdot\mathbf{U}}}\frac{d}{dt}(\mathbf{U}\cdot\mathbf{U}) = \frac{\mathbf{U}'\cdot\mathbf{U} + \mathbf{U}\cdot\mathbf{U}'}{2\sqrt{\mathbf{U}\cdot\mathbf{U}}} = \frac{\mathbf{U}\cdot\mathbf{U}'}{|\mathbf{U}|} ,$$

as claimed.

9. $\mathbf{v} = \frac{d\mathbf{R}}{dt} = \mathbf{i} + \mathbf{j}t^{-1}$, velocity vector ;

$\mathbf{a} = \frac{d\mathbf{v}}{dt} = -\mathbf{j}t^{-2}$, acceleration vector ;

$|\mathbf{v}| = \frac{ds}{dt} = \left(1+t^{-2}\right)^{1/2} = \frac{\sqrt{1+t^2}}{t}$, speed ;

$\frac{d^2s}{dt^2} = -t^{-2}\sqrt{1+t^2} + t^{-1}\left[\frac{t}{\sqrt{1+t^2}}\right]$ or, $a_T = \frac{-1}{t^2\sqrt{1+t^2}}$, tangential component of acceleration;

$$a_N = \sqrt{|\mathbf{a}|^2 - a_T^2} = \sqrt{\frac{1}{t^4} - \frac{1}{t^4(1+t^2)}} = \sqrt{\frac{t^2}{t^4(1+t^2)}} = \frac{1}{t\sqrt{1+t^2}} ,$$

normal component of acceleration

10. $\mathbf{v} = \mathbf{i}\,e^t(\cos t - \sin t) + \mathbf{j}\,e^t(\cos t + \sin t) + \mathbf{k}\,e^t$ and

$\mathbf{a} = -\mathbf{i}\,2e^t \sin t + \mathbf{j}\,2e^t \cos t + \mathbf{k}\,e^t$.

Thus, when $t = 0$ $\mathbf{v}(0) = \mathbf{i} + \mathbf{j} + \mathbf{k}$ and $\mathbf{a}(0) = \mathbf{j}2 + \mathbf{k}$. Then,

$$\mathbf{v}(0) \times \mathbf{a}(0) = \begin{vmatrix} \mathbf{i} & \mathbf{j} & \mathbf{k} \\ 1 & 1 & 1 \\ 0 & 2 & 1 \end{vmatrix} = -\mathbf{i} - \mathbf{j} + \mathbf{k}2$$

and $|\mathbf{v}(0) \times \mathbf{a}(0)| = \sqrt{6}$, $|\mathbf{v}(0)| = \sqrt{3}$. Therefore, the curvature is given by

$$\kappa = \frac{|\mathbf{v}(0) \times \mathbf{a}(0)|}{|\mathbf{v}(0)|^3} = \frac{\sqrt{6}}{3\sqrt{3}} = \frac{\sqrt{2}}{3} .$$

11. $\dfrac{dr}{dt} = -b \sin\theta \dfrac{d\theta}{dt}$ and $r\dfrac{d\theta}{dt} = b\cos\theta \dfrac{d\theta}{dt}$.

Since $\mathbf{v} = \mathbf{u}_r \dfrac{dr}{dt} + \mathbf{u}_\theta\, r \dfrac{d\theta}{dt}$, we have

$$|\mathbf{v}| = v_0 = \sqrt{b^2 \sin^2\theta + b^2 \cos^2\theta}\;\left|\frac{d\theta}{dt}\right| = b\frac{d\theta}{dt} ,$$

since $b > 0$ and the particle is moving counterclockwise.

Therefore, $\dfrac{d\theta}{dt} = \dfrac{v_0}{b}$ is constant. Next,

$$\frac{d^2r}{dt^2} = \frac{d\dot{r}}{d\theta}\frac{d\theta}{dt} = -b\cos\theta\left(\frac{d\theta}{dt}\right)^2 = -\frac{v_0^2}{b}\cos\theta , \text{ so}$$

$$\frac{d^2r}{dt^2} - r\left(\frac{d\theta}{dt}\right)^2 = -\frac{v_0^2}{b}\cos\theta - b\cos\theta\left(\frac{v_0}{b}\right)^2 = -\frac{2v_0^2}{b}\cos\theta ,$$

$\mathbf{u}_r$-component of $\mathbf{a}$;

$\dfrac{d^2\theta}{dt^2} = 0$, and we find

$$2\frac{dr}{dt}\frac{d\theta}{dt} = -2b\sin\theta\left(\frac{d\theta}{dt}\right)^2 = -\frac{2v_0^2}{b}\sin\theta ,$$ $\mathbf{u}_\theta$-component of $\mathbf{a}$.

Hence, the acceleration vector is given by

$$\mathbf{a} = \mathbf{u}_r\left(-\frac{2v_0^2}{b}\cos\theta\right) + \mathbf{u}_\theta\left(-\frac{2v_0^2}{b}\sin\theta\right) ..$$

13-1 Functions of Two or More Variables.

OBJECTIVE A: Given a function of two independent variables x and y, modifying Example 2 in this article of the textbook, express the function in polar coordinates and determine if it has a limit as (x,y) approaches $(0,0)$.

1. Consider the function $u(x,y) = \dfrac{x+y}{\sqrt{x^2+y^2}}$. Using polar coordinates

   $x =$ ________ and $y =$ ________, $(r \neq 0)$, we have
   $u =$ ____________. No matter how small r may be, as the point (x,y), or (r,θ), moves around the origin u takes on all values b between $-\sqrt{2}$ (when $\theta =$ ___) and ___ (when $\theta = \frac{\pi}{4}$). Thus we cannot make u close to any particular number by keeping (x,y) close to $(0,0)$. Therefore, the function u ________ have a limit as (x,y) approaches $(0,0)$.

OBJECTIVE B: Given an elementary function of two variables x and y, find its limit as (x,y) approaches the point (a,b), if the limit exists.

2. $\lim\limits_{\substack{x\to 0\\ y\to 0}} \dfrac{e^x + e^{-y}}{xy+1} = \dfrac{1 + \_\_\_}{0+1} = \_\_\_$.

3. $\lim\limits_{\substack{x\to 3\\ y\to 3}} \tan^{-1}\dfrac{y}{x} = \tan^{-1}\dfrac{3}{\_\_\_} = \tan^{-1} \_\_\_ = \_\_\_$.

OBJECTIVE C: Given a function $w = f(x,y)$, find the largest possible domain for f if w is to be a real variable.

4. For the function $w = \sqrt{x^2 - y^2}$, in order that w be a real number, the expression, $x^2 - y^2$ must be ____________ or, $x^2 \geq$ ___.

---

1. $r \cos\theta$, $r \sin\theta$, $\cos\theta + \sin\theta$, $\frac{5\pi}{4}$, $\sqrt{2}$, does not

2. 1, 2

3. 3, 1, $\frac{\pi}{4}$

Equivalently, $|x| \geq$ ____ .

Sketch a graph of the domain set in the coordinate system at the right.

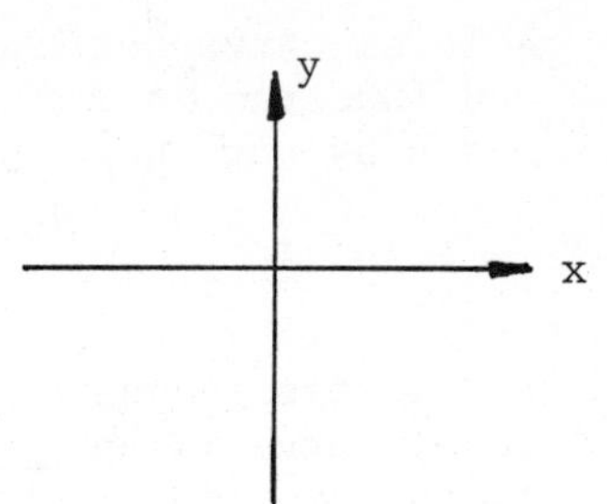

5. Consider the function $w = \frac{x}{y} e^{-x^2}$. Now $\frac{x}{y}$ is a real number for any value of x and any value of $y \neq$ ___ . Also, e^{-x^2} is a real number for x satisfying ______________ . Therefore, w is a real number except when ________ ; that is, the domain of w contains all points in the xy-plane except the __________ .

13-2 The Directional Derivative: Special Cases.

OBJECTIVE A: Given a function $w = f(x,y)$, represent the function
(a) by sketching a surface in xyw-space, and
(b) by drawing a family of level curves, $f(x,y) = \text{constant}$.

6. Let $f(x,y) = 4x^2 + 9y^2$.
The equation

$$w = 4x^2 + 9y^2$$

describes an elliptic ______________ in xyw-space. The section cut out from the surface by the yw-plane is

$x = 0$, ___________ which is a parabola with vertex at the origin, opening upward. When $y = 0$,

___________ is also a parabola opening upward. Sketch the graph of the surface.

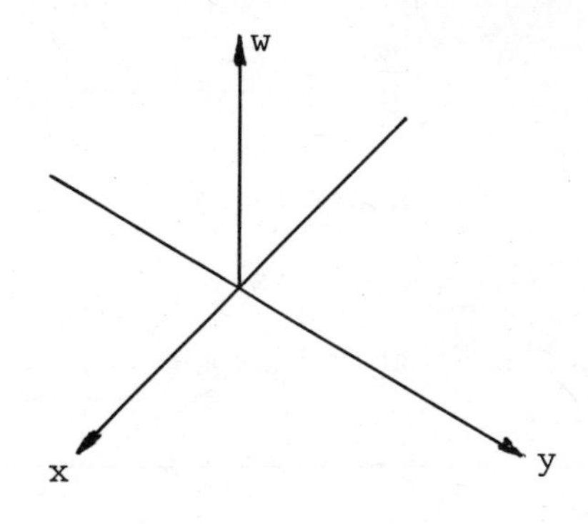

---

4. nonnegative, y^2 , $|y|$, shaded portion in figure gives (x,y) points where $|x| \geq |y|$

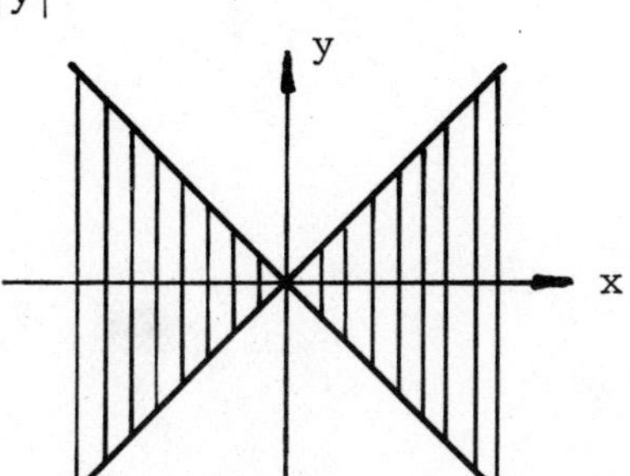

5. 0, $-\infty < x < \infty$, $y = 0$, x-axis

6. paraboloid, $w = 9y^2$, $w = 4x^2$

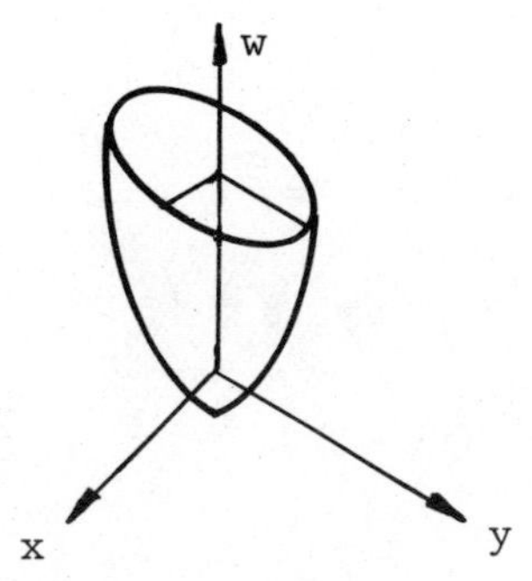

7. A level curve for the constant c, and function in Problem 6, is given by the equation

$c = \underline{\hspace{3cm}}$.

For a fixed value of c, the level curve is an ________ in the xy-plane. Sketch a family of level curves.

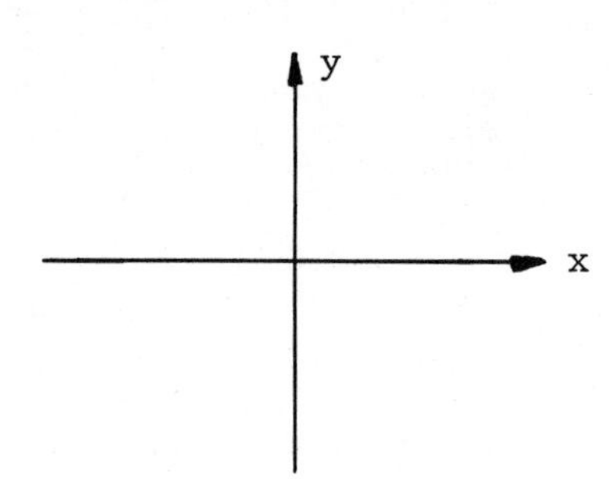

OBJECTIVE B: Given an equation of a real-valued function of several variables, find the partial derivatives with respect to each variable.

8. $f(x,y) = xe^{-y} + y \cos x$

$f_x(x,y) = e^{-y} + \underline{\hspace{3cm}}$ and $f_y(x,y) = \underline{\hspace{3cm}} + \cos x$.

9. $w = xy\, e^{-y^2/2}$

$\frac{\partial w}{\partial x} = \underline{\hspace{3cm}}$ and $\frac{\partial w}{\partial y} = xe^{-y^2/2} - \underline{\hspace{3cm}}$.

10. $g(x,y,z) = \frac{z}{y} + \tan(xy-1)$

$g_x(x,y,z) = \underline{\hspace{4cm}}$,

$g_y(x,y,z) = \underline{\hspace{1.5cm}} + x \sec^2(xy-1)$, and $g_z(x,y,z) = \underline{\hspace{1cm}}$.

---

7. $4x^2 + 9y^2$, ellipse

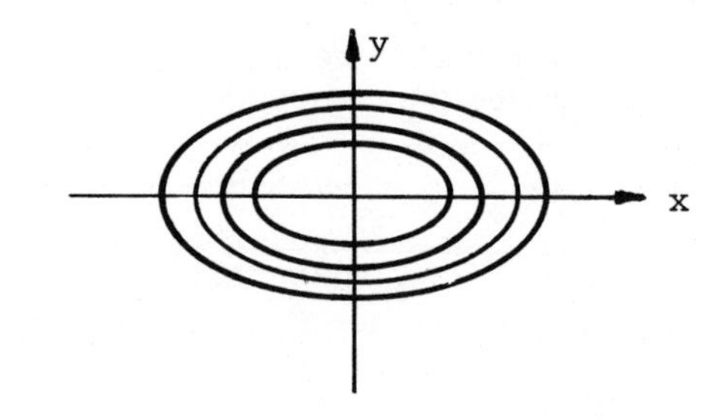

8. $-y \sin x$, $-xe^{-y}$

9. $ye^{-y^2/2}$, $xy^2\, e^{-y^2/2}$

10. $y \sec^2(xy-1)$, $-\frac{z}{y^2}$, $\frac{1}{y}$

11. $h(x,y,z,w,\lambda,\mu) = (x-w)^2 + (y-z)^2 - \lambda(y-x-1) - \mu(w-z)^2$

$h_x(x,y,z,w,\lambda,\mu) = 2(x-w) +$ ___ ,
$h_y(x,y,z,w,\lambda,\mu) =$ ________ $- \lambda$,
$h_z(x,y,z,w,\lambda,\mu) = -2(y-z) +$ _____ ,
$h_w(x,y,z,w,\lambda,\mu) =$ ________ $- \mu$,
$h_\lambda(x,y,z,w,\lambda,\mu) =$ __________ , and
$h_\mu(x,y,z,w,\lambda,\mu) =$ ________ .

OBJECTIVE C: Express the spherical coordinates ρ, ϕ, θ as functions of the cartesian coordinates x, y, z and calculate partial derivatives such as $\partial\rho/\partial x$, $\partial\phi/\partial z$, and $\partial\theta/\partial y$.

12. The relationships between cartesian and spherical coordinates are given by $x = \rho \sin\phi \cos\theta$, $y =$ ____________ , and $z =$ ________ . Thus, $x^2 + y^2 + z^2 =$ ____ , or $0 \le \rho =$ ____________ . Also, $\frac{y}{x} =$ ________ so that $\theta =$ ________ . Finally,

$\frac{z}{\rho} =$ ________ so that $\phi =$ ____________ .

13. From the equations derived in Problem 22 above,

$$\frac{\partial\rho}{\partial z} = \frac{\_\_\_\_}{\sqrt{x^2+y^2+z^2}} ;$$

$$\frac{\partial\phi}{\partial y} = \frac{-1}{\sqrt{1 - \_\_\_\_\_\_}} \cdot \left(-\frac{1}{2}z\right)(2y)\left(x^2+y^2+z^2\right)^{-3/2}$$

$$= \frac{\_\_\_\_\_\_\_\_}{\_\_\_\_\_\_\_\_} ; \quad \text{and} \quad \frac{\partial\theta}{\partial z} = \_\_\_ .$$

---

11. λ, $2(y-z)$, $2z\mu$, $-2(x-w)$, $-y+x+1$, $-w+z^2$

12. $\rho\sin\phi\sin\theta$, $\rho\cos\phi$, ρ^2, $\sqrt{x^2+y^2+z^2}$, $\tan\theta$, $\tan^{-1}\frac{y}{x}$, $\cos\phi$, $\cos^{-1}\frac{z}{\sqrt{x^2+y^2+z^2}}$

13. z, $\frac{z^2}{x^2+y^2+z^2}$, $\frac{yz}{\left(x^2+y^2+z^2\right)\sqrt{x^2+y^2}}$, 0

14. Using the relationships expressed in Problem 12 above,

$\frac{\partial x}{\partial \rho}$ = ________ ; $\frac{\partial y}{\partial \phi}$ = ________ ; and $\frac{\partial z}{\partial \phi}$ = ________.

13-3 Tangent Plane and Normal Line.

OBJECTIVE A: Given a surface $w = f(x,y)$, find an equation of the tangent plane to the surface at a specified point P_o, if the tangent plane exists.

15. The point $P_o(1,-2,3)$ lies on the surface $w = \sqrt{3x^2 - xy + y^2}$.
To find an equation of the tangent plane, calculate

$$f_x(1,-2) = \frac{1}{2}(6x - y)\left(3x^2 - xy + y^2\right)^{-1/2}\Big|_{(1,-2)} = \underline{\quad} ;$$

$$f_y(1,-2) = \underline{\qquad\qquad}\Big|_{(1,-2)} = -\frac{5}{6} ;$$

Therefore, an equation of the tangent plane is

$\frac{4}{3}(x - 1) + \underline{\quad}(y + 2) + \underline{\quad}(w - 3) = 0$ or, $8x - 5y - 6w = \underline{\quad}$.

OBJECTIVE B: Given a surface $w = f(x,y)$, find the normal line (if it exists) to the surface at a specified point P_o.

16. For the surface and point in Problem 15 above, for a normal vector $\mathbf{N}$ to the surface at P_o we may take $\mathbf{N} = 8\mathbf{i} +$ ________ . Thus, equations for the normal line are given by

$x = \underline{\quad} + 8t$, $y = -2 - \underline{\quad}$, $w = $ ________ .

Alternatively, the normal line is

$\frac{x - 1}{8} = \frac{y + 2}{\underline{\quad}} = \underline{\qquad}$.

13-4 Approximate Value of Δw.

OBJECTIVE A: Given a surface $w = f(x,y)$, find the change along the plane tangent to the surface at some given point (x_o, y_o, w_o) for specified increments Δx and Δy.

17. Consider the surface $w = f(x,y) = y \cos(x + y)$. For the point $\left(\frac{\pi}{2}, \frac{\pi}{2}, \underline{\quad}\right)$, the change along the tangent plane is given by

---

14. $\sin \phi \cos \theta$, $\rho \cos \phi \sin \theta$, $-\rho \sin \phi$

15. $\frac{4}{3}$, $\frac{1}{2}(2y - x)\left(3x^2 - xy + y^2\right)^{-1/2}$, $-\frac{5}{6}$, -1, 0 16. $-5\mathbf{j} - 6\mathbf{k}$, 1, $5t$, $3 - 6t$, -5, $\frac{w - 3}{-6}$

$\Delta w_{tan} = f_x\left(\frac{\pi}{2}, \frac{\pi}{2}\right) \Delta x + $ ______________ .

Now, $\frac{\partial w}{\partial x} =$ ______________ and $\frac{\partial w}{\partial y} =$ ______________________ .

Thus, $f_x\left(\frac{\pi}{2}, \frac{\pi}{2}\right) = 0$ and $f_y\left(\frac{\pi}{2}, \frac{\pi}{2}\right) =$ ____ . For the increments $\Delta x = -0.1$ and $\Delta y = 0.05$ we have $\Delta w_{tan} =$ ________ as the change along the tangent plane. On the other hand, the actual change along the surface itself is

$$\Delta w = f\left(\frac{\pi}{2} - 0.1, \frac{\pi}{2} + 0.05\right) - f\left(\frac{\pi}{2}, \frac{\pi}{2}\right)$$

$$\approx f(1.471, 1.621) + \underline{\quad}$$

$$\approx \underline{\qquad\quad} \cos(3.092) + 1.571$$

$$\approx 1.621(-0.999) + 1.571 \approx -0.048$$

with the calculations done on a calculator.

OBJECTIVE B: Given a function $w = f(x,y)$, find a linear approximation to it near a specified point.

18. Given the function $w = f(x,y)$, a linear approximation to it near the point (x_o,y_o) is

$w =$ ______________________________ .

The right side of the previous equation represents the ________ that is tangent to the surface $w = f(x,y)$ at the point (x_o,y_o) .

19. For the surface $f(x,y) = 3xy + ye^x - x \sin y$ we have

$f_x(x,y) =$ ______________ and $f_y(x,y) = 3x + e^x - x \cos y$.

Hence, near the point $\left(0, \frac{\pi}{2}\right)$ the surface $w = f(x,y)$ can be approximated linearly by the tangent plane

$w = \underline{\quad} + $ ________ $(x - 0) + (1)$ $\underline{\qquad\quad}$ = ______________ .

---

17. $-\frac{\pi}{2}$, $f_y\left(\frac{\pi}{2}, \frac{\pi}{2}\right) \Delta y$, $-y \sin(x+y)$, $\cos(x+y) - y \sin(x+y)$, -1, -0.05, $\frac{\pi}{2}$, 1.621

18. $f(x_o,y_o) + f_x(x_o,y_o)(x - x_o) + f_y(x_o,y_o)(y - y_o)$, plane

19. $3y + ye^x - \sin y$, $\frac{\pi}{2}$, $(2\pi - 1)$, $y - \frac{\pi}{2}$, $(2\pi - 1)x + y$

OBJECTIVE C: Use the approximation

$$\Delta w \approx \Delta w_{tan} = f_x(x_o,y_o)\Delta x + f_y(x_o,y_o)\Delta y$$

to discuss the sensitivity of $w = f(x,y)$ to changes in its variables near a given point (x_o,y_o).

20. The surface area of a cone of radius r and height h is given by the formula

$$S = \pi r\sqrt{r^2+h^2} .$$

To calculate ΔS_{tan}, we determine

$\frac{\partial S}{\partial r} = \pi\sqrt{r^2+h^2} +$ ________________ , and $\frac{\partial S}{\partial h} =$ ________________ .

Therefore, near the point $(r_o,h_o) = (3,4)$,

$\Delta S \approx \Delta S_{tan} = \frac{\partial S}{\partial r}(r_o,h_o)\Delta r + \frac{\partial S}{\partial h}(r_o,h_o)\Delta h =$ _____ $\Delta r +$ _____ Δh .

Hence, a one-unit change in r will change S by about $34\pi/5$ units. A one-unit change in h will change S by about ________ units. The surface area of a cone of radius 3 and height 4 is nearly ________ times more sensitive to a small change in r than it is to a change of the same size in h.

13-5 The Directional Derivative: General Case.

OBJECTIVE: Given a function f of two or three variables, find the directional derivative of f at a given point, and in the direction of a given vector **A**.

21. Let us find the directional derivative of $f = e^{x \tan y}$ at the point $(1, \pi/4)$, and in the direction $\mathbf{A} = \frac{1}{e}\mathbf{i} + \frac{2}{e}\mathbf{j}$. Now,

$\frac{\partial f}{\partial x} =$ ________________ and $\frac{\partial f}{\partial y} =$ ________________ .

Thus, we obtain the gradient vector

$\mathbf{v} = \mathbf{i}f_x\left(1,\frac{\pi}{4}\right) + \mathbf{j}f_y\left(1,\frac{\pi}{4}\right) =$ ____________ . Next, $|\mathbf{A}| =$ ____ ,

and hence a unit vector in the direction of **A** is given by

$\mathbf{u} = \frac{\mathbf{A}}{|\mathbf{A}|} =$ ________________ .

---

20. $\pi r^2\left(r^2+h^2\right)^{-1/2}$, $\pi rh\left(r^2+h^2\right)^{-1/2}$, $\frac{34\pi}{5}$, $\frac{12\pi}{5}$, $\frac{12\pi}{5}$, three

21. $\tan y\, e^{x\tan y}$, $x\sec^2 y\, e^{x\tan y}$, $e\mathbf{i} + 2e\mathbf{j}$, $\frac{\sqrt{5}}{e}$, $\frac{1}{\sqrt{5}}\mathbf{i} + \frac{2}{\sqrt{5}}\mathbf{j}$, e, $\frac{2}{\sqrt{5}}$, $\sqrt{5}\, e$

The derivative of f in the direction **A** can now be calculated from the vectors **u** and **v** as

$$\mathbf{u}\cdot\mathbf{v} = \frac{1}{\sqrt{5}}(\_\_\_) + \frac{\quad}{\_\_\_}(2e) = \_\_\_\_\_ .$$

22. Consider the function $f = x^2 + 5y \sin z$. To find the directional derivative of f at the point $P_0(1, 1, \pi/2)$ in the direction from P_0 toward the point $P_1(4, 5, \pi/2)$, we first calculate the partial derivatives of f at P_0. Thus,

$$f_x = \_\_\_ \Big|_{(1,\ 1,\ \pi/2)} = \_\_\_ , \qquad f_y = 5 \sin z \Big|_{(1,\ 1,\ \pi/2)} = \_\_\_ ,$$

$$f_z = \_\_\_\_\_\_\_ \Big|_{(1,\ 1,\ \pi/2)} = 0 .$$

Therefore, the gradient of f at $(1, 1, \pi/2)$ is

$$\mathbf{v} = \_\_\_\_\_\_\_\_\_\_ .$$

The vector $\mathbf{A} = \overrightarrow{P_0P_1} = 3\mathbf{i} + 4\mathbf{j} + 0\mathbf{k}$ has magnitude

$|\mathbf{A}| = \sqrt{\_\_\_\_\_\_\_\_} = \_\_\_$. Thus, the direction of **A** is

$$\mathbf{u} = \frac{\mathbf{A}}{|\mathbf{A}|} = \_\_\_\_\_\_\_\_\_\_ .$$

The derivative of f in the direction **A** is

$$\mathbf{u}\cdot\mathbf{v} = \frac{3}{5}(\_\_\_) + \frac{\quad}{\_\_\_}(5) + 0(0) = \frac{\quad}{\_\_\_} .$$

13-6 The Gradient.

OBJECTIVE A: Given a function $w = f(x,y,z)$, find the gradient vector $(\text{grad } f)_0$ at a specified point $P_0(x_0,y_0,z_0)$.

23. For the function $f(x,y,z) = y \ln x^2 + x \cos z$ and the point $P_0(-1, 2, \pi/2)$, the gradient vector is obtained by calculating the partial derivatives of f at P_0:

---

22. $2x$, 2, 5, $5y \cos z$, $2\mathbf{i} + 5\mathbf{j} + 0\mathbf{k}$, $3^2 + 4^2 + 0^2$, 5, $\frac{3}{5}\mathbf{i} + \frac{4}{5}\mathbf{j} + 0\mathbf{k}$, 2, $\frac{4}{5}$, $\frac{26}{5}$

23. $\frac{2y}{x} + \cos z$, 0, $-x \sin z$, 1, $-4\mathbf{i} + 0\mathbf{j} + \mathbf{k}$

$f_x = \underline{\qquad\qquad}\Big|_0 = -4$, $f_y = \ln x^2\Big|_0 = \underline{\quad}$,

$f_z = \underline{\qquad\qquad}\Big|_0 = \underline{\quad}$. Thus, $(\text{grad } f)_0 = \underline{\qquad\qquad\qquad}$.

OBJECTIVE B: Given a function f of two or three variables, determine the direction one should travel, starting from a given point P_0 , to obtain the most rapid rate of increase or decrease (whichever is specified) of the function.

24. The function $w = f(x,y,z)$ changes most rapidly from the point $P_0(x_0,y_0,z_0)$ in the direction of the vector ____________ . Moreover, the directional derivative in this direction equals the ____________ of the gradient vector. In this case the gradient vector lies in 3-space which contains the domain of the function f .

25. In which direction should one travel, starting from the point $P_0(1, 1, \pi/2)$, in order to obtain the most rapid rate of decrease of the function $f = x^2 + 5y \sin z$? What is the instantaneous rate of change of f per unit of distance in this direction?

<u>Solution</u>. In Problem 22 we calculated $(\text{grad } f)_0 = \underline{\qquad\qquad\qquad}$. Since the directional derivative of f in the direction $\mathbf{u}$ is given by

$$\mathbf{u} \cdot (\text{grad } f)_0 = |(\text{grad } f)_0| \cos \theta ,$$

where θ is the angle between the unit vector $\mathbf{u}$ and the vector __________ , the value of the directional derivative is smallest when $\cos \theta = \underline{\quad}$ or $\theta = \underline{\quad}$. That is, $\mathbf{u}$ points in the negative direction of the gradient.

Thus, $\mathbf{u} = \underline{\qquad\qquad\qquad\qquad}$. The value of the directional derivative in this direction is

$$\mathbf{u} \cdot (\text{grad } f)_0 = \underline{\qquad\qquad} .$$

This value is the negative of the magnitude of the gradient.

26. The function $w = f(x,y)$ changes most rapidly from the point $P_0(x_0,y_0)$ in the direction of the gradient vector

$$(\text{grad } f)_0 = \underline{\qquad\qquad\qquad\qquad\qquad}$$

---

24. $(\text{grad } f)_0$, magnitude

25. $2\mathbf{i} + 5\mathbf{j} + 0\mathbf{k}$, $(\text{grad } f)_0$, -1, π, $-\frac{2}{\sqrt{29}}\mathbf{i} - \frac{5}{\sqrt{29}}\mathbf{j} + 0\mathbf{k}$, $-\sqrt{29}$

26. $\mathbf{i}f_x(x_0,y_0) + \mathbf{j}f_y(x_0,y_0)$, $(\text{grad } f)_0$

in the xy-plane (which contains the domain of the function). The directional derivative in this direction equals the magnitude of the vector ____________ .

27. Consider the function $f = x^y$. At the point $P_0(e,0)$ the partial derivatives of f are

$$f_x = \underline{\hspace{2cm}}\Big|_{(e,0)} = \underline{\hspace{1cm}}, \quad f_y = \underline{\hspace{2cm}}\Big|_{(e,0)} = \underline{\hspace{1cm}}$$

so that $(\text{grad } f)_0 = $ __________ . The maximum value of the directional derivative at P_0 occurs in the direction $\mathbf{u} =$ ___ , and this value equals $|(\text{grad } f)_0| =$ ___ .

Remark. Notice that the gradient vector for a function of two variables lies in the xy-plane, whereas for a function of three variables it lies in space.

OBJECTIVE C: Find the plane which is tangent to the surface $f(x,y,z) = \text{constant}$ at a specified point $P_0(x_0,y_0,z_0)$.

28. The gradient vector $(\text{grad } f)_0$ is __________ to the surface

$$f(x,y,z) = \text{constant}$$

at the point $P_0(x_0,y_0,z_0)$ on the surface.

29. Consider the surface $z = e^{3x} \sin 3y + 2$. Let $w = f(x,y,z) = e^{3x} \sin 3y - z$ so the equation of the surface has the form

$$f(x,y,z) = \text{constant} .$$

In this case the constant is equal to ___ . The point $P_0(0, \pi/6, 3)$ lies on the surface, and the vector

$$(\text{grad } f)_0 = \left(\mathbf{i}\, f_x + \mathbf{j}\, f_y + \mathbf{k}\, f_z\right)_0$$

$$= \left(\mathbf{i}\ 3e^{3x} \sin 3y + \mathbf{j}\ \underline{\hspace{3cm}} - \mathbf{k}\right)_0$$

$$= \underline{\hspace{3cm}}$$

is normal to the surface at P_0 . Thus, an equation of the tangent plane to the surface at P_0 is

$$\underline{\hspace{1cm}}(x-0) + 0\left(\underline{\hspace{1.5cm}}\right) + \underline{\hspace{1cm}}(z-3) = 0 , \quad \text{or } z = \underline{\hspace{2cm}} .$$

---

27. yx^{y-1}, 0, $x^y \ln x$, 1, $0\mathbf{i} + 1\mathbf{j}$, $\mathbf{j}$, 1

28. normal

29. −2, $3e^{3x} \cos 3y$, $3\mathbf{i} + 0\mathbf{j} - \mathbf{k}$, 3, $y - \frac{\pi}{6}$, −1, $3(x+1)$

30. Since the gradient vector $3\mathbf{i} + 0\mathbf{j} - \mathbf{k}$ is normal to the surface in Problem 29 at the point $P_0(0, \pi/6, 3)$, equations of the line normal to the surface at P_0 are

$x = 0 - 3t$, $y =$ __________, and $z =$ ______.

These equations also may be written in the form $\frac{x}{-3} =$ ______, $y =$ ___.

OBJECTIVE D: Given a function $w = f(x,y,z)$, calculate its <u>Laplacian</u> $\frac{\partial^2 f}{\partial x^2} + \frac{\partial^2 f}{\partial y^2} + \frac{\partial^2 f}{\partial z^2}$, provided it exists.

31. Consider the function $f = x^2y^3 + xyz$. Then the partial derivatives of f are

$\frac{\partial f}{\partial x} =$ __________, $\frac{\partial f}{\partial y} = 3x^2y^2 + xz$, $\frac{\partial f}{\partial z} =$ ____.

Taking the partial derivatives again,

$\frac{\partial^2 f}{\partial x^2} = \frac{\partial}{\partial x}\left(\frac{\partial f}{\partial x}\right) =$ _____, $\frac{\partial^2 f}{\partial y^2} = \frac{\partial}{\partial y}\left(\frac{\partial f}{\partial y}\right) =$ _______, and

$\frac{\partial^2 f}{\partial z^2} = 0$. Thus, the Laplacian of f is given by

$\frac{\partial^2 f}{\partial x^2} + \frac{\partial^2 f}{\partial y^2} + \frac{\partial^2 f}{\partial z^2} =$ ____________.

13-7 <u>The Chain Rule for Partial Derivatives</u>.

OBJECTIVE: Let w be a differentiable function of the variables $x,y,z,\ldots,v$ and let these in turn be differentiable functions of a second set of variables $p,q,r,\ldots,t$. Calculate the derivative of w with respect to any one of the variables in the second set by use of the chain rule for partial derivatives.

32. To express dw/dt as a function of t if $w = x^2 - xy$ and $x = e^t$, $y = \ln t$ we first differentiate w with respect to each of the variables x, y in the first set:

$\frac{\partial w}{\partial x} =$ ________ and $\frac{\partial w}{\partial y} =$ ____.

---

30. $\frac{\pi}{6} - 0t$, $3 - t$, $\frac{z-3}{-1}$, $\frac{\pi}{6}$ 31. $2xy^3 + yz$, xy, $2y^3$, $6x^2y$, $2y^3 + 6x^2y$

32. $2x - y$, $-x$, e^t, $\frac{1}{t}$, $\frac{\partial w}{\partial y}\frac{dy}{dt}$, $(-x)\frac{1}{t}$, $\frac{1}{t}e^t$

Next, we differentiate each of the variables x,y with respect to the one variable t of the second set:

$$\frac{dx}{dt} = \_\_\_\_ \quad \text{and} \quad \frac{dy}{dt} = \_\_\_ .$$

Then we form the products of corresponding derivatives, add these products together, and substitute as follows:

$$\frac{dw}{dt} = \frac{\partial w}{\partial x}\frac{dx}{dt} + \_\_\_\_\_\_ = (2x-y)e^t + \_\_\_\_\_\_\_ = (2e^t - \ln t)e^t - \_\_\_\_ .$$

33. To express $\partial w/\partial u$ and $\partial w/\partial v$ as functions of u and v if $w = x^y + z$ and $x = 1+u^2$, $y = e^v$, $z = uv$ we first differentiate w with respect to each of the variables x,y,z in the first set:

$$\frac{\partial w}{\partial x} = \_\_\_\_\_\_\_ , \quad \frac{\partial w}{\partial y} = \_\_\_\_\_\_\_\_ , \quad \frac{\partial w}{\partial z} = \_\_ .$$

Next, we differentiate each of those variables with respect to the first variable u of the second set:

$$\frac{\partial x}{\partial u} = \_\_\_\_ , \quad \frac{\partial y}{\partial u} = 0 , \quad \frac{\partial z}{\partial u} = \_\_ .$$

Then,

$$\frac{\partial w}{\partial u} = \frac{\partial w}{\partial x}\frac{\partial x}{\partial u} + \frac{\partial w}{\partial y}\frac{\partial y}{\partial u} + \_\_\_\_\_\_$$

$$= \left(yx^{y-1}\right)(2u) + \left(x^y \ln x\right)(0) + \_\_\_\_\_ = e^v\left(1+u^2\right)^{e^v-1}(2u) + v .$$

To calculate $\partial w/\partial v$ we differentiate each of the variables x,y,z in the first set with respect to the second variable v of the second set:

$$\frac{\partial x}{\partial v} = 0 , \quad \frac{\partial y}{\partial v} = \_\_\_\_ , \quad \frac{\partial z}{\partial v} = \_\_\_ .$$

Then,

$$\frac{\partial w}{\partial v} = \frac{\partial w}{\partial x}\frac{\partial x}{\partial v} + \_\_\_\_\_\_\_\_\_\_\_\_$$

$$= \left(yx^{y-1}\right)(0) + \_\_\_\_\_\_\_\_ + (1)(u) = \_\_\_\_\_\_\_\_\_\_\_\_\_\_ .$$

34. Suppose that $w = f\left(\frac{y}{x}\right)$ for a differentiable function f. Show that

$$\frac{\partial w}{\partial x} = -\frac{y}{x^2} f'\left(\frac{y}{x}\right) \quad \text{and} \quad \frac{\partial w}{\partial y} = \frac{1}{x} f'\left(\frac{y}{x}\right) .$$

---

33. yx^{y-1}, $x^y \ln x$, 1, $2u$, v, $\frac{\partial w}{\partial z}\frac{\partial z}{\partial u}$, $1(v)$, e^v, u, $\frac{\partial w}{\partial y}\frac{\partial y}{\partial v} + \frac{\partial w}{\partial z}\frac{\partial z}{\partial v}$, $(x^y \ln x)e^v$, $\left(1+u^2\right)^{e^v} \ln(1+u^2)e^v + u$

Solution. Let $w = f(u)$ where $u = y/x$. To find the partial derivatives of w, we first differentiate w with respect to the single variable u of the first set: $\frac{dw}{du} = f'(u)$. To find $\partial w/\partial x$ we next calculate the derivative of the variable u from the first set with respect to the first variable x of the second set: $\frac{\partial u}{\partial x} =$ ______ .

Then, the chain rule gives

$$\frac{\partial w}{\partial x} = \frac{dw}{du}\frac{\partial u}{\partial x} = \underline{\qquad\qquad\qquad} \quad \text{as claimed.}$$

To find $\partial w/\partial y$ we calculate the derivative of the variable u from the first set with respect to the second variable y of the second set:

$\frac{\partial u}{\partial y} =$ ______ . Then, the chain rule gives

$$\frac{\partial w}{\partial y} = \underline{\qquad\qquad\qquad} = \underline{\qquad\qquad\qquad} .$$

35. A point moves on the surface $w = x^3 + 2xy^2$ with x increasing at the rate of 4 cm/sec and y increasing at 6 cm/sec. At what rate is w/ x changing at the moment when $x = 2$ cm and $y = 5$ cm ?

Solution. Let $u = \partial w/\partial x =$ ____________ . We want to calculate du/dt. Now, by the chain rule,

$$\frac{du}{dt} = \underline{\qquad\qquad\qquad\qquad\qquad} = (6x)\frac{dx}{dt} + \underline{\qquad\qquad\qquad} .$$

At the specified moment,

$$\frac{du}{dt} = (12)(4) + \underline{\qquad\qquad\qquad} = \underline{\qquad} \ \text{cm}^2/\text{sec} .$$

13-8 The Total Differential.

OBJECTIVE A: Find the total differential of a function of the variables $x, y, z, \ldots, v$ if each of these variables is in turn a differentiable function of the variables in a second set $p, q, r, \ldots, t$.

36. Suppose $w = \sin(x^2 - 2xy + y^2)$ where $x = uv$ and $y = u/v$. Then,

$$dw = (2x - 2y)\cos(x^2 - 2xy + y^2)\,dx + \underline{\qquad\qquad\qquad\qquad\qquad\qquad}$$
$$= 2(x - y)\cos(x^2 - 2xy + y^2)\,[\underline{\qquad\qquad}]$$

---

34. $-\frac{y}{x^2}$, $f'(u)\left(-\frac{y}{x^2}\right)$, $\frac{1}{x}$, $\frac{dw}{du}\frac{\partial u}{\partial y}$, $f'(u)\left(\frac{1}{x}\right)$

35. $3x^2 + 2y^2$, $\frac{\partial u}{\partial x}\frac{dx}{dt} + \frac{\partial u}{\partial y}\frac{dy}{dt}$, $(4y)\frac{dy}{dt}$, $(20)(6)$, 168

with $dx = v\,du + u\,dv$ and $dy =$ ______________ .

Hence, substitution gives,

$$dw = 2\left(uv - \frac{u}{v}\right)\cos\left(u^2v^2 - 2u^2 + \frac{u^2}{v^2}\right)\left[\_\_\_\_ du + \_\_\_\_ dv\right].$$

Here $du = \Delta u$ and $dv = \Delta v$ are new independent variables.

37. From the calculation in Problem 36,

$$\frac{\partial w}{\partial v} = \_\_\_\_\_\_\_\_\_\_\_\_\_\_\_\_\_\_\_\_\_\_\_\_\_\_\_\_\_\_ .$$

OBJECTIVE B: Use differentials to approximate increments.

38. A closed metal can in the shape of a right circular cylinder is to have an inside height of 6 in., an inside radius of 2 in., and a thickness Of 0.1 in. If the cost of the metal to be used is 10 cents per cubic inch, find an approximate cost of the metal to be used in the manufacture of the can.

Solution. The volume of a right circular cylinder is

$$V = V(r,h) = \pi r^2 h ,$$

where r is the radius and h is the height. The exact volume of metal in the can is the increment $\Delta V = V(2.1, 6.2) - V(2,6)$. We will approximate this increment by the total differential dV. Thus,

$dV =$ ________________ $= (2\pi rh)\,dr +$ __________ .

When $r = 2$, $h = 6$, $dr = \Delta r = 0.1$, and $dh = \Delta h =$ _____, we have

$dV = 2\pi(2)(6)(0.1) +$ ____________ $=$ _______ .

Because the cost is 10 cents per cubic inch, the approximation $\Delta V \approx dV$ gives $(10)(3.2\pi) \approx 100.53$ cents as the approximate cost of the metal to be used in the manufacture of the can.

---

36. $(2y - 2x)\cos(x^2 - 2xy + y^2)\,dy$, $dx - dy$, $\frac{1}{v}\,du - \frac{u}{v^2}\,dv$, $\left(v - \frac{1}{v}\right)$, $\left(u + \frac{u}{v^2}\right)$

37. $2\left(uv - \frac{u}{v}\right)\cos\left(u^2v^2 - 2u^2 + \frac{u^2}{v^2}\right)\left(u + \frac{u}{v^2}\right)$

38. $\frac{\partial V}{\partial r}\,dr + \frac{\partial V}{\partial h}\,dh$, $(\pi r^2)\,dh$, 0.2, $\pi(2)^2(0.2)$, 3.2π

OBJECTIVE C: Given a set of variables $x, y, z, \ldots, v$, each of which is in turn a differentiable function of the variables in a second set $p, q, r, \ldots, t$, find the differentials (if possible) $dp, dq, dr, \ldots, dt$ in terms of the differentials $dx, dy, dz, \ldots, dv$.

39. To find dp and dq in terms of dx and dy, where

$$x = p + q^2 \quad \text{and} \quad y = q - p^2 \quad , \quad pq \neq -\frac{1}{4} ,$$

we first calculate dx and dy : $dx = dp + 2q\,dq$ and dy = ____________ .
We solve these equation simultaneously for dp and dq by the method of determinants:

$$dp = \frac{\begin{vmatrix} dx & 2q \\ dy & 1 \end{vmatrix}}{\begin{vmatrix} 1 & 2q \\ -2p & 1 \end{vmatrix}} = \frac{1}{1+4pq}(dx - 2q\,dy) \quad \text{and}$$

$$dq = \frac{\begin{vmatrix} \_\_ & \_\_ \\ \_\_ & \_\_ \end{vmatrix}}{1+4pq} = \frac{1}{1+4pq}(\_\_\_\_\_\_\_\_\_\_) .$$

From the right-hand sides of the expressions for dp and dq, we may read off $\frac{\partial p}{\partial x}$, $\frac{\partial p}{\partial y}$, $\frac{\partial q}{\partial x}$, and $\frac{\partial q}{\partial y}$. For instance,

$$\frac{\partial p}{\partial x} = \frac{1}{1+4pq} \quad \text{and} \quad \frac{\partial q}{\partial x} = \_\_\_\_\_\_\_\_ .$$

13-9 Maxima and Minima of Functions of Two Independent Variables.

OBJECTIVE: Given the surface $z = f(x,y)$ defined by a function f which has continuous partial derivatives over some region R, examine the surface for high and low points.

40. A necessary condition that must be satisfied if $z = f(x,y)$ has a maximum (or minimum) value occurring at the point (a,b) that is not on the boundary of the region R is that

______________________________ .

---

39. $-2p\,dp + dq$, $\begin{vmatrix} 1 & dx \\ -2p & dy \end{vmatrix}$, $2p\,dx + dy$, $\frac{2p}{1+4pq}$

40. $\frac{\partial f}{\partial x} = 0$ and $\frac{\partial f}{\partial y} = 0$ at (a,b)

41. Consider the surface $f(x,y) = x^2 + 2y^2 + 2x - y$. To find the high and low points on the surface we apply the first necessary condition for a maximum or minimum, namely

$$f_x = \_\_\_\_\_\_ = 0 \quad \text{and} \quad f_y = \_\_\_\_\_\_ = 0 .$$

Solving these equations simultaneously gives $x = -1$ and $y =$ _____ . Thus, the point ____________ is a candidate at which a high or low point may occur. The value for f is $f(-1, 1/4) = -9/8$. To examine the behavior of the difference

$$D = f(x,y) - f(-1, 1/4)$$

let $x = -1 + h$ and $y = \frac{1}{4} + k$. Then

$$D = (-1+h)^2 + 2\left(\tfrac{1}{4} + k\right)^2 + 2(-1+h) - \left(\tfrac{1}{4} + k\right) - \left(-\tfrac{9}{8}\right) = \_\_\_\_\_\_\_\_ .$$

Since D is ____________ for all values of h, k except $h = k = 0$, the surface has a _______ point at $(-1, 1/4, -9/8)$. The given function has an absolute minimum $-9/8$.

42. Consider the surface $f(x,y) = x - y + xy$. To examine the surface for high and low points, apply the first necessary condition:

$$f_x = \_\_\_\_\_\_ = 0 \quad \text{and} \quad f_y = \_\_\_\_\_\_ = 0 .$$

Solving these equations simultaneously gives the point _______ as a candidate. The corresponding value for f is ___ . Let $x = 1 + h$ and $y = -1 + k$, and examine the difference

$$D = f(1+h,-1+k) - f(1,-1) = (1+h) - (-1+k) + (1+h)(-1+k) - 1 = \_\_\_\_ .$$

For values of h and k of the same sign the difference D is ____________ ; for values when h and k are of opposite sign the difference D is ____________ . Thus, the surface has a __________ point occurring at $(1,-1,1)$.

13-10 The Method of Least Squares.

OBJECTIVE: Apply the method of least squares to obtain the line $y = mx + b$ that best fits a given set of data points $(x_1,y_1), (x_2,y_2), \ldots, (x_n,y_n)$.

43. According to Problem 1, page 615 of the textbook, the least squares computations may be summarized as follows:

$$m\left(\sum x_i\right) + nb = \_\_\_\_\_\_ ; \quad m\left(\sum x_i^2\right) + b\left(\sum x_i\right) = \_\_\_\_\_\_ ,$$

where all sums run from $i = 1$ to $i = n$ (the number of data points).

---

41. $2x + 2$, $4y - 1$, $1/4$, $(-1, 1/4)$, $h^2 + 2k^2$, positive, low

42. $1 + y$, $-1 + x$, $(1,-1)$, 1, hk, positive, negative, saddle

43. $\sum y_i$, $\sum x_i y_i$

44. Let us apply the least squares technique to find an equation of the line that best fits the data points (0,1), (1,3), (2,2), (3,4), (4,5). Thus,

$$\sum x_i = 0 + 1 + 2 + 3 + 4 = \_\_\_\_$$

$$\sum y_i = 1 + 3 + 2 + 4 + 5 = \_\_\_\_$$

$$\sum x_i y_i = 0 + 3 + \_\_\_\_\_\_\_\_\_\_ = \_\_\_\_$$

$$\sum x_i^2 = 0 + 1 + 4 + \_\_\_\_\_\_\_\_ = \_\_\_\_ .$$

These give rise to the pair of simultaneous equations

$30m + 10b = \_\_\_\_$ and $\_\_\_\_\_\_\_\_ = 15$.

Solving simultaneously by the method of determinants,

$$m = \frac{\begin{vmatrix} 39 & 10 \\ 15 & 5 \end{vmatrix}}{\begin{vmatrix} 30 & 10 \\ 10 & 5 \end{vmatrix}} = \frac{\_\_\_}{50} = \_\_\_\_ , \text{ and}$$

$$b = \frac{\begin{vmatrix} 30 & 39 \\ 10 & 15 \end{vmatrix}}{50} = \frac{\_\_\_}{50} = \_\_\_\_ .$$

Therefore, an equation for the best fitting line is $y = \_\_\_\_\_\_\_\_\_\_$.

13-11 Maxima and Minima of Functions of Several Independent Variables. Lagrange Multipliers.

OBJECTIVE: Solve extremal problems for functions of several independent variables subject to one or more constraint equations, using the method of Lagrange multipliers.

45. To determine the minimum distance from the origin to the hyperbola $xy = 1$, we would want to minimize the function $f(x,y) = x^2 + y^2$ subject to the constraint $g(x,y) = xy - 1 = 0$. We let

$H(x,y,\lambda) = x^2 + y^2 - \lambda(xy - 1)$. Then, $H_x = \_\_\_\_\_\_\_\_ = 0$,
$H_y = 2y - \lambda x = 0$, $H_\lambda = \_\_\_\_\_\_\_\_ = 0$.

---

44. 10, 15, 4 + 12 + 20, 39, 9 + 16, 30, 39, 10m + 5b, 45, 0.9, 60, 1.2, 0.9x + 1.2

From the first two of these equations we obtain $x = -\frac{1}{2}\lambda y$ and $y = \underline{\qquad}$, and subsequent substitution into the last equation $xy = 1$ gives $\underline{\qquad\qquad} = 1$. Since $xy = 1$ this yields $\lambda^2 = 4$ or, $\lambda = \pm 2$. Now, $0 = xy - 1 = x\left(\frac{1}{2}\lambda x\right) - 1 = \frac{\lambda x^2}{2} - 1$ implies $x^2 = \frac{2}{\lambda}$. Thus, λ cannot be negative for real x. Hence, $\lambda = \underline{\quad}$. Then $x^2 = \frac{2}{\lambda} = 1$ or $x = \pm 1$. Since $xy = 1$ this provides the points $(1,1)$ and $\underline{\qquad\qquad}$. From the geometry of the problem it is clear there is a minimum distance from the origin to the hyperbola, and it has the value

$$D = \sqrt{f(1,1)} = \sqrt{f(-1,-1)} = \underline{\quad}.$$

Notice that there is no maximum distance.

46. Determine the radius r and the height h of the cylinder of maximum surface area which can be inscribed in a sphere of radius a.

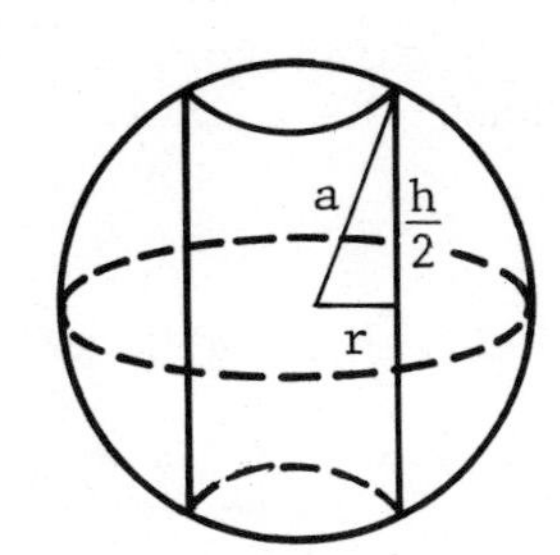

Solution. We want to maximize the surface area

$$f(r,h) = 2\pi rh$$

of the inscribed cylinder subject to the constraint

$$r^2 + \left(\frac{h}{2}\right)^2 = a^2,$$

as in the figure at the right. Assume that both $r > 0$ and $h > 0$ so we do have an inscribed cylinder. Let

$$H(r,h,\lambda) = 2\pi rh - \lambda\left(r^2 + \frac{h^2}{4} - a^2\right).$$

Then,

$H_r = \underline{\qquad\qquad} = 0$, $H_h = 2\pi r - \frac{1}{2}h\lambda = 0$, and

$H_\lambda = \underline{\qquad\qquad\qquad} = 0$. If $\lambda = 0$, the equations $H_r = H_h = 0$ would require both $r = 0$ and $h = 0$, contrary to the assumption that we do have an inscribed cylinder. Thus, neither r, h, nor λ is zero. Also, from the equations $H_r = 0$ and $H_h = 0$, $\lambda r = \pi h$ and $\lambda h = 4\pi r$. Division of these last equations gives

$$\frac{\lambda r}{\lambda h} = \frac{\pi h}{\underline{\qquad}} \text{ or, } 4r^2 = \underline{\quad}.$$

---

45. $2x - \lambda y$, $-xy + 1$, $\frac{1}{2}\lambda x$, $\frac{1}{4}\lambda^2 xy$, 2, $(-1,-1)$, $\sqrt{2}$

46. $2\pi h - 2r\lambda$, $-r^2 - \frac{h^2}{4} + a^2$, $4\pi r$, h^2, r^2, $\frac{a}{\sqrt{2}}$, $\sqrt{2}\,a$, $2\pi a^2$

Substitution into the constraint equation then yields

$r^2 + \_\_\_\_\_ = a^2$ or, $r = \_\_\_\_\_$ (since $r > 0$).

Then, $h = 2r = \_\_\_\_\_$. The geometric nature of the problem ensures that the point $(r,h) = \left(\frac{a}{\sqrt{2}}, \sqrt{2}\,a\right)$ yields a maximum surface area

$$f\left(\frac{a}{\sqrt{2}}, \sqrt{2}\,a\right) = \_\_\_\_\_\_\_ .$$

47. To maximize the function $f(x,y,z) = x^2 + 2y - z^2$ subject to the two constraints $2x - y = 0$ and $y + z = 0$, form the function

$$H(x,y,z,\lambda,\mu) = x^2 + 2y - z^2 - \lambda(2x - y) - \mu(y + z) .$$

Then,

$H_x = \_\_\_\_\_\_\_\_ = 0$, $H_y = \_\_\_\_\_\_\_\_\_\_ = 0$, $H_z = \_\_\_\_\_\_\_\_\_ = 0$,

$H_\lambda = -2x + y = 0$, and $H_\mu = -y - z = 0$.

From $H_x = 0$ we have $x = \_\_\_$. Substitution into $H_\lambda = 0$ and $H_\mu = 0$ gives $y = 2x = \_\_\_\_$ and $z = -y = \_\_\_\_$. Substitution of this expression for z into the equation $H_z = 0$ gives $\mu = \_\_\_\_$. Thus, the equation for $H_y = 0$ becomes

$$2 + \lambda - \mu = 2 + \lambda - \_\_\_\_ = 0 \quad \text{or,} \quad \lambda = \_\_\_ .$$

Then, this value of λ gives

$$x = \frac{2}{3}, \quad y = \_\_\_, \quad z = \_\_\_, \quad \text{and} \quad \mu = \_\_\_ .$$

Therefore, the point $P_0\left(\frac{2}{3}, \frac{4}{3}, -\frac{4}{3}\right)$ gives $f\left(\frac{2}{3}, \frac{4}{3}, -\frac{4}{3}\right) = \_\_\_$.

This is a maximum value for f because when $x = y/2$ and $z = -y$, from the constraint equations, are substituted into $f(x,y,z)$ we find,

$$F(y) = \frac{y^2}{4} + 2y - y^2 = 2y - \frac{3}{4}y^2 \quad \text{and} \quad F''(y) < 0.$$

Thus, f does not have a ___________ value subject to the constraints.

13-12 Higher-Order Derivatives.

OBJECTIVE A: Given a function of several independent variables, calculate all partial derivatives of the second-order.

48. For the function $f(x,y) = xe^{-y} + y\cos x$ we found in Problem 8 above that,

$$f_x = e^{-y} - y\sin x \quad \text{and} \quad f_y = -xe^{-y} + \cos x .$$

---

47. $2x - 2\lambda$, $2 + \lambda - \mu$, $-2z - \mu = 0$, λ, 2λ, -2λ, 4λ, 4λ, $\frac{2}{3}$, $\frac{4}{3}$, $-\frac{4}{3}$, $\frac{8}{3}$, $\frac{4}{3}$, minimum

Hence, the second-order partials are given by

$f_{xx} = \frac{\partial}{\partial x}(f_x) = $ __________ , $f_{xy} = \frac{\partial}{\partial y}(f_x) = $ __________ ,

$f_{yx} = $ __________ $= -e^{-y} - \sin x$, $f_{yy} = $ __________ $=$ ______ .

49. Let $g(x,y,z) = x^3 + xyz + 3yz^2$. Then,

$g_x = 3x^2y + yz$, $g_y = $ __________ , $g_z = xy + 6yz$.

The second-order partials are given by

$g_{xx} = $ ______ , $g_{xy} = $ ________ , $g_{xz} = y$,

$g_{yx} = 3x^2 + z$, $g_{yy} = $ ________ , $g_{yz} = $ ________ ,

$g_{zx} = $ ______ , $g_{zy} = x + 6z$, $g_{zz} = $ ________ .

OBJECTIVE B: Suppose $w = h(x,y) = f(u) + g(v)$, where $f(u)$ and $g(v)$ are twice-differentiable functions of u and v , and $u = u(x,y)$, $v = v(x,y)$ are twice-differentiable functions of x and y . Calculate the second-order partials w_{xx} , w_{xy} , w_{yz} , and w_{yy} .

50. Consider $w = h(x,y) = f(u)$, where $u = x \cos y$. Then,

$w_x = \frac{dw}{du}\frac{\partial u}{\partial x}$ __________ ,

$w_y = \frac{dw}{du}\frac{\partial u}{\partial y}$ __________ ,

$w_{xx} = \frac{\partial}{\partial x}(w_x) = f''(u)\frac{\partial u}{\partial x}\cos y = $ __________ ,

$w_{xy} = \frac{\partial}{\partial y}(w_x) = f''(u)\frac{\partial u}{\partial y}\cos y + ($ __________ $)$

$=$ ________________ ,

$w_{yx} = \frac{\partial}{\partial x}(w_y) = -f'(u)\sin y + ($ __________ $)$

$=$ ________________ ,

---

48. $-y \cos x$, $-e^{-y} - \sin x$, $\frac{\partial}{\partial x}(f_y)$, $\frac{\partial}{\partial y}(f_y)$, xe^{-y}

49. $x^3 + xz + 3z^2$, $6xy$, $3x^2 + z$, 0, $x + 6z$, y, $6y$

50. $f'(u)\cos y$, $-x f'(u) \sin y$, $f''(u)\cos^2 y$, $f'(u)(-\sin y)$,

$-x f''(u) \sin y \cos y - f'(u) \sin y$, $-x f''(u)\frac{\partial u}{\partial x}\sin y$,

$-f'(u)\sin y - x f''(u) \cos y \sin y$, $-x f''(u)\frac{\partial u}{\partial y}\sin y - x f'(u)\cos y$

$$w_{yy} = \frac{\partial}{\partial y}(w_y) = \underline{\hspace{6cm}}$$

$$= x^2 f''(u) \sin^2 y - x f'(u) \cos y .$$

13-13 Exact Differentials.

OBJECTIVE: Given an expression of the form

$$M(x,y)\,dx + N(x,y)\,dy$$

determine whether it is or is not an exact differential. If the expression is the differential of a function $f(x,y)$, find f .

51. The expression $M(x,y)\,dx + N(x,y)\,dy$ is called an exact ____________ if and only if there is a function $w = f(x,y)$ with

________________________ .

52. A necessary and sufficient condition that the expression $M(x,y)\,dx + N(x,y)\,dy$ be exact is that

____________ ,

provided that the functions $M(x,y)$ and $N(x,y)$ and their partial derivatives M_x, M_y, N_x, N_y are __________ for all real values of x and y in a ______________ region R .

53. To test the expression

$$(y + x \sin y)\,dx + (x + y \sin x)\,dy$$

for exactness, we find

$$\frac{\partial M}{\partial y} = \frac{\partial}{\partial y}(y + x \sin y) = \underline{\hspace{3cm}} , \text{ and}$$

$$\frac{\partial N}{\partial x} = \frac{\partial}{\partial x}(x + y \sin x) = \underline{\hspace{3cm}} .$$

Therefore, the expression ________ exact.

54. Consider the expression

$$(xe^y - e^{2y})\,dy + (e^y + x)\,dx .$$

$$\text{Now, } \frac{\partial M}{\partial y} = \frac{\partial}{\partial y}\left(\underline{\hspace{2cm}}\right) = \underline{\hspace{1cm}} , \text{ and}$$

$$\frac{\partial N}{\partial x} = \frac{\partial}{\partial x}\left(\underline{\hspace{2cm}}\right) = e^y .$$

---

51. differential, $df(x,y) = M(x,y)\,dx + N(x,y)\,dy$

52. $\frac{\partial M}{\partial y} = \frac{\partial N}{\partial x}$, continuous, simply connected

53. $1 + x \cos y$, $1 + y \cos x$, is not

Therefore, the expression ____ exact. Thus, we seek $f(x,y)$ such that

$$\frac{\partial f}{\partial x} = e^y + x \quad \text{and} \quad \frac{\partial f}{\partial y} = xe^y - e^{2y} .$$

Integrating the first of these with respect to x while holding y constant and adding $g(y)$ as our "constant of integration," we have

$f(x,y) = $ ____________ $+ g(y)$.

Differentiating this with respect to y with x held constant and equating the result to $xe^y - e^{2y}$ gives

$xe^y + g'(y) = xe^y - e^{2y}$, or $g'(y) = $ ______ .

Therefore, $g(y) = $ ________ $+ C$ and

$f(x,y) = $ ______________________ ,

where C is an arbitrary constant. Thus we have found infinitely many functions with the differential $(xe^y - e^{2y})\,dy + (e^y + x)\,dx$, one for each value of C .

13-14 Derivatives of Integrals.

OBJECTIVE A: Find the derivative, with respect to x , of an integral of the form

$$\int_{u(x)}^{v(x)} f(t)\,dt .$$

55. If f is continuous on $a \le t \le b$, and if u and v are ____________ functions of x such that $u(x)$ and $v(x)$ lie between a and b , then

$$\frac{d}{dx} \int_{u(x)}^{v(x)} f(t)\,dt = \text{\_\_\_\_\_\_\_\_\_\_\_\_\_\_\_\_} .$$

56. Assuming $x > 0$,

$$\frac{d}{dx} \int_{\ln x}^{\sqrt{x}} e^t\,dt = e^{\sqrt{x}} \frac{d}{dx} (\text{\_\_\_\_}) - \text{\_\_\_\_\_} \frac{d}{dx} (\ln x)$$

$$= \text{\_\_\_\_\_\_\_\_\_\_\_\_\_\_\_\_} .$$

---

54. $e^y + x$, e^y, $xe^y - e^{2y}$, is, $xe^y + \frac{1}{2}x^2$, $-e^{2y}$, $-\frac{1}{2}e^{2y}$, $xe^y + \frac{1}{2}x^2 - \frac{1}{2}e^{2y} + C$

55. differentiable, $f\big(v(x)\big)\frac{dv}{dx} - f\big(u(x)\big)\frac{du}{dx}$

56. $\sqrt{x}$, $e^{\ln x}$, $\frac{1}{2\sqrt{x}} e^{\sqrt{x}} - x\left(\frac{1}{x}\right)$ or $\frac{1}{2\sqrt{x}} e^{\sqrt{x}} - 1$

OBJECTIVE B: Find the derivative, with respect to x, of a power function $\left(u(x)\right)^{v(x)}$, where $u(x) > 0$, and u and v are differentiable functions of x.

57. $\frac{d}{dx}\left(u(x)^{v(x)}\right) =$ ______________________ .

58. Using the result in Problem 57 above,

$$\frac{d}{dx}\left(x^4+1\right)^{e^x} = \left(x^4+1\right)^{e^x}\left[\frac{e^x}{x^4+1}\cdot\frac{d}{dx}\left(\_\_\_\_\_\_\right) + \frac{d}{dx}(e^x)\cdot\_\_\_\_\_\_\_\_\right]$$

$$= \_\_\_\_\_\_\_\_\_\_\_\_\_\_\_\_\_\_\_\_\_\_ .$$

59. To obtain the result in Problem 57 above, let $F(u,v) = u^v$, where $u = u(x)$ and $v = v(x)$ are differentiable functions of x, and $u(x) > 0$. Then

$$\frac{dF}{dx} = \frac{\partial F}{\partial u}\frac{du}{dx} + \frac{\partial F}{\partial v}\frac{dv}{dx} = \_\_\_\_\_\_\_\_\frac{du}{dx} + \_\_\_\_\_\_\_\_\frac{dv}{dx}$$

$$= u^v\left[\_\_\_\_\_\_\_\_\_\_\_\_\_\_\_\_\_\_\right].$$

---

57. $u^v\left(\frac{v}{u}\cdot\frac{du}{dx} + \frac{dv}{dx}\cdot\ln u\right)$

58. x^4+1, $\ln(x^4+1)$, $e^x\left(x^4+1\right)^{e^x}\left[\frac{4x^3}{x^4+1} + \ln(x^4+1)\right]$

59. vu^{v-1}, $u^v \ln u$, $\frac{v}{u}\frac{du}{dx} + \ln u\,\frac{dv}{dx}$

CHAPTER 13 OBJECTIVE - PROBLEM KEY

Objective	Problems in Thomas/Finney Text	Objective	Problems in Thomas/Finney Text
13-1 A	p. 573, 4-7	13-6 D	p. 600, 16-23;
B	p. 573, 8,9		p. 632, 7-13
C	p. 573, 10-13	13-7	p. 604, 1-14
13-2 A	p. 580, 1-5	13-8 A	p. 609, 1
B	p. 580, 6-16	B	p. 609, 2,3
C	p. 581, 19-24	C	p. 609, 4,5,6
13-3 A,B	p. 583, 1-5	13-9	p. 612, 1-7
13-4 A	p. 592, 1,2	13-10	p. 615, 1-8
B	p. 592, 5-7	13-11	p. 627, 1-9, 11, 12
C	p. 592, 8,9,11	13-12 A	p. 632, 7-17
13-5	p. 596, 1-4, 6,8	B	p. 632, 1-6
13-6 A	p. 599, 1-7	13-13	p. 637, 1-7
B	p. 596, 5,9;	13-14 A	p. 639, 1-5
	p. 599, 12,15	B	p. 639, 6-10
C	p. 599, 10,11; 8,9		

CHAPTER 13 SELF-TEST

1. Determine, if possible, the following limits. If the limit exists, say what it is; if it does not exist, say why not.

   (a) $\displaystyle\lim_{\substack{x\to -1\\ y\to 1}} \frac{4x^2 - 2y + 3}{x^2 - y^2 - 1}$ (b) $\displaystyle\lim_{\substack{x\to 0\\ y\to 0}} \frac{y^2}{x^2 + y^2}$ (c) $\displaystyle\lim_{\substack{x\to 0\\ y\to 0}} \frac{x - x\cos y}{2y\sin x}$

2. (a) Sketch the surface $w = f(x,y) = x^2 - 2$ in xyw-space.
   (b) Sketch the level curves of the function

   $f(x,y) = y - x^2 + x$ for $w = 0$, $w = 2$, and $w = -2$.

   (c) Find the largest possible xy-domain on which $w = \ln \frac{x}{y}$ is a real variable.

3. Find all first- and second-order partial derivatives.

   (a) $w = 3e^{2x} \cos y$ (b) $f(x,y,z) = zx^y$, $x > 0$

4. Find the normal line to the surface $f(x,y) = e^x \sin y$ at the point $P_0(1, \pi/2, e)$.

5. Find an equation of the tangent plane to the hyperboloid $4x^2 - 9y^2 - 9z^2 - 36 = 0$ at $P_0\left(3\sqrt{3}, 2, 2\right)$.

6. Find the change along the plane tangent to the surface

   $$w = \frac{x - y}{x + y} \quad \text{at} \quad P_0(-1,2,-3),$$

   when $\Delta x = -0.1$ and $\Delta y = 0.02$. Compare this value with the increment Δw.

7. Find a linear approximation to the function

$$f(x,y) = xy\,e^{x+y}$$

near the point $\left(\frac{1}{2}, -\frac{1}{2}\right)$.

8. (a) Find the directional derivative of the function $f = z\tan^{-1}\frac{x}{y}$ at the point $P_0(1,1,2)$, and in the direction $\mathbf{A} = -6\mathbf{i} + 2\mathbf{j} - 3\mathbf{k}$.

(b) What is the maximum value of the directional derivatives of f at P_0, and the direction of the maximum rate of change?

9. Find the indicated derivatives using the chain rule.

(a) $\frac{dw}{dt}$, for $w = t^2 + x\sin y$, $x = \frac{1}{t}$, $y = \tan^{-1} t$

(b) $\frac{\partial^2 w}{\partial v^2}$, for $w = \sin(4x+5y)$, $x = u+v$, $y = u-v$

10. If $w = x + f(xy)$, where f is a differentiable function, show that

$$x\frac{\partial w}{\partial x} - y\frac{\partial w}{\partial y} = x.$$

11. Use differentials to estimate the change in the volume

$$V = \frac{1}{3}\pi h\left(R^2 + Rr + r^2\right)$$

of the frustrum of a cone if the upper radius r is decreased from 4 to 3.8 cm, the base radius R is increased from 6 to 6.1 cm, and the height is increased from 8 to 8.3 cm.

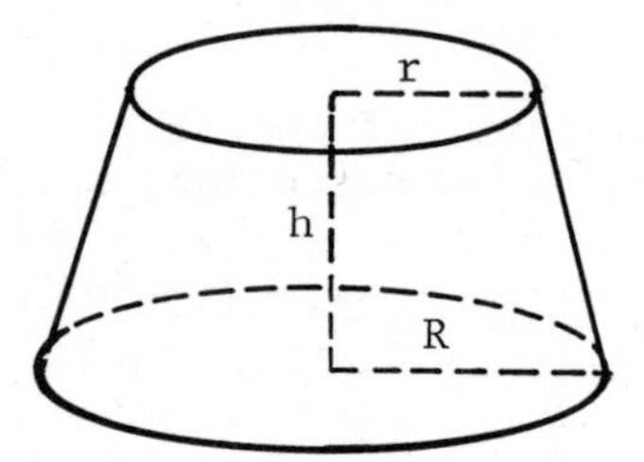

12. Find the total differential df.

(a) $f(x,y) = x\tan y + y\cot x$

(b) $f(x,y,z) = xe^{z^2-y^2}$

13. Given that $x = uv$ and $y = u^2+v^2$, find du and dv in terms of dx and dy.

14. Examine the surface $z = f(x,y) = 2x^2 + y^2 - 4x + 2y + 2xy + 4$ for high and low points.

15. Use the method of least squares to find an equation of the line that best fits the data points $(-1,2)$, $(0,1)$, $(3,-1)$.

16. Find the maximum value of $w = xyz$ among all points lying on the intersection of the planes $x+y+z = 40$ and $x+y-z = 0$.

17. Show that the expression

$$\left(\frac{y}{x} + \ln y\right) dx + \left(\frac{x}{y} + \ln x\right) dy$$

is an exact differential df, and find $f(x,y)$.

18. Find the derivative, with respect to x, of each of the following:

(a) $\int_{x^2}^{x} \sin^2 t \, dt$ (b) $y = (x)^{x^2}$, $x > 0$

SOLUTIONS TO CHAPTER 13 SELF-TEST

1. (a) $$\lim_{\substack{x \to -1 \\ y \to 1}} \frac{4x^2 - 2y + 3}{x^2 - y^2 - 1} = \frac{4(-1)^2 - 2(1) + 3}{(-1)^2 - (1)^2 - 1} = \frac{5}{-1} = -5$$

(b) Let $x = r \cos \theta$, $y = r \sin \theta$, $(r \neq 0)$ so that $g = \frac{y^2}{x^2 + y^2}$ becomes

$g = \sin^2 \theta$. Thus, g takes on all values between 0 and 1 as the point (x,y), or (r,θ), moves around the origin, no matter how small r may be. Thus, the limit does not exist as (x,y) approaches $(0,0)$.

(c) $$\lim_{\substack{x \to 0 \\ y \to 0}} \frac{x - x \cos y}{2y \sin x} = \lim_{\substack{x \to 0 \\ y \to 0}} \frac{x(1 - \cos y)}{2 \sin x \cdot y}$$

$$= \lim_{x \to 0} \frac{x}{2 \sin x} \cdot \lim_{y \to 0} \frac{1 - \cos y}{y} = \frac{1}{2} \cdot 0 = 0$$

2. (a) For every value of y, f is the parabola

$$w = x^2 - 2 .$$

The graph of this cylinder is sketched at the right.

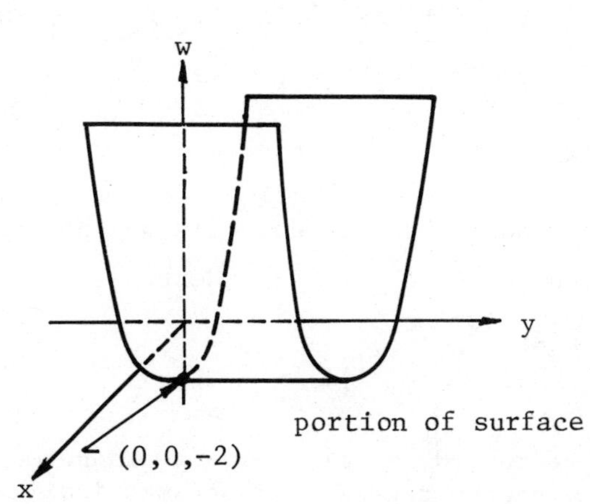

(b) For any fixed $w = c$,

$y - x^2 + x = c$ or,

$y - c = x^2 - x$ gives

$$y - \left(c - \frac{1}{4}\right) = \left(x - \frac{1}{2}\right)^2 ,$$

which is a parabola. The level curves $w = 0, 2, -2$ are sketched at the right.

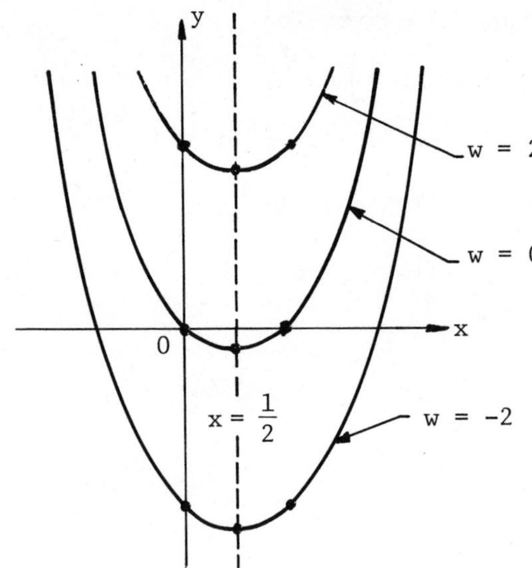

(c) For $w = \ln \frac{x}{y}$ to be real-valued it is necessary that $y \neq 0$ and that $\frac{x}{y} > 0$. Therefore, the domain is the set of points (x,y) where x and y are both positive, or both negative (i.e., the first and third quadrants of the xy-plane excluding the axes).

3. (a) $\frac{\partial w}{\partial x} = 6e^{2x} \cos y$, $\frac{\partial w}{\partial y} = -3e^{2x} \sin y$, $\frac{\partial^2 w}{\partial x \partial y} = \frac{\partial^2 w}{\partial y \partial x} = -6e^{2x} \sin y$,

$\frac{\partial^2 w}{\partial x^2} = 12e^{2x} \cos y$, $\frac{\partial^2 w}{\partial y^2} = -3e^{2x} \cos y$

(b) $f_x = yzx^{y-1}$, $f_y = zx^y \ln x$, $f_z = x^y$, $f_{xx} = y(y-1)zx^{y-2}$,

$f_{xy} = zx^{y-1} + yzx^{y-1} \ln x$, $f_{xz} = yx^{y-1}$, $f_{yx} = zx^{y-1} + yzx^{y-1} \ln x$,

$f_{yy} = zx^y (\ln x)^2$, $f_{yz} = x^y \ln x$, $f_{zx} = yx^{y-1}$, $f_{zy} = x^y \ln x$,

$f_{zz} = 0$

4. Notice that the point P_0 does lie on the surface since its coordinates satisfy the equation $w = e^x \sin y$. Next,

$$f_x\left(1, \frac{\pi}{2}\right) = e^x \sin y \Big|_{\left(1, \frac{\pi}{2}\right)} = e , \quad \text{and} \quad f_y\left(1, \frac{\pi}{2}\right) = e^x \cos y \Big|_{\left(1, \frac{\pi}{2}\right)} = 0.$$

Thus, the vector $\mathbf{N} = e\mathbf{i} + 0\mathbf{j} - \mathbf{k}$ is normal to the surface at P_0. The normal line can be represented parametrically by

$$x = 1 + et , \quad y = \frac{\pi}{2} , \quad z = e - t .$$

5. Let $g(x,y,z) = 4x^2 - 9y^2 - 9z^2$. Notice that P_0 does lie on the given surface since $g(3\sqrt{3}, 2, 2) = 36$. Next,

$$g_x(3\sqrt{3}, 2, 2) = 8x\Big|_{P_0} = 24\sqrt{3},$$

$$g_y(3\sqrt{3}, 2, 2) = -18y\Big|_{P_0} = -36,$$

$$g_z(3\sqrt{3}, 2, 2) = -18z\Big|_{P_0} = -36.$$

The vector grad $g(P_0) = 24\sqrt{3}\,\mathbf{i} - 36\mathbf{j} - 36\mathbf{k}$ is normal to the surface at P_0. Hence,

$24\sqrt{3}\left(x - 3\sqrt{3}\right) - 36(y-2) - 36(z-2) = 0$ or,

$2\sqrt{3}\,x - 3y - 3z = 6$ is an equation of the tangent plane.

6. $$\Delta w_{tan} = \frac{\partial w}{\partial x}\Delta x + \frac{\partial w}{\partial y}\Delta y = \frac{2y}{(x+y)^2}\Delta x - \frac{2x}{(x+y)^2}\Delta y$$

Thus for the point $(x_0, y_0) = (-1,2)$, $\Delta x = -0.1$ and $\Delta y = 0.02$, we find

$$\Delta w_{tan} = \frac{4}{1}(-0.1) + \frac{2}{1}(0.02) = -0.36.$$

On the other hand, the increment giving the change along the surface $w = f(x,y)$ itself is

$$\Delta w = f(-1.1, 2.02) - f(-1,2) = \frac{-1.1 - 2.02}{-1.1 + 2.02} - \frac{-1-2}{-1+2} = \frac{-3.12}{0.92} + \frac{3}{1} = -0.39$$

7. Calculating the partial derivatives

$$f_x(1,-1) = ye^{x+y} + xye^{x+y}\Big|_{\left(\frac{1}{2}, -\frac{1}{2}\right)} = -\frac{1}{2} - \frac{1}{4} = -\frac{3}{4}, \text{ and}$$

$$f_y(1,-1) = xe^{x+y} + xye^{x+y}\Big|_{\left(\frac{1}{2}, -\frac{1}{2}\right)} = \frac{1}{2} - \frac{1}{4} = \frac{1}{4}.$$

Thus, a linear approximation near $\left(\frac{1}{2}, -\frac{1}{2}\right)$ is given by

$$w = f\left(\frac{1}{2}, -\frac{1}{2}\right) + f_x\left(\frac{1}{2}, -\frac{1}{2}\right)\left(x - \frac{1}{2}\right) + f_y\left(\frac{1}{2}, -\frac{1}{2}\right)\left(y + \frac{1}{2}\right)$$

$$= -\frac{1}{4} + \left(-\frac{3}{4}\right)\left(x - \frac{1}{2}\right) + \frac{1}{4}\left(y + \frac{1}{2}\right), \text{ or } 4w = 1 - 3x + y.$$

8. (a) $f_x = z \cdot \dfrac{1}{1+(x/y)^2} \cdot \dfrac{1}{y} = \dfrac{zy}{x^2+y^2}$,

$f_y = z \cdot \dfrac{1}{1+(x/y)^2} \cdot \left(\dfrac{-x}{y^2}\right) = \dfrac{-zx}{x^2+y^2}$, and $f_z = \tan^{-1}\dfrac{x}{y}$

Then, $\text{grad } f(P_0) = f_x(1,1,2)\mathbf{i} + f_y(1,1,2)\mathbf{j} + f_z(1,1,2)\mathbf{k} = \mathbf{i} - \mathbf{j} + \frac{\pi}{4}\mathbf{k}$.

A unit vector in the direction of $\mathbf{A}$ is

$$\mathbf{u} = \frac{\mathbf{A}}{|\mathbf{A}|} = \frac{\mathbf{A}}{\sqrt{36+4+9}} = -\frac{6}{7}\mathbf{i} + \frac{2}{7}\mathbf{j} - \frac{3}{7}\mathbf{k} .$$

Thus, the directional derivative of f in the direction $\mathbf{A}$ at P_0 is

$$\text{grad } f(P_0)\cdot\mathbf{u} = -\frac{6}{7} - \frac{2}{7} - \frac{3\pi}{28} = -\frac{32+3\pi}{28} \approx -1.48 .$$

(b) The direction of the maximum rate of change is in the direction of the gradient, $\text{grad } f(P_0)$. The value of the directional derivative in that direction is

$$|\text{grad } f(P_0)| = \sqrt{1+1+(\pi/4)^2} = \frac{1}{4}\sqrt{32+\pi^2} \approx 1.62 .$$

9. (a) Let $z = t$ so that $w = z^2 + x\sin y$. Then,

$$\frac{dw}{dt} = \frac{\partial w}{\partial x}\frac{dx}{dt} + \frac{\partial w}{\partial y}\frac{dy}{dt} + \frac{\partial w}{\partial z}\frac{dz}{dt}$$

$$= (\sin y)\left(-t^{-2}\right) + (x\cos y)\left(\frac{1}{1+t^2}\right) + (2z)(1)$$

$$= -t^{-2}\sin(\tan^{-1} t) + \frac{1}{t(1+t^2)}\cos(\tan^{-1} t) + 2t .$$

(b) $\dfrac{\partial w}{\partial v} = \dfrac{\partial w}{\partial x}\dfrac{\partial x}{\partial v} + \dfrac{\partial w}{\partial y}\dfrac{\partial y}{\partial v} = 4\cos(4x+5y) - 5\cos(4x+5y)$

Then, setting $r = \partial w/\partial v = -\cos(4x+5y)$,

$$\frac{\partial^2 w}{\partial v^2} = \frac{\partial}{\partial v}\left(\frac{\partial w}{\partial v}\right) = \frac{\partial r}{\partial v} = \frac{\partial r}{\partial x}\frac{\partial x}{\partial v} + \frac{\partial r}{\partial y}\frac{\partial y}{\partial v}$$

$$= 4\sin(4x+5y) - 5\sin(4x+5y) = -\sin(4x+5y) = -\sin(9u-v) .$$

10. Let $u = xy$, so that $w = x + f(u)$. Then,

$$\frac{\partial w}{\partial x} = 1 + f'(u)\frac{\partial u}{\partial x} = 1 + y f'(xy) , \text{ and}$$

$$\frac{\partial w}{\partial y} = f'(u)\frac{\partial u}{\partial y} = x f'(xy) .$$

Therefore,

$$x\frac{\partial w}{\partial x} - y\frac{\partial w}{\partial y} = x\left(1 + y f'(xy)\right) - y\left(x f'(xy)\right) = x , \text{ as claimed.}$$

11. $dV = \frac{1}{3}\pi h(R+2r)\,dr + \frac{1}{3}\pi h(2R+r)\,dR + \frac{1}{3}\pi\left(R^2+Rr+r^2\right)dh$.

At $r = 4$, $R = 6$, $h = 8$, $dr = -0.2$, $dR = 0.1$, and $dh = 0.3$, we find

$$dV = \left(\frac{112\pi}{3}\right)(-0.2) + \left(\frac{128\pi}{3}\right)(0.1) + \left(\frac{76\pi}{3}\right)(0.3) = 4.4\pi$$

The volume increases by about $4.4\pi \approx 13.8$ cubic centimeters.

12. (a) $df = f_x\,dx + f_y\,dy$

$$= (\tan y - y\csc^2 x)\,dx + (x\sec^2 y + \cot x)\,dy$$

(b) $df = f_x\,dx + f_y\,dy + f_z\,dz$

$$= e^{z^2-y^2}\,dx - 2xy\,e^{z^2-y^2}\,dy + 2xz\,e^{z^2-y^2}\,dz$$

13. $dx = v\,du + u\,dv$ and $dy = 2u\,du + 2v\,dv$.

Thus,

$$du = \frac{\begin{vmatrix} dx & u \\ dy & 2v \end{vmatrix}}{\begin{vmatrix} v & u \\ 2u & 2v \end{vmatrix}} = \frac{2v\,dx - u\,dy}{2(v^2-u^2)} = \frac{v}{v^2-u^2}\,dx - \frac{u}{2(v^2-u^2)}\,dy \; ;$$

$$dv = \frac{\begin{vmatrix} v & dx \\ 2u & dy \end{vmatrix}}{2(v^2-u^2)} = -\frac{u}{v^2-u^2}\,dx + \frac{v}{2(v^2-u^2)}\,dy \; .$$

14. From the first necessary condition for a maximum or minimum,

$$\frac{\partial z}{\partial x} = 4x - 4 + 2y = 0 \quad \text{and} \quad \frac{\partial z}{\partial y} = 2y + 2 + 2x = 0 \quad \text{or,}$$

$2x + y = 2$ and $x + y = -1$. Solving these equations simultaneously gives $x = 3$, $y = -4$. For small values of h and k we examine the behavior of the difference

$$\begin{aligned} D &= f(3+h, -4+k) - f(3,-4) \\ &= 2(3+h)^2 + (k-4)^2 - 4(3+h) + 2(k-4) + 2(3+h)(k-4) + 10 \\ &= 2(6h+h^2) + (k^2-8k) - 4h + 2k + 2(3k-4h+hk) \\ &= (h+k)^2 > 0 \end{aligned}$$

Thus the surface has a <u>low</u> point at $(3,-4,-6)$.
The given function has an absolute minimum -6 .

15. $x_1 + x_2 + x_3 = -1 + 0 + 3 = 2$

$x_1^2 + x_2^2 + x_3^2 = 1 + 0 + 9 = 10$

$y_1 + y_2 + y_3 = 2 + 1 - 1 = 2$

$x_1y_1 + x_2y_2 + x_3y_3 = -2 + 0 - 3 = -5$

This gives the least squares equations $10m + 2b = -5$ and $2m + 3b = 2$. Solving simultaneously by determinants, $m = -\frac{19}{26}$ and $b = \frac{30}{26}$. Thus,

$$y = -\frac{19}{26}x + \frac{30}{26}$$

is an equation of the best fitting line.

16. Form the function

$$H(x,y,z,\lambda,\mu) = xyz - \lambda(x+y+z-40) - \mu(x+y-z) .$$

Setting $H_x = H_y = H_z = H_\lambda = H_\mu = 0$ gives the equations

$$yz - \lambda - \mu = 0 \qquad (1) \qquad\qquad xz - \lambda - \mu = 0 \qquad (2)$$

$$xy - \lambda + \mu = 0 \qquad (3) \qquad\qquad -x - y - z + 40 = 0 \qquad (4)$$

$$-x - y + z = 0 \qquad (5)$$

Subtracting equation (4) from (5) gives $2z - 40 = 0$ or $z = 20$. Substitution into (1) and (2) yields

$$\left.\begin{array}{l} 20y = \lambda + \mu \\ 20x = \lambda + \mu \end{array}\right\} \quad \text{thus,} \quad x = y .$$

Substitution into (5) gives $-2x + 20 = 0$ or $x = 10$ and $y = 10$. Thus, $(10,10,20)$ is the unique extremal point subject to the two constraints. Thus, $w = 10 \cdot 10 \cdot 20 = 2000$ is the <u>maximum value</u>.

To see that this is a <u>maximum</u>, note that the intersection of the two planes defined by the constraint equations is the line $x + y = 20$, $z = 20$. That is we can think of maximizing $w = 20xy$ for points on the line $x + y = 20$. For $y > 20$ or $x > 20$, the line passes into the second or fourth quadrants, respectively, whence $w = 20xy$ would be negative. Thus, there can be <u>no minimum</u>. The maximum occurs precisely when $x = y = 10$ along that line.

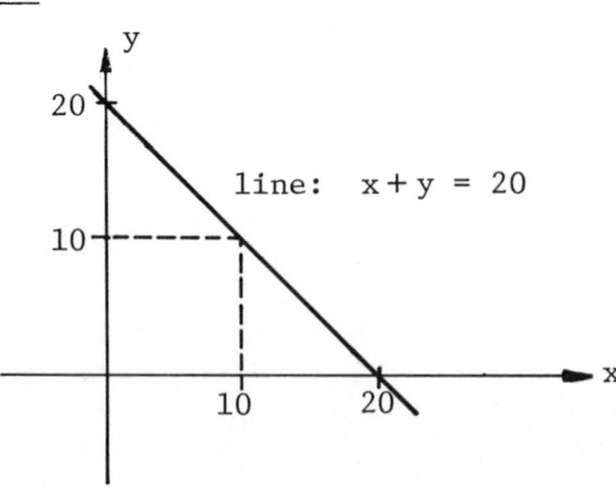

17. $\frac{\partial M}{\partial y} = \frac{\partial}{\partial y}\left(\frac{y}{x} + \ln y\right) = \frac{1}{x} + \frac{1}{y} = \frac{\partial}{\partial x}\left(\frac{x}{y} + \ln x\right) = \frac{\partial N}{\partial x}$

so the expression is an exact differential df . Then, $\frac{\partial f}{\partial x} = \frac{y}{x} + \ln y$ so that

$f(x,y) = y \ln x + x \ln y + g(y)$. Differentiating this with respect to y with x held constant,

$$\frac{\partial f}{\partial y} = \ln x + \frac{x}{y} + g'(y) = N(x,y) = \frac{x}{y} + \ln x .$$

Thus, $g'(y) = 0$ so that $g(y) = C$ is constant. Then,

$f(x,y) = y \ln x + x \ln y + C$, where C is an arbitrary constant.

18. (a) $\frac{d}{dx}\int_{x^2}^{x} \sin^2 t\, dt = \sin^2 x \frac{d}{dx}(x) - \sin^2(x^2)\frac{d}{dx}(x^2)$

$= \sin^2 x - 2x \sin^2(x^2)$

(b) $\frac{d}{dx}(x)^{x^2} = (x)^{x^2}\left[\frac{x^2}{x}\cdot\frac{d}{dx}(x) + \frac{d}{dx}(x^2)\cdot \ln x\right] = (x)^{x^2+1}(1 + 2\ln x)$

14-1 Double Integrals.

OBJECTIVE A: Evaluate a given (double) iterated integral, and sketch the region over which the integration extends.

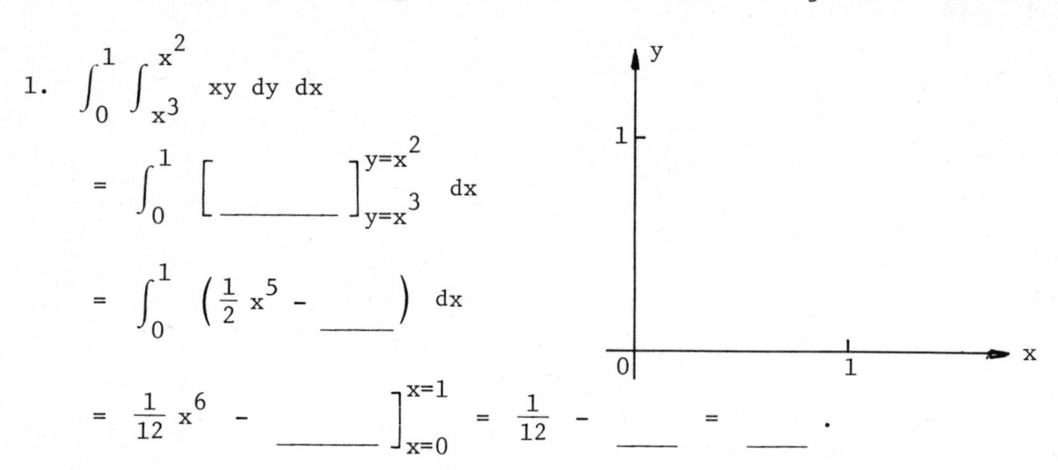

1. $\int_0^1 \int_{x^3}^{x^2} xy \, dy \, dx$

$= \int_0^1 \left[\underline{\qquad\qquad} \right]_{y=x^3}^{y=x^2} dx$

$= \int_0^1 \left(\frac{1}{2} x^5 - \underline{\qquad} \right) dx$

$= \frac{1}{12} x^6 - \underline{\qquad\qquad} \Big]_{x=0}^{x=1} = \frac{1}{12} - \underline{\qquad} = \underline{\qquad}$.

Sketch the region of integration in the coordinate system provided above.

2. $\int_0^{\pi} \int_0^{y^2} \sin \frac{x}{y} \, dx \, dy$

$= \int_0^{\pi} \left[\underline{\qquad\qquad} \right]_{x=0}^{x=y^2} dy$

$= \int_0^{\pi} (-y \cos y + \underline{\qquad}) \, dy$

$= \left(-y \sin y \Big]_{y=0}^{y=\pi} + \int_0^{\pi} \sin y \, dy \right) + \underline{\qquad\qquad} \Big]_{y=0}^{y=\pi}$

$= 0 + (\underline{\qquad\qquad}) \Big]_{y=0}^{y=\pi} + \frac{\pi^2}{2} = \underline{\qquad\qquad} \approx 6.93$.

Sketch the region of integration.

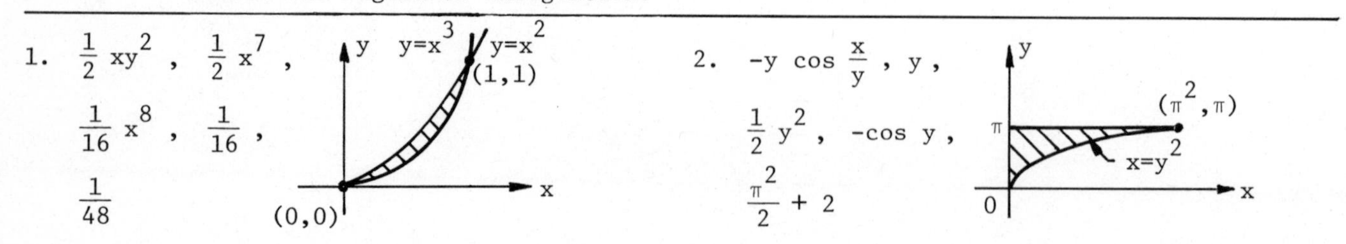

1. $\frac{1}{2} xy^2$, $\frac{1}{2} x^7$, $\frac{1}{16} x^8$, $\frac{1}{16}$, $\frac{1}{48}$

2. $-y \cos \frac{x}{y}$, y , $\frac{1}{2} y^2$, $-\cos y$, $\frac{\pi^2}{2} + 2$

OBJECTIVE B: Given a (double) iterated integral, write an equivalent double iterated integral with the order of integration reversed. Sketch the region over which the integration takes place and evaluate the new integral.

3. 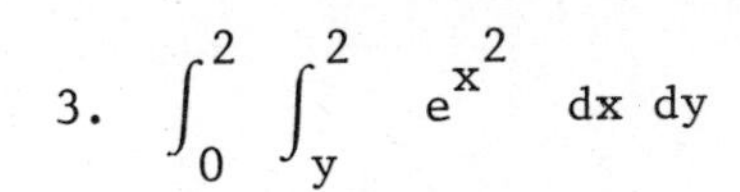

$\int_0^2 \int_y^2 e^{x^2}\, dx\, dy$

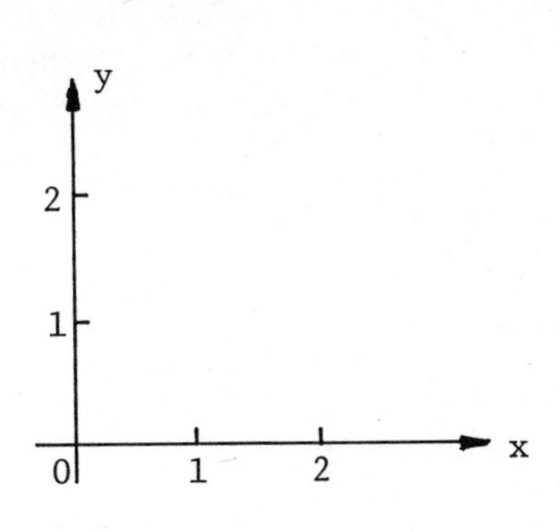

First sketch the region of integration in the coordinate system at the right: as y varies from $y = 0$ to $y = 2$, x varies from ________________ . Reversing the order of integration: as x varies from $x = 0$ to $x = 2$, y varies from ________________ . Thus,

$$\int_0^2 \int_y^2 e^{x^2}\, dx\, dy = \int_{\_\_}^{\_\_} \int_{\_\_}^{\_\_} e^{x^2}\, dy\, dx$$

$$= \int_0^2 \_\_\_\_\_\_ \Big]_{y=0}^{y=x} dx = \int_0^2 \_\_\_\_\_\_ \, dx$$

$$= \_\_\_\_\_\_ \Big]_{x=0}^{x=2} = \_\_\_\_\_\_\_\_ \approx 26.80 .$$

4. $\int_0^2 \int_y^{6-y^2} 2xy^2\, dx\, dy$

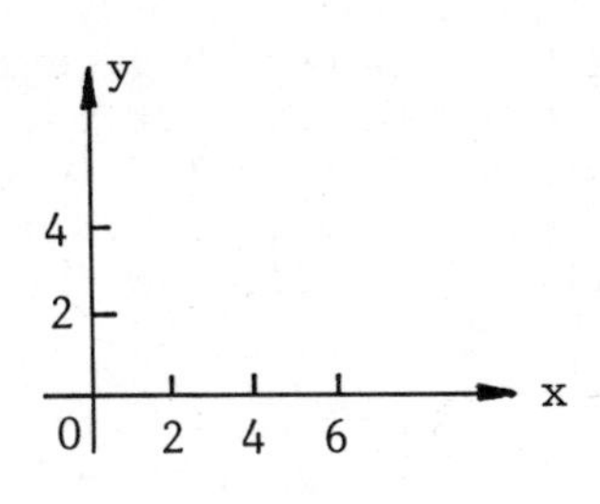

First sketch the region of integration in the coordinate system at the right: as y varies from $y = 0$ to $y = 2$, x varies from ________________ . Reversing the order of integration, as x varies from $x = 0$ to $x = 2$, y varies from ________________ ; but as x varies from $x = 2$ to $x = 6$, y varies from ________________ .

---

3. $x = y$ to $x = 2$, $y = 0$ to $y = x$,

$\int_0^2 \int_0^x$, ye^{x^2} , xe^{x^2} , $\frac{1}{2} e^{x^2}$,

$\frac{1}{2}(e^4 - 1)$

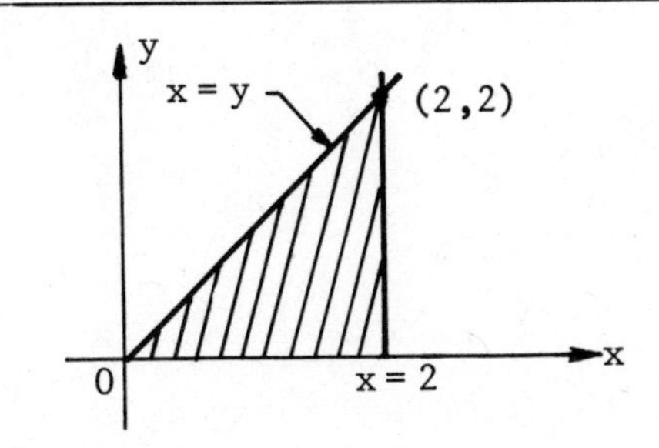

Thus, the iterated integral becomes the sum of two iterated integrals when the order of integration is reversed:

$$\int_0^2 \int_y^{6-y^2} 2xy^2 \, dx\, dy = \int_{\_\_}^{\_\_} \int_{\_\_}^{\_\_} 2xy^2 \, dy\, dx + \int_{\_\_}^{\_\_} \int_{\_\_}^{\_\_} 2xy^2 \, dy\, dx$$

$$= \int_0^2 \underline{\qquad} \Big]_{y=0}^{y=x} dx + \int_2^6 \frac{2}{3} xy^3 \Big]_{\_\_}^{\_\_} dx$$

$$= \int_0^2 \frac{2}{3} x^4 \, dx + \int_2^6 \frac{2}{3} x(6-x)^{3/2} \, dx$$

$$= \underline{\qquad} \Big]_{x=0}^{x=2} + \int_4^0 -\frac{2}{3} (6-u)u^{3/2} \, du$$

(where $u = 6 - x$)

$$= \underline{\quad} - \frac{2}{3}\left(\frac{12}{5} u^{5/2} - \frac{2}{7} u^{7/2}\right)\Big]_{u=4}^{u=0} \approx 31.09 \, .$$

OBJECTIVE C: Find the volume of a solid whose base is a specified region A in the xy-plane, and whose top is a given surface $z = F(x,y)$

5. The base of a solid is the region in the xy-plane that is bounded by the ellipse $x^2 + 9y^2 = 36$ for $y \leq 0$. Sketch this region in the coordinate system at the right. The top of the solid is the plane $z = -y$. Thus, the volume of the solid is computed as the following double integral:

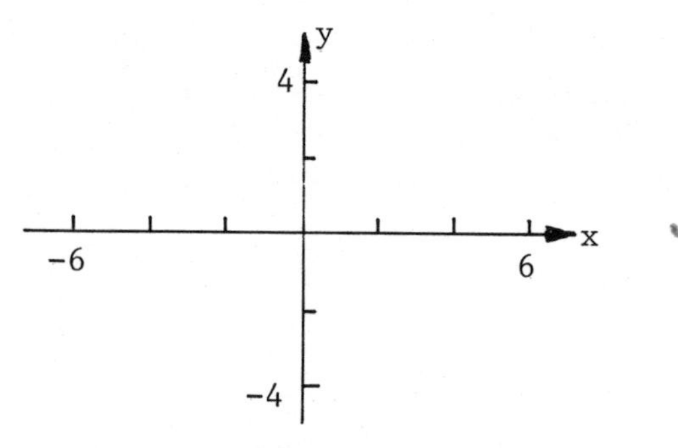

---

4. $x = y$ to $x = 6 - y^2$, $y = 0$ to $y = x$,

$y = 0$ to $y = \sqrt{6-x}$,

$\int_0^2 \int_0^x + \int_2^6 \int_0^{\sqrt{6-x}}$,

$\frac{2}{3} xy^3$, $\Big]_{y=0}^{y=\sqrt{6-x}}$, $\frac{2}{15} x^5$, $\frac{64}{15}$

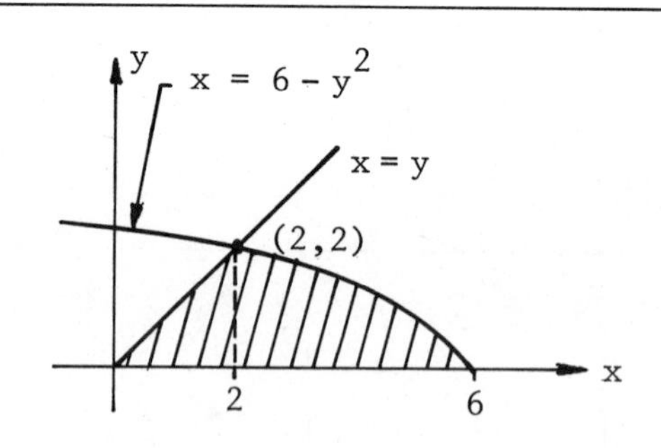

$$V = \int_{\_\_}^{\_\_} \int_{\_\_\_\_}^{\_\_\_\_} - y \, dy \, dx = \int_{-6}^{6} \underline{\qquad\qquad} \Big]_{y=\_\_\_\_}^{y=\_\_\_} dx$$

$$= \frac{1}{2} \int_{-6}^{6} \underline{\qquad\qquad\qquad} dx = \frac{1}{18} \Big(\underline{\qquad\qquad\qquad} \Big)\Big]_{x=-6}^{x=6} = \underline{\qquad} .$$

14-2 Area by Double Integration.

OBJECTIVE: For a specified region A in the xy-plane
- (a) sketch the region A;
- (b) label each bounding curve of the region with its equation, and find the coordinates of the boundary points where the curves intersect; and
- (c) find the area of the region by evaluating an appropriate (double) iterated integral.

6. Let us find the area of the region A in the xy-plane bounded by the circle $x^2 + y^2 = 16$ and the parabola $y^2 = 6x$. First sketch the region A and label the bounding curves in the coordinate system at the right. Notice that we seek the area inside both curves because x must be positive in order that $y^2 = 6x$ hold for real numbers. Next we calculate the points of intersection of the circle and parabola. Substituting $y^2 = 6x$ into the equation for the circle gives

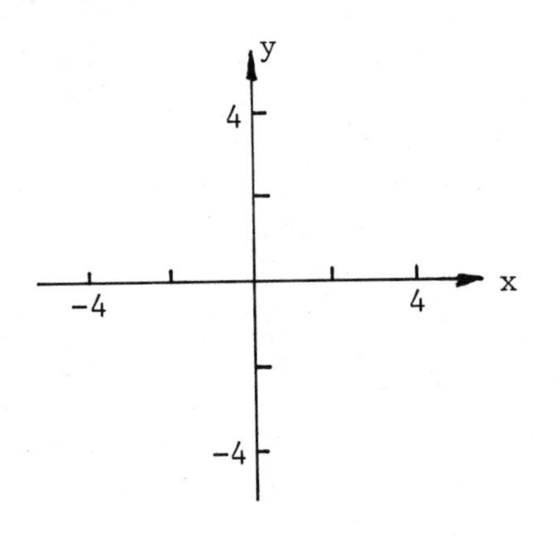

$\underline{\qquad\qquad} = 16$ or, $(x+8)(\underline{\qquad\quad}) = 0$. Thus, $x = \underline{\quad}$ since x cannot be negative. Then $y^2 = 6x = \underline{\quad}$. This yields the two points $\underline{\qquad}$ and $\underline{\qquad}$ as the boundary points where the two curves intersect. We now want to decide what order of integration we should choose. We see that vertical strips sometimes go from the lower branch of the parabola to the upper branch (if $0 \le x \le 2$), but sometimes from the lower semi-circle to the upper semi-circle (if $2 \le x \le 4$). Thus, integration in the order of first y and then x requires that the area be taken in two separate pieces.

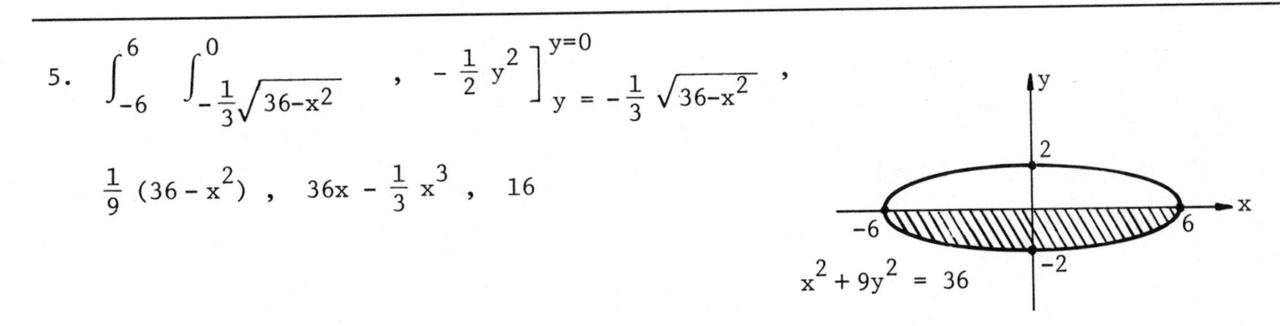

On the other hand, horizontal strips always go from the parabola as the left boundary over to the circle on the right, and the area is given by the integral

$$A = \int_{\_\_\_}^{\_\_\_} \int_{\_\_\_\_}^{\_\_\_\_} dx\, dy \ .$$

Evaluation of this integral gives the result

$$A = \int_{-2\sqrt{3}}^{2\sqrt{3}} \left(\_\_\_\_\_\_\_\_\_\_ \right) dy$$

$$= \frac{1}{2}\left(y\sqrt{16-y^2} + 16 \sin^{-1} \frac{y}{4} \right) - \frac{1}{18} y^3 \Big]_{y=-2\sqrt{3}}^{y=2\sqrt{3}}$$

$$= \frac{4\sqrt{3}}{3} + 8 \sin^{-1} \frac{\sqrt{3}}{2} - 8 \sin^{-1} \left(-\frac{\sqrt{3}}{2} \right)$$

$$= \frac{4\sqrt{3}}{3} + 8\left(\frac{\pi}{3}\right) - 8\left(-\frac{\pi}{3}\right) = \frac{4\sqrt{3} + 16\pi}{3} \approx 19.06 \ .$$

7. Find the area of the region in the first quadrant that is bounded by the curves $x + y = 4$ and $y = 3/x$.

Solution. Sketch the region in the coordinate system at the right, and label the bounding curves. To calculate the points of intersection of the line and the hyperbola, substitute $y = 4 - x$ into $xy = 3$ which gives the result ____________ , or $(x-3)($ ______ $) = 0$. This yields the two points $(3,1)$ and ______ as the boundary points where the two curves intersect. In this example, either order of integration leads to a single double integral giving the area. If we use horizontal strips, the area is given by

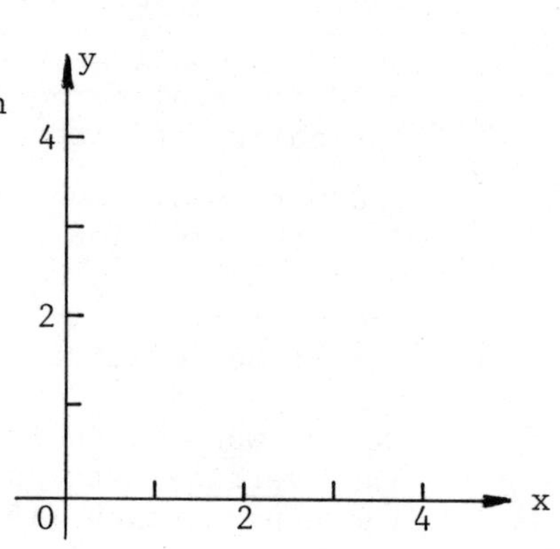

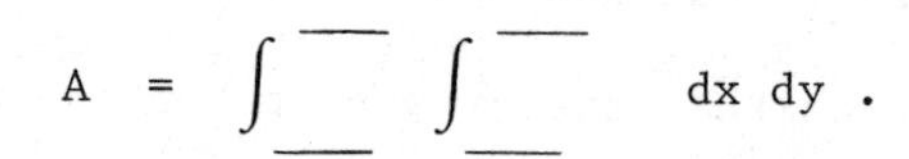

---

6. $x^2 + 6x$, $x - 2$, 2, 12 , $\left(2, -2\sqrt{3}\right)$,

$\left(2, 2\sqrt{3}\right)$, $\displaystyle\int_{-2\sqrt{3}}^{2\sqrt{3}} \int_{y^2/6}^{\sqrt{16-y^2}}$,

$\sqrt{16-y^2} - \frac{1}{6} y^2$

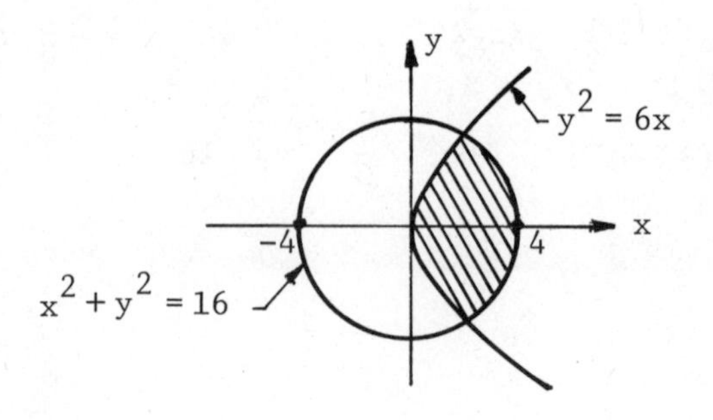

On the other hand, using vertical strips the area is given by

$$A = \int_{\_\_\_}^{\_\_\_} \int_{\_\_\_}^{\_\_\_} dy\ dx\ .$$

Evaluation of either integral (we take the latter) gives

$$A = \int_1^3 \left(\_\_\_\_\_\_\_\_ \right) dx$$

$$= 4x - \frac{1}{2}x^2 - 3\ \ln x \Big]_{x=1}^{x=3} = 4 - 3\ \ln 3 \approx 0.704\ .$$

14-3 Physical Applications.

OBJECTIVE A: Given a plane region A in the xy-plane, find its center of gravity.

8. Consider the region A in the xy-plane bounded by the curves $y^2 = x$ and $y = x$. The region A is sketched in the figure at the right. To find the center of gravity we assume the region has constant uniform density $\delta(x,y) \equiv c$. First, calculate the mass of the region. This is given by the double integral

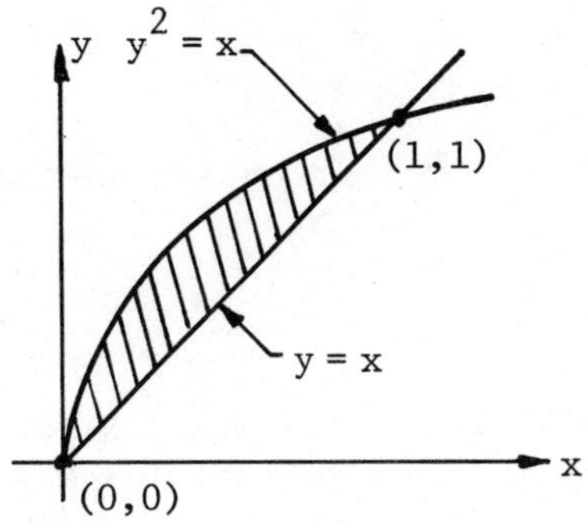

$$M = \iint \delta(x,y)\ dA = \int_{\_\_\_}^{\_\_\_} \int_{\_\_\_}^{\_\_\_} c\ \ dy\ dx$$

$$= \int_0^1 \_\_\_\_\_\_\_\_\_\_ \ dx = c\left(\frac{2}{3}x^{3/2} - \frac{1}{2}x^2\right)\Big]_{x=0}^{x=1} = \_\_\_\ .$$

---

7. $x(4-x) = 3$, $x-1$, $(1,3)$,

$$\int_1^3 \int_{3/y}^{4-y} dx\ dy\ , \quad \int_1^3 \int_{3/x}^{4-x} dy\ dx\ ,$$

$$4 - x - \frac{3}{x}$$

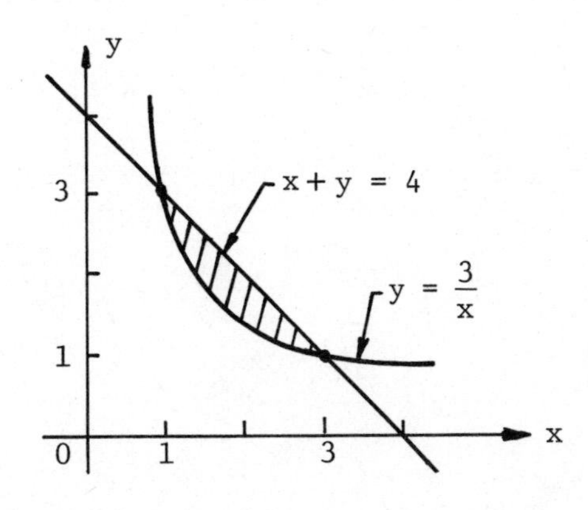

Next, calculate the moment of the mass with respect to the x-axis,

$$M_x = \iint \_\_\_\_\_\_\_\_ = \int_0^1 \int_x^{\sqrt{x}} cy\, dy\, dx$$

$$= \frac{c}{2} \int_0^1 \_\_\_\Big]_{y=x}^{y=\sqrt{x}} dx = \frac{c}{2} \int_0^1 (x - x^2)\, dx$$

$$= \frac{c}{2} \left(\frac{1}{2} x^2 - \frac{1}{3} x^3\right)\Big]_{x=0}^{x=1} = \_\_\_\_ .$$

The moment of the mass with respect to the y-axis is,

$$M_y = \iint x\, \delta(x,y)\, dA = \int_{\_\_}^{\_\_} \int_{\_\_}^{\_\_} cx\, dy\, dx$$

$$= c \int_0^1 x\left(\sqrt{x} - x\right) dx = c\left(\_\_\_\_\_\_\_\_\_\_\right)\Big]_{x=0}^{x=1} = \_\_\_\_ .$$

We then obtain the coordinates of the center of gravity,

$$\bar{x} = \frac{M_y}{M} = \_\_\_ \quad \text{and} \quad \bar{y} = \frac{M_x}{M} = \_\_\_ .$$

OBJECTIVE B: Given a plane region A in the xy-plane, find its moments of inertia about various axes.

9. Find the moment of inertia about the x-axis and about the y-axis for the plane region bounded by the parabola $y = kx^2$, $k > 0$, and the line $y = b$, $b > 0$.

<u>Solution</u>. A sketch of the region is given in the figure on the next page. The x-coordinates of the points of intersection of the line $y = b$ and the parabola $y = kx^2$ are $x = -\sqrt{b/k}$ and $x = \_\_\_\_$.

---

8. $\int_0^1 \int_x^{\sqrt{x}}$, $c\left(\sqrt{x} - x\right)$, $\frac{c}{6}$, $y\, \delta(x,y)\, dA$, y^2, $\frac{c}{12}$, $\int_0^1 \int_x^{\sqrt{x}}$,

$\frac{2}{5} x^{5/2} - \frac{1}{3} x^3$, $\frac{c}{15}$, $\frac{2}{5}$, $\frac{1}{2}$

9. $\sqrt{b/k}$, $\int_0^b \int_{-\sqrt{y/k}}^{\sqrt{y/k}}$, $y^{5/2}$, $\frac{4b^{7/2}}{7\sqrt{k}}$, $x^2 \delta(x,y)\, dA$, $y^{3/2}$, $\frac{4b^{5/2}}{15k^{3/2}}$,

$\frac{b}{5k} M$

We assume that the region is of constant uniform density $\delta(x,y) \equiv 1$. Then the moment of inertia about the x-axis is given by

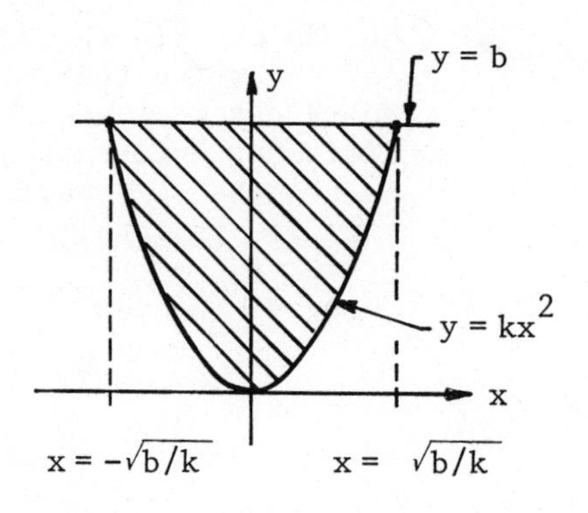

$$I_x = \iint y^2 \, \delta(x,y) \, dA$$

$$= \int_{\_\_}^{\_\_} \int_{\_\_\_\_}^{\_\_\_\_} y^2 \, dx \, dy$$

$$= \frac{2}{\sqrt{k}} \int_0^b \_\_\_\_\_ \, dy = \_\_\_\_\_ .$$

The moment of inertia about the y-axis is

$$I_y = \iint \_\_\_\_\_\_\_\_ = \int_0^b \int_{-\sqrt{y/k}}^{\sqrt{y/k}} x^2 \, dx \, dy$$

$$= \frac{2}{3k^{3/2}} \int_0^b \_\_\_\_\_ \, dy = \_\_\_\_\_ .$$

If we calculate the mass of the plane region,

$$M = \iint \delta(x,y) \, dA = \int_0^b \int_{-\sqrt{y/k}}^{\sqrt{y/k}} dx \, dy = 2 \int_0^b \sqrt{y/k} \, dy = \frac{4b^{3/2}}{4\sqrt{k}} ,$$

then we can express the moments of inertia in terms of the mass as follows:

$$I_x = \frac{3b^2}{7} M \quad \text{and} \quad I_y = \_\_\_\_\_ .$$

14-4 Polar Coordinates.

OBJECTIVE A: Given a (double) iterated integral in cartesian coordinates, change it to an equivalent double integral in polar coordinates and then evaluate the integral thus obtained.

10. $\displaystyle\int_0^a \int_y^{\sqrt{a^2 - y^2}} \sin(x^2 + y^2) \, dx \, dy$

---

10. 0, a, $\displaystyle\int_0^{\pi/4} \int_0^a \sin(r^2) \, r \, dr \, d\theta$,

$-\frac{1}{2}\cos(r^2)$, $\frac{\pi}{4}\left(\frac{1}{2} - \frac{1}{2}\cos a^2\right)$

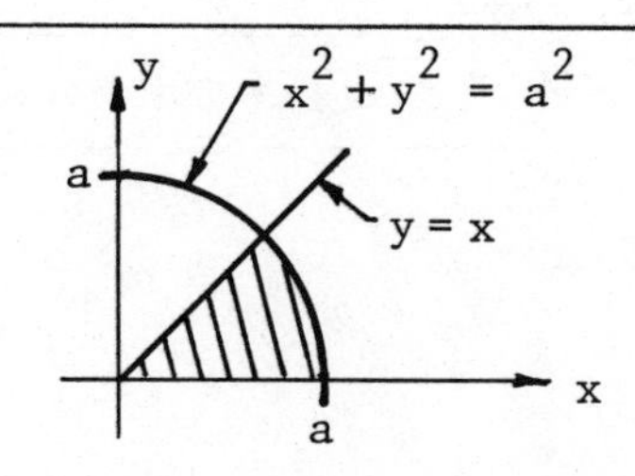

First sketch the region of integration in the coordinate system at the right. In converting the integral to polar coordinates we need to find the polar limits of integration for the region. We see that θ varies from $\theta =$ ___ to $\theta = \pi/4$ radians, and r varies from $r = 0$ to $r =$ ___ units. Thus, the integral becomes

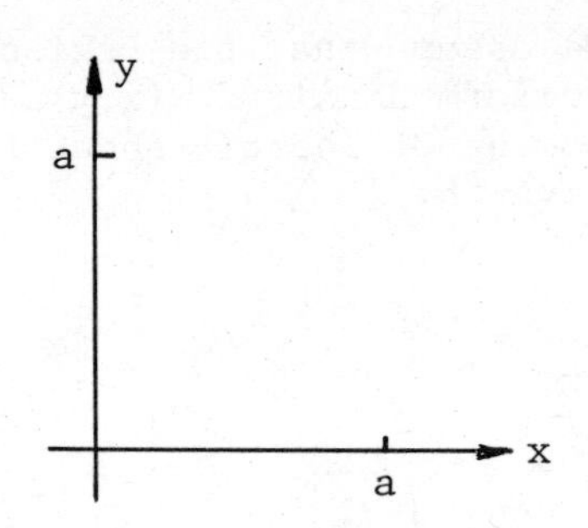

<u>In cartesian coordinates</u> — <u>In polar coordinates</u>

$$\int_0^a \int_y^{\sqrt{a^2 - y^2}} \sin(x^2 + y^2)\, dx\, dy = \int_{\_\_\_}^{\pi/4} \int_0^{\_\_\_} \_\_\_\_\_\_\_\_\_\_ \, dr\, d\theta$$

$$= \int_0^{\pi/4} \Big[\_\_\_\_\_\_\_\_\_\_ \Big]_{r=0}^{r=a} d\theta$$

$$= \int_0^{\pi/4} \left(\frac{1}{2} - \frac{1}{2} \cos a^2 \right) d\theta$$

$$= \_\_\_\_\_\_\_\_\_\_\_\_ .$$

11. $$\int_0^1 \int_{\sqrt{1-x^2}}^{\sqrt{2-x^2}} \frac{1}{\sqrt{x^2 + y^2}}\, dy\, dx$$

A sketch of the region of integration is shown at the right. Notice that the line $y = x$ intersects the circle $x^2 + y^2 = 2$ at the point $(1,1)$. Let us find the polar limits of integration for this region. The region can be described as the union of two regions: as θ varies from $\theta = 0$ to $\theta = \pi/4$, r varies from $r =$ ___ (on the circle $x^2 + y^2 = 1$) to the line $x = 1$ or $r = \sec\theta$ (since $x = r\cos\theta = 1$); as θ varies from $\theta = \pi/4$ to $\theta = \pi/2$, r varies from ____________ (from the circle $x^2 + y^2 = 1$ to the circle $x^2 + y^2 = 2$). Therefore, the integral becomes

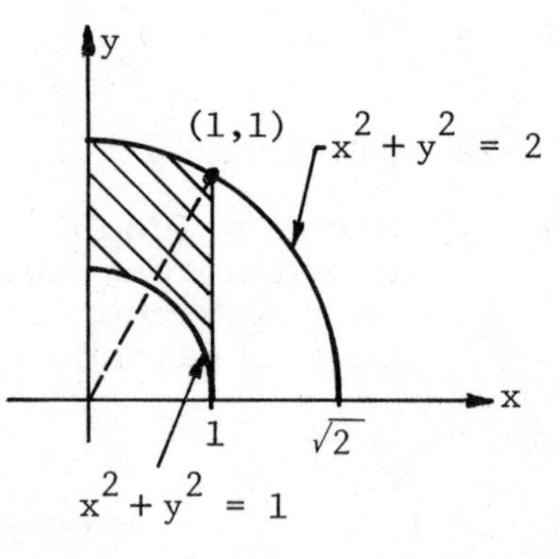

---

11. 1, $r = 1$ to $r = \sqrt{2}$, $\int_0^{\pi/4} \int_1^{\sec\theta} + \int_{\pi/4}^{\pi/2} \int_1^{\sqrt{2}}$, $\sec\theta - 1$, $\left(\sqrt{2} - 1\right)\theta$, $\frac{\pi}{4}\left(\sqrt{2} - 1\right)$

In cartesian coordinates

$$\int_0^1 \int_{\sqrt{1-x^2}}^{\sqrt{2-x^2}} \frac{1}{\sqrt{x^2+y^2}}\, dy\, dx$$

In polar coordinates

$$= \int_0^{\pi/4} \int_{\_\_\_}^{\_\_\_} \left(\frac{1}{r}\right) r\, dr\, d\theta \quad + \quad \int_{\_\_\_}^{\_\_\_} \int_{\_\_\_}^{\_\_\_} \left(\frac{1}{r}\right) r\, dr\, d\theta$$

$$= \int_0^{\pi/4} \left(\_\_\_\_\_\_\_\_\_\_\right) d\theta \quad + \quad \int_{\pi/4}^{\pi/2} \left(\sqrt{2} - 1\right) d\theta$$

$$= \ln\ |\sec\theta + \tan\theta| - \theta\ \Big]_{\theta=0}^{\theta=\pi/4} \quad + \quad \_\_\_\_\_\_\_\_\_\_ \Big]_{\theta=\pi/4}^{\theta=\pi/2}$$

$$= \left(\ln\ \left|\sec\frac{\pi}{4} + \tan\frac{\pi}{4}\right| - \frac{\pi}{4} - 0\right) + \_\_\_\_\_\_\_\_\_\_$$

$$= \ln\left(\sqrt{2}+1\right) + \frac{\pi}{4}\left(\sqrt{2}-1\right) \approx 1.21 \ .$$

OBJECTIVE B: In an applied problem involving double integration (e.g., finding an area, volume, center of gravity or moment of inertia), express the double integral in polar coordinates (when appropriate, to make the integrations easier), and then evaluate the integral thus obtained.

12. Let us find the area of one petal of the rose $r = a \sin 3\theta$ sketched in the figure at the right. The petal in the first quadrant is traced out as θ varies from $\theta = 0$ to $\theta = \pi/3$: r starts at $r = 0$, reaches its maximum value $r = a$ when $\theta =$ ____ , and returns to $r = 0$ when $\theta = \pi/3$. Thus, as θ varies from $\theta = 0$ to $\theta = \pi/3$, r varies from __________ __________ . The area of the petal is given in terms of polar coordinates by

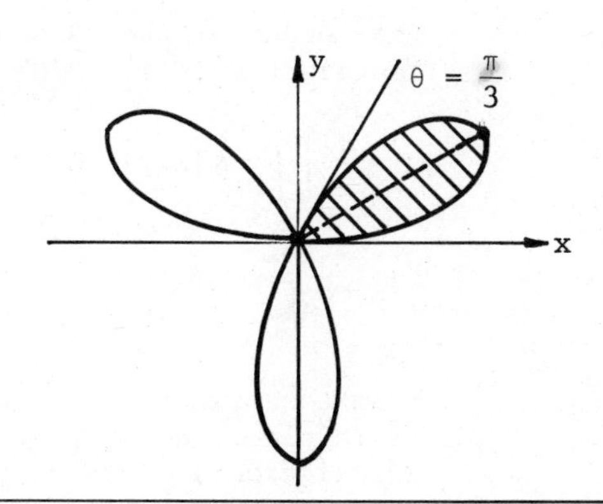

---

12. $\pi/6$, $r = 0$ to $r = a \sin 3\theta$ (the boundary of the petal) , $\int_0^{\pi/3} \int_0^{a \sin 3\theta}$

$\frac{1}{2} r^2$, $\frac{1}{2} a^2 \sin^2 3\theta$, $\theta - \frac{1}{6} \sin 6\theta$, $\frac{\pi a^2}{12}$

$$A = \iint dA = \int_{\_\_\_}^{\_\_\_} \int_{\_\_\_}^{\_\_\_} r\ dr\ d\theta$$

$$= \int_0^{\pi/3} \_\_\_\_\_ \Big]_{r=0}^{r=a \sin\theta} d\theta = \int_0^{\pi/3} \_\_\_\_\_\_\_\_\_\_ d\theta$$

$$= \frac{a^2}{4} \int_0^{\pi/3} (1 - \cos 6\theta)\ d\theta = \frac{a^2}{4} \Big[\_\_\_\_\_\_\_\_\_\_ \Big]_{\theta=0}^{\theta=\pi/3} = \_\_\_\_ .$$

13. Let us find the center of mass of the semi-circular region bounded by the x-axis and the curve $y = \sqrt{a^2 - x^2}$, when the density of the region is proportional to the square of its distance from the center of the semi-circular arc: $\delta(x,y) = k(x^2 + y^2)$ for some proportionality constant k. The region is sketched in the figure at the right. In terms of polar coordinates, the mass of the region is given by

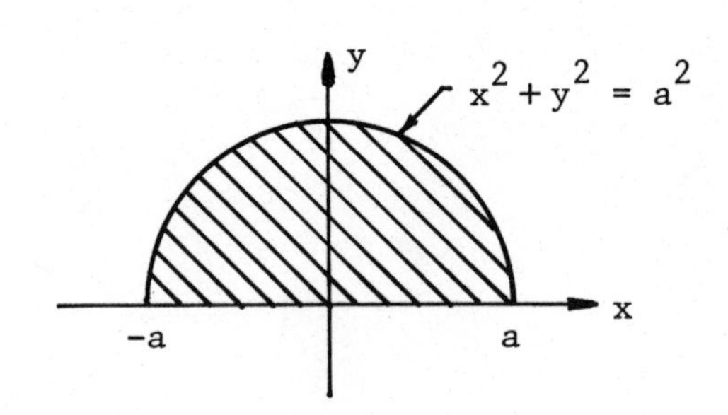

$$M = \iint \delta(x,y)\ dA = \int_{\_\_\_}^{\_\_\_} \int_{\_\_\_}^{\_\_\_} kr^2 \cdot r\ dr\ d\theta$$

$$= \int_0^{\pi} \frac{k}{4} r^4 \Big]_{r=0}^{r=a} d\theta = \_\_\_\_\_\_\_\_ .$$

The moment of the mass with respect to the x-axis in terms of polar coordinates is

$$M_x = \iint y\ \delta(x,y)\ dA = \int_0^{\pi} \int_0^{a} \_\_\_\_\_\_\_\_\_\_ \cdot r\ dr\ d\theta$$

$$= \int_0^{\pi} \_\_\_\_\_\_\_\_ d\theta = \frac{2ka^5}{5} .$$

From the symmetry of the plate and its density function across the y-axis, it is clear that M_y is zero. (This may also be verified by direct calculation.) Thus,

$\bar{x} = \_\_\_$ and $\bar{y} = \_\_\_\_$

give the coordinates for the center of mass.

---

13. $\int_0^{\pi} \int_0^{a}$, $\frac{\pi k a^4}{4}$, $kr^2(r \sin\theta)$, $\frac{ka^5}{5} \sin\theta$, 0, $\frac{8a}{5\pi}$

14. Suppose we wish to find the volume of the ellipsoid $9x^2 + 9y^2 + z^2 = 9$. In cartesian coordinates the volume is given by

$$V = 2 \int_{-1}^{1} \int_{-\sqrt{1-x^2}}^{\sqrt{1-x^2}} z \, dy \, dx \quad , \quad \text{where } z = 3\sqrt{1-x^2-y^2}$$

(the factor 2 occurs because we selected $z \geq 0$, and the ellipsoid is symmetric across the plane $z = 0$). We see that this iterated integral would be very cumbersome to calculate. Let us try the integration in polar coordinates: $x = r \cos \theta$ and $y = r \sin \theta$ transforms the equation $9x^2 + 9y^2 + z^2 = 9$ into the polar equation ______________. Thus, when $z = 0$ (in the xy-plane), $r =$ ___ . Therefore, in terms of polar coordinates the volume is given by

$$V = 2 \int_{\_\_}^{\_\_} \int_{\_\_}^{\_\_} 3\sqrt{1-r^2} \cdot r \, dr \, d\theta$$

$$= 2 \int_{0}^{2\pi} \_\_\_\_\_\_\_\_\_\_ \Big]_{r=0}^{r=1} d\theta = 2 \int_{0}^{2\pi} d\theta = 4\pi .$$

14-5 Triple Integrals, Volume.

OBJECTIVE: By triple integration, find the volume of a specified region V in xyz-space.

15. Let us calculate the volume of the region in space bounded by the cylinders $z = x^2$ and $z = 4 - x^2$, and the planes $y = 0$, $y = 5$. A sketch of the region is shown at the right.

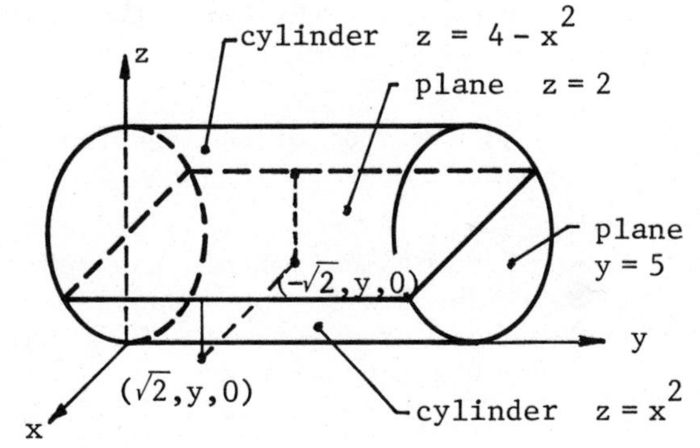

The two surfaces intersect when $4 - x^2 = x^2$, or $x^2 = 2$; that is they intersect on the plane $z = 2$. The volume projects into the region A in the xy-plane described by $-\sqrt{2} \leq x \leq \sqrt{2}$ and $0 \leq y \leq 5$. Thus, the volume is given by the triple iterated integral,

---

14. $z^2 + 9r^2 = 9$, 1, $\int_0^{2\pi} \int_0^1$, $-(1-r^2)^{3/2}$

15. $\int_{-\sqrt{2}}^{\sqrt{2}} \int_0^5 \int_{x^2}^{4-x^2}$, $4 - 2x^2$, $y(4-2x^2)$, $5(4-2x^2)$, $20x - \frac{10}{3}x^3$, $\frac{80\sqrt{2}}{3}$

$$V = \int_{\_\_\_}^{\_\_\_} \int_{\_\_\_}^{\_\_\_} \int_{\_\_\_}^{\_\_\_} dz\, dy\, dx$$

$$= \int_{-\sqrt{2}}^{\sqrt{2}} \int_0^5 \_\_\_\_\_\_ \, dy\, dx = \int_{-\sqrt{2}}^{\sqrt{2}} \_\_\_\_\_\_ \Big]_{y=0}^{y=5} dx$$

$$= \int_{-\sqrt{2}}^{\sqrt{2}} \_\_\_\_\_\_ \, dx = \_\_\_\_\_\_ \Big]_{x=-\sqrt{2}}^{x=\sqrt{2}} = \_\_\_\_\_\_ .$$

14-6 Cylindrical Coordinates.

OBJECTIVE: Find by triple integration, in cylindrical coordinates, the volume of a specified region in space. (The region may be described in cartesian coordinates.)

16. Consider the region in space that lies inside the sphere $x^2 + y^2 + z^2 = 9$, but outside the cylinder $x^2 + y^2 = 1$. The region is sketched in the figure at the right. The z-axis is an axis of symmetry so that cylindrical coordinates should be useful for finding the volume of the region by triple integration.

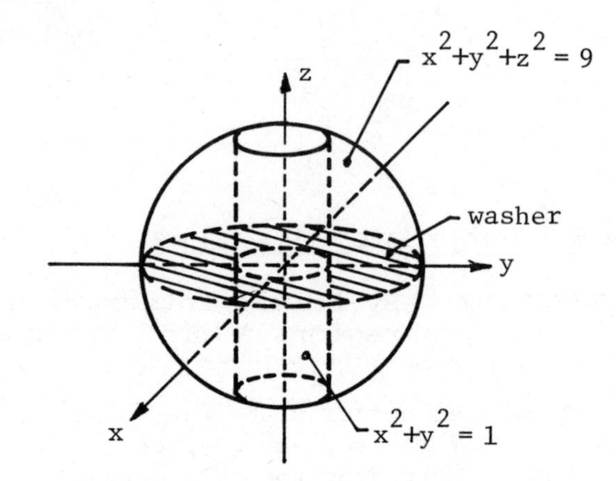

The equation of the sphere in cylindrical coordinates is $z^2 = \_\_\_\_\_\_$. The projection of the volume within the sphere and outside the cylinder in the xy-plane is a washer with inner radius $r = 1$ and outer radius $r = \_\_\_$. This is the shaded area shown in the diagram, and in polar coordinates it is described by the inequalities $\_\_\_ \le r \le \_\_\_$ and $\_\_\_ \le \theta \le \_\_\_$. For a point (x,y), or (r,θ), in this washer, the altitude z of the solid region varies from $z = -\sqrt{9-r^2}$ (the surface of the sphere below the xy-plane) to $z = \_\_\_\_\_\_$ (that portion of the surface above the xy-plane). Then the volume of the solid region in cylindrical coordinates is given by

---

16. $9-r^2$, 3, $1 \le r \le 3$, $0 \le \theta \le 2\pi$, $\sqrt{9-r^2}$, $\int_1^3 \int_0^{2\pi} \int_{-\sqrt{9-r^2}}^{\sqrt{9-r^2}} dz\, r\, d\theta\, dr$,

$2r\sqrt{9-r^2}$, $-\frac{4\pi}{3}\left(9-r^2\right)^{3/2}$

$$V = \int_1^3 \int_{\_\_\_}^{2\pi} \int_{-\sqrt{9-r^2}}^{\_\_\_\_\_} \_\_\_\_\_\_\_\_ = \int_1^3 \int_0^{2\pi} \_\_\_\_\_\_\_\_ \, d\theta \, dr$$

$$= \int_1^3 4\pi r \sqrt{9-r^2} \, dr = \_\_\_\_\_\_\_\_\_\_ \Big]_{r=1}^{r=3} = \frac{64\pi\sqrt{2}}{3} \approx 94.78 .$$

14-7 Physical Applications of Triple Integration.

OBJECTIVE: By triple integration, find the mass, the center of gravity, or the moments of inertia of a mass M distributed over a region V of xyz-space and having density $\delta = \delta(x,y,z)$ at the point (x,y,z) of V. Use cartesian or cylindrical coordinates to evaluate your integrals, whichever is more convenient.

17. Find the center of gravity of the homogenous solid inside the paraboloid $x^2 + y^2 = z$ and outside the cone $x^2 + y^2 = z^2$.

Solution. The solid region is sketched at the right. A slice through the solid by a plane parallel to the xy-plane produces a washer-shaped area like that shaded area shown in the figure. The cone and the paraboloid intersect when $z = z^2$, or $z = 0$ (the base) or $z =$ ___ (the top rim of the solid). Since the region is geometrical we take the density as unity. Because the z-axis is an axis of symmetry it will be convenient to use cylindrical coordinates. In cylindrical coordinates, the equation $x^2 + y^2 = z^2$ of the cone becomes ________ and the equation for the paraboloid becomes ________ . The projection of the solid into the xy-plane is the interior of a circle centered at the origin of radius $r =$ ___ (since $r = z$ for the top rim of the solid). In polar coordinates this circular disk is expressed by the inequalities ___ $\le r \le$ ___ and ___ $\le \theta \le$ ___ . For a point (x,y), or (r,θ), in this disk the altitude of the solid region varies from the paraboloid $z =$ ___ to the surface of the cone $z =$ ___ , both expressed in terms of cylindrical coordinates.

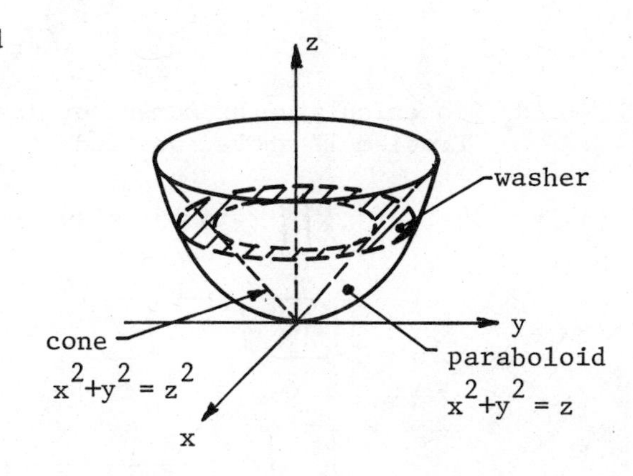

---

17. 1, $r^2 = z^2$, $r^2 = z$, 1, $0 \le r \le 1$, $0 \le \theta \le 2\pi$, r^2, r, $\int_0^{2\pi} \int_0^1 \int_{r^2}^{r}$,

$r(r - r^2)$, $\frac{1}{3} r^3 - \frac{1}{4} r^4$, 0, 0, $\int_0^{2\pi} \int_0^1 \int_{r^2}^{r}$, $\frac{1}{2} rz^2$, $\frac{1}{2}$

Therefore, the mass of the solid is given by the triple integral

$$M = \iiint dV = \int_{\_\_}^{\_\_} \int_{\_\_}^{\_\_} \int_{\_\_}^{\_\_} dz\ r\ dr\ d\theta$$

$$= \int_0^{2\pi} \int_0^1 \_\_\_\_\_\_\_\_\ dr\ d\theta = \int_0^{2\pi} \_\_\_\_\_\_\_\_ \Big]_{r=0}^{r=1} d\theta = \frac{\pi}{6} .$$

From the symmetry of the solid, it is clear that the center of gravity must lie on the z-axis. Thus, $\bar{x} = \_\_\_$ and $\bar{y} = \_\_\_$. We calculate $\bar{z}$:

$$\iiint z\ dV = \int_{\_\_}^{\_\_} \int_{\_\_}^{\_\_} \int_{\_\_}^{\_\_} z\ dz\ r\ dr\ d\theta$$

$$= \int_0^{2\pi} \int_0^1 \_\_\_\_\_\_ \Big]_{z=r^2}^{z=r} dr\ d\theta = \frac{1}{2} \int_0^{2\pi} \int_0^1 (r^3 - r^5)\ dr\ d\theta$$

$$= \frac{1}{2} \int_0^{2\pi} \left(\frac{1}{4} r^4 - \frac{1}{6} r^6 \right) \Big]_{r=0}^{r=1} d\theta = \frac{\pi}{12} . \quad \text{Therefore, } \bar{z} = \_\_\_ .$$

18. To calculate the moment of inertia about the z-axis of the solid in Problem 17 above, we find

$$I_z = \iiint (x^2 + y^2)\ \delta\ dV \qquad (\text{where } \delta = 1)$$

$$= \int_{\_\_}^{\_\_} \int_{\_\_}^{\_\_} \int_{\_\_}^{\_\_} \_\_\_\ dz\ r\ dr\ d\theta$$

$$= \int_0^{2\pi} \int_0^1 \_\_\_\_\_\_\_\_\ dr\ d\theta$$

$$= \int_0^{2\pi} \_\_\_\_\_\_\_\_ \Big]_{r=0}^{r=1} d\theta = \_\_\_ .$$

---

18. $\int_0^{2\pi} \int_0^1 \int_{r^2}^{r}\ r^2$, $r^3(r - r^2)$, $\frac{1}{5} r^5 - \frac{1}{6} r^6$, $\frac{\pi}{15}$

14-8 Spherical Coordinates.

OBJECTIVE A: Find by triple integration, in spherical coordinates, the volume of a specified region in space. (The region may be described in cartesian coordinates.)

19. Consider the region in space that lies inside the sphere $x^2 + y^2 + z^2 = 4$, but outside the cone $x^2 + y^2 = z^2$. The region is sketched in the figure at the right. The origin is a point of symmetry of the solid so that spherical coordinates should be useful for finding the volume of the region by triple integration.

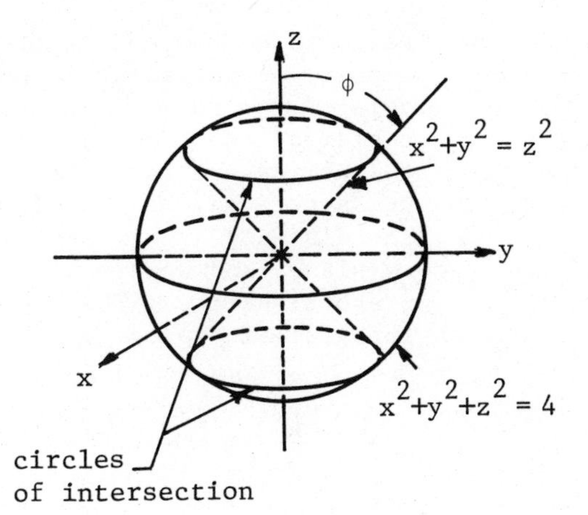

To find the intersection of the sphere and the cone, we substitute $x^2 + y^2 = z^2$ into the equation of the sphere obtaining $z^2 + z^2 = 4$ or $z^2 =$ ____ . Thus, $z = -\sqrt{2}$, or $z = \sqrt{2}$. The plane $z = -\sqrt{2}$ gives the circle of intersection $x^2 + y^2 = 2$ on the sphere which lies $\sqrt{2}$ units below the xy-plane; and $z = \sqrt{2}$ provides the corresponding circle of intersection above the xy-plane. In spherical coordinates, $z = \rho \cos\phi$, so the plane $z = \sqrt{2}$ corresponds to $\sqrt{2} = \rho \cos\phi$. Since the circles of intersection lie on the sphere where $\rho = 2$, the angle ϕ (see figure) for the upper circle satisfies the equation $\sqrt{2} = 2\cos\phi$. Thus, $\phi =$ ____ radians. Similarly, the angle for the lower circle satisfies $-\sqrt{2} = 2\cos\phi$, or $\phi = 3\pi/4$ radians.

Therefore, the volume of the region inside the sphere but outside the cone is given in spherical coordinates by

$$V = \int_{\_\_}^{\_\_} \int_{\_\_}^{\_\_} \int_{\_\_}^{\_\_} \rho^2 \sin\phi \; d\rho \, d\theta \, d\phi$$

$$= \int_{\pi/4}^{3\pi/4} \int_0^{2\pi} \frac{1}{3}\rho^3 \sin\phi \Big]_{\rho=0}^{\rho=2} d\theta \, d\phi = \int_{\pi/4}^{3\pi/4} \int_0^{2\pi} \_\_\_\_ \; d\theta \, d\phi$$

$$= \int_{\pi/4}^{3\pi/4} \_\_\_\_\_\_ \; d\phi = \frac{16\pi\sqrt{2}}{3} \approx 23.70 .$$

---

19. 2, $\pi/4$, $\int_{\pi/4}^{3\pi/4} \int_0^{2\pi} \int_0^{2}$, $\frac{8}{3}\sin\phi$, $\frac{16\pi}{3}\sin\phi$

OBJECTIVE B: By triple integration, using spherical coordinates, find the mass, the center of gravity, or the moments of inertia of a mass M distributed over a region V of xyz-space and having density $\delta = \delta(x,y,z)$ at the point (x,y,z) of V.

20. Let us find the center of gravity of a solid hemisphere of radius a if the density at any point is proportional to the distance of the point from the axis of the solid. We choose coordinates so that the origin is at the center of the sphere, the z-axis is the axis of the hemisphere, and consider the hemisphere that lies above the xy-plane. In spherical coordinates the hemisphere may be described by $\rho = a$, $0 \le \phi \le \pi/2$, and $\_\_\_ \le \theta \le \_\_\_$. The distance $r = \rho \sin \phi$ is measured from the z-axis parallel to the xy-plane, and we are given that $\delta = k\rho \sin \phi$ for some constant k of proportionality. Therefore, the mass of the solid is given by the triple integral (in spherical coordinates),

$$M = \int_{\_\_}^{\_\_} \int_{\_\_}^{\_\_} \int_{\_\_}^{\_\_} (k\rho \sin \phi)\, \rho^2 \sin \phi \; d\rho \; d\phi \; d\theta$$

$$= \int_0^{2\pi} \int_0^{\pi/2} \underline{\qquad\qquad\qquad} \; d\phi \; d\theta$$

$$= \frac{ka^4}{4} \int_0^{2\pi} \left[\frac{\phi}{2} - \frac{1}{4} \sin 2\phi \right]_{\phi=0}^{\phi=\pi/2} d\theta = \underline{\qquad\qquad} .$$

Because of the symmetry of the solid about the z-axis in both shape and density, $\bar{x} = \bar{y} = 0$. We calculate $\bar{z}$:

$$\iiint z\delta \; dV = \int_0^{2\pi} \int_0^{\pi/2} \int_0^{a} \underline{\qquad\qquad\qquad\qquad\qquad} \; d\rho \; d\phi \; d\theta$$

$$= \int_0^{2\pi} \int_0^{\pi/2} \underline{\qquad\qquad\qquad\qquad} \; d\phi \; d\theta$$

$$= \int_0^{2\pi} \frac{ka^5}{15} \sin^3 \phi \Big]_{\phi=0}^{\phi=\pi/2} d\theta = \underline{\qquad\qquad} .$$

Therefore, $\bar{z} = \dfrac{2\pi ka^5}{15M} = \underline{\qquad\qquad}$.

---

20. $0 \le \theta \le 2\pi$, $\int_0^{2\pi} \int_0^{\pi/2} \int_0^{a}$, $\frac{1}{4} ka^4 \sin^2 \phi$, $\frac{ka^4 \pi^2}{8}$,

$(\rho \cos \phi)(k\rho \sin \phi)\rho^2 \sin \phi$, $\frac{1}{5} ka^5 \sin^2 \phi \cos \phi$, $\frac{2\pi ka^5}{15}$, $\frac{16a}{15\pi}$

14-9 Surface Area.

OBJECTIVE: Find, by integration, the area of a specified surface in space.

21. Let G be a region of the xy-plane and let the function $z = f(x,y)$, $(x,y) \in G$, together with its first partial derivatives, be ____________ in G. The area of the surface described by f may then be computed as the integral,

$$S = \iint_G \underline{\qquad\qquad\qquad} \, dx\, dy .$$

22. Another integral expression for surface area is obtained from the following information. Let $\mathbf{N}$ be an upward normal to the surface, and let $\mathbf{p}$ denote some upward vector perpendicular to the ground plane over which the surface lies. Let γ be the angle between $\mathbf{N}$ and $\mathbf{p}$. Then, the area of the surface can be computed by the integral

$$S = \iint_G \underline{\qquad} \, dA = \iint_G \underline{\qquad} \, dA .$$

23. Suppose we wish to find the surface area of that part of the cylinder $z^2 + x^2 = a^2$, $a > 0$, which lies above the triangle in the xy-plane bounded by $x = 0$, $y = 0$, and $x + y = a$. A sketch of the surface is shown at the right. Since $z \geq 0$, we can take $z = \sqrt{a^2 - x^2}$ as the function $z = f(x,y)$ describing the surface over the triangular region G in the xy-plane. Then,

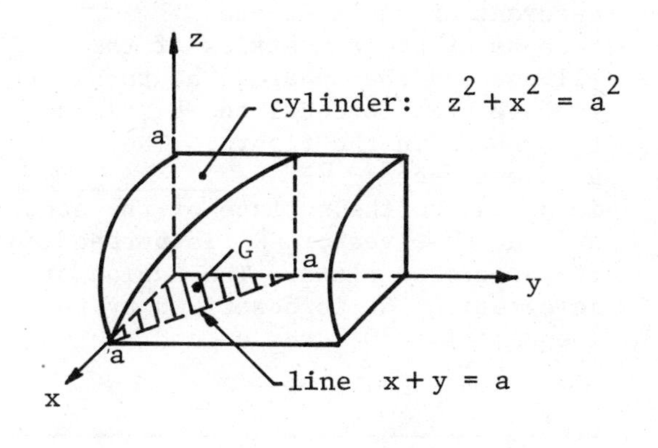

$$\frac{\partial z}{\partial x} = \underline{\qquad\qquad} \quad \text{and} \quad \frac{\partial z}{\partial y} = \underline{\quad} .$$

Thus, $\sqrt{\left(\frac{\partial z}{\partial x}\right)^2 + \left(\frac{\partial z}{\partial y}\right)^2 + 1} = \underline{\qquad\qquad}$.

---

21. continuous, $\sqrt{(f_x)^2 + (f_y)^2 + 1}$

22. $\frac{1}{\cos \gamma}$, $\frac{|\mathbf{N}|\,|\mathbf{p}|}{|\mathbf{N} \cdot \mathbf{p}|}$

The region of integration G is described by the inequalities $\_\_\_ \le x \le \_\_\_$ and $\_\_\_ \le y \le \_\_\_\_\_$. Therefore, the surface area is given by the integral

$$S = \int_{\_\_\_}^{\_\_\_} \int_{\_\_\_}^{\_\_\_} \_\_\_\_\_\_\_\_\_\_ \, dy\, dx = a \int_0^a \frac{a - x}{\sqrt{a^2 - x^2}} \, dx$$

$$= a^2 \int_0^a \frac{dx}{\sqrt{a^2 - x^2}} - a \int_0^a \frac{x\, dx}{\sqrt{a^2 - x^2}}$$

$$= a^2 \left(\_\_\_\_\_\_\_\_ \right) + a\sqrt{a^2 - x^2} \Big]_{x=0}^{x=a} = \_\_\_\_\_\_\_\_\_\_ .$$

24. Let us find the area of the surface of that portion of the sphere $x^2 + y^2 + z^2 = a^2$ that lies inside the elliptic cylinder $4y^2 + x^2 = a^2$. The figure at the right shows that portion which lies above the first quadrant of the xy-plane. Because of the symmetries of the ellipse and the sphere, the total surface area is equal to 8 times that shown in the figure. The gradient vector $\nabla F = \_\_\_\_\_\_\_\_\_\_$ is normal to the surface of the sphere, and the unit vector $\mathbf{k}$ is perpendicular to the ground plane. The region of integration G is described by the inequalities $0 \le x \le a$ and $\_\_\_ < y \le \_\_\_\_\_\_\_\_$.

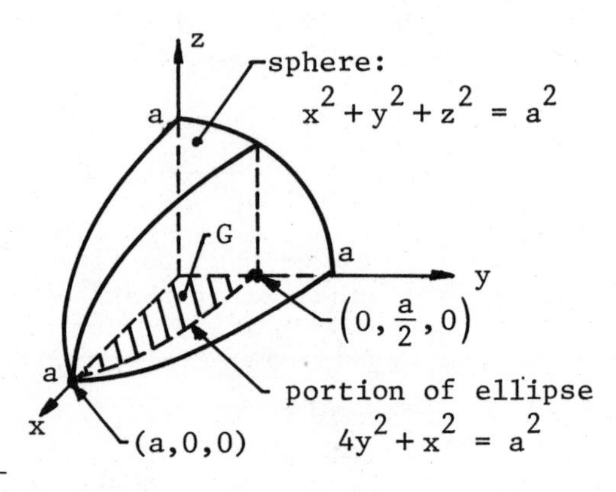

---

23. $\dfrac{-x}{\sqrt{a^2 - x^2}}$, 0, $\dfrac{a}{\sqrt{a^2 - x^2}}$, $0 \le x \le a$, $0 \le y \le a - x$, $\displaystyle\int_0^a \int_0^{a-x} \frac{a}{\sqrt{a^2 - x^2}} \, dy\, dx$,

$\sin^{-1} \dfrac{x}{a}$, $\dfrac{a^2}{2}(\pi - 2)$

24. $2x\mathbf{i} + 2y\mathbf{j} + 2z\mathbf{k}$, $0 \le y \le \frac{1}{2}\sqrt{a^2 - x^2}$, $\dfrac{\sqrt{4x^2 + 4y^2 + 4z^2}}{2z}$, $\displaystyle\int_0^a \int_0^{\frac{1}{2}\sqrt{a^2 - x^2}}$,

$\dfrac{\pi}{6}$, $\dfrac{4\pi a^2}{3}$

Therefore, the total surface area is given by

$$S = 8 \iint_G \frac{|\nabla F|}{|\nabla F \cdot \mathbf{k}|}\, dA = 8 \iint_G \underline{\hspace{4cm}}\, dA$$

$$= 8 \int_{\_\_\_}^{\_\_\_} \int_{\_\_\_}^{\_\_\_} \frac{a}{\sqrt{a^2 - x^2 - y^2}}\, dy\, dx$$

(since $x^2 + y^2 + z^2 = a^2$, $z \geq 0$)

$$= 8a \int_0^a \sin^{-1} \frac{y}{\sqrt{a^2 - x^2}} \Bigg]_{y=0}^{\frac{1}{2}\sqrt{a^2 - x^2}} dx$$

$$= 8a \int_0^a \underline{\hspace{1cm}}\, dx = \underline{\hspace{2cm}} .$$

CHAPTER 14 OBJECTIVE - PROBLEM KEY

Objective	Problems in Thomas/Finney Text	Objective	Problems in Thomas/Finney Text
14-1 A	p. 652, 1-4	14-4 B	p. 663, 7-12
B	p. 652, 5-8	14-5	p. 666, 1-10
C	p. 652, 9,10	14-6	p. 667, 1-6
14-2	p. 654, 1-12	14-7	p. 669, 1-10
14-3 A	p. 658, 1,4,6	14-8 A	p. 670, 1,3
B	p. 658, 2,3,5,7	B	p. 670, 2,4
14-4 A	p. 663, 1-6	14-9	p. 675, 2-10

CHAPTER 14 SELF-TEST

1. Evaluate the following double integrals, and sketch the region over which the integration extends.

   (a) $\int_1^3 \int_{-y}^{2y} xe^{y^3}\, dx\, dy$ (b) $\int_0^{\pi/2} \int_0^{\cos x} e^y \sin x\, dy\, dx$

2. Write an equivalent double integral with the order of integration reversed for each of the following. Sketch the region over which the integration extends and evaluate the new integral.

   (a) $\int_1^2 \int_{x^2}^{x^3} dy\, dx$ (b) $\int_0^2 \int_0^{\sqrt{4-y^2}} \left(4 - x^2\right)^{1/2} dx\, dy$

3. Find, by double integration, the volume of the solid under the paraboloid $z = x^2 + y^2$ and lying above the region in the xy-plane bounded by the curves $y = x$ and $y = x^2$.

4. Find, by double integration, the area of the region in the xy-plane bounded by the curves $y = x^2$ and $y = 4x - x^2$.

5. Find the center of mass of the region in the first quadrant of the xy-plane bounded by the curve $y = \sqrt{x}$ and the line $x = 4$, if the density at (x,y) is $\delta = xy$.

6. Find the moments of inertia about the x- and y-axes for the region in Problem 5 (with the same density).

7. Change the double integral $\int_0^{2a} \int_0^{\sqrt{2ay - y^2}} x \, dx \, dy$ to an equivalent double integral in polar coordinates, and then evaluate.

8. By double integration, find the volume of the solid under the hemisphere $x^2 + y^2 + z^2 = 4$, $z \geq 0$, and lying above the region in the xy-plane within the cylinder $(x - 1)^2 + y^2 = 1$.

9. By triple integration, find the volume of the solid within the sphere $x^2 + y^2 + z^2 = z$ and above the cone $z^2 = x^2 + y^2$.

10. Find the center of gravity of the wedge cut out of the cylinder $x^2 + y^2 = 1$ by the plane $z = y$ above, and the plane $z = 0$ below.

11. Find the moment of inertia of a right circular cone of base radius a , altitude h , and constant density $\delta = 1$, about its axis (i.e., the axis through the vertex and perpendicular to the base).

12. Find the surface area of that part of the cylinder $x^2 + z^2 = a^2$ that lies inside the cylinder $x^2 + y^2 = a^2$.

SOLUTIONS TO CHAPTER 14 SELF-TEST

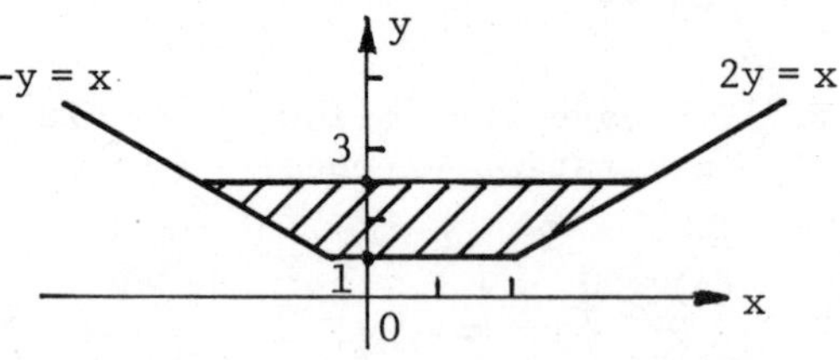

1. (a) The region of integration is shown in the figure at the right. Evaluation of the integral leads to,

$$\int_1^3 \int_{-y}^{2y} xe^{y^3}\, dx\, dy = \int_1^3 \frac{1}{2} x^2 e^{y^3} \Big]_{x=-y}^{x=2y} dy$$

$$= \int_1^3 \frac{1}{2}\left(4y^2 - y^2\right) e^{y^3}\, dy = \int_1^3 \frac{3}{2} y^2 e^{y^3}\, dy$$

$$= \frac{1}{2} e^{y^3} \Big]_{y=1}^{y=3} = \frac{1}{2}\left(e^{27} - e\right) .$$

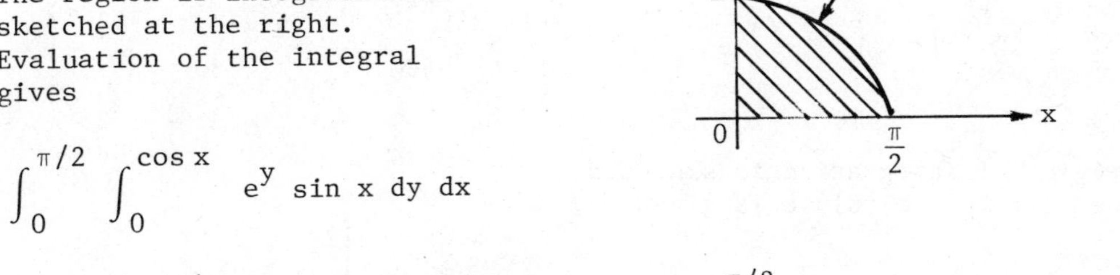

(b) The region if integration is sketched at the right. Evaluation of the integral gives

$$\int_0^{\pi/2} \int_0^{\cos x} e^y \sin x\, dy\, dx$$

$$= \int_0^{\pi/2} e^y \sin x \Big]_{y=0}^{y=\cos x} dx = \int_0^{\pi/2} \left(\sin x\, e^{\cos x} - \sin x\right) dx$$

$$= -e^{\cos x} + \cos x \Big]_{x=0}^{x=\pi/2} = e - 2 \approx 0.718 .$$

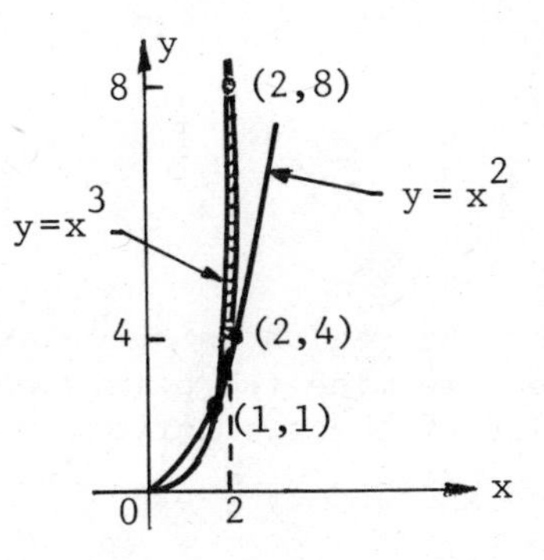

2. (a) The region of integration is sketched at the right. As y varies from $y = 1$ to $y = 4$, x varies from $x = y^{1/3}$ to $x = y^{1/2}$; but as y varies from $y = 4$ to $y = 8$ x varies from $x = y^{1/3}$ to $x = 2$. Thus, the integral becomes the sum of two integrals when the order of integration is reversed:

$$\int_1^2 \int_{x^2}^{x^3} dy\, dx = \int_1^4 \int_{y^{1/3}}^{y^{1/2}} dx\, dy + \int_4^8 \int_{y^{1/3}}^{2} dx\, dy$$

$$= \int_1^4 \left(y^{1/2} - y^{1/3}\right) dy + \int_4^8 \left(2 - y^{1/3}\right) dy = \frac{17}{12} .$$

(b) The region of integration is sketched at the right. Reversing the order of integration gives

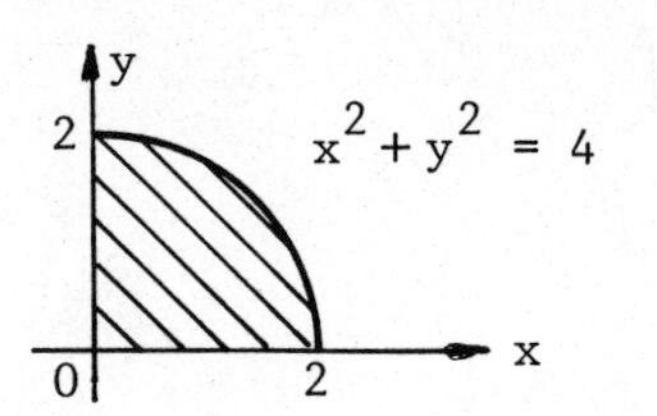

$$\int_0^2 \int_0^{\sqrt{4-y^2}} \left(4-x^2\right)^{1/2} \, dx\, dy$$

$$= \int_0^2 \int_0^{\sqrt{4-x^2}} \left(4-x^2\right)^{1/2} \, dy\, dx = \int_0^2 y\left(4-x^2\right)^{1/2} \Big]_{y=0}^{y=\sqrt{4-x^2}} dx$$

$$= \int_0^2 \left(4-x^2\right) dx = 4x - \frac{1}{3}x^3 \Big]_{x=0}^{x=2} = \frac{16}{3} .$$

3. The region of integration is sketched at the right. The volume is given by,

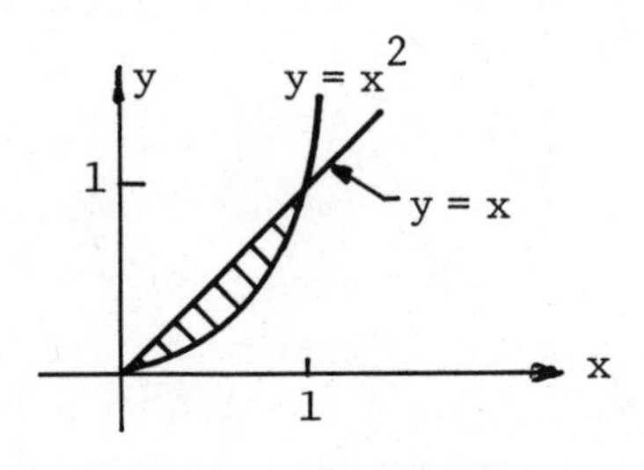

$$V = \iint z \, dA = \int_0^1 \int_{x^2}^{x} \left(x^2 + y^2\right) dy\, dx$$

$$= \int_0^1 \left(x^2 y + \frac{1}{3} y^3\right) \Big]_{y=x^2}^{y=x} dx$$

$$= \int_0^1 \left(\frac{4}{3}x^3 - x^4 - \frac{1}{3}x^6\right) dx = \frac{1}{3}x^4 - \frac{1}{5}x^5 - \frac{1}{21}x^7 \Big]_{x=0}^{x=1} = \frac{9}{105} \approx 0.086 .$$

4. A sketch of the region is shown at the right. To determine the points of intersection of the two curves, we equate $y = x^2$ and $y = 4x - x^2$:

$$x^2 = 4x - x^2 \quad \text{or,} \quad 2x(x-2) = 0 .$$

The points of intersection are therefore (0,0) and (2,4) . The area of the region is given by

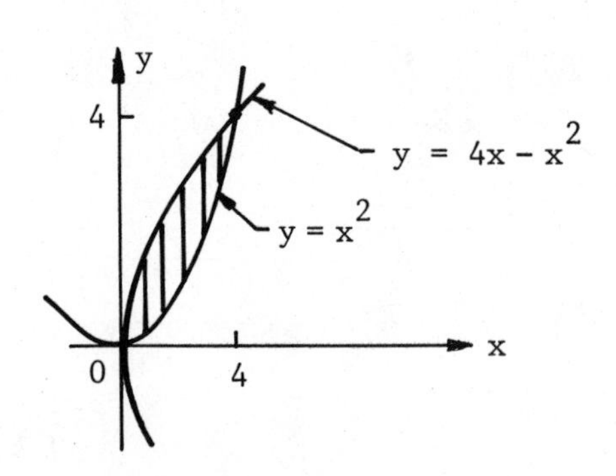

$$A = \iint dA = \int_0^2 \int_{x^2}^{4x-x^2} dy\, dx$$

$$= \int_0^2 \left(4x - x^2 - x^2\right) dx = \frac{8}{3} .$$

5. The mass of the region is given by,

$$M = \iint \delta \, dA = \int_0^4 \int_0^{\sqrt{x}} xy \, dy \, dx$$

$$= \int_0^4 \frac{1}{2} xy^2 \Big]_{y=0}^{y=\sqrt{x}} dx$$

$$= \frac{1}{2} \int_0^4 x^2 \, dx = \frac{32}{3} .$$

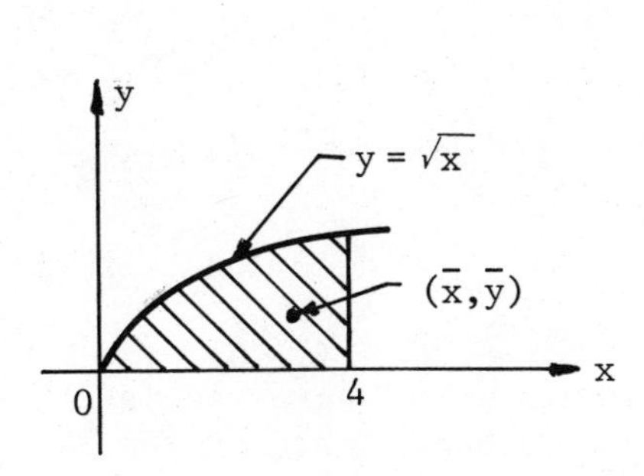

Calculation of the first moments gives,

$$M_x = \iint y \, \delta \, dA = \int_0^4 \int_0^{\sqrt{x}} xy^2 \, dy \, dx = \int_0^4 \frac{1}{3} xy^3 \Big]_{y=0}^{y=\sqrt{x}} dx$$

$$= \frac{1}{3} \int_0^4 x^{5/2} \, dx = \frac{256}{21} ,$$

and

$$M_y = \iint x \, \delta \, dA = \int_0^4 \int_0^{\sqrt{x}} x^2 y \, dy \, dx = \int_0^4 \frac{1}{2} x^2 y^2 \Big]_{y=0}^{y=\sqrt{x}} dx$$

$$= \frac{1}{2} \int_0^4 x^3 \, dx = \frac{256}{8} = 32 .$$

Therefore, $\bar{x} = \frac{M_y}{M} = 3$ and $\bar{y} = \frac{M_x}{M} = \frac{8}{7}$.

6. The moment of inertia about the x-axis is given by,

$$I_x = \iint y^2 \, \delta \, dA = \int_0^4 \int_0^{\sqrt{x}} xy^3 \, dy \, dx$$

$$= \int_0^4 \frac{1}{4} y^4 x \Big]_{y=0}^{y=\sqrt{x}} dx = \frac{1}{4} \int_0^4 x^3 \, dx = 16 .$$

The moment of inertia about the y-axis is given by,

$$I_y = \iint x^2 \, \delta \, dA = \int_0^4 \int_0^{\sqrt{x}} x^3 y \, dy \, dx$$

$$= \int_0^4 \frac{1}{2} x^3 y^2 \Big]_{y=0}^{y=\sqrt{x}} dx = \frac{1}{2} \int_0^4 x^4 \, dx = 102.4 .$$

(From Problem 8, page 658 in the text, the polar moment of inertia is given by

$I_0 = I_x + I_y = 118.4$.)

7. For the region of integration, as y varies from $y = 0$ to $y = 2a$, x varies from $x = 0$ to $x = \sqrt{2ay - y^2}$. The equation $x = \sqrt{2ay - y^2}$ describes the right semi-circle $x^2 + (y - a)^2 = a^2$, $x \geq 0$. The region of integration is the shaded portion in the figure at the right. A polar equation for the circle is $r = 2a \sin \theta$. Then the region of integration may be described in terms of polar coordinates as follows: θ varies from $\theta = 0$ to $\theta = \pi/2$, and r varies from $r = 0$ to $r = 2a \sin \theta$. Thus, the double integral becomes,

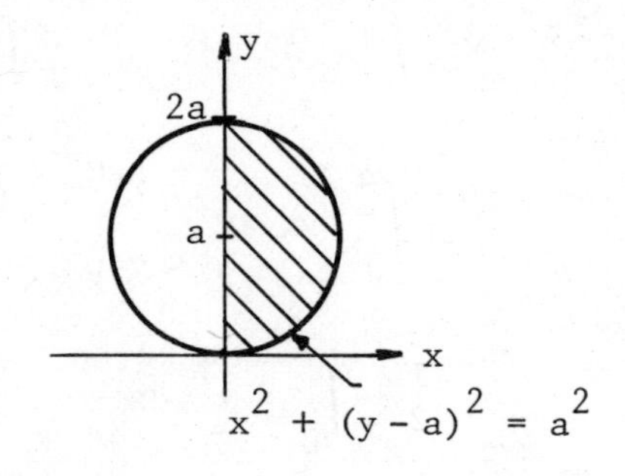

$$\int_0^{2a} \int_0^{\sqrt{2ay-y^2}} x \, dx \, dy = \int_0^{\pi/2} \int_0^{2a \sin \theta} r \cos \theta \cdot r \, dr \, d\theta$$

$$= \int_0^{\pi/2} \frac{8a^3}{3} \sin^3 \theta \cos \theta \, d\theta = \frac{2a^3}{3} \sin^4 \theta \Big]_{\theta=0}^{\theta=\pi/2} = \frac{2a^3}{3} .$$

8. The cartesian equation $z = \sqrt{4 - x^2 - y^2}$ can be written as $z = \sqrt{4 - r^2}$ using polar coordinates for x and y . The region of integration in the xy-plane is the interior of the circle $(x - 1)^2 + y^2 = 1$. In terms of polar coordinates this region is described by the inequalities $-\frac{\pi}{2} \leq \theta \leq \frac{\pi}{2}$ and $0 \leq r \leq 2 \cos \theta$. Because of the symmetry of the solid we need only compute the volume lying above the interior of the semi-circle in the first quadrant of the xy-plane, and then multiply by 2 . Then the total volume is given by

$$V = \iint z \, dA = 2 \int_0^{\pi/2} \int_0^{2 \cos \theta} \sqrt{4 - r^2} \cdot r \, dr \, d\theta$$

$$= 2 \int_0^{\pi/2} -\frac{1}{3} \left(4 - r^2\right)^{3/2} \Big]_{r=0}^{r=2 \cos \theta} d\theta = -\frac{16}{3} \int_0^{\pi/2} \left(\sin^3 \theta - 1 \right) d\theta$$

$$= -\frac{16}{3} \left[-\frac{1}{3} \cos \theta (\sin^2 \theta + 2) - \theta \right]_{\theta=0}^{\theta=\pi/2} = \frac{8\pi}{3} - \frac{32}{9} \approx 4.82 .$$

9. The solid region resembles an ice cream cone, and is sketched on the next page. The sphere is centered at $(0, 0, 1/2)$ and has radius $a = 1/2$. The cone and the sphere intersect when $z^2 + z^2 = z$, or $z = 0$ (the bottom tip of the cone), or $z = 1/2$ (at the rim of the cone). The equation of the cone can be written in

spherical coordinates as $\rho^2 \cos^2 \phi = \rho^2 \sin^2 \phi$ or, for $\rho \neq 0$, $\phi = \pi/4$ (since $0 \le \phi \le \pi$ by definition). In spherical coordinates the equation of the sphere is $\rho^2 = \rho \cos \phi$, or $\rho = \cos \phi$ (since $\rho \neq 0$ for the top of the region). Therefore, the volume of the region is given in spherical coordinates by the triple integral,

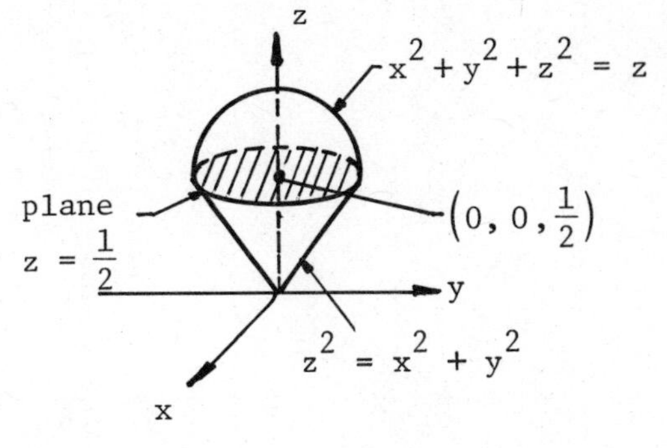

$$V = \iiint dV = \int_0^{2\pi} \int_0^{\pi/4} \int_0^{\cos\phi} \rho^2 \sin\phi \, d\rho \, d\phi \, d\theta$$

$$= \int_0^{2\pi} \int_0^{\pi/4} \frac{1}{3} \cos^3 \phi \sin \phi \, d\phi \, d\theta$$

$$= -\frac{1}{12} \int_0^{2\pi} \cos^4 \phi \Big]_{\phi=0}^{\phi=\pi/4} d\theta = \frac{\pi}{8} \approx 0.39 .$$

10. A sketch of the solid is shown in the figure at the right. We will use cylindrical coordinates. The plane $z = y$ becomes $z = r \sin \theta$ in cylindrical coordinates. The region of integration in the xy-plane is the interior of the semi-circle $y = \sqrt{1 - x^2}$. In cylindrical coordinates this region is given by $0 \le \theta \le \pi$ and $0 \le r \le 1$. Since the region is geometrical we take the density as unity. Then the mass is given by,

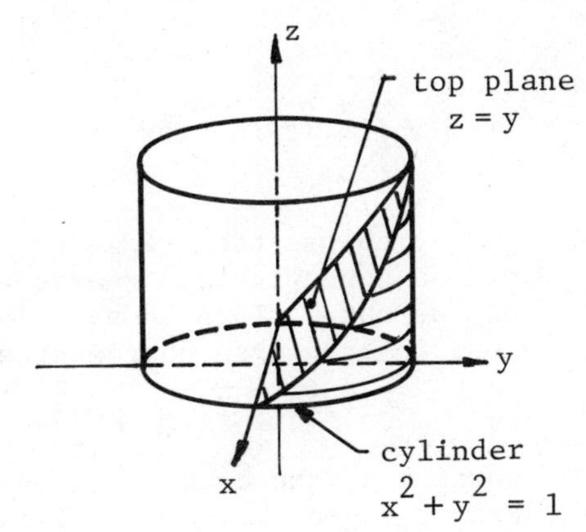

$$M = \iiint \delta \, dV = \int_0^{\pi} \int_0^1 \int_0^{r \sin\theta} dz \, r \, dr \, d\theta$$

$$= \int_0^{\pi} \int_0^1 r^2 \sin \theta \, dr \, d\theta = \frac{1}{3} \int_0^{\pi} \sin \theta \, d\theta = \frac{2}{3} .$$

From the symmetry of the solid across the yz-plane, $\bar{x} = 0$. To find $\bar{y}$ and $\bar{z}$ we calculate the first moments:

$$\iiint y\,\delta\,dV = \int_0^{\pi}\int_0^{1}\int_0^{r\sin\theta} r\sin\theta\,dz\,r\,dr\,d\theta$$

$$= \int_0^{\pi}\int_0^{1} r^3\sin^2\theta\,dr\,d\theta = \frac{1}{4}\int_0^{\pi}\sin^2\theta\,d\theta$$

$$= \frac{1}{4}\left(\frac{\theta}{2} - \frac{1}{4}\sin 2\theta\right)\Big]_{\theta=0}^{\theta=\pi} = \frac{\pi}{8}$$

so that $\bar{y} = \frac{\pi}{8} \div \frac{2}{3} = \frac{3\pi}{16}$; and

$$\iiint z\,\delta\,dV = \int_0^{\pi}\int_0^{1}\int_0^{r\sin\theta} z\,dz\,r\,dr\,d\theta$$

$$= \int_0^{\pi}\int_0^{1} \frac{1}{2}z^2\Big]_{z=0}^{z=r\sin\theta} r\,dr\,d\theta$$

$$= \int_0^{\pi}\int_0^{1} \frac{1}{2}r^3\sin^2\theta\,dr\,d\theta = \frac{1}{8}\int_0^{\pi}\sin^2\theta\,d\theta = \frac{\pi}{16}$$

from which it follows that $\bar{z} = \frac{\pi}{16} \div \frac{2}{3} = \frac{3\pi}{32}$.

11. We may choose the origin at the vertex of the cone with the z-axis as the axis of the cone. Then an equation for the cone in cartesian coordinates is given by $z = \sqrt{x^2 + y^2}$, $0 \le z \le h$. A portion of the cone is shown at the right. From the geometry of the figure we see that the vertex angle α of the cone satisfies $\tan\alpha = a/h$. In spherical coordinates, the cartesian equation $z = h$ for the base of the cone becomes $\rho\cos\phi = h$ or, $\rho = h\sec\phi$. Thus, the moment of inertia about the z-axis is given in spherical coordinates by the triple integral,

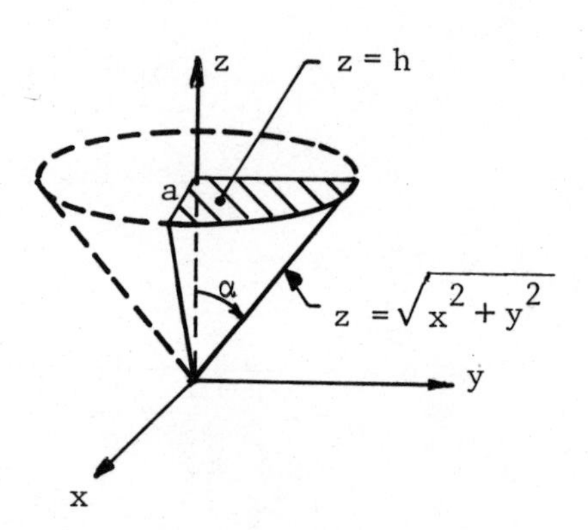

$$I_z = \iiint (x^2 + y^2)\ \delta\ dV \qquad (\delta \equiv 1)$$

$$= \int_0^{2\pi} \int_0^{\alpha} \int_0^{h \sec \phi} (\rho^2 \sin^2 \phi)\ \rho^2 \sin \phi\ d\rho\ d\phi\ d\theta$$

$$= \frac{h^5}{5} \int_0^{2\pi} \int_0^{\alpha} \tan^3 \phi \sec^2 \phi\ d\phi\ d\theta$$

$$= \frac{h^5}{5} \int_0^{2\pi} \frac{1}{4} \tan^4 \alpha\ d\theta = \frac{\pi h^5}{10} \tan^4 \alpha = \frac{\pi a^4 h}{10}$$

since $\alpha = \tan^{-1}(a/h)$.

12. A sketch of the intersecting cylinders is at the right. By the symmetry of the surface we need only calculate the area of the top surface (above the xy-plane), and multiply this by 2 .

In that case, $z = \sqrt{a^2 - x^2}$ and this yields,

$$\frac{\partial z}{\partial x} = \frac{-x}{\sqrt{a^2 - x^2}} \quad \text{and} \quad \frac{\partial z}{\partial y} = 0 .$$

Therefore,

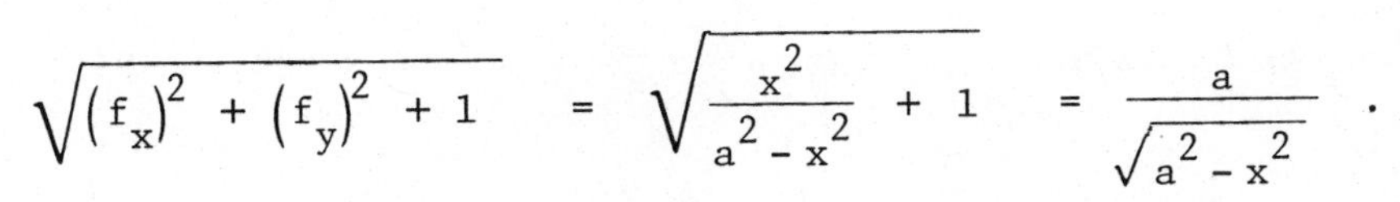

$$\sqrt{(f_x)^2 + (f_y)^2 + 1} = \sqrt{\frac{x^2}{a^2 - x^2} + 1} = \frac{a}{\sqrt{a^2 - x^2}} .$$

The region in the xy-plane over which the surface lies is the interior of the circle $x^2 + y^2 = a^2$. Thus, the total surface area is given by the double integral,

$$S = 2 \int_{-a}^{a} \int_{-\sqrt{a^2 - x^2}}^{\sqrt{a^2 - x^2}} \frac{a}{\sqrt{a^2 - x^2}}\ dy\ dx$$

$$= 2a \int_{-a}^{a} 2\ dx = 8a^2 .$$

CHAPTER 15 VECTOR ANALYSIS

15-1 Introduction: Vector Fields.

OBJECTIVE A: Define the term vector field and give several examples of vector fields.

1. Let G be a region of 3-space. If, to each point P in G, a ______ $\mathbf{F}(P)$ is assigned, the collection of all such ______ is called a vector ______ .

2. For a fluid flowing with steady-state flow in a region G, a vector field is given by the assignment of a ______ vector $\mathbf{v}$ indicating the flow at each point of G. This vector field may be described by writing a vector equation for the velocity at each point $P(x,y,z)$ of G.

3. The gravitational force field induced at the point $P(x,y,z)$ in space by a mass M that is taken to lie at the origin O is described by the equation

$$\mathbf{F} = \underline{\hspace{6cm}},$$

where G is the gravitational constant. The direction of $\mathbf{F}$ is from ___ toward ___ . At points on the surface of the sphere $|\overrightarrow{OP}| = a$, the vectors in the gravitational field all have the same ______ , and all point toward the ______ of the sphere.

OBJECTIVE B: Given a differentiable scalar function $w = f(x,y,z)$, find the gradient field $\mathbf{F}(x,y,z) = \nabla f$.

4. Consider the function $f(x,y,z) = \dfrac{GM}{\left(x^2+y^2+z^2\right)^{1/2}}$. From the definition of grad f we have

$$\mathbf{F} = \nabla f = \mathbf{i}\frac{\partial f}{\partial x} + \mathbf{j}\,\underline{\hspace{1cm}} + \mathbf{k}\,\underline{\hspace{1cm}}$$

$$= \mathbf{i}\frac{-GMx}{\left(x^2+y^2+z^2\right)^{3/2}} + \mathbf{j}\,\underline{\hspace{3cm}} + \mathbf{k}\frac{-GMz}{\left(x^2+y^2+z^2\right)^{3/2}}$$

$$= \underline{\hspace{3cm}}\,(\mathbf{i}x + \mathbf{j}y + \mathbf{k}z).$$

---

1. vector, vectors, field

2. velocity

3. $\dfrac{-GM(\mathbf{i}x + \mathbf{j}y + \mathbf{k}z)}{\left(x^2+y^2+z^2\right)^{3/2}}$, P, O, length, center

Therefore, we observe that the graviational field in Problem 3 above (Example 4, page 682 of the text) is the gradient field of the function f, where G is the gravitational constant and M is the mass located at the origin.

15-2 Surface Integrals.

OBJECTIVE A: Let S be a given surface defined by $z = f(x,y)$ for (x,y) in some specified closed, bounded region R of the xy-plane. Assume that f and its first partial derivatives f_x and f_y are continuous functions throughout R and on its boundary. If $h(x,y,z)$ is a given continuous function on S, find

$$\iint_S h(x,y,z) \; d\sigma$$

5. A practical procedure for evaluating the surface integral described in the above Objective is to replace z by its value ________ on the surface S, and replace $d\sigma$ by

$$d\sigma = \underline{\hspace{4cm}},$$

and evaluate the resulting ________ integral over the region R in the xy-plane into which S projects.

6. Consider the surface integral

$\iint_S z^2 \, d\sigma$, where S is the portion of the cone $z = \sqrt{x^2 + y^2}$ between the planes $z = 2$ and $z = 5$. The surface S and its projection R onto the xy-plane are sketched at the right. Observe that R is the ring ___ $\le x^2 + y^2 \le$ ___ . The function

$f(x,y) =$ ____________ defines the surface S over R.

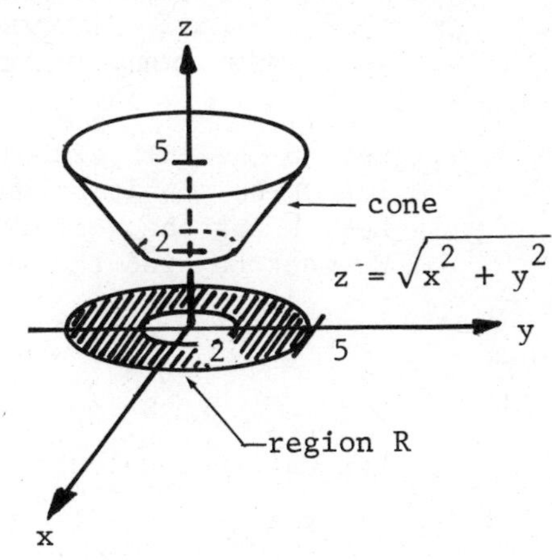

---

4. $\dfrac{\partial f}{\partial y}$, $\dfrac{\partial f}{\partial z}$, $\dfrac{-GMy}{\left(x^2 + y^2 + z^2\right)^{3/2}}$, $\dfrac{-GM}{\left(x^2 + y^2 + z^2\right)^{3/2}}$

5. $f(x,y)$, $\sqrt{1 + f_x^{\,2} + f_y^{\,2}}\; dA$, double

6. $4 \le x^2 + y^2 \le 25$, $\sqrt{x^2 + y^2}$, $\dfrac{x}{\sqrt{x^2 + y^2}}$, $\dfrac{y}{\sqrt{x^2 + y^2}}$, $\dfrac{y}{z}$, y^2/z^2, $\sqrt{2}$, $\int_0^{2\pi} \int_2^5 \sqrt{2}\; r^3 \, dr \, d\theta$, $\dfrac{\sqrt{2}\;\pi}{2}$

Thus,

$$f_x = \underline{\qquad\qquad} = \frac{x}{z} \quad \text{and} \quad f_y = \underline{\qquad\qquad} = \underline{\quad} .$$

Hence,

$$d\sigma = \sqrt{1 + \left(x^2/z^2\right) + \left(\underline{\qquad}\right)} \quad dA = \underline{\qquad} \; dA$$

because $x^2 + y^2 + z^2 = 2z^2$ on S. Therefore,

$$\iint_S z^2 \, d\sigma = \iint_R (x^2 + y^2) \sqrt{2} \; dA = \int_{\_}^{\_} \int_{\_}^{\_} \underline{\qquad\qquad} \; dr \, d\theta$$

$$= \underline{\qquad} (5^4 - 2^4) = \frac{609\pi \sqrt{2}}{2} \approx 1352.86 .$$

OBJECTIVE B: Evaluate integrals of the form

$$\iint_S \mathbf{F} \cdot \mathbf{n} \; d\sigma ,$$

for a specified vector field $\mathbf{F}$, and a given surface S defined as in Objective A above, where $\mathbf{n}$ is a unit vector normal to S and pointing in a consistent way (e.g., always outward or always inward). Assume that the components of $\mathbf{F}$ are continuous on S.

7. Let S be the paraboloid $z = 1 - x^2 - y^2$ above the xy-plane, and let $\mathbf{n}$ be a unit normal to S at each point, which is directed upward. Let $\mathbf{F}$ be the vector field on S given by $\mathbf{F} = \mathbf{i}x + \mathbf{j}y + \mathbf{k}z$. We want to find the surface integral

$$\iint_S \mathbf{F} \cdot \mathbf{n} \; d\sigma .$$

Let us first calculate the vector $\mathbf{n}$. If we define $g(x,y,z) = z - (1 - x^2 - y^2)$, then the gradient vector $\nabla g = \mathbf{i}\, 2x + \mathbf{j} \underline{\qquad} + \mathbf{k}$ is perpendicular to the surface S at each point. Moreover, it is directed upward and outward so it is pointing in the direction specified for $\mathbf{n}$. The unit normal $\mathbf{n}$ can be written as

---

7. $2y$, $\dfrac{\mathbf{i}\,2x + \mathbf{j}\,2y + \mathbf{k}}{\sqrt{4x^2 + 4y^2 + 1}}$, $\dfrac{2x^2 + 2y^2 + z}{\sqrt{4x^2 + 4y^2 + 1}}$, $\sqrt{4x^2 + 4y^2 + 1}$, $0 \le x^2 + y^2 \le 1$,

$1 + x^2 + y^2$, $\displaystyle\int_0^{2\pi} \int_0^1 (1 + r^2)\, r \, dr \, d\theta$, $\dfrac{3\pi}{2}$

$$\mathbf{n} = \frac{\nabla g}{|\nabla g|} = \underline{\hspace{4cm}} . \quad \text{Therefore,}$$

$$\mathbf{F} \cdot \mathbf{n} = \underline{\hspace{4cm}} = \frac{1 + x^2 + y^2}{\sqrt{4x^2 + 4y^2 + 1}}$$

because $z = 1 - x^2 - y^2$ on the surface S .

Next, $d\sigma = \sqrt{1 + z_x^2 + z_y^2} \; dA = \underline{\hspace{3cm}} \; dA$.

The surface S projects onto the region R in the xy-plane described by $\underline{\hspace{4cm}}$. Therefore

$$\iint_S \mathbf{F} \cdot \mathbf{n} \, d\sigma = \iint_R \left(\underline{\hspace{4cm}} \right) dA$$

$$= \int_{\_\_}^{\_\_} \int_{\_\_}^{\_\_} \underline{\hspace{3cm}} \; dr \, d\theta = \underline{\hspace{1cm}} \approx 4.71 .$$

(↑ in polar coordinates)

15-3 Line Integrals.

OBJECTIVE A: Let C be a directed curve in space from a given point A to a given point B , expressed parametrically by $x = f(t)$, $y = g(t)$, $z = h(t)$, for $t_A \le t \le t_B$. Assume that the functions f, g, h are continuous and have bounded piecewise-continuous first derivatives on $[t_A, t_B]$. Let $w = w(x,y,z)$ be a specified scalar function of position that is continuous in a region D that contains C . Then evaluate the line integral,

$$\int_C w \, ds .$$

8. Consider the line integral $\int_C y \, ds$, where C is the curve given by $x = t$ and $y = t^3$, $-1 \le t \le 0$. Thus, the curve C lies in the xy-plane. Now,

$$ds = \sqrt{\left(\frac{dx}{dt}\right)^2 + \left(\frac{dy}{dt}\right)^2} \; dt = \underline{\hspace{4cm}} .$$

---

8. $\sqrt{1 + 9t^4} \; dt$, $t^3\sqrt{1 + 9t^4}$, $\frac{1}{54}\left(1 + 9t^4\right)^{3/2}$, $\frac{1 - 10\sqrt{10}}{54} \approx -0.57$

Therefore,

$$\int_C y\,ds = \int_{-1}^{0} \underline{\hspace{3cm}}\, dt = \underline{\hspace{3cm}}\Big]_{-1}^{0}$$

$$= \underline{\hspace{2.5cm}} .$$

9. Let us evaluate $\int_C w\,ds$ for $w = \left(1 + \frac{9}{4} z^{2/3}\right)^{1/4}$, where C is given by $x = \cos t$, $y = \sin t$, $z = t^{3/2}$, $0 \le t \le 20/3$. Now,

$$ds = \sqrt{\left(\frac{dx}{dt}\right)^2 + \left(\frac{dy}{dt}\right)^2 + \left(\frac{dz}{dt}\right)^2}\, dt = \underline{\hspace{3cm}} .$$

Thus, we obtain

$$\int_C w\,ds = \int_{\_\_}^{\_\_} \underline{\hspace{4cm}}\, dt = \int_0^{20/3} \left(1 + \frac{9}{4}t\right)^{3/4} dt$$

$$= \underline{\hspace{3.5cm}}\Big]_0^{20/3} = \frac{16}{23}(2^7 - 1) \approx 32.25 .$$

10. Evaluate $\int_C w\,ds$ for $w = x^2 - y^2$, where C is the circle $x^2 + y^2 = 4$.

Solution. We can parameterize C by $x = 2\cos\theta$ and $y = 2\sin\theta$, $0 \le \theta \le 2\pi$. (Any parameterization satisfying the hypotheses of Objective A will give the same value for the line integral.) Then,

$$ds = \sqrt{\left(\frac{dx}{d\theta}\right)^2 + \left(\frac{dy}{d\theta}\right)^2}\, d\theta = \underline{\hspace{5cm}} = 2\,d\theta ,$$

and we get

$$\int_C w\,ds = \int_{\_\_}^{\_\_} \underline{\hspace{2.5cm}}\, d\theta = \underline{\hspace{2.5cm}}\Big]_{\_\_}^{\_\_} = \underline{\hspace{1cm}} .$$

---

9. $\sqrt{1 + \frac{9}{4}t}\; dt$, $\int_0^{20/3} \left(1 + \frac{9}{4}t\right)^{1/4} \sqrt{1 + \frac{9}{4}t}\; dt$, $\frac{16}{63}\left(1 + \frac{9}{4}t\right)^{7/4}$

10. $\sqrt{4\sin^2\theta + 4\cos^2\theta}\; d\theta$, $\int_0^{2\pi} 8\cos 2\theta\, d\theta$, $4\sin 2\theta\Big]_0^{2\pi}$, 0

OBJECTIVE B: For a specified curve C as in Objective A, evaluate the line integral

$$\int_C \mathbf{F} \cdot d\mathbf{R}$$

where $\mathbf{F} = \mathbf{i}M(x,y,z) + \mathbf{j}N(x,y,z) + \mathbf{k}P(x,y,z)$ is a given vector field with continuous components throughout some connected region D that contains C. Assume $\mathbf{R} = \mathbf{i}x + \mathbf{j}y + \mathbf{k}z$ is the vector from the origin to the point (x,y,z).

11. If $\mathbf{F}$ is interpreted as a force whose point of application moves along the curve C from a point A to a point B, then the line integral

$$\int_C \mathbf{F} \cdot d\mathbf{R}$$

is the ________ done by the _________ .

12. If we calculate the dot product of the vectors $\mathbf{F}$ and $d\mathbf{R}$, an alternate form for the line integral in the Objective B is

______________________________ .

13. Let us calculate $\int_C \mathbf{F} \cdot d\mathbf{R}$ for the force $\mathbf{F} = \mathbf{i}(x-z) + \mathbf{j}(1-xy) + \mathbf{k}y$ as its point of application moves from the origin to the point $A(1,1,1)$
(a) along the straight line $x = y = z$; and
(b) along the curve

$$x = t^2, \quad y = t, \quad z = t^3, \quad 0 \leq t \leq 1 .$$

Solution. (a) The integral to be evaluated is

$$\int_C \mathbf{F} \cdot d\mathbf{R} = \int_C (x-z)\,dx + (1-xy)\,dy + y\,dz ,$$

which, for the straight line path $x = y = z$, becomes

$$\int_C \mathbf{F} \cdot d\mathbf{R} = \int_{\_\_}^{\_\_} \_\_\_\_\_\_\_\_\_\_\_\_ \, dy = \_\_\_ .$$

---

11. work, force

12. $\int_C M\,dx + N\,dy + P\,dz$

(b) Along the curve $x = t^2$, $y = t$, $z = t^3$, $0 \leq t \leq 1$ we get

$$\int_C \mathbf{F} \cdot d\mathbf{R} = \int_{\_\_\_}^{\_\_\_} \_\_\_\_\_\_\_\_\_\_\_\_\_\_\_\_\_\_\_\_ dt$$

$$= \int_0^1 (1 + 4t^3 - 2t^4) \, dt = \_\_\_ .$$

Notice that the line integral depends on the path C joining the origin to A(1,1,1) since the values obtained in parts (a) and (b) are different. Therefore, the field **F** __________ conservative.

14. Suppose a particle moves upward along the helix whose vector equation is given by

$$\mathbf{R}(t) = \mathbf{i} \cos t + \mathbf{j} \sin t + \mathbf{k} t , \quad 0 \leq t \leq 2\pi .$$

Find the work done on the particle by the force

$$\mathbf{F} = \mathbf{i}(-zy) + \mathbf{j}(zx) + \mathbf{k}(xy)$$

as the point of application moves along C from the point A(1,0,0) to the point B(1,0,2π) .

Solution. The work done is given by the integral

$$W = \int_C \mathbf{F} \cdot d\mathbf{R} = \int_{t_1}^{t_2} \mathbf{F}\big(x(t), y(t), z(t)\big) \cdot \frac{d\mathbf{R}}{dt} \, dt .$$

Now, in terms of the parameter t , the force **F** with point of application on the helix C is given by

$$\mathbf{F}\big(x(t), y(t), z(t)\big) = \_\_\_\_\_\_\_\_\_\_\_\_\_\_\_\_\_\_\_\_ .$$

Also, $\frac{d\mathbf{R}}{dt} = \mathbf{i}(-\sin t) + \mathbf{j} \, \_\_\_\_\_ + \mathbf{k}$ so that

$$\mathbf{F} \cdot \frac{d\mathbf{R}}{dt} = t \sin^2 t + \_\_\_\_\_\_\_\_\_\_\_\_ = \_\_\_\_\_\_\_\_\_\_\_\_ .$$

---

13. (a) $\int_0^1 (1 + y - y^2) \, dy$, $\frac{7}{6}$

(b) $\int_0^1 (t^2 - t^3) \, 2t \, dt + (1 - t^3) \, dt + t \cdot 3t^2 \, dt$, $\frac{8}{5}$, is not

14. $\mathbf{i}(-t \sin t) + \mathbf{j}(t \cos t) + \mathbf{k}(\sin t \cos t)$, $\cos t$, $t \cos^2 t + \sin t \cos t$,

$t + \sin t \cos t$, A(1,0,0), B(1,0,2π), $\int_0^{2\pi} (t + \sin t \cos t) \, dt$, $\frac{1}{2} t^2 + \frac{1}{2} \sin^2 t \Big]_0^{2\pi}$

As t varies from $t = 0$ to $t = 2\pi$, the point $P(x,y,z)$ on C varies from ____________ to ____________ . Therefore, the work done by $\mathbf{F}$ from A to B is given by,

$$W = \int_{\_\_}^{\_\_} \_\_\_\_\_\_\_\_\_\_\_\_ \, dt = \Big[\_\_\_\_\_\_\_\_\_\_\_\_\Big]_{\_\_}^{\_\_} = 2\pi^2 .$$

15. Find the work done by the force

$$\mathbf{F} = \mathbf{i}y + \mathbf{j}z^2 + \mathbf{k}x ,$$

as the point of application moves from $A(0,0,0)$ along the y-axis to $(0,-5,0)$, then in a straight line to the point $B(0,1,1)$.

Solution. Let C_1 denote the line segment from $A(0,0,0)$ to $(0,-5,0)$, and let C_2 denote the straight line segment from $(0,-5,0)$ to $B(0,1,1)$. The path is sketched at the right. The total work done is the sum of two line integrals,

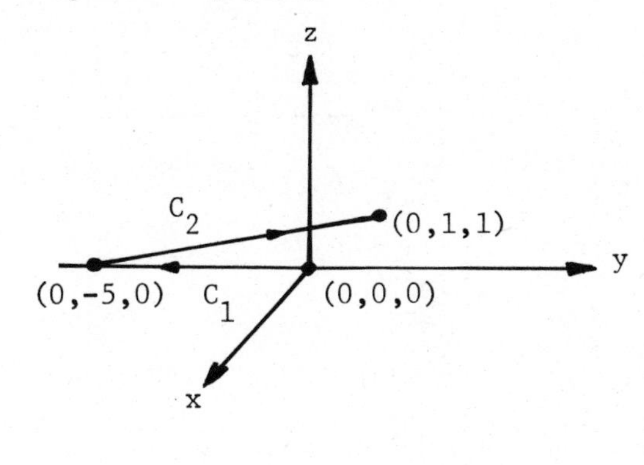

$$W = \int_{C_1} \mathbf{F} \cdot d\mathbf{R} + \int_{C_2} \mathbf{F} \cdot d\mathbf{R} .$$

We want to find parametric representations for C_1 and C_2. The line segment C_1 is given by $x = 0$, $y = -t$, $z = 0$ for $0 \le t \le 5$. Notice that as t varies from $t = 0$ to $t = 5$, we move from the point __________ to the point __________ . A parametric representation for C_2 is given by $x = 0$, $y = 1 + 6t$, $z =$ ________ . As t varies from $t =$ ___ to $t =$ ___ , we move from the point $(0,-5,0)$ to the point $(0,1,1)$. Thus,

on C_1 : $\quad \mathbf{R} = \mathbf{R}_1(t) = -t\mathbf{j}$, $\quad \dfrac{d\mathbf{R}_1}{dt} =$ ___ ,

$\mathbf{F} = \mathbf{F}_1(t) =$ _____ , and $\mathbf{F}_1 \cdot \dfrac{d\mathbf{R}_1}{dt} = 0$;

on C_2 : $\quad \mathbf{R} = \mathbf{R}_2(t) = \mathbf{j}(1+6t) + \mathbf{k}$ ______ , $\quad \dfrac{d\mathbf{R}_2}{dt} =$ _______ ,

$\mathbf{F} = \mathbf{F}_2(t) =$ ______________________ ,

and $\mathbf{F}_2 \cdot \dfrac{d\mathbf{R}_2}{dt} =$ __________ .

---

15. $(0,0,0)$, $(0,-5,0)$, $1+t$, -1, 0,
$-\mathbf{j}$, $-t\mathbf{i}$, $1+t$, $6\mathbf{j}+\mathbf{k}$, $\mathbf{i}(1+6t) + \mathbf{j}(1+t)^2 + 0\mathbf{k}$, $6(1+t)^2$, 0, $\int_{-1}^{0}$, 2

Therefore, the work done is,

$$W = \int_0^5 \_\_\_ \, dt + \int_{\_\_\_}^{\_\_\_} 6(1+t)^2 \, dt = \_\_\_ .$$

OBJECTIVE C: For a given vector field $\mathbf{F}$, find a differentiable scalar function f, if possible, such that $\mathbf{F} = \text{grad } f$.

16. If there does exist a scalar function f with $\mathbf{F} = \text{grad } f = \nabla f$, then the field $\mathbf{F}$ is said to be ________________ .

17. Whenever a field $\mathbf{F}$ is conservative, the integrand in the work integral,

$$\mathbf{F} \cdot d\mathbf{R} = M\,dx + N\,dy + P\,dz$$

is an ______________________ .

18. If $M(x,y,z)$, $N(x,y,z)$, and $P(x,y,z)$ are continuous, together with their ________________ partial derivatives, then a <u>necessary</u> condition for the expression

$$M\,dx + N\,dy + P\,dz$$

to be an exact differential is that the following equations all be satisfied:

_____________ , _____________ , _____________ .

19. If $\mathbf{F}$ is a vector field with continuous components throughout some connected region D containing the points A and B, then a necessary and sufficient condition that the integral

$$\int_A^B \mathbf{F} \cdot d\mathbf{R}$$

be independent of the path joining A and B in D is that $\mathbf{F}$ be a ________________ field.

20. Consider the vector field

$$\mathbf{F}(x,y,z) = \mathbf{i}(2xyz) + \mathbf{j}(x^2 z) + \mathbf{k}(x^2 y + 1) .$$

Let us determine if $\mathbf{F}$ is conservative.

---

16. conservative

17. exact differential

18. first-order, $\frac{\partial M}{\partial y} = \frac{\partial N}{\partial x}$, $\frac{\partial M}{\partial z} = \frac{\partial P}{\partial x}$, $\frac{\partial N}{\partial z} = \frac{\partial P}{\partial y}$

19. conservative

20. $x^2 z$, $2xz$, P, may, $x^2 z$, $x^2 y + 1$, $x^2 yz$, $x^2 z$, 0, $x^2 yz$, $x^2 y$, 1, z, $x^2 yz + z + C_1$, is

First, $\mathbf{F} \cdot d\mathbf{R} = (2xyz)\,dx + (x^2z)\,dy + (x^2y+1)\,dz$. Then, applying the test in Problem 18 above, with $M = 2xyz$, $N =$ _____ , and $P = x^2y+1$, we obtain

$$\frac{\partial M}{\partial z} = 2xy = \frac{\partial P}{\partial x} \ , \quad \frac{\partial M}{\partial y} = \_\_\_\_ = \frac{\partial N}{\partial x} \ , \text{ and } \frac{\partial \_\_}{\partial y} = x^2 = \frac{\partial N}{\partial z} \ .$$

Therefore, we conclude that there _______ be a function $f(x,y,z)$ such that $\mathbf{F} \cdot d\mathbf{R} = df$. We want to find f . If f does exist, then

$$\frac{\partial f}{\partial x} = 2xyz \ , \quad \frac{\partial f}{\partial y} = \_\_\_\_\_ \ , \text{ and } \frac{\partial f}{\partial z} = \_\_\_\_\_\_\_\_\_\_$$ all must hold. To find f we integrate the first of these equations with respect to x , holding y and z constant, obtaining

$$f(x,y,z) = \_\_\_\_\_\_\_\_ + g(y,z) \ ,$$

where $g(y,z)$ is an arbitrary function acting as the "constant of integration." Next, we differentiate the last equation with respect to y , holding x and z constant, equate this to $\partial f/\partial y$, and solve for $\partial g/\partial y$:

$$\_\_\_\_\_ + \frac{\partial g(y,z)}{\partial y} = x^2 z \ , \text{ so that } \frac{\partial g(y,z)}{\partial y} = \_\_\_\_\_ \ .$$

Therefore, $g(y,z) = h(z)$ is a function of z alone. The expression for f then becomes

$$f(x,y,z) = \_\_\_\_\_ + h(z) \ .$$

We differentiate this last equation with respect to z , holding x and y constant, equate this to $\partial f/\partial z$, and solve for $h'(z)$:

$$\_\_\_\_\_ + h'(z) = x^2y + 1 \ , \text{ so that } h'(z) = \_\_\_ \ . \text{ Therefore,}$$

$h(z) =$ ___ $+ C_1$ for some arbitrary constant C_1 . Hence we may write f as

$$f(x,y,z) = \_\_\_\_\_\_\_\_\_\_\_\_\_\_\_\_ \ .$$

It is easy to verify that $\mathbf{F} = \nabla f$, so we conclude that $\mathbf{F}$ ____ a conservative field.

21. Suppose it is required to find the line integral $\int_C \mathbf{F} \cdot d\mathbf{R}$ for the force field $\mathbf{F}$ in Problem 20 above, where C is composed of the line segments $A(0,0,0)$ to $(0,-1,-3)$ and from $(0,-1,-3)$ to $B(-1,1,2)$. Then,

$$\int_C \mathbf{F} \cdot d\mathbf{R} = f(\_\_\_\_\_\_\_\_) - f(\_\_\_\_\_\_\_\_) = \_\_\_ \ .$$

---

21. (-1,1,2), (0,0,0), 4

22. Let C be a directed curve that is continuous and has bounded and piecewise-continuous first derivatives from A(-2,1,0) to B(1,0,1) . Let **F** be the gravitational field of a mass M located at the origin. Find the work W done by **F** on a unit point mass that traverses C .

Solution. We know from Problem 4 of this chapter that the gravitational field **F** is the gradient of the function

$$f(x,y,z) = \_\_\_\_\_\_\_\_\_\_\_\_\_\_\_\_ ,$$

where G is the gravitational constant. Consequently,

$$W = \int_C \mathbf{F} \cdot d\mathbf{R} = f(\_\_\_\_\_\_) - f(\_\_\_\_\_\_) = \_\_\_\_\_\_\_\_\_\_\_\_\_\_ .$$

15-4 Two-Dimensional Fields: Line Integrals in the Plane and their Relation to Surface Integrals on Cylinders.

OBJECTIVE A: Given a continuous two-dimensional vector flow field $\mathbf{F} = \delta\mathbf{v}$, find the rate of mass transport (i.e., the flux of **F**) across a specified curve C in the xy-plane. Assume that C is piecewise smooth enough to have a tangent and that the direction across C is specified.

23. If **n** is a unit normal vector indicating the direction of flow across C , then the flux of **F** across C is given as a line integral

$$\_\_\_\_\_\_\_\_\_\_\_\_\_\_ .$$

24. If C is a simple closed curve in the xy-plane, and if the counter-clockwise direction on C is taken as the positive direction (as the direction of increasing arc length), then the flux of the field $\mathbf{F}(x,y) = \mathbf{i}M(x,y) + \mathbf{j}N(x,y)$ outward across C can be given by the line integral

$$\_\_\_\_\_\_\_\_\_\_\_\_\_\_\_\_\_\_ .$$

This integral has the advantage that it can be evaluated using ______ parameterization of C , provided we integrate in the positive direction along C .

---

22. $\dfrac{GM}{\left(x^2+y^2+z^2\right)^{1/2}}$, (1,0,1), (-2,1,0), $GM\left(\dfrac{1}{\sqrt{2}} - \dfrac{1}{\sqrt{5}}\right) \approx 0.26\ GM$

23. $\int_C \mathbf{F}\cdot\mathbf{n}\ ds$

24. $\int_C (M\,dy - N\,dx)$, any

25. Find the flux of the field $\mathbf{F} = \mathbf{i}xy^2 + \mathbf{j}x^2y$ outward across the circle $x^2 + y^2 = a^2$.

Solution. A parameterization of the circle is $x = a \cos t$, $y =$ ________, $0 \le t \le$ ____. Using the line integral in Problem 24 for the flux we have,

$$\int_C (M\,dy - N\,dx) = \int_C \text{\_\_\_\_\_\_\_\_\_\_\_\_}$$

$$= \int_0^{2\pi} \text{\_\_\_\_\_\_\_\_\_\_\_\_}\,dt$$

$$= 2a^4 \int_0^{2\pi} \frac{1}{4}(1 + \cos 2t)\ \text{\_\_\_\_\_\_\_\_}\,dt$$

$$= 2a^4 \int_0^{2\pi} \frac{1}{4}\left(1 - \cos^2 2t\right)\,dt$$

$$= 2a^4 \int_0^{2\pi} \frac{1}{4}\left[1 - \frac{1}{2}(1 + \cos 4t)\right]\,dt$$

$$= 2a^4 \int_0^{2\pi} \text{\_\_\_\_\_\_\_\_\_\_\_\_}\,dt$$

$$= 2a^4 \left[\frac{t}{8} - \frac{1}{32}\sin 4t\right]_0^{2\pi} = \text{\_\_\_\_\_\_}.$$

The positive answer means that the net flux is __________.

OBJECTIVE B: Given a continuous three-dimensional vector flow field $\mathbf{F} = \delta\mathbf{v}$, find the rate of mass transport outward through a specified smooth closed surface S bounding a region D in its interior.

26. In Exercise 4, page 709 of the text, you are asked to explain why the surface integral

____

can be interpreted as the rate of mass transport outward through S if $\mathbf{n}$ is the outward-pointing unit vector normal to S. We will use this result in the next problem.

---

25. $a \sin t$, 2π, $xy^2\,dy - x^2y\,dx$, $2a^4 \cos^2 t \sin^2 t$, $1 - \cos 2t$, $\frac{1}{8} - \frac{1}{8}\cos 4t$, $\pi a^4/2$, outward

26. $\iint_S \delta(\mathbf{v}\cdot\mathbf{n})\,d\sigma$

27. Consider the flow field $\mathbf{F} = \mathbf{i}y - \mathbf{j}x + \mathbf{k}(x^2+y^2)$ through the portion of the paraboloid $z = 9-x^2-y^2$ above the xy-plane, and directed outward through the surface. We want to find the rate of mass transport. Therefore, we seek to evaluate the surface integral

_______________ ,

where $\mathbf{n}$ is the outward-pointing unit normal to the paraboloid. Let us first calculate this vector $\mathbf{n}$. If we define $g(x,y,z) = z-(9-x^2-y^2)$, then the vector ________ is a unit vector normal to the surface at each point. We find this vector to be $\frac{\nabla g}{|\nabla g|} = \frac{\mathbf{i}2x + \mathbf{j}2y + \mathbf{k}}{\underline{\qquad\qquad}}$, and it is directed upward from the surface of the paraboloid. Thus we can take $\mathbf{n} = \nabla g/|\nabla g|$. Then,

$$\mathbf{F}\cdot\mathbf{n} = \frac{\underline{\qquad\qquad}}{\sqrt{4x^2+4y^2+1}} . \text{ Next,}$$

$$d\sigma = \sqrt{1+\left(\frac{\partial z}{\partial x}\right)^2+\left(\frac{\partial z}{\partial y}\right)^2}\ dA = \underline{\qquad\qquad\qquad}\ dx\,dy .$$

The paraboloid S projects onto the region R in the xy-plane bounded by the circle $x^2+y^2 = 9$. Therefore, the rate of mass transport is,

$$\iint_S \mathbf{F}\cdot\mathbf{n}\,d\sigma = \iint_R \left(\frac{x^2+y^2}{\sqrt{4x^2+4y^2+1}}\right)\sqrt{4x^2+4y^2+1}\ dx\,dy$$

$$= \int_{\_\_}^{\_\_}\int_{\_\_}^{\_\_} \underline{\qquad}\ dr\,d\theta = \frac{\pi 3^4}{2} \approx 127.23 .$$

↑ in polar coordinates

---

27. $\iint_S \mathbf{F}\cdot\mathbf{n}\,d\sigma$, $\frac{\nabla g}{|\nabla g|}$, $\sqrt{4x^2+4y^2+1}$, $2xy-2xy+x^2+y^2$, $\sqrt{4x^2+4y^2+1}$,

$\int_0^{2\pi}\int_0^3 r^3\,dr\,d\theta$

15-5 Green's Theorem.

OBJECTIVE A: Use Green's Theorem to evaluate a given line integral around a simple closed curve C. Assume all hypotheses of the Theorem are satisfied, and that the curve C is oriented in the counterclockwise direction in the xy-plane.

28. If C is a suitable simple __________ curve in the xy-plane containing the region R as its interior, and if M, N, $\partial M/\partial y$, and $\partial N/\partial x$ are ____________ functions of (x,y) in R and on C, then

$$\oint_C M\,dx + N\,dy = \underline{\hspace{4cm}}.$$

29. Give some examples of "suitable" regions R with a piecewise smooth boundary C in the xy-plane for which Green's Theorem applies. (see pages 712-713 of the Thomas/Finney text.)

(a) ____

____ ,

(b) ______________________ ,

(c) ______________________ ,

(d) ____

____ .

30. Let C be the closed curve bounded by

$x = 0$, $y = \sqrt{1-x^2}$ and $y = 0$,

oriented counterclockwise.

Evaluate the line integral

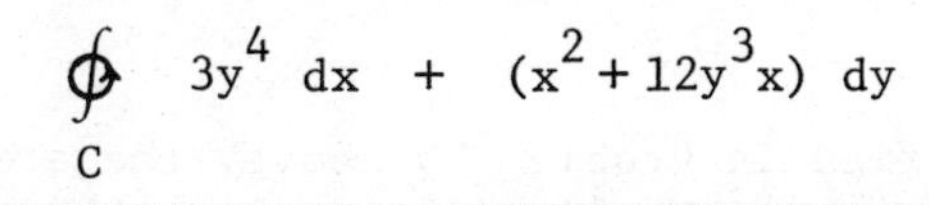

$$\oint_C 3y^4\,dx + (x^2 + 12y^3x)\,dy$$

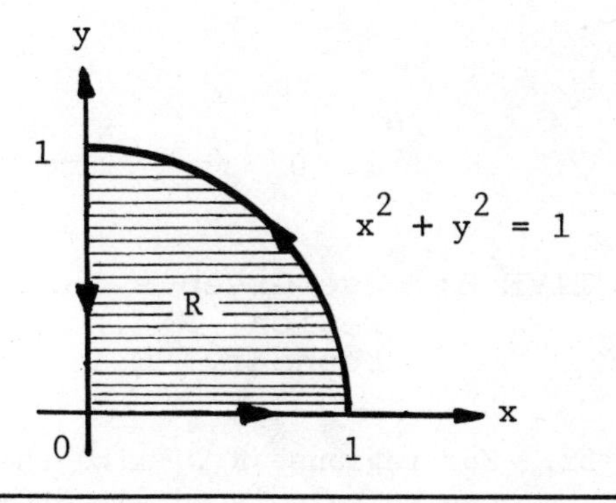

---

28. closed, continuous, $\iint_R \left[\frac{\partial N}{\partial x} - \frac{\partial M}{\partial y}\right] dx\,dy$

29. (a) R is bounded by a simple closed curve C such that a line parallel to either axis cuts C in at most two points

(b) R is a rectangle

(c) R is an annulus

(d) R combines regions like those in (a), (b), or (c) (see figures 15-17 and 15-18 in the text)

Solution. We will use Green's Theorem. Thus, for $M = 3y^4$ and $N = x^2 + 12y^3x$, $\frac{\partial N}{\partial x} - \frac{\partial M}{\partial y} = (\underline{\qquad\qquad}) - 12y^3 = \underline{\qquad}$. Hence,

$$\oint_C 3y^4\,dx + (x^2 + 12y^3x)\,dy = \int_{\_\_}^{\_\_}\int_{\_\_}^{\_\_\_\_} \underline{\qquad}\;dy\,dx$$

$$= \int_0^1 \underline{\qquad\qquad\qquad}\;dx = \underline{\qquad}.$$

31. Let us evaluate the line integral

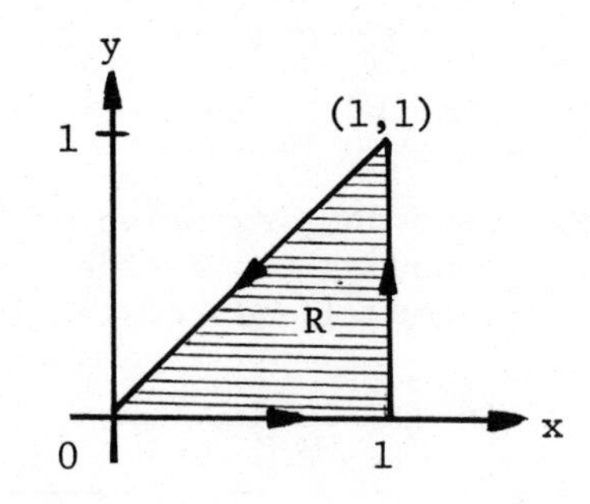

$$\oint_C xy\,dx + \left(x^{3/2} + y^{3/2}\right)dy,$$

where C is the triangle with vertices $(0,0)$, $(1,0)$, $(1,1)$ oriented counterclockwise. Using Green's Theorem,

$$\oint_C xy\,dx + \left(x^{3/2} + y^{3/2}\right)dy = \iint_R \underline{\qquad\qquad\qquad}\;dA$$

$$= \int_{\_\_}^{\_\_}\int_{\_\_}^{\_\_}\left(\frac{3}{2}x^{1/2} - x\right)dy\,dx$$

$$= \int_0^1 \underline{\qquad\qquad\qquad}\;dx = \frac{3}{5}x^{5/2} - \underline{\qquad}\Big]_0^1 = \underline{\qquad}.$$

OBJECTIVE B: Use Green's Theorem to find the area of a given suitable region R in the xy-plane with a piecewise smooth boundary C.

32. For regions R, like those specified in Problem 29 above, the area of R can be calculated via Green's Theorem as the line integral

$$\underline{\qquad\qquad\qquad\qquad\qquad\qquad}.$$

---

30. $2x + 12y^3$, $2x$, $\int_0^1 \int_0^{\sqrt{1-x^2}} 2x\,dy\,dx$, $2x\sqrt{1-x^2}$, $\frac{2}{3}$

31. $\frac{3}{2}x^{1/2} - x$, $\int_0^1 \int_0^x$, $x\left(\frac{3}{2}x^{1/2} - x\right)$, $\frac{1}{3}x^3$, $\frac{4}{15}$ 32. $\frac{1}{2}\oint_C x\,dy - y\,dx$

33. Let us find the area of the region bounded below by the x-axis and above by one arch of the cycloid parameterized by

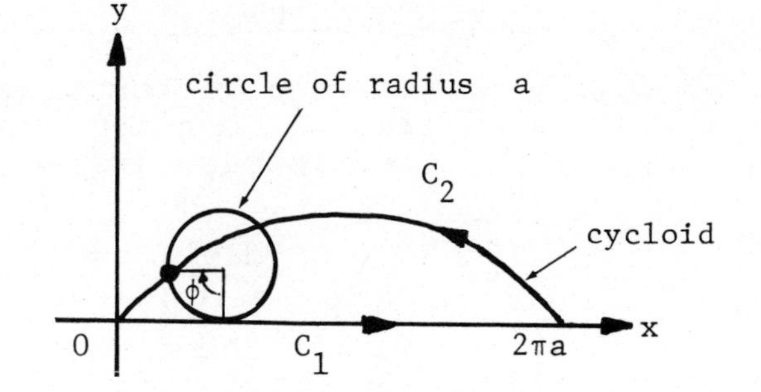

$$x = a(\phi - \sin\phi) ,$$

$$y = a(1 - \cos\phi) ,$$

$0 \leq \phi \leq 2\pi$. (See the figure at the right.) We take the counterclockwise orientation on the curve C bounding the region. The curve C is made up of two smooth curves: the x-axis C_1 , and the cycloid arch C_2 . Now,

on C_1 : $x\,dy - y\,dx =$ __________ $= 0$, and

on C_2 : $x\,dy - y\,dx = a(\phi - \sin\phi)(a\sin\phi\,d\phi)$

$-$ ____________________

$=$ ________ .

Therefore, the area of the region is given by (use the correct orientation),

$$A = \frac{1}{2}\oint_C x\,dy - y\,dx$$

$$= \frac{a^2}{2}\int_{\_\_}^{\_\_} \_\_\_\_\_\_\_\_\_\_\_\_\_\_\_\_\_\_ \,d\phi$$

$$= \frac{a^2}{2}\Big[\sin\phi - \phi\cos\phi + \_\_\_\_\_\_\_\_\Big]_{2\pi}^{0} = \_\_\_\_ .$$

---

33. $x \cdot 0 - 0 \cdot dx$, $a(1 - \cos\phi)(a - a\cos\phi)\,d\phi$, $a^2(\phi\sin\phi + 2\cos\phi - 2)$, $\int_{2\pi}^{0} (\phi\sin\phi + 2\cos\phi - 2)\,d\phi$, $2\sin\phi - 2\phi$, $3\pi a^2$

15-6 The Divergence Theorem.

OBJECTIVE A: Given a vector field $\mathbf{F} = \mathbf{i}M + \mathbf{j}N + \mathbf{k}P$, with M, N, and P continuous functions of (x,y,z) that have continuous first-order partial derivatives:

$$\operatorname{div}\mathbf{F} = \frac{\partial M}{\partial x} + \frac{\partial N}{\partial y} + \frac{\partial P}{\partial z};$$

verify the divergence theorem for the cube with center at the origin and faces in the planes $x = \pm a$, $y = \pm a$, and $z = \pm a$.

34. For the cube, the divergence theorem gives

$$\iiint_D \operatorname{div}\mathbf{F}\ dV = \underline{\qquad\qquad\qquad},$$

where $\mathbf{n}\,d\sigma$ is a vector element of surface area directed along the unit outer normal vector $\mathbf{n}$, and S is the surface of the cube enclosing the region D.

35. Let us verify the divergence theorem for the field $\mathbf{F} = \mathbf{i}(2xyz) + \mathbf{j}(x^2z) + \mathbf{k}(x^2y+1)$ and the cube with center at the origin and faces in the planes $x = \pm 1$, $y = \pm 1$, $z = \pm 1$. First,

$$\operatorname{div}\mathbf{F} = \frac{\partial M}{\partial x} + \frac{\partial N}{\partial y} + \frac{\partial P}{\partial z} = 2yz + \underline{\quad} + 0 = \underline{\qquad}.$$

Thus, if D is the region enclosed by the cube,

$$\iiint_D \operatorname{div}\mathbf{F}\ dV = \int_{-1}^{1}\int_{-1}^{1}\int_{-1}^{1} 2yz\ dz\ dy\ dx$$

$$= \int_{-1}^{1}\int_{-1}^{1} \underline{\qquad}\Big]_{z=-1}^{z=1} dy\ dx = \underline{\quad}.$$

Next, we compute $\iint \mathbf{F}\cdot\mathbf{n}\ d\sigma$ as the sum of the integrals over the six faces separately. We begin with the two faces perpendicular to the x-axis.

---

34. $\iint_S \mathbf{F}\cdot\mathbf{n}\ d\sigma$

35. 0, $2yz$, yz^2, 0, $-1 \le y \le 1$, $-1 \le z \le 1$, $\mathbf{i}$, $\mathbf{i}(2yz) + \mathbf{j}z + \mathbf{k}(y+1)$, $dy\,dz$, $\int_{-1}^{1}\int_{-1}^{1} 2yz\ dz\,dy$, 0, $-1 \le x \le 1$, $-1 \le z \le 1$, $\mathbf{j}$, $\mathbf{i}(2xz) + \mathbf{j}(x^2z) + \mathbf{k}(x^2+1)$, $dz\ dx$, $\int_{-1}^{1}\int_{-1}^{1} x^2z\ dz\ dx$, 0, 0

For the face $x = -1$ and the face $x = +1$, respectively, we have the first and second lines of the following table (complete the table):

range of integration	outward unit normal	field **F**
$-1 \le y \le 1, \quad -1 \le z \le 1$	$-\mathbf{i}$	$\mathbf{i}(-2yz) + \mathbf{j}z + \mathbf{k}(y+1)$
____________	____	____________

For each of these planes, $x = x(y,z)$ is a constant function of (y,z) so that

$$d\sigma = \sqrt{1 + \left(\frac{\partial x}{\partial y}\right)^2 + \left(\frac{\partial x}{\partial z}\right)^2}\ dy\, dz = \underline{\qquad\quad} .$$

Therefore, the sum of the integrals over these two faces is

$$\iint \mathbf{F}\cdot\mathbf{n}\, d\sigma = \int_{-1}^{1}\int_{-1}^{1} 2yz\, dz\, dy + \underline{\qquad\qquad\qquad\qquad}$$

$$= 2\int_{-1}^{1} yz^2\Big]_{z=-1}^{z=1} dy = \underline{\qquad\quad} .$$

Now consider the two faces perpendicular to the y-axis. For the face $y = -1$ and the face $y = +1$, respectively, complete the first and second lines of the following table:

range of integration	outward unit normal	field **F**
$-1 \le x \le 1, \quad -1 \le z \le 1$	$-\mathbf{j}$	$\mathbf{i}(-2xz) + \mathbf{j}(x^2z) + \mathbf{k}(-x^2+1)$
____________	____	____________

For each of these planes $d\sigma = \underline{\qquad\quad}$. Therefore, the sum of the surface integrals over these two faces is

$$\iint \mathbf{F}\cdot\mathbf{n}\, d\sigma = \int_{-1}^{1}\int_{-1}^{1} (-x^2z)\, dz\, dx + \underline{\qquad\qquad\qquad\qquad}$$

$$= \underline{\quad\ } .$$

Finally, we take the two faces perpendicular to the z-axis. For the face $z = -1$ and the face $z = +1$, respectively, we obtain the first and second lines of the following table:

range of integration	outward unit normal	field $\mathbf{F}$
$-1 \le x \le 1, \quad -1 \le y \le 1$	$-\mathbf{k}$	$\mathbf{i}(-2xy) + \mathbf{j}(-x^2) + \mathbf{k}(x^2y+1)$
$-1 \le x \le 1, \quad -1 \le y \le 1$	$\mathbf{k}$	$\mathbf{i}(2xy) + \mathbf{j}x^2 + \mathbf{k}(x^2y+1)$

For each of these planes $d\sigma = dx\,dy$, and the sum of the surface integrals over these two faces is

$$\iint \mathbf{F}\cdot\mathbf{n}\, d\sigma = \int_{-1}^{1}\int_{-1}^{1} -(x^2y+1)\, dx\, dy + \int_{-1}^{1}\int_{-1}^{1} (x^2y+1)\, dx\, dy$$

$$= 0 .$$

Therefore, the surface integral over the six faces of the cube is

$$\iint_S \mathbf{F}\cdot\mathbf{n}\, d\sigma = \underline{\qquad} = \iiint_D \operatorname{div}\mathbf{F}\, dV .$$

OBJECTIVE B: Given a vector field $\mathbf{F} = \mathbf{i}M + \mathbf{j}N + \mathbf{k}P$ as in Objective A, and given a suitable piecewise smooth surface S enclosing the region D in space, verify the divergence theorem. Assume that the projections of D into the coordinate planes are simply connected regions.

36. Give some examples of "suitable" regions D enclosed by a piecewise smooth bounding surface S for which the divergence theorem applies. One example is when D is a convex region with no holes such that when D is projected into a coordinate plane it produces a simply connected region R, with the property that any line perpendicular to that coordinate plane at an interior point of R intersects the bounding surface S in at most ______ points. Other examples are:

(a) ____________ ,

(b) ______________________________ ,

(c) ____
______________________________ .

---

36. two, (a) cubes
(b) two concentric spheres
(c) regions that can be split up into a finite number of simple regions of the type described above

37. Let us verify the divergence theorem for the vector field $\mathbf{F} = \mathbf{i}x + \mathbf{j}y + \mathbf{k}z$, where D is the region inside the cylinder $x^2 + y^2 = 1$ and between the planes $z = 0$ and $z = 2$. Now,

$$\text{div } \mathbf{F} = \frac{\partial M}{\partial x} + \frac{\partial N}{\partial y} + \frac{\partial P}{\partial z} = \_\_\_ \,,$$ so that

$$\iiint_D \text{div } \mathbf{F} \; dV = 3 \iiint_D dV = \_\_\_\_\_\_\_\_\_\_ = 6\pi .$$

The surface integral $\iint \mathbf{F} \cdot \mathbf{n} \, d\sigma$ is the sum of three integrals: one over the top, another one over the bottom, and a third one over the side of the cylinder. We calculate each of these separately.

The top: $z = 2$, $0 \le x^2 + y^2 \le 1$, $\mathbf{n} = \_\_\_$ is the outer normal, and

$\mathbf{F} = \_\_\_\_\_\_\_\_\_\_$. Then,

$$\iint \mathbf{F} \cdot \mathbf{n} \, d\sigma = \iint \_\_\_ \, d\sigma = 2 \times \text{area of top} = \_\_\_ .$$

The bottom: $z = 0$, $0 \le x^2 + y^2 \le 1$, $\mathbf{n} = \_\_\_$ is the outer normal, and

$\mathbf{F} = \_\_\_\_\_\_\_\_$. Then,

$$\iint \mathbf{F} \cdot \mathbf{n} \, d\sigma = \iint \_\_\_ \, d\sigma = 0 .$$

The side: $x^2 + y^2 = 1$, $0 \le z \le 2$, $\mathbf{F} = \_\_\_\_\_\_\_\_\_\_$ and

$$\mathbf{n} = \frac{\mathbf{i}x + \mathbf{j}y}{\sqrt{x^2 + y^2}} = \mathbf{i}x + \mathbf{j}y$$ is the outer normal. Hence,

$$\iint \mathbf{F} \cdot \mathbf{n} \, d\sigma = \iint \_\_\_\_\_\_ \, d\sigma = \iint d\sigma = \_\_\_\_\_\_\_\_ = 4\pi .$$

Therefore, the surface integral over the entire cylindrical can is

$$\iint_S \mathbf{F} \cdot \mathbf{n} \, d\sigma = \_\_\_ + 0 + 4\pi = \iiint_D \text{div } \mathbf{F} \; dV .$$

---

37. 3, $3\pi(1)^2(2)$, $\mathbf{k}$, $\mathbf{i}x + \mathbf{j}y + \mathbf{k}2$, 2, 2π, $-\mathbf{k}$, $\mathbf{i}x + \mathbf{j}y$, 0, $\mathbf{i}x + \mathbf{j}y + \mathbf{k}z$, $x^2 + y^2$, $2\pi(1)(2)$, 2π

15-7 Stokes's Theorem.

OBJECTIVE: Given a vector field $\mathbf{F} = \mathbf{i}M + \mathbf{j}N + \mathbf{k}P$, where M, N, and P are continuous functions of (x,y,z), together with their first-order partial derivatives, throughout a region D containing a specified smooth, simply connected, orientable surface S bounded by a simple closed curve C; verify Stokes's Theorem. Assume that the positive direction around C is the one induced by the positive orientation of S.

38. Under the hypotheses stated in the above Objective, if $\mathbf{n}$ is a positive unit vector normal to S, then

$$\oint_C \mathbf{F} \cdot d\mathbf{R} = \underline{\qquad\qquad} .$$

39. Consider the oriented triangle shown in the figure at the right. Let

$$\mathbf{F} = \mathbf{i}\,y^2 + \mathbf{j}\,xy - \mathbf{k}\,2xz .$$

We will verify Stokes's Theorem for the triangular plane bounded by the sides C_1, C_2, C_3 directed as shown. First, let us find the positive unit vector normal to the plane that agrees with the orientation around the triangle. A vector normal to the plane is $2\mathbf{i} \times (2\mathbf{j} + \mathbf{k}) = \underline{\qquad\qquad}$. Notice that this vector points upward (positive $\mathbf{k}$ component) and toward the left in the figure (negative $\mathbf{j}$ component), so it points in the positive direction relative to the counterclockwise direction around the boundary. Therefore, the unit vector $\mathbf{n}$ is given by $\mathbf{n} = \underline{\qquad\qquad}$.

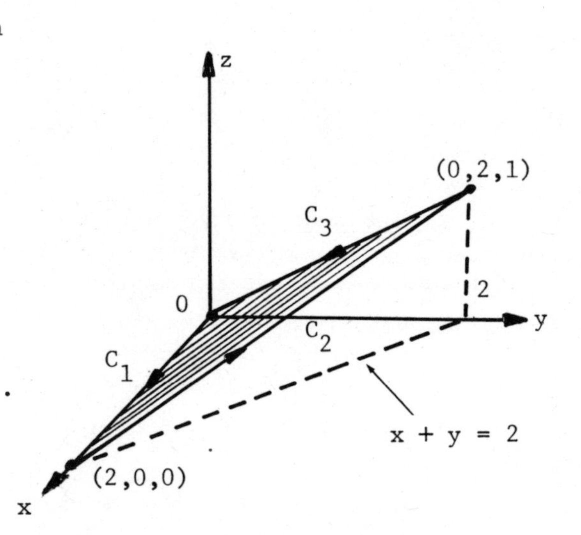

---

38. $\displaystyle\iint_S \text{curl}\,\mathbf{F} \cdot \mathbf{n}\; d\sigma$

39. $-2\mathbf{j} + 4\mathbf{k}$, $\dfrac{-\mathbf{j} + 2\mathbf{k}}{\sqrt{5}}$, $\dfrac{y}{2}$, $\mathbf{j}2z - \mathbf{k}y$, $\sqrt{1 + 0 + \frac{1}{4}}$, $-\frac{3}{2}y$, $\displaystyle\int_0^2 \int_0^{2-x}$, -2,

$y^2\,dx + xy\,dy - 2xz\,dz$, 0, $(2 - 2t)(2t)(2\,dt)$, $4(-3t^2 + t)$, $\displaystyle\int_0^1$, -2, 0, -2

An equation of the plane is

$-2(y-0) \ + \ 4(z-0) \ = \ 0$, or $z = \underline{\qquad}$. Also,

$$\text{curl } \mathbf{F} = \begin{vmatrix} \mathbf{i} & \mathbf{j} & \mathbf{k} \\ \frac{\partial}{\partial x} & \frac{\partial}{\partial y} & \frac{\partial}{\partial z} \\ y^2 & xy & -2xz \end{vmatrix} = \underline{\qquad\qquad} .$$

For the element of surface area we use,

$$d\sigma = \sqrt{1 + \left(\frac{\partial z}{\partial x}\right)^2 + \left(\frac{\partial z}{\partial y}\right)^2} \ dx\,dy = \underline{\qquad\qquad} \ dx\,dy$$

or $d\sigma = \frac{1}{2}\sqrt{5} \ dx\,dy$. Therefore,

$$\text{curl } \mathbf{F} \cdot \mathbf{n} \ d\sigma = (\mathbf{j}2z - \mathbf{k}y) \cdot \left(-\frac{1}{2}\mathbf{j} + \mathbf{k}\right) dx\,dy = (-z-y)\ dx\,dy$$

$$= \underline{\qquad} \ dx\,dy \quad \text{on the plane.}$$

Hence,

$$\iint_S \text{curl } \mathbf{F} \cdot \mathbf{n} \ d\sigma = \int_{\underline{\quad}}^{\overline{\quad}} \int_{\underline{\quad}}^{\overline{\quad}} -\frac{3}{2} y \ dy\,dx = \int_0^2 -\frac{3}{4}(2-x)^2 \ dx$$

$$= \underline{\qquad} .$$

Now, to find the line integral $\oint_C \mathbf{F} \cdot d\mathbf{R}$, we will calculate three integrals, one for C_1, C_2, and C_3 separately, and sum them. The integrand in the line integral is

$$\mathbf{F} \cdot (\mathbf{i}dx + \mathbf{j}dy + \mathbf{k}dz) = \underline{\qquad\qquad\qquad} .$$

On C_1 : $z = 0$, $y = 0$, $0 \le x \le 2$, and thus

$$\mathbf{F} \cdot d\mathbf{R} = \underline{\quad} . \quad \text{Hence} \quad \int_{C_1} \mathbf{F} \cdot d\mathbf{R} = 0 .$$

On C_2 : The line segment from $(2,0,0)$ to $(0,2,1)$ can be represented parametrically by $x = 2-2t$, $y = 2t$, $z = t$ for $0 \le t \le 1$. Then, $\mathbf{F} \cdot d\mathbf{R}$ on the segment can be expressed in terms of t by,

$$\mathbf{F} \cdot d\mathbf{R} = (2t)^2(-2\,dt) + \underline{\qquad\qquad\qquad} - 2(2-2t)t\,dt$$

$$= \underline{\qquad\qquad\quad} dt \text{ , and therefore}$$

$$\int_{C_2} \mathbf{F} \cdot d\mathbf{R} = \int_{\_\_}^{\_\_} 4(-3t^2+t)\,dt = \underline{\quad} .$$

(check the orientation for C_2 here)

On C_3 : $x = 0$, $z = \frac{y}{2}$, $2 \geq y \geq 0$, and thus

$$\mathbf{F} \cdot d\mathbf{R} = y^2 \cdot 0 + 0\,dy - 0\,dz = \underline{\quad} \text{ and } \int_{C_3} \mathbf{F} \cdot d\mathbf{R} = 0 .$$

Therefore, summing the three line integrals,

$$\oint_C \mathbf{F} \cdot d\mathbf{R} = \underline{\quad} = \iint_S \text{curl } \mathbf{F} \cdot \mathbf{n}\,d\sigma .$$

CHAPTER 15 OBJECTIVE - PROBLEM KEY

Objective	Problems in Thomas/Finney Text	Objective	Problems in Thomas/Finney Text
15-1 A	See page 680 in the text	15-4 A	p. 709, 3
B	p. 684, 6-10	B	p. 709, 4-6
15-2 A	p. 690, 5,6	15-5 A	p. 718, 3-7,11,12
B	p. 690, 7-10	B	p. 711, Example 2
15-3 A	p. 701, 1,2	15-6 A	p. 725, 1-5
B	p. 702, 3-5,9-13,19-21	B	p. 726, 6-10
C	p. 702, 14-18	15-7	p. 734, 1-4,6

CHAPTER 15 SELF-TEST

1. Find the gradient fields for the given functions.

   (a) $f(x,y,z) = 5x^2 - 3y^3 + 2xyz$

   (b) $f(x,y,z) = e^{xy}(x^2 + z^2)$

2. Let S be the portion of the cylinder $z = \frac{1}{2}y^2$ for which $x \geq 0$, $y \geq 0$, and $x + y \leq 1$. Find the surface integral

$$\iint_S \sqrt{x(1+2z)}\,d\sigma .$$

3. Evaluate the integral

$$\iint_S \mathbf{F} \cdot \mathbf{n} \; d\sigma ,$$

for $\mathbf{F} = \mathbf{i}x + \mathbf{j}y + \mathbf{k}z$, given that S is the plane of the triangle with vertices $(a,0,0)$, $(0,a,0)$, $(0,0,a)$, $a > 0$, lying in the first octant, and $\mathbf{n}$ is the upper normal unit vector.

4. Evaluate the following line integrals.

(a) $\int_C \sqrt{x+z} \; ds$, where C is the line segment from the point $A(0,0,0)$ to $B(4,5,6)$;

(b) $\int_C y^2 \, dx + x^2 \, dy$, where C is the right semi-circle $x = \sqrt{1-y^2}$ directed from the point $A(0,-1)$ to $B(0,1)$.

5. Find the work done by the force

$$\mathbf{F} = \mathbf{i}\left(\frac{1}{x+3y+2z} - 5x\right) + \mathbf{j}\left(\frac{3}{x+3y+2z} + 3y^2\right) + \mathbf{k}\left(\frac{2}{x+3y+2z}\right)$$

as the point of application moves from $A(0,2,0)$ along the y-axis to $(0,1,0)$, and from there in a straight line to $B(1,1,1)$. <u>Hint</u>: Is the field conservative?

6. Given the vector $\mathbf{F} = \mathbf{i}(2xz+1) + \mathbf{j}\,2y(z+1) + \mathbf{k}(x^2+y^2+3z^2)$, find a function $f(x,y,z)$, if possible, such that $\mathbf{F} = \text{grad } f$.

7. Find the rate of mass transport outward through the sphere $x^2+y^2+z^2 = 4$ if the vector flow field is $\mathbf{F} = \delta\mathbf{v} = -\mathbf{i}y + \mathbf{j}x + \mathbf{k}z^4$.

8. Use Green's Theorem to evaluate the line integral $\oint_C y \, dx - x \, dy$,

where C is the cardioid $r = 1 - \cos\theta$ directed in the counterclockwise direction.

9. Use Green's Theorem to find the area of the region bounded by the hypocycloid parameterized by $x = a\cos^3 t$, $y = a\sin^3 t$, for $0 \le t \le 2\pi$. Assume the boundary is directed in the counterclockwise direction.

10. Use the divergence theorem to find the rate of mass transport outward through the region inside the sphere $x^2+y^2+z^2 = 4$ and outside the cylinder $x^2+y^2 = 1$, if the vector flow field is $\mathbf{F} = \mathbf{i}6x - \mathbf{j}13y + \mathbf{k}12z$.

11. Let C be the intersection of the sphere $x^2+y^2+z^2 = 1$ and the cone $z = \sqrt{x^2+y^2}$, directed in the counterclockwise sense around the z-axis. For the field $\mathbf{F} = \mathbf{i}(x^2+z) + \mathbf{j}(y^2+2x) + \mathbf{k}(z^2-y)$, use Stokes's Theorem to calculate the line integral

$$\oint_C \mathbf{F}\cdot d\mathbf{R} \quad , \quad \mathbf{R} = \mathbf{i}x + \mathbf{j}y + \mathbf{k}z \ .$$

12. A force is given by $\mathbf{F} = \mathbf{i}(x^2 - xy^3) + \mathbf{j}(y^2 - 2xy)$, and its point of application moves counterclockwise around the square in the xy-plane with vertices (0,0) , (3,0) , (3,3) , and (0,3) . Find the work done.

SOLUTIONS TO CHAPTER 15 SELF-TEST

1. (a) $\nabla f = \mathbf{i}\frac{\partial f}{\partial x} + \mathbf{j}\frac{\partial f}{\partial y} + \mathbf{k}\frac{\partial f}{\partial z} = \mathbf{i}(10x+2yz) + \mathbf{j}(2xz-9y^2) + \mathbf{k}\,2xy$

   (b) $\nabla f = \mathbf{i}\,e^{xy}\left(2x+yx^2+yz^2\right) + \mathbf{j}\ xe^{xy}(x^2+y^2) + \mathbf{k}\ 2ze^{xy}$

2. The portion of the cylindrical surface is sketched below. For the element of surface area $d\phi$ we use

$$d\sigma = \sqrt{1 + \left(\frac{\partial z}{\partial x}\right)^2 + \left(\frac{\partial z}{\partial y}\right)^2}\ dx\,dy = \sqrt{1+0+y^2}\ dx\,dy\ .$$

The surface integral becomes

$$\iint_S \sqrt{x(1+2z)}\quad d\sigma$$

$$= \int_0^1 \int_0^{1-x} \sqrt{x}\ (1+y^2)\ dy\,dx$$

$$= \int_0^1 \sqrt{x}\ \left[(1-x) + \frac{1}{3}(1-x)^3\right] dx$$

$$= \int_0^1 \left(\frac{4}{3}x^{1/2} - 2x^{3/2} + x^{5/2} - \frac{1}{3}x^{7/2}\right) dx$$

$$= \frac{8}{9} - \frac{4}{5} + \frac{2}{7} - \frac{2}{27} \approx 0.30\ .$$

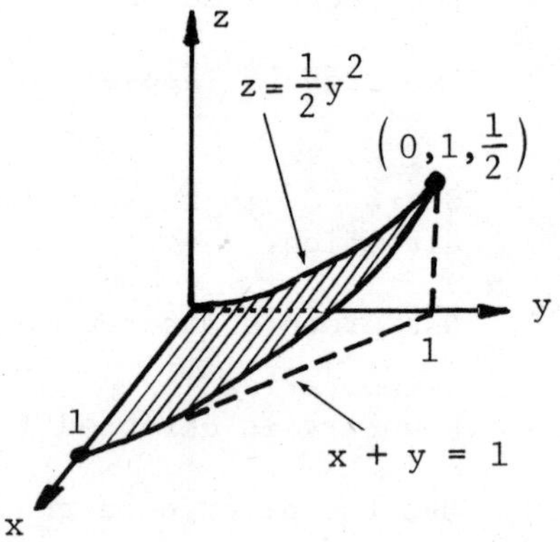

z
$z = \frac{1}{2}y^2$
$\left(0,1,\frac{1}{2}\right)$
y
1
1
x + y = 1
x

3. The triangular shaped surface S is shown at the right. An equation of the plane is

$$x + y + z = a, \quad \text{or} \quad z = a - x - y .$$

A vector normal to the plane is

$\mathbf{N} = \mathbf{i} + \mathbf{j} + \mathbf{k}$, and this vector points in the upward direction. Thus,

$$\mathbf{n} = \frac{\mathbf{i} + \mathbf{j} + \mathbf{k}}{\sqrt{3}} .$$

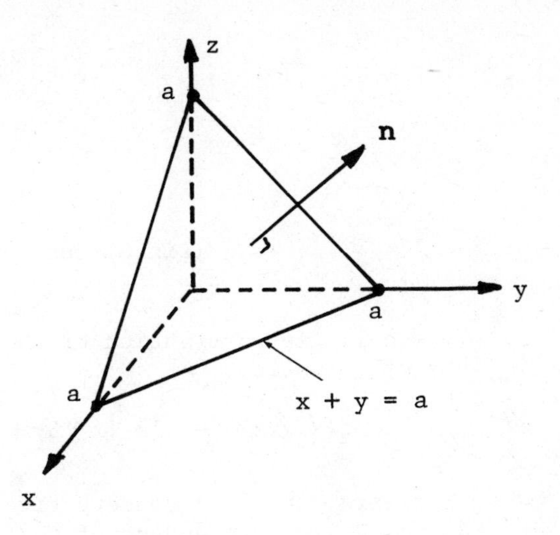

For the element of surface area we have,

$$d\sigma = \sqrt{1 + \left(\frac{\partial z}{\partial x}\right)^2 + \left(\frac{\partial z}{\partial y}\right)^2}\; dx\, dy = \sqrt{3}\; dx\, dy .$$

The projection of the surface onto the xy-plane is the region described by $x \geq 0$, $y \geq 0$, and $x + y \leq a$ (see figure). Thus,

$$\iint_S \mathbf{F} \cdot \mathbf{n}\, d\sigma = \iint_S \frac{1}{\sqrt{3}} (x + y + z)\, d\sigma$$

$$= \int_0^a \int_0^{a-x} (x + y + a - x - y)\, dy\, dx$$

$$= a \int_0^a (a - x)\, dx = \frac{1}{2} a^3 .$$

4. (a) A parametric representation for the line segment C is $x = 4t$, $y = 5t$, $z = 6t$ with $0 \leq t \leq 1$. Thus,

$$\int_C \sqrt{x + z}\; ds = \int_0^1 \sqrt{4t + 6t} \cdot \sqrt{\left(\frac{dx}{dt}\right)^2 + \left(\frac{dy}{dt}\right)^2 + \left(\frac{dz}{dt}\right)^2}\; dt$$

$$= \int_0^1 \sqrt{10t}\; \sqrt{77}\; dt = \frac{2}{3} \sqrt{770} \approx 18.50 .$$

(b) A parametric representation for the semi-circle is $x = \cos \theta$, $y = \sin \theta$ with $-\pi/2 \leq \theta \leq \pi/2$. (Note the orientation.) Thus,

$$\int_C y^2\, dx + x^2\, dy = \int_{-\pi/2}^{\pi/2} \left[\sin^2 \theta(-\sin \theta) + \cos^2 \theta (\cos \theta) \right] d\theta$$

$$= \int_{-\pi/2}^{\pi/2} (\cos^3\theta - \sin^3\theta)\, d\theta = \int_{-\pi/2}^{\pi/2} \cos^3\theta\, d\theta$$

(since $\sin^3\theta$ is an odd function)

$$= \frac{1}{3}(\sin\theta)(\cos^2\theta + 2)\Big]_{-\pi/2}^{\pi/2} = \frac{4}{3}.$$

5. From a little observation of the components of F, it is easy to see that $F = \nabla f$, where

$$f(x,y,z) = \ln|x + 3y + 2z| - \frac{5}{2}x^2 + y^3 .$$

Therefore, F is a conservative force field so that the work done by F from A to B is independent of the path joining them. The work done is

$$W = \int_A^B F \cdot dR = f(1,1,1) - f(0,2,0)$$

$$= \left(\ln 6 - \frac{5}{2} + 1\right) - (\ln 6 - 0 + 8) = -\frac{19}{2}.$$

6. If $M = 2xz + 1$, $N = 2y(z+1)$, and $P = x^2 + y^2 + 3z^2$, then

$$\frac{\partial M}{\partial z} = 2x = \frac{\partial P}{\partial x}, \quad \frac{\partial N}{\partial z} = 2y = \frac{\partial P}{\partial y}, \quad \frac{\partial M}{\partial y} = 0 = \frac{\partial N}{\partial x}.$$

Thus, there may be a function $f(x,y,z)$ with $F = \nabla f$. Now, if f does exist,

$$\frac{\partial f}{\partial x} = 2xz + 1, \quad \frac{\partial f}{\partial y} = 2y(z+1), \quad \text{and} \quad \frac{\partial f}{\partial z} = x^2 + y^2 + 3z^2 .$$

Integration of the first of these equations with respect to x, holding y and z fixed, gives

$$f(x,y,z) = x^2 z + x + \underbrace{g(y,z)}_{\text{"constant of integration"}} .$$

Then,

$$2y(z+1) = \frac{\partial f}{\partial y} = \frac{\partial g(y,z)}{\partial y} \quad \text{so} \quad g(y,z) = y^2(z+1) + h(z),$$

where $h(z)$ is the "constant of integration" obtained when $\partial g/\partial y = 2y(z+1)$ is integrated with respect to y with z fixed. Then,

$$f(x,y,z) = x^2 z + x + y^2(z+1) + h(z), \quad \text{and}$$

$$x^2 + y^2 + 3z^2 = \frac{\partial f}{\partial z} = x^2 + y^2 + h'(z) \quad \text{so} \quad h'(z) = 3z^2 .$$

Hence, $h(z) = z^3 + C$ and $f(x,y,z) = x(xz+1) + y^2(z+1) + z^3 + C$,

where C is an arbitrary constant.

7. We divide the sphere into the upper hemisphere S_1: $z = \sqrt{4 - x^2 - y^2}$ and the lower hemisphere S_2: $z = -\sqrt{4 - x^2 - y^2}$. In either case, a unit outer normal is given by

$$\mathbf{n} = \frac{\mathbf{i}x + \mathbf{j}y + \mathbf{k}z}{\sqrt{x^2 + y^2 + z^2}} = \frac{\nabla w}{|\nabla w|} \quad \text{for } w = x^2 + y^2 + z^2 - 4 .$$

Thus, $\mathbf{n} = \frac{1}{2}(\mathbf{i}x + \mathbf{j}y + \mathbf{k}z)$ since $x^2 + y^2 + z^2 = 4$ on the sphere. Notice that $\mathbf{n}$ points in the outward direction (upward for $z > 0$ and downward for $z < 0$) .

An element of surface area for S_1 or S_2 is given by

$$d\sigma = \sqrt{1 + \left(\frac{\partial z}{\partial x}\right)^2 + \left(\frac{\partial z}{\partial y}\right)^2}\, dx\, dy = \sqrt{1 + \frac{x^2}{z^2} + \frac{y^2}{z^2}}\, dx\, dy$$

$$= \sqrt{\frac{x^2 + y^2 + z^2}{z^2}}\, dx\, dy = \frac{2}{|z|} .$$

Thus, the rate of mass transport outward through the sphere S is,

$$\iint_S \mathbf{F} \cdot \mathbf{n}\, d\sigma = \iint_{S_1} \mathbf{F} \cdot \mathbf{n}\, d\sigma + \iint_{S_2} \mathbf{F} \cdot \mathbf{n}\, d\sigma$$

$$= \iint_{S_1} \frac{1}{2} z^5 \cdot \frac{2}{|z|}\, dx\, dy + \iint_{S_2} \frac{1}{2} z^5 \cdot \frac{2}{|z|}\, dx\, dy$$

$$= \iint_{x^2+y^2 \leq 4} z^4\, dx\, dy + \iint_{x^2+y^2 \leq 4} -z^4\, dx\, dy = 0 .$$

8. Let R denote the region that lies inside the cardioid. Then,

$$\oint_C y\, dx - x\, dy = -2 \iint_R dx\, dy = -2 \int_0^{2\pi} \int_0^{1 - \cos\theta} r\, dr\, d\theta$$

↑ note the orientation and polar coordinates

$$= \int_{2\pi}^{0} (1 - 2\cos\theta + \cos^2\theta)\, d\theta$$

$$= \theta - 2\sin\theta + \frac{\theta}{2} + \frac{1}{4}\sin 2\theta \Big]_{2\pi}^{0} = -3\pi \; .$$

9. A sketch of the region is given at the right. From the symmetry of the region we can calculate the area of the region in the first quadrant and multiply by 4. We use the Corollary to Green's Theorem (p. 712). For the parameterization

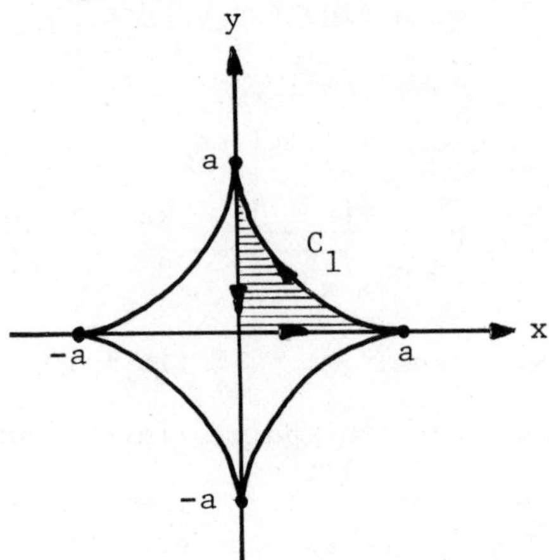

$$x = a\cos^3 t \quad \text{and} \quad y = a\sin^3 t \; ,$$

$$dx = -3a\sin t\cos^2 t\, dt \quad \text{and}$$

$$dy = 3a\cos t\sin^2 t\, dt \; .$$

Thus, the area of the entire region is given by

$$A = \oint \frac{1}{2}(x\,dy - y\,dx) = 4\int_{C_1} \frac{1}{2}(x\,dy - y\,dx)$$

$$= 4\int_0^{\pi/2} \frac{1}{2}(3a^2\cos^4 t\sin^2 t + 3a^2\sin^4 t\cos^2 t)\, dt$$

$$= 6a^2\int_0^{\pi/2} \cos^2 t\sin^2 t\, dt = 6a^2\left[\frac{t}{8} - \frac{1}{32}\sin 4t\right]_0^{\pi/2} = \frac{3\pi}{8}a^2 \; .$$

(as in Problem 25 of this chapter)

10. Evaluation of $\iint_S \mathbf{F}\cdot\mathbf{n}\, d\sigma$ itself would involve finding two surface integrals, one for the sphere and one for the inside cylinder. Instead we use the divergence theorem:

$$\iint_S \mathbf{F}\cdot\mathbf{n}\, d\sigma = \iiint_D \text{div}\,\mathbf{F}\, dV = \iiint_D (6 - 13 + 12)\, dV$$

$$= 5\int_0^{2\pi}\int_1^2\int_{-\sqrt{4-r^2}}^{\sqrt{4-r^2}} dz\, r\, dr\, d\theta$$

cylindrical coordinates

$$= 5\int_0^{2\pi}\int_1^2 2r\sqrt{4-r^2}\, dr\, d\theta$$

$$= 5\int_0^{2\pi} -\frac{2}{3}\left(4-r^2\right)^{3/2}\Big]_{r=1}^{r=2} d\theta = 20\pi\sqrt{3} \approx 108.83 \; .$$

11. The intersection of the sphere and the cone is shown at the right. The sphere and cone intersect when

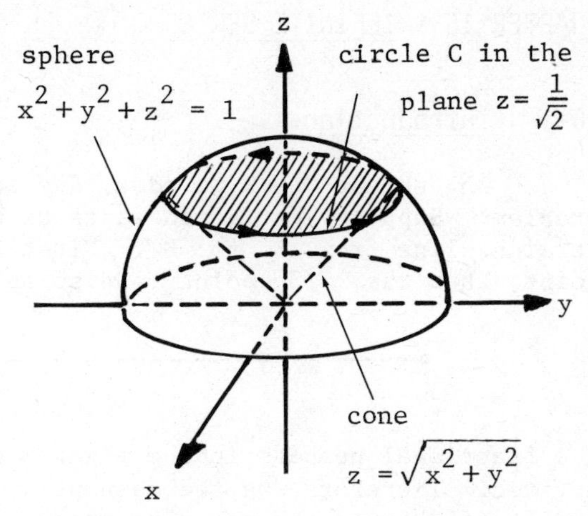

$$x^2 + y^2 + \left(\sqrt{x^2 + y^2}\right)^2 = 1$$

or

$$x^2 + y^2 = \frac{1}{2} \quad \text{and} \quad z = \frac{1}{\sqrt{2}} .$$

We can apply Stokes's Theorem using the circle $C : x^2 + y^2 = \frac{1}{2}$ considered in the plane surface $z = \frac{1}{\sqrt{2}}$ (rather than using the spherical cap above it for S.) The reason for this choice of the surface is that its unit normal is the constant vector $\mathbf{n} = \mathbf{k}$ (rather than the varying unit normal on the cap). Then,

$$\text{curl } \mathbf{F} = \begin{vmatrix} \mathbf{i} & \mathbf{j} & \mathbf{k} \\ \frac{\partial}{\partial x} & \frac{\partial}{\partial y} & \frac{\partial}{\partial z} \\ x^2 + z & y^2 + 2x & z^2 - y \end{vmatrix} = -\mathbf{i} + \mathbf{j} + 2\mathbf{k} ,$$

and Stokes's Theorem gives

$$\oint_C \mathbf{F} \cdot d\mathbf{R} = \iint_S \text{curl } \mathbf{F} \cdot \mathbf{n} \, d\sigma$$

$$= \iint_S 2 \, d\sigma = 2 \text{ times the area of circle } C$$

$$= 2 \cdot \pi \left(\frac{1}{\sqrt{2}}\right)^2 = \pi .$$

12. According to Green's Theorem, the work done is

$$W = \oint_C \mathbf{F} \cdot d\mathbf{R} = \int_0^3 \int_0^3 \left[\frac{\partial}{\partial x}(y^2 - 2xy) - \frac{\partial}{\partial y}(x^2 - xy^3)\right] dy \, dx$$

$$= \int_0^3 \int_0^3 (-2y + 3xy^2) \, dy \, dx = \int_0^3 (-9 + 27x) \, dx = \frac{189}{2} .$$

CHAPTER 16 INFINITE SERIES

16-1 Introduction.

An ancient Greek paradox, due to the mathematician Zeno, concerned the following problem. Suppose that a man wants to walk a certain distance, say two miles, along a straight line from A to B. First he must pass the half-way point, then the 3/4 point, then the 7/8 point, and so on as illustrated in the following figure.

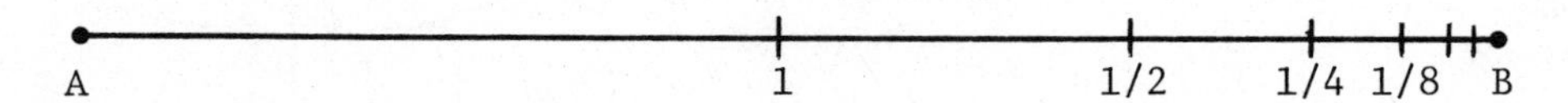

The fractional numbers in the figure indicate the distance in miles remaining to be covered. Therefore, on the assumption that a finite length contains an infinite number of points, the man must pass an infinite number of distance markers along the way. But that is impossible. The paradox is that the man does get to B, and in a finite amount of time, assuming he walks at some steady pace.

An analysis of the problem is not difficult. The total distance s from A to B is an infinite sum expressible as,

$$s = 1 + \left(\frac{1}{2}\right) + \left(\frac{1}{2}\right)^2 + \left(\frac{1}{2}\right)^3 + \ldots,$$

and the paradox is dispelled if this infinite sum equals the finite number of 2 miles. That turns out to be exactly the case, as you will see further on in this chapter when it is established that

$$\frac{1}{1-x} = 1 + x + x^2 + \ldots + x^n + \ldots, \quad \text{if } |x| < 1 .$$

For then,

$$2 = \frac{1}{1-\frac{1}{2}} = 1 + \frac{1}{2} + \left(\frac{1}{2}\right)^2 + \left(\frac{1}{2}\right)^3 + \ldots + \left(\frac{1}{2}\right)^n + \ldots .$$

16-2 Sequences.

OBJECTIVE A: Given a defining rule for the sequence $\{a_n\}$, write the first few items of the sequence.

1. A sequence is a ____________ whose domain is the set of ____________________ .

2. The numbers in the range of a sequence are called the _________ of the sequence. The number a_n is called the ____________ of the sequence, or the term with _________ n.

3. For the sequence whose defining rule is $a_n = 2 + \frac{1}{n}$, the first four terms are

   $a_1 = 3$, $a_2 =$ _____ , $a_3 =$ _____ , and $a_4 =$ _____ .

---

1. function, positive integers 2. terms, nth term, index 3. $\frac{5}{2}$, $\frac{7}{3}$, $\frac{9}{4}$

4. For the sequence whose defining rule is $a_n = \frac{4^n}{n!}$, the first four terms are

$a_1 = 4$, $a_2 =$ ___ , $a_3 =$ ______ , and $a_4 =$ ______ .

5. For the sequence whose defining rule is $a_n = (-1)^{n+1}\left(\frac{n+1}{n^3}\right)$, the first four terms are

$a_1 = 2$, $a_2 =$ _____ , $a_3 =$ ______ , and $a_4 =$ ______ .

OBJECTIVE B: Given a sequence $\{a_n\}$ determine if it converges or diverges. Find the limit if the sequence does converge.

6. The sequence $\{a_n\}$ converges to the number L if to every ________________ ε there corresponds an _________ N such that

___________________________________ .

If no such limit L exists, we say that a_n ____________ .

7. If $0 < b < 1$, then $\{b^n\}$ converges to 0 . To see why, consider the inequality

$$|b^n - 0| = b^n < \varepsilon .$$

Thus, we seek an integer N such that

__________ for all $n > N$.

Since the natural logarithm $y = \ln x$ is an increasing function for all x ,

$b^n < \varepsilon$ is equivalent to $n \ln b < \ln \varepsilon$.

Also, because $\ln b$ is a negative number for $0 < b < 1$, this latter inequality is equivalent to

$n >$ ______________ .

Therefore, we need only choose an integer N satisfying

__________________ ,

and the criterion set forth in Problem 6 for convergence to 0 is satisfied.

---

4. 8, 32/3, 32/3

5. -3/8, 4/27, -5/64

6. positive number, index, $|a_n - L| < \varepsilon$ for all $n > N$, diverges

7. $b^n < \varepsilon$, $\ln \varepsilon/\ln b$, $N > \ln \varepsilon/\ln b$

8. Consider the sequence defined by $a_n = \frac{4^n}{n!}$. Thus,

$$a_n = \frac{4 \cdot 4 \cdot 4 \cdot 4 \cdots 4}{1 \cdot 2 \cdot 3 \cdot 4 \cdots n} = \_\_\_ \left(\frac{4 \cdot 4 \cdots 4}{5 \cdot 6 \cdots n}\right) \quad \text{if} \quad n > 5$$

$$\leq \frac{32}{3}\left(\frac{4}{5}\right)^{\_\_\_\_} = \left(\frac{32}{3}\right)\left(\frac{5}{4}\right)^4\left(\frac{4}{5}\right)^n$$

Since $0 \leq a_n$ for all n, the Sandwich Theorem 2 on page 746 of the Thomas/Finney text gives $a_n \to$ ___ because $(4/5)^n \to 0$ from Problem 7.

9. For the sequence $\left\{\frac{n^3+5}{n^2-1}\right\}$, $\lim_{n\to\infty} \frac{n^3+5}{n^2-1} = \lim_{n\to\infty} \frac{n + (5/n^2)}{1 - (1/n^2)} =$ ___ .

Therefore, the sequence ____________ .

10. Let $a_n = \left(1 - \frac{1}{2^2}\right)\left(1 - \frac{1}{3^2}\right)\left(1 - \frac{1}{4^2}\right) \cdots \left(1 - \frac{1}{n^2}\right)$.

Thus, $\ln a_n = \sum_{k=2}^{n} \ln\left(1 - \frac{1}{k^2}\right) = \sum_{k=2}^{n} \ln\left(\frac{k^2-1}{k^2}\right)$

$$= \sum_{k=2}^{n} \left[\ln(k+1) + \ln(k-1) - \_\_\_\_\_\_\right]$$

$$\begin{aligned} = -\ln 2 &+ \left[\ln 3 + \ln 1 - \ln 2\right] \\ &+ \left[\ln 4 + \ln 2 - 2\ln 3\right] \\ &+ \left[\_\_\_\_\_\_\_\_\_\_\right] \\ &+ \left[\ln 6 + \ln 4 - 2\ln 5\right] \\ &\ \ \vdots \\ &+ \left[\ln n + \ln(n-2) - 2\ln(n-1)\right] \\ &+ \left[\ln(n+1) + \ln(n-1) - 2\ln n\right] \end{aligned}$$

---

8. $\frac{32}{3}$, $n-4$, 0

9. ∞, diverges

10. $2\ln k$, $\ln 5 + \ln 3 - 2\ln 4$, $\ln(n+1) - \ln n$, $\ln\frac{1}{2}$, $\frac{1}{2}$

Now, look closely at the expanded sum on the right. Notice that two terms contain $+\ln 3$ and one term contains $-2\ln 3$, two terms contain $+\ln 4$ and one term contains $-2\ln 4$, and so forth. Thus, most of the terms cancel each other out and we are left with,

$$\ln a_n = -\ln 2 + \underline{\qquad\qquad\qquad} .$$

Therefore,

$$\lim_{n\to\infty} \ln a_n = \ln\frac{1}{2} + \lim_{n\to\infty} \ln\left(\frac{n+1}{n}\right) = \underline{\qquad\qquad} .$$

By applying Theorem 3 on page 747 in the text to $f(x) = e^x$, we have

$$a_n = e^{\ln a_n} \to e^{\ln(1/2)} = \underline{\qquad} \quad \text{as} \quad n\to\infty .$$

11. For the sequence defined by $a_n = (-1)^{n+1}\dfrac{n+1}{n^3}$,

$$0 \le |a_n| = \underline{\qquad} = \frac{1}{n^2} + \underline{\qquad} .$$

Therefore,

$$(-1)^{n+1}\frac{n+1}{n^3} \to \underline{\qquad} \quad \text{as} \quad n\to\infty ,$$

by the sequence version of the Sandwich Theorem.

16-3 Limits That Arise Frequently.

OBJECTIVE: Given a sequence $\{a_n\}$, determine if it converges or diverges. If it converges, use the limits calculated in this article of the text, or logarithms or l'Hôpital's rule, to find its limit.

12. Let $a_n = \left(\dfrac{3n-1}{5n+1}\right)^n$. Then,

$$\frac{3n-1}{5n+1} < \frac{3n}{5n+1} < \underline{\qquad} \quad \text{implies} \quad 0 \le \left(\frac{3n-1}{5n+1}\right)^n < \underline{\qquad} .$$

Therefore $a_n \to \underline{\qquad}$ because $\left(\frac{3}{5}\right)^n \to 0$.

13. Let $a_n = \left(\dfrac{3n+1}{5n-1}\right)^{1/n}$. Then,

$$\ln a_n = \frac{1}{n}\ln(3n+1) - \underline{\qquad\qquad\qquad} .$$

---

11. $\dfrac{n+1}{n^3}$, $\dfrac{1}{n^3}$, 0

12. $\dfrac{3}{5}$, $\left(\dfrac{3}{5}\right)^n$, 0

By l'Hôpital's rule,

$$\lim_{n\to\infty} \frac{\ln(3n+1)}{n} = \lim_{n\to\infty} \frac{\rule{5em}{0.4pt}}{1} = \underline{\quad}, \quad \text{and}$$

$$\lim_{n\to\infty} \frac{\ln(5n-1)}{n} = \lim_{n\to\infty} \frac{\rule{5em}{0.4pt}}{1} = \underline{\quad}.$$

Then, $\ln a_n \to 0$ so that $a_n = e^{\ln a_n} \to$ ________

by Theorem 3 of Article 16-2, page 747, in the text.

14. Consider the sequence defined by $a_n = \left(1 + e^{-n}\right)^n$. Then,

$$\ln a_n = \underline{\qquad\qquad\qquad},$$

and by l'Hôpital's rule,

$$\lim_{n\to\infty} \frac{\ln(1+e^{-n})}{1/n} = \lim_{n\to\infty} \frac{\rule{6em}{0.4pt}}{-1/n^2} \qquad (0/0)$$

$$= \lim_{n\to\infty} \frac{n^2}{e^n + 1} \qquad (\infty/\infty)$$

$$= \lim_{n\to\infty} \frac{\rule{2em}{0.4pt}}{\rule{2em}{0.4pt}} \qquad (\text{still } \infty/\infty)$$

$$= \lim_{n\to\infty} \frac{2}{e^n} = \underline{\quad}.$$

Therefore,

$$a_n = e^{\ln a_n} \to \underline{\qquad\quad}.$$

15. Let $a_n = \sqrt[n]{n^3}$. Then, $a_n = \left(n^3\right)^{1/n} = n^{\rule{1em}{0.4pt}} = \left(\sqrt[n]{n}\right)^{\rule{1em}{0.4pt}}$.

Now, $\sqrt[n]{n} \to$ ___ by limit two in the text (see page 749),

and if $f(x) = x^3$, then

$$a_n = f\left(\sqrt[n]{n}\right) \to \underline{\qquad\quad}$$

by Theorem 3 of Article 16-2, page 747, in the text.

---

13. $\frac{1}{n}\ln(5n-1)$, $3/(3n+1)$, 0, $5/(5n-1)$, 0, $e^0 = 1$

14. $n\ln\left(1+e^{-n}\right)$, $\left(\frac{1}{1+e^{-n}}\right)\left(-e^{-n}\right)$, $\frac{2n}{e^n}$, 0, $e^0 = 1$

15. $3/n$, 3, 1, $1^3 = 1$

16-4 Infinite Series.

16. If $\{a_n\}$ is a sequence, and

$$s_n = a_1 + a_2 + \dots + a_n ,$$

then the sequence $\{s_n\}$ is called an ________________ .

17. The number s_n is called the ________________ of the series.

18. Instead of $\{s_n\}$ we usually use the notation ________ for the series.

19. The series $\sum_{n=1}^{\infty} a_n$ is said to converge if the sequence _____ converges to a finite limit L. In that case we write ______________ or $a_1 + a_2 + \dots + a_n + \dots = L$. If no such limit exists, the series is said to ________ .

OBJECTIVE A: For a given geometric series $\sum_{n=0}^{\infty} ar^n$, determine if the series converges or diverges. If it does converge, then compute the sum of the series. The indexing of the series may be changed for a given problem.

20. Consider the series $\sum_{n=0}^{\infty} \frac{3}{5^n}$. This is a geometric series with $a =$ ___ and $r =$ _____ . Since $|r| < 1$, the geometric series __________ , and its sum is given by

$$\sum_{n=0}^{\infty} \frac{3}{5^n} = \text{\_\_\_\_\_\_} = \text{\_\_\_} .$$

---

16. infinite series

17. nth partial sum

18. $\sum_{n=1}^{\infty} a_n$

19. $\{s_n\}$, $\sum_{n=1}^{\infty} a_n = L$, diverge

20. 3, $\frac{1}{5}$, converges, $\frac{3}{1 - \frac{1}{5}}$, 15/4

21. The series $\sum_{n=2}^{\infty} (-1)^n \frac{4}{3^n}$ is a geometric series with $a =$ ___ and

$r =$ ___ . Since $|r| < 1$, the geometric series ________ . However,

the index begins with $n = 2$ instead of $n = 0$. Now,

$$\sum_{n=2}^{\infty} (-1)^n \frac{4}{3^n} = \sum_{n=0}^{\infty} (-1)^n \frac{4}{3^n} - \left(\_\_\_\_\_\_\right)$$

$$= \_\_\_\_\_\_ - \frac{8}{3} = \_\_\_ .$$

22. The series $\sum_{n=3}^{\infty} \frac{2^n}{7}$ is a geometric series with $a =$ ___ and $r =$ ___ .

Since ________ the series diverges.

23. The repeating decimal 0.15 15 15 ... is a geometric series in disguise. It can be written as

$$0.15\ 15\ 15 \ldots = \frac{15}{100} + \frac{15}{\_\_\_} + \frac{15}{\_\_\_} + \ldots$$

$$= \sum_{n=1}^{\infty} \_\_\_\_ = \sum_{n=0}^{\infty} \frac{15}{100^n} - \_\_\_$$

$$= \_\_\_\_\_\_ - 15 = \frac{\_\_\_\_\_\_}{99} .$$

24. Sometimes the terms of a given series are a sum or difference of terms, each of which belongs to a geometric series. For example,

$$\sum_{n=0}^{\infty} \left(\frac{7}{3^n} - \frac{1}{2^n}\right) = \sum_{n=0}^{\infty} \frac{7}{3^n} - \sum_{n=0}^{\infty} \frac{1}{2^n}$$

$$= \_\_\_\_ - \_\_\_\_ = \_\_\_ - \frac{4}{2} = \_\_\_ .$$

---

21. 4, $-\frac{1}{3}$, converges, $4 - \frac{4}{3}$, $\frac{4}{1 + \frac{1}{3}}$, $\frac{1}{3}$

22. $\frac{1}{7}$, 2, $|r| > 1$

23. 100^2, 100^3, $\frac{15}{100^n}$, 15, $\frac{15}{1 - \frac{1}{100}}$, $1500 - 1485$ or 15

24. $\frac{7}{1 - \frac{1}{3}}$, $\frac{1}{1 - \frac{1}{2}}$, $\frac{21}{2}$, $\frac{17}{2}$

OBJECTIVE B: Given an elementary series $\sum_{n=1}^{\infty} a_n$, determine whether it converges or diverges. If it converges, find the sum.

25. Consider the series $\sum_{n=1}^{\infty} \frac{1}{(2n-1)(2n+1)}$. This is not a geometric series.

However, we can use partial fractions to re-write the kth term:

$$\frac{1}{(2k-1)(2k+1)} = \frac{1}{2}\left[\underline{\qquad} - \frac{1}{2k+1}\right].$$

This permits us to write the partial sum

$$\sum_{n=1}^{k} \frac{1}{(2n-1)(2n+1)} = \frac{1}{1\cdot 3} + \frac{1}{3\cdot 5} + \cdots + \frac{1}{(2k-1)(2k+1)}$$

as

$$s_k = \frac{1}{2}\left(\frac{1}{1} - \frac{1}{3}\right) + \frac{1}{2}\left(\underline{\qquad}\right) + \frac{1}{2}\left(\underline{\qquad}\right) + \cdots + \frac{1}{2}\left(\frac{1}{2k-1} - \frac{1}{2k+1}\right).$$

By removing parentheses on the right, and combining terms, we find that

$$s_k = \underline{\qquad\qquad}.$$

Therefore, $s_k \to \underline{\quad}$ and the series ________ converge. Hence,

$$\sum_{n=1}^{\infty} \frac{1}{(2n-1)(2n+1)} = \underline{\quad}.$$

26. For the series $\sum_{n=1}^{\infty} \frac{n^n}{n!}$, we have for every index n,

$$a_n = \frac{n^n}{n!} = \frac{n\cdot n\cdot n\cdots n}{1\cdot 2\cdot 3\cdots n} = \left(\frac{n}{1}\right)\left(\frac{n}{2}\right)\left(\underline{\quad}\right)\cdots\left(\frac{n}{n}\right) > \underline{\quad}.$$

Therefore, $\lim_{n\to\infty} a_n \neq 0$. We conclude that the series ____________.

---

25. $\frac{1}{2k-1}$, $\frac{1}{3} - \frac{1}{5}$, $\frac{1}{5} - \frac{1}{7}$, $\frac{1}{2}\left(1 - \frac{1}{2k+1}\right)$, $\frac{1}{2}$, does, $\frac{1}{2}$

26. $\frac{n}{3}$, 1, diverges

27. Consider the series $\sum_{n=1}^{\infty} (-1)^{n+1}\left(1 + \frac{1}{3^n}\right)$. For large values of the index n , the absolute value of the nth term,

$$|a_n| = 1 + 3^{-n}$$

is close to ___ . Therefore, the limit $\lim_{n\to\infty} (-1)^{n+1}\left(1 + \frac{1}{3^n}\right)$ __________ exist. We conclude that the series __________ .

28. For the series $\sum_{n=1}^{\infty} \left(\frac{n+2}{n}\right)^n$, we have

$\lim_{n\to\infty} \left(\frac{n+2}{n}\right)^n = \lim_{n\to\infty} \left(1 + \frac{2}{n}\right)^n =$ ____ . Thus, the series __________ because the limit of the nth term is not zero.

29. The kth partial sum of the series $\sum_{n=1}^{\infty} \left[n^3 - (n+1)^3\right]$ is

$s_k = (1 - 2^3) +$ ______________________________ .

By removing parentheses on the right, and combining terms, we find that $s_k =$ ______________ . Therefore, the number s_k is less than or equal to $-k^3$ at each stage, so we conclude that the series __________ .

16-5 Tests for Convergence of Series with Nonnegative Terms.

OBJECTIVE: Use one of the five tests for divergence and convergence of infinite series, listed on page 776 of the Thomas/Finney text, to determine whether a given series converges or diverges.

30. $\sum_{n=1}^{\infty} \frac{n+5}{n^2 - 3n + 5}$

For every index n , $5n > -3n + 5$ because n is a positive integer.

---

27. 1, does not, diverges

28. e^2, diverges

29. $(2^3 - 3^3) + \ldots + \left[k^3 - (k+1)^3\right]$, $1 - (k+1)^3$, diverges

Then, $n^2 + 5n >$ ____________ , and since $n^2 - 3n + 5$ is positive for every index it follows that

$$\frac{n+5}{n^2 - 3n + 5} > \underline{\quad} .$$

We conclude that the given series ____________ by comparison with the series $\sum \frac{1}{n}$.

31. $\sum_{n=2}^{\infty} \frac{1}{n(\ln n)^2}$

We apply the integral test: $\int_2^{\infty} \frac{dx}{x(\ln x)^2} = \lim_{b \to \infty} \underline{\qquad} \Big]_b^2 = \underline{\qquad}$.

Therefore, the integral ____________ and hence the series ____________ .

32. $\sum_{n=1}^{\infty} \frac{(\ln n)^2}{n^3}$

The following argument shows that $\ln n < \sqrt{n}$ for every index n . First, define the function $g(x) = \sqrt{x} - \ln x$. Now $g'(x) =$ ____________ is nonnegative if $x \geq$ ___ , and it follows that g is an increasing function of x for $x \geq 4$. Also, $g(4) = 2 - \ln 4 \approx 0.613$ is positive. Therefore, $g(x) > 0$ for $x \geq 4$. A simple verification using tables, or a calculator, shows that $g(1)$, $g(2)$, and $g(3)$ are positive. Hence, we have established that $\ln n < \sqrt{n}$ for every positive integer n . Using this fact,

$$\frac{(\ln n)^2}{n^3} < \frac{(\sqrt{n})^2}{n^3} = \underline{\quad} .$$

We conclude that the series $\sum_{n=1}^{\infty} \frac{(\ln n)^2}{n^3}$ ____________ .

33. $\sum_{n=1}^{\infty} \frac{n!}{3^n}$

We try the ratio test. Thus, $\frac{a_{n+1}}{a_n} = \frac{(n+1)!/3^{n+1}}{n!/3^n} =$ ____________ .

Hence, $\rho = \lim_{n \to \infty} \frac{a_{n+1}}{a_n} =$ ____ , and the series ____________ .

---

30. $n^2 - 3n + 5$, $\frac{1}{n}$, diverges

31. $-\frac{1}{\ln x}$, $\frac{1}{\ln 2}$, converges, converges

32. $\frac{1}{2\sqrt{x}} - \frac{1}{x}$, 4, $\frac{1}{n^2}$, converges

33. $\frac{1}{3} \cdot (n+1)$, $+\infty$, diverges

34. $\displaystyle\sum_{n=1}^{\infty} \left(\sqrt[n]{n} - 1\right)^n$

We try the root test. Thus, for $a_n = \left(\sqrt[n]{n} - 1\right)^n$,

$\sqrt[n]{a_n}$ = ____________ → ___ .

Because ρ = ___ we conclude that the series ____________ according to the root test.

OBJECTIVE B: Given a series whose nth term is an elementary expression containing some power of the variable x , find all values of x for which the series will converge. Begin with the ratio test or the root test, and then apply other tests as needed.

35. $\displaystyle\sum_{n=1}^{\infty} \frac{|x|^n}{\sqrt{n}\ 3^n}$

The nth term of the series is $a_n = \dfrac{|x|^n}{\sqrt{n}\ 3^n}$. If we apply the root test,

$\sqrt[n]{a_n} = \dfrac{|x|}{\sqrt[n]{\sqrt{n}} \cdot 3}$ → ______ .

The root test therefore tells us that the series converges if $|x|$ is less than ___ . We don't know what happens when $|x| = 3$. However, in that case the series becomes

$$\sum_{n=1}^{\infty} \frac{3^n}{\sqrt{n}\ 3^n} = \sum_{n=1}^{\infty} \frac{1}{n^{1/2}} .$$

This is a p-series with p = ___ , and therefore the series ____________ . Thus, the original series

$\displaystyle\sum_{n=1}^{\infty} \frac{|x|^n}{\sqrt{n}\ 3^n}$, converges for all values of x satisfying __________ .

OBJECTIVE C: Discuss the cardinal principle governing increasing sequences and how it applies to infinite series of nonnegative terms.

36. An increasing sequence $\{s_n\}$ is a sequence with the property that __________ for every n .

---

34. $\sqrt[n]{n} - 1$, 0, 0, converges

35. $\dfrac{|x|}{3}$, 3, $\dfrac{1}{2}$, diverges, $|x| < 3$

36. $s_n \leq s_{n+1}$

37. A sequence $\{s_n\}$ is said to be __________ from above if there is a finite constant M such that $s_n \leq M$ for every n.

38. If $\{s_n\}$ is an increasing sequence that is bounded from above, then it __________.

39. If an increasing sequence $\{s_n\}$ fails to be bounded, then it __________ __________.

40. Applying the principle given in Problems 38 and 39 (see Theorem 1, page 764 of the text), if $\sum a_n$ is a series of nonnegative terms, then it converges if and only if the sequence of partial sums $\{s_n\}$ satisfies what property?

16-6 Absolute Convergence.

41. A series $\sum_{n=1}^{\infty} a_n$ is said to converge absolutely if __________.

42. True or False:
(a) If a series converges absolutely, then it converges.
(b) If a series converges, then it converges absolutely.

OBJECTIVE: Given an infinite series containing (possibly) negative terms, determine whether it converges absolutely. Use the tests for convergence and divergence listed on page 776 of the text.

43. $\sum_{n=1}^{\infty} \frac{(-1)^{n+1} n \ln n}{3^n}$

The absolute value of the nth term of the series is $|a_n| = \frac{n \ln n}{3^n}$.

Applying the ratio test, $\frac{|a_{n+1}|}{|a_n|} = \frac{(n+1)\ln(n+1)/3^{n+1}}{n \ln n/3^n} =$ __________.

---

37. bounded

38. converges

39. diverges to plus infinity

40. having an upper bound, as in Problem 37

41. $\sum_{n=1}^{\infty} |a_n|$ converges

42. (a) True (b) False

43. $\frac{1}{3}\left(\frac{n+1}{n}\right)\frac{\ln(n+1)}{\ln n}$, $\frac{1/(n+1)}{1/n}$, 1, $\frac{1}{3}$, does

By l'Hôpital's rule,

$\lim_{n \to \infty} \frac{\ln (n+1)}{\ln n} = \lim_{n \to \infty}$ ____________ = ___ .

Therefore,

$\lim_{n \to \infty} \frac{|a_{n+1}|}{|a_n|} =$ ___ , so we conclude that $\sum_{n=1}^{\infty} \frac{(-1)^{n+1} n \ln n}{3^n}$

________ converge absolutely.

44. $\sum_{n=1}^{\infty} \frac{2n - n^2}{n^3}$

The nth term of the series is $a_n = \frac{2n - n^2}{n^3} = \frac{n(2 - n)}{n^3}$ which is negative if $n >$ ___ . Thus $-a_n = \frac{n - 2}{n^2}$ is positive for $n > 2$. Now, $\frac{n - 2}{n^2} > \frac{1}{2n}$ whenever $n >$ ___ . Since the series $\sum_{n=1}^{\infty} \frac{1}{2n}$ ____________ ,

we conclude that the series $\sum_{n=1}^{\infty} |a_n| = \sum_{n=1}^{\infty} \frac{n - 2}{n^2}$ ____________ by the comparison test. Therefore, the original series __________ converge absolutely.

16-7 Alternating Series, Conditional Convergence.

OBJECTIVE A: Given an alternating series, determine if it converges or diverges. A test for convergence is Leibniz's Theorem on page 781 of the text, but another test may be required.

45. $\sum_{n=1}^{\infty} (-1)^{n+1} \frac{n}{n^2 + 1}$

First we see that $a_n = \frac{n}{n^2 + 1}$ is positive for every n . Also, $\lim_{n \to \infty} a_n =$ ___ . Next, we compare a_{n+1} with a_n for arbitrary n . Now,

$$\frac{n+1}{(n+1)^2 + 1} \le \frac{n}{n^2 + 1} \text{ if and only if } (n+1)(n^2+1) \le n\left[(n+1)^2 + 1\right] .$$

This last inequality is equivalent to $n^3 + n^2 + n + 1 \le$ ______________ or, $1 \le n^2 + n$ which is true. We conclude that the alternating series ________ converge by Leibniz's Theorem.

---

44. 2, 4, diverges, diverges, does not   45. 0, $n^3 + 2n^2 + 2n$, does

46. $\sum_{n=1}^{\infty} (-1)^{n+1} \frac{1}{n^{1+1/n}}$

It is clear that $a_n = \frac{1}{n^{1+1/n}} = \frac{1}{n \cdot \sqrt[n]{n}}$ is positive for every n.

Also, $\lim_{n\to\infty} a_n = \lim_{n\to\infty} \frac{1}{n} \cdot \lim_{n\to\infty} \frac{1}{n \cdot \sqrt[n]{n}} =$ ______ .

We would like to show that $\{a_n\}$ is a ____________ sequence. One way is to replace n by the continuous variable x and show that the resultant function

$$y = f(x) = x^{1+1/x}, \quad x > 0,$$

which is the reciprocal of a_n for $x = n$, is an increasing function of x for every x. If we take the logarithm of both sides of this last equation, and differentiate implicitly with respect to x, we obtain

$\ln y = \left(1 + \frac{1}{x}\right) \ln x$, and $\frac{y'}{y} = \frac{1}{x^2}$(__________) $+ \frac{1}{x}$.

Simplifying algebraically, $y' = \frac{y}{x^2}$(1 + ___ − ln x) .

Thus, since $x > \ln x$ for all $x > 0$, we find that y' is positive, and $y = f(x)$ is increasing for all x (so the reciprocal is ____________). Therefore we have established that

$$\frac{1}{(n+1) \cdot \sqrt[(n+1)]{n+1}} < \frac{1}{n \cdot \sqrt[n]{n}} \quad \text{for every index } n.$$

We conclude that the alternating series ____________ .

OBJECTIVE B: Given an infinite series, use the tests studied in this chapter of the text to determine if the series is absolutely convergent, conditionally convergent, or divergent.

47. If a series $\sum a_n$ converges, but the series of absolute values $\sum |a_n|$ diverges, we say that the original series $\sum a_n$ is ________________ .

48. In Problem 46 above, the series $\sum_{n=1}^{\infty} (-1)^{n+1} \frac{1}{n^{1+1/n}}$ was shown to be convergent. We want to know if the series converges absolutely.

---

46. $0 \cdot 1 = 0$, decreasing, $1 - \ln x$, x, decreasing, converges

47. conditionally convergent

Using the same technique as in Problem 46, it is easy to establish that

$$\sqrt[n+1]{n+1} < \sqrt[n]{n} \quad \text{if } n \geq 3:$$

we define the function $y = x^{1/x}$, $x \geq 3$, and show that y' is always negative; whence we conclude that y is a ______________ function of x. In particular,

$$\sqrt[n]{n} < \sqrt[3]{3} \approx 1.44 < \frac{3}{2} \quad \text{if } n \geq \_\_\_ .$$

It follows that

$$\frac{1}{n \cdot \sqrt[n]{n}} > \_\_\_\_ \quad \text{if } n \geq 4 .$$

Since the harmonic series $\sum (1/n)$ diverges, we find that the series

$$\sum_{n=1}^{\infty} \frac{1}{n^{1+1/n}}$$ ____________ by the comparison test. Therefore, the original series $$\sum_{n=1}^{\infty} (-1)^{n+1} \frac{1}{n^{1+1/n}}$$ is ______________________________.

49. $$\sum_{n=1}^{\infty} (-1)^{n+1} \left(\frac{n+1}{n}\right)^n$$

In this case, $a_n = \left(\frac{n+1}{n}\right)^n = \left(1 + \frac{1}{n}\right)^n \to \_\_\_$.

Therefore the series ____________ .

50. $$\sum_{n=2}^{\infty} \frac{\cos n\pi}{n\sqrt{\ln n}}$$

Since $\cos n\pi$ is 1 when n is even, and -1 when n is odd, the series is alternating in sign. Let us see if the series converges absolutely. Now,

$$|a_n| = \left|\frac{\cos n\pi}{n\sqrt{\ln n}}\right| = \_\_\_\_\_\_\_\_\_\_\_\_ .$$ Applying the integral test,

$$\int_1^{\infty} \frac{dx}{x\sqrt{\ln x}} = \lim_{b \to \infty} \_\_\_\_\_\_\_\_\_\_\_\_ \Big]_1^b = \_\_\_ .$$

---

48. decreasing, 4, $\frac{2}{3n}$, diverges, conditionally convergent 49. e, diverges

50. $\frac{1}{n\sqrt{\ln n}}$, $2\sqrt{\ln x}$, $+\infty$, diverges, 0, increasing, conditionally convergent

Therefore, the series $\sum |a_n|$ ____________ . To see if the original series converges we check the three conditions of Leibniz's Theorem: (remember that the numerator $\cos n\pi$ simply determines the sign of the nth term of the series)

$\frac{1}{n\sqrt{\ln n}}$ is positive for all n, and converges to ____ . Also, it is clear that $\frac{1}{(n+1)\sqrt{\ln(n+1)}} < \frac{1}{n\sqrt{\ln n}}$ because $y = \ln x$ is an ______________ function of x. Therefore, the series

$$\sum_{n=1}^{\infty} \frac{\cos n\pi}{n\sqrt{\ln n}} \text{ is } \underline{\hspace{6cm}} .$$

OBJECTIVE C: Use the Alternating Series Estimation Theorem to estimate the magnitude of the error if the first k terms, for some specified number k, are used to approximate a given alternating series.

51. It can be shown, with a little work, that the alternating harmonic series

$$\sum_{n=1}^{\infty} \frac{(-1)^{n+1}}{n}$$

converges to $\ln 2$. If we wish to approximate $\ln 2$ correct to four decimal places using this series, the alternating series error estimation gives

$$\underline{\quad} < 0.5 \times 10^{-5} , \quad \text{or} \quad n > \underline{\hspace{3cm}} .$$

Therefore, we would need to sum the first 200,000 terms of the alternating harmonic series to ensure four decimal place accuracy in approximating $\ln 2$. This does not mean that fewer terms would not provide that accuracy. A more efficient approximation for $\ln 2$, accurate to four decimal places, uses Simpson's rule with $n = 6$ to estimate

$$\int_1^2 \frac{dx}{x}$$

(see Chapter 4, Test Problem 11 in this manual).

---

51. $\frac{1}{n}$, 2×10^5

16-8 Power Series for Functions.

OBJECTIVE: Find the Taylor series at $x = a$, or the Maclaurin series, for a given function $y = f(x)$. Assume that $x = a$ is specified and that f has finite derivatives of all orders at $x = a$.

52. A series of the form $\sum_{n=0}^{\infty} a_n x^n$ is called a ______________ .

53. If $y = f(x)$ has finite derivatives of all orders at $x = a$, the particular power series

$$f(a) + f'(a)(x-a) + \frac{f''(a)}{2!}(x-a)^2 + \dots + \frac{f^{(n)}(a)}{n!}(x-a)^n + \dots$$

is called the ______________________________ . If $a = 0$, the series is known as the ____________________ for f . The Taylor series for a function may or may not converge to the function. This problem is investigated in the next article.

54. If $y = f(x)$ has finite derivatives of order up to and including n , then the polynomial

$$f_n(x) = f(a) + f'(a)(x-a) + \frac{f''(a)}{2!}(x-a)^2 + \dots + \frac{f^{(n)}(a)}{n!}(x-a)^n$$

is called the nth-degree ____________________________________ . The graph of this polynomial passes through the point __________ , and its first n derivatives match the first n derivatives of ________ at ______ . Each nonnegative integer n corresponds to a Taylor polynomial for f at $x = a$, provided the first n derivatives of f exist at $x = a$.

55. Let us find the Taylor polynomials $f_3(x)$ and $f_4(x)$ for the function $f(x) = a^x$, $a > 0$, at $x = 1$. To do this we need to complete the following table:

n	$f^{(n)}(x)$	$f^{(n)}(1)$
0	a^x	a
1	$a^x \ln a$	$a \ln a$
2	__________	__________
3	__________	__________
4	__________	__________

---

52. power series

53. Taylor series for f at $x = a$, Maclaurin series

54. Taylor polynomial of f at $x = a$, $(a, f(a))$, $y = f(x)$, $x = a$

Then,

$$f_3(x) = a + a(\ln a)(x-1) + \frac{a(\ln a)^2}{2!}(x-1)^2 + \_\_\_\_\_\_\_\_\_\_,$$

$$f_4(x) = \_\_\_\_\_\_\_\_\_\_\_\_\_\_\_\_\_\_\_\_ .$$

56. For the function $f(x) = a^x$ in Problem 55, the Taylor polynomial at $x = 1$ is

$$\sum_{n=0}^{\infty} \_\_\_\_\_\_\_\_\_\_ .$$

57. Let us find the Maclaurin series for the function $f(x) = x^5 + 4x^4 + 3x^3 + 2x + 1$. We need to find the derivatives of f of all orders, and evaluate them at $x = 0$:

$f'(x) = \_\_\_\_\_\_\_\_$, $f'(0) = 2$

$f^{(2)}(x) = \_\_\_\_\_\_\_\_$, $f^{(2)}(0) = \_\_\_$

$f^{(3)}(x) = \_\_\_\_\_\_\_\_$, $f^{(3)}(0) = \_\_\_$

$f^{(4)}(x) = 120x + 96$, $f^{(4)}(0) = 96$

$f^{(5)}(x) = \_\_\_$, $f^{(5)}(0) = \_\_\_$

$f^{(6)}(x) = 0$, $f^{(6)}(0) = 0$

In general, $f^{(k)}(0) = \_\_\_$ if $k \geq 6$. Thus, the Maclaurin series is

$\_\_\_\_\_\_\_\_\_\_\_\_\_\_\_\_\_\_\_\_$,

which simplifies to $1 + 2x + 3x^3 + 4x^4 + x^5$. Therefore, the Maclaurin series for a polynomial expressed in powers of x is the polynomial itself.

---

55. for $n = k$, $f^{(k)}(1) = a(\ln a)^k$; $\frac{a(\ln a)^3}{3!}(x-1)^3$,

$$a + a(\ln a)(x-1) + \frac{a(\ln a)^2}{2!}(x-1)^2 + \frac{a(\ln a)^3}{3!}(x-1)^3 + \frac{a(\ln a)^4}{4!}(x-1)^4$$

56. $\frac{a(\ln a)^n}{n!}(x-1)^n$

57. $5x^4 + 16x^3 + 9x^2 + 2$, $20x^3 + 48x^2 + 18x$, 0, $60x^2 + 96x + 18$, 18, 120, 120, 0,

$$1 + 2x + 0x^2 + \frac{18}{3!}x^3 + \frac{96}{4!}x^4 + \frac{120}{5!}x^5$$

58. Suppose we want to express the polynomial in Problem 57 in powers of $(x+1)$ instead of powers of x. We find the Taylor series of f at $x =$ ___ . From our previous calculations of the derivatives, we find that

$f(-1) = -1$, $f'(-1) = 0$, $f^{(2)}(-1) =$ ____ , $f^{(3)}(-1) =$ ____ ,
$f^{(4)}(-1) =$ ____ , $f^{(5)}(-1) =$ ____ , and $f^{(k)}(-1) =$ ___ if $k \geq 6$.
Thus, the Taylor series of f at $x = -1$ is

____ ,

which simplifies to $-1 + 5(x+1)^2 - 3(x+1)^3 - (x+1)^4 + (x+1)^5$.

16-9 Taylor's Theorem With Remainder: Sines, Cosines, and e^x .

59. The statement of Taylor's Theorem in the text gives the remainder term in integral form as

$R_n(x,a) =$ ________________________ .

This remainder term measures the error in the approximation of $y = f(x)$ by the nth-degree Taylor polynomial at ________ . Thus, the Taylor-series expansion for $f(x)$ will converge to $f(x)$ provided that

____________________ .

60. An alternate form for the remainder term, employing differentiation is known as Lagrange's Form of the remainder, and is given by

$R_n(x,a) =$ ____________________ ,

where the number c lies between ____________ . This form is very useful because often we can bound the derivative $f^{(n+1)}(c)$ by some constant M : $|f^{(n+1)}(c)| \leq M$. This ensures that $R_n(x,a) \to 0$.

---

58. -1, 10, -18, -24, 120, 0,

$$-1 + 0(x+1) + \frac{10}{2!}(x+1)^2 - \frac{18}{3!}(x+1)^3 - \frac{24}{4!}(x+1)^4 + \frac{120}{5!}(x+1)^5$$

59. $\displaystyle\int_a^x \frac{(x-t)^n}{n!} f^{(n+1)}(t)\,dt$, $x = a$, $\displaystyle\lim_{n\to\infty} R_n(x,a) = 0$

60. $f^{(n+1)}(c)\dfrac{(x-a)^{n+1}}{(n+1)!}$, a and x

OBJECTIVE A: Using the Maclaurin series for the functions e^x , $\sin x$, and $\cos x$, write the Maclaurin series for functions which are combinations of sines, cosines, exponentials, or powers of x .

61. The Maclaurin series for e^x , $\sin x$, and $\cos x$ are e^x = __________ ,

$\sin x$ = ______________ , and $\cos x$ = ______________ .

Each of these series converges to its respective function for every value of x .

62. Let us find the Maclaurin series for $\sin^3 x$. A trigonometric identity gives

$$\sin^3 x = \frac{1}{4}(3 \sin x - \sin 3x) .$$

We use Maclaurin series for the terms on the right side:

$$3 \sin x = 3x - \frac{3x^3}{3!} + \frac{3x^5}{5!} - \frac{3x^7}{7!} + \ldots ,$$

$\sin 3x$ = ______________________ ,

$$3 \sin x - \sin 3x = 4x^3 - 2x^5 + \frac{52}{5!}x^7 - \ldots$$

Therefore, $\sin^3 x$ = ______________

$$= \sum_{n=0}^{\infty} \frac{(-1)^n\left(3 - 3^{2n+1}\right)}{4(2n+1)!} x^{2n+1} .$$

OBJECTIVE B: Use the remainder Estimation Theorem, on page 795 of the Thomas/Finney text, to estimate the truncation error when a Taylor polynomial is used to approximate a given function. Assume that the function has derivatives of all orders.

63. We will calculate $\cos \sqrt{2}$ with an error less than 10^{-6} . By Taylor's Theorem,

$\cos \sqrt{2}$ = ______________________ + $R_{2k}(x,0)$

---

61. $\sum_{n=0}^{\infty} \frac{x^n}{n!}$, $\sum_{n=0}^{\infty} \frac{(-1)^n x^{2n+1}}{(2n+1)!}$, $\sum_{n=0}^{\infty} \frac{(-1)^n x^{2n}}{(2n)!}$

62. $3x - \frac{(3x)^3}{3!} + \frac{(3x)^5}{5!} - \frac{(3x)^7}{7!} + \ldots$, $x^3 - \frac{1}{2}x^5 + \frac{13}{5!}x^7 - \ldots$

The Remainder Estimation Theorem, with $M =$ ___ and $x =$ ____ , gives $|R_{2k}| \leq 1 \cdot$ ____________ . By trial we find that

$$\frac{(\sqrt{2})^{11}}{11!} = 0.0000011337 > 10^{-6} \quad \text{and} \quad \frac{(\sqrt{2})^{13}}{13!} = 0.0000000145 < 10^{-6}.$$

Thus, we should take $(2k+1)$ to be at least ____ , or k to be at least 6 . With an error less than 10^{-6} ,

$$\cos\sqrt{2} = 1 - \frac{2}{2!} + \frac{4}{4!} - \frac{8}{6!} + \ldots + \underline{\quad\quad}$$

(↑ last term)

$$\approx 0.155944 .$$

64. Let us determine for what values of x we can replace e^x by $1 + x + (x^2/2) + (x^3/3!)$ with an error of magnitude less than 5×10^{-5} . In Example 6, page 797 of the text, the Remainder Estimation Theorem gave

$$|R_4| \leq \underline{\qquad\qquad} ,$$

and we desire $|R_4| < 5 \times 10^{-5}$. This will be the case if

$$|x|^4 < \underline{\qquad} \quad \text{or ,} \quad |x| < \underline{\qquad\qquad} \approx 0.14142 .$$

For instance,

$$e^{0.1} = 1 + (0.1) + \frac{(0.1)^2}{2} + \frac{(0.1)^3}{6} \approx 1.10517$$

is correct to five decimal places.

16-10 Further Computations, Logarithms, Arctangents, and π.

OBJECTIVE: Use a suitable series to calculate a given quantity to three decimal places. Show that the remainder term does not exceed 5×10^{-4} . (Assume the quantity is the value of a function whose series expansion has been studied in this chapter of the text.)

65. To calculate $\ln 0.75$, we use the identity

$$\ln 0.75 = \ln\frac{3}{4} = \ln\frac{3}{2} - \underline{\qquad} .$$

---

63. $1 - \frac{2}{2!} + \frac{4}{4!} - \frac{8}{6!} + \ldots + (-1)^k \frac{2^k}{(2k)!}$, 1, $\sqrt{2}$, $\frac{(\sqrt{2})^{2k+1}}{(2k+1)!}$, 13, 6, $\frac{64}{12!}$

64. $3 \cdot \frac{|x|^4}{4!}$, 40×10^{-5}, $\sqrt[4]{0.0004}$

We can use the calculation for $\ln 2$ obtained on page 803 of the text: $\ln 2 \approx 0.69315$. To obtain the first term on the right side of the previous equation we use the series

$$\ln \frac{N+1}{N} = 2\left(\underline{\qquad\qquad\qquad\qquad\qquad\qquad} \right) \quad \text{with } N = 2 .$$

Then,

$$\ln \frac{3}{2} = 2\left(\frac{1}{5} + \frac{1}{3(5)^3} + \frac{1}{5(5)^5} + \frac{1}{7(5)^7} + \cdots\right)$$

$$\approx 2(0.2 + 0.00266667 + 0.000064 + 0.00000183 + \ldots)$$

$$\approx 0.40547 .$$

The error satisfies

$$|R_7(x,0)| \leq \frac{1}{8} \cdot \frac{|x|^8}{1 - |x|} , \quad \text{where} \quad x = \frac{1}{2N+1} = \underline{\quad} .$$

Thus,

$$|R_7(x,0)| \leq \frac{5}{32}\left|\frac{1}{5}\right|^8 < 5 \times 10^{-6} .$$

It follows that

$$\ln \frac{3}{4} \approx 0.40547 - 0.69315 = -0.28768 ,$$ accurate to five decimal places.

16-11 A Second Derivative Test for Maxima and Minima of Functions of Two Independent Variables.

OBJECTIVE: Use the second derivative test as stated in Problem 1(b), page 811 of the text, to test a given surface $z = f(x,y)$ for maxima, minima, and saddle points. Assume the function f and its first- and second-order partial derivatives are continuous throughout some neighborhood of each point $P(a,b)$ where $f_x(a,b) = f_y(a,b) = 0$.

66. Let $A = f_{xx}(a,b)$, $B = f_{xy}(a,b)$, and $C = f_{yy}(a,b)$. Suppose $f_x(a,b) = f_y(a,b) = 0$ and $A \neq 0$. Then at (a,b) the function $f(x,y)$ has

 (i) a relative minimum if __________________,

 (ii) a relative maximum if __________________,

 (iii) a saddle point if __________.

---

65. $\ln 2$, $\frac{1}{2N+1} + \frac{1}{3(2N+1)^3} + \frac{1}{5(2N+1)^5} + \cdots$, $\frac{1}{5}$

66. (i) $AC - B^2 > 0$ and $A > 0$,

 (ii) $AC - B^2 > 0$ and $A < 0$,

 (iii) $AC - B^2 < 0$

67. Consider the function $f(x,y) = x^2 - xy + y^3 - x$. We set

$\frac{\partial f}{\partial x} = 2x - y - 1 = 0$ and $\frac{\partial f}{\partial y} =$ ____________ $= 0$.

The first equation is equivalent to $y = 2x - 1$, and substitution of this into the second equation gives $-x + 3(2x - 1)^2 = 0$, or $12x^2 - 13x + 3 = 0$.

Thus, $x = \frac{1}{3}$ or $x = \frac{3}{4}$. It follows that the points ________ and ________ are critical points where f_x and f_y are simultaneously 0. Next we find, $f_{xx} =$ ___, $f_{xy} =$ ___, and $f_{yy} = 6y$.

Let us test the critical point $\left(\frac{1}{3}, -\frac{1}{3}\right)$. If

$A = f_{xx}\left(\frac{1}{3}, -\frac{1}{3}\right) =$ ___, $B = f_{xy}\left(\frac{1}{3}, -\frac{1}{3}\right) =$ ___, and

$C = f_{yy}\left(\frac{1}{3}, -\frac{1}{3}\right) =$ ___, then $AC - B^2 = -5$. We conclude that the surface has a ________________ at $\left(\frac{1}{3}, -\frac{1}{3}\right)$.

To test the point $\left(\frac{3}{4}, \frac{1}{2}\right)$, let

$A = f_{xx}\left(\frac{3}{4}, \frac{1}{2}\right) = 2$, $B = f_{xy}\left(\frac{3}{4}, \frac{1}{2}\right) =$ ___, $C = f_{yy}\left(\frac{3}{4}, \frac{1}{2}\right) = 3$.

Then, $AC - B^2 =$ ___ and $A > 0$. We conclude that the surface has a ____________________ at $\left(\frac{3}{4}, \frac{1}{2}\right)$.

68. For the function $f(x,y) = x^3 - 3x^2y^2 + y^4$, we find

$\frac{\partial f}{\partial x} =$ ______________ $= 3x($ __________ $)$ and

$\frac{\partial f}{\partial y} =$ ______________ $= 2y($ __________ $)$.

From $\frac{\partial f}{\partial x} = 0$ and $\frac{\partial f}{\partial y} = 0$, $x = 0$ if and only if $y = 0$ also. Hence ______ is one critical point. If $x \neq 0$, then $\frac{\partial f}{\partial x} = 3x(x - 2y^2) = 0$ demands $x = 2y^2$. Since $y \neq 0$, $2y(2y^2 - 3x^2) = 0$ gives $2y^2 - 3x^2 = 0$, and substitution of $x = 2y^2$ into this last equation gives __________ $= 0$

---

67. $-x + 3y^2$, $\left(\frac{1}{3}, -\frac{1}{3}\right)$, $\left(\frac{3}{4}, \frac{1}{2}\right)$, 2, -1, 2, -1, -2, saddle point, -1, 5, relative minimum

68. $3x^2 - 6xy^2$, $x - 2y^2$, $4y^3 - 6x^2y$, $2y^2 - 3x^2$, (0,0), $x - 3x^2$, $\frac{1}{3}$, $\left(\frac{1}{3}, -\frac{1}{\sqrt{6}}\right)$, $-12xy$, 1, $-\frac{4}{\sqrt{6}}$, $-\frac{4}{3}$, saddle point, $\frac{4}{3}$, $-\frac{4}{3}$, saddle point

or, $x(1-3x) = 0$. It follows that $x =$ ____ . From $x = 2y^2$ we find that $\left(\frac{1}{3}, \frac{1}{\sqrt{6}}\right)$ and __________ are also critical points of f . Next we test these critical points. First, we compute the second partial derivatives:

$f_{xx} = 6(x - y^2)$, $f_{xy} =$ ____ , and $f_{yy} = 6(2y^2 - x^2)$.

(a) The critical point $(0,0)$:

$A = f_{xx}(0,0) = 0$, $B = f_{xy}(0,0) = 0$, and $C = f_{yy}(0,0) = 0$.

Therefore, the second derivative test yields no information concerning the surface at the point $(0,0)$. However, a little further analysis shows that the surface has a saddle point at $(0,0)$.

(b) The critical point $\left(\frac{1}{3}, \frac{1}{\sqrt{6}}\right)$:

$A = f_{xx}\left(\frac{1}{3}, \frac{1}{\sqrt{6}}\right) =$ ____ , $B = f_{xy}\left(\frac{1}{3}, \frac{1}{\sqrt{6}}\right) =$ ____ , and

$C = f_{yy}\left(\frac{1}{3}, \frac{1}{\sqrt{6}}\right) = \frac{4}{3}$. Then, $AC - B^2 =$ ____ and we conclude that the surface has a ______________ at $\left(\frac{1}{3}, \frac{1}{\sqrt{6}}\right)$.

(c) The critical point $\left(\frac{1}{3}, -\frac{1}{\sqrt{6}}\right)$:

$A = f_{xx}\left(\frac{1}{3}, -\frac{1}{\sqrt{6}}\right) = 1$, $B = f_{xy}\left(\frac{1}{3}, -\frac{1}{\sqrt{6}}\right) = \frac{4}{\sqrt{6}}$, and

$C = f_{yy}\left(\frac{1}{3}, -\frac{1}{\sqrt{6}}\right) =$ ____ . Thus, $AC - B^2 =$ ____ and we conclude that the surface has a ______________ at $\left(\frac{1}{3}, -\frac{1}{\sqrt{6}}\right)$.

16-12 Indeterminate Forms.

OBJECTIVE: Use series to evaluate the limit $\lim_{x \to a} \frac{f(x)}{g(x)}$, at a point a where $f(x)$ and $g(x)$ are both zero. Assume that the functions f and g have series expansions in powers of $x - a$ that converge in some interval $|x - a| < \delta$.

69. $\lim_{x \to 0} \frac{e^{2x} - 1}{x}$

The Maclaurin series for e^{2x} , to terms in x^3 , is

$$e^{2x} = \sum_{n=0}^{\infty} \frac{(2x)^n}{n!} = \underline{\hspace{6cm}} .$$

Hence, $e^{2x} - 1 = 2x\left(\underline{\hspace{4cm}}\right)$,
and

$$\lim_{x \to 0} \frac{e^{2x} - 1}{x} = \lim_{x \to 0} 2\left(1 + x + \frac{2}{3}x^2 + \ldots\right) = \underline{\hspace{1cm}} .$$

70. $\lim_{x \to 0} \dfrac{\tan x - x}{x - \sin x}$

The Maclaurin series for $\tan x$ and $\sin x$, to terms in x^5, are

$$\tan x = x + \frac{x^3}{3} + \frac{2x^5}{15} + \ldots , \qquad \sin x = \underline{\hspace{5cm}} .$$

Hence,

$$\tan x - x = \frac{x^3}{3}\left(1 + \frac{2}{5}x^2 + \ldots\right) \quad \text{and}$$

$$x - \sin x = \frac{x^3}{3}\left(\underline{\hspace{3.5cm}}\right) .$$

Therefore,

$$\lim_{x \to 0} \frac{\tan x - x}{x - \sin x} = \lim_{x \to 0} \frac{\left(\underline{\hspace{3.5cm}}\right)}{\left(\frac{1}{2} - \frac{1}{40}x^2 + \ldots\right)} = \underline{\hspace{1cm}} .$$

16-13 Convergence of Power Series; Integration and Differentiation.

OBJECTIVE A: Given a power series $\sum_{n=0}^{\infty} a_n x^n$, find its interval of convergence. If the interval is finite, determine whether the series converges at each endpoint.

71. $\sum_{n=1}^{\infty} \dfrac{1}{\sqrt{n}\ 3^n} x^n$

We apply the ratio test to the series of absolute values, and find

$$\rho = \lim_{n \to \infty} \left| \frac{x^{n+1}}{\sqrt{n+1}\ 3^{n+1}} \cdot \underline{\hspace{2cm}} \right| = \lim_{n \to \infty} \frac{\sqrt{n}}{\underline{\hspace{2cm}}} |x| = \underline{\hspace{2cm}} .$$

---

69. $1 + 2x + \dfrac{4x^2}{2!} + \dfrac{8x^3}{3!} + \ldots$, $\quad 1 + x + \dfrac{2}{3}x^2 + \ldots$, $\quad 2$

70. $x - \dfrac{x^3}{3!} + \dfrac{x^5}{5!} - \ldots$, $\quad \dfrac{1}{2} - \dfrac{1}{40}x^2 + \ldots$, $\quad 1 + \dfrac{2}{5}x^2 + \ldots$, $\quad 2$

Therefore, the original series converges absolutely if $|x| <$ ___ and diverges if __________ . When $x = 3$, the series becomes

$$\sum_{n=1}^{\infty} \underline{\qquad}\ , \quad \text{the p-series with } p = \underline{\quad}\ ;$$

this series __________ . When $x = -3$, the series becomes

$$\sum_{n=1}^{\infty} \underline{\qquad\quad}\ ,$$

and this series ____________ , by Leibniz's Theorem. Therefore, the interval of convergence of the original power series is ____________ .

72. $\displaystyle\sum_{n=1}^{\infty} \frac{2^n}{n\left(3^{n+2}\right)}\, x^{n+1}$

The power series converges for $x = 0$. For $x \neq 0$, we apply the root test to the series of absolute values, and find

$$\rho = \lim_{n\to\infty} \sqrt[n]{\frac{2^n\ |x|^n\ |x|}{n \cdot 3^n \cdot 3^2}} = \lim_{n\to\infty} \underline{\qquad\qquad\qquad}$$

$$= \frac{2\ |x|\ \cdot 1}{\underline{\qquad}} < 1\ , \quad \text{if} \quad |x| < \underline{\quad}\ .$$

Therefore, the original series converges absolutely if $|x| < 3/2$ and diverges if $|x| > 3/2$. When $x = 3/2$, the series becomes

$$\sum_{n=1}^{\infty} \frac{2^n}{n\left(3^{n+2}\right)} \left(\frac{3}{2}\right)^{n+1} = \sum_{n=1}^{\infty} \underline{\quad}\ ,$$

and this series ____________ . When $x = -3/2$, the series becomes

$$\sum_{n=1}^{\infty} \frac{(-1)^{n+1}}{6n}\ ,$$

and this series ____________ , by Leibniz's Theorem. Therefore, the interval of convergence of the original power series is ____________ .

---

71. $\dfrac{\sqrt{n}\ 3^n}{x^n}$, $3\sqrt{n+1}$, $\frac{1}{3}|x|$, 3, $|x| > 3$, $\dfrac{1}{\sqrt{n}}$, $\frac{1}{2}$, diverges, $\dfrac{(-1)^n}{\sqrt{n}}$, converges, $-3 \leq x < 3$

72. $\dfrac{2\ |x|\ \sqrt[n]{|x|}}{\sqrt[n]{n} \cdot 3 \cdot \sqrt[n]{9}}$, $1 \cdot 3 \cdot 1$, $\frac{3}{2}$, $\dfrac{1}{6n}$, diverges, converges, $-\frac{3}{2} \leq x < \frac{3}{2}$

73. $\sum_{n=1}^{\infty} \sin(5n)(x-\pi)^n$

For every value of x, $|\sin(5n)(x-\pi)^n| \leq |x-\pi|^n$.

The geometric series $\sum_{n=1}^{\infty} |x-\pi|^n$ converges if ______________ and diverges if ______________. Therefore, by the comparison test, the original series converges absolutely if ______________. Suppose $|x-\pi| = 1$. Then the series becomes,

$$\sum_{n=1}^{\infty} \sin(5n) \qquad \text{or} \qquad \sum_{n=1}^{\infty} \text{\_\_\_\_\_\_\_\_\_\_\_\_\_\_} .$$

However, $\lim_{n \to \infty} \sin(5n)$ fails to exist, so neither of these series can converge. We conclude that the interval of convergence of the original series is ________________.

OBJECTIVE B: Given a power series $f(x) = \sum a_n x^n$, find the power series for $f'(x)$.

74. In Example 7, on page 820 of the text, it is given that

$$\frac{1}{1+t^2} = 1 - t^2 + t^4 - t^6 + \ldots, \qquad \text{for } -1 < t < 1 .$$

Therefore, using the term-by-term differentiation theorem,

$$\frac{-2t}{\left(1+t^2\right)^2} = \text{\_\_\_\_\_\_\_\_\_\_\_\_\_\_\_\_\_\_\_\_}, \quad \text{for } \text{\_\_\_\_\_\_\_\_\_\_} .$$

OBJECTIVE C: If f is a function having a known power series $f(x) = \sum a_n x^n$, use the series and a calculator to estimate the integral $\int_0^b f(x)\,dx$, assuming that b lies within the interval of convergence.

75. Let us find $\int_0^{0.2} \cos\sqrt{x}\,dx$ accurate to five decimal places.

---

73. $|x-\pi| < 1$, $|x-\pi| \geq 1$, $|x-\pi| < 1$, $(-1)^n \sin(5n)$, $\pi - 1 < x < \pi + 1$

74. $-2t + 4t^3 - 6t^5 + \ldots$, $-1 < t < 1$

Now,

$$\cos x = 1 - \frac{x^2}{2!} + \frac{x^4}{4!} - \frac{x^6}{6!} + \frac{x^8}{8!} - \cdots ,$$

so the power series for $\cos \sqrt{x}$ is given by

$$\cos \sqrt{x} = \underline{\hspace{6cm}} , \quad x \geq 0 .$$

Thus, using term-by-term integration,

$$\int_0^{0.2} \cos \sqrt{x} \, dx = \underline{\hspace{8cm}} \Big]_0^{0.2}$$

$$= 0.2 - \frac{0.04}{4} + \frac{0.008}{72} - \frac{0.0016}{2880} + \frac{0.00032}{201600} - \cdots$$

$$\approx 0.2 - 0.01 + 0.00011 - 0.00000056 + \cdots$$

Hence, $\int_0^{0.2} \cos \sqrt{x} \, dx \approx$ ________

with an error of less than 5×10^{-6} .

CHAPTER 16 OBJECTIVE - PROBLEM KEY

Objective	Problems in Thomas/Finney Text	Objective	Problems in Thomas/Finney Text
16-2 A,B	p. 748, 1-55	16-8	p. 791, 1-13, 15-20
16-3	p. 752, 1-30	16-9 A	p. 799, 1-7
16-4 A	p. 761, 3-7,10-18,23,24,39,40	B	p. 799, 13-25
B	p. 761, 1,2,25-38	16-10	p. 806, 1-8
16-5 A	p. 776, 1-31	16-11	p. 811, 1-8
B	p. 777, 32-34	16-12	p. 814, 1-20
16-6	p. 780, 1-18	16-13 A	p. 822, 1-20
16-7 A	p. 785, 1-10	B	p. 823, 23-26
B	p. 785, 11-28	C	p. 823, 27-34
C	p. 785, 29-34		

CHAPTER 16 SELF-TEST

1. Determine if each sequence $\{a_n\}$ converges or diverges. Find the limit of the sequence if it does converge.

(a) $a_n = \sqrt{n+1} - \sqrt{n}$

(b) $a_n = \dfrac{1 + (-1)^n}{\sqrt[n]{n}}$

(c) $a_n = \left(\dfrac{n - 0.05}{n}\right)^n$

(d) $a_n = \dfrac{2^n}{5^{3 + 1/n}}$

---

75. $1 - \frac{x}{2!} + \frac{x^2}{4!} - \frac{x^3}{6!} + \frac{x^4}{8!} - \cdots$, $x - \frac{x^2}{2\cdot 2!} + \frac{x^3}{3\cdot 4!} - \frac{x^4}{4\cdot 6!} + \frac{x^5}{5\cdot 8!} - \cdots$, 0.21011

2. Find the sum of each series.

(a) $\sum_{n=0}^{\infty} (-1)^n \frac{3}{5^n}$ (b) $\sum_{n=4}^{\infty} \frac{2}{(4n-3)(4n+1)}$ (c) $\sum_{n=0}^{\infty} \left(\frac{5}{3^n} - \frac{2}{7^n}\right)$

(d) $\frac{127}{1000} + \frac{127}{1000^2} + \frac{127}{1000^3} + \dots + \frac{127}{1000^n} + \dots$

In Problems 3 - 8, determine whether the given series converges or diverges. In each case, give a reason for your answer.

3. $\sum_{n=1}^{\infty} \frac{\sqrt{n}}{n^2+3}$ 4. $\sum_{n=1}^{\infty} \frac{n!\,3^n}{10^n}$ 5. $\sum_{n=1}^{\infty} \sin\left(\frac{n\pi - 2}{3n}\right)$

6. $\sum_{n=1}^{\infty} \left(\frac{n}{2n+5}\right)^n$ 7. $\sum_{n=1}^{\infty} \frac{1}{n+\sqrt{n}}$ 8. $\sum_{n=1}^{\infty} \frac{\tan^{-1} n}{n^2+1}$

9. Find all values of x for which the given series converge.

(a) $\sum_{n=1}^{\infty} (2x-1)^{n!}$ (b) $\sum_{n=2}^{\infty} \frac{\ln n}{n} x^n$

In Problems 10 - 13, determine whether the series are absolutely convergent, conditionally convergent, or divergent.

10. $\sum_{n=1}^{\infty} (-1)^{n+1} \frac{\sin n}{n^2+1}$ 11. $\sum_{n=1}^{\infty} (-1)^{n+1} \frac{1}{(n+1)^{1/n}}$

12. $\sum_{n=2}^{\infty} (-1)^n \frac{1}{(\ln n)^2}$ 13. $\sum_{n=1}^{\infty} (-1)^{n+1} \frac{n+1}{7n-2}$

14. Estimate the magnitude of the error if the first five terms are used to approximate the series,

$$\sum_{n=1}^{\infty} (-1)^{n+1} \frac{2^n}{3^n} .$$

Sum the first five terms, and state whether your approximation underestimates or overestimates the sum of the series.

15. Find the Taylor series of $f(x) = \sqrt{x}$ at $a = 9$. Do not be concerned with whether the series converges to the given function f.

16. Find the Maclaurin series for the function $f(x) = x \ln(1+x^2)$ using series that have already been obtained in the Thomas/Finney text.

17. Use series to estimate the number $e^{-1/3}$ with an error of magnitude less than 0.001 .

18. Test the surface $z = xy^2 - 2xy + 2x^2 - 15x$ for maxima, minima, and saddle points.

19. Use series to evaluate the following limits.

(a) $\lim_{x \to 0} \frac{\sin x - x \cos x}{x^3}$ (b) $\lim_{x \to 0} \frac{\ln(1-2x)}{\tan \pi x}$

20. Find the first three nonzero terms in the Maclaurin series for the function $f(x) = \sec^2 x$ using the Maclaurin series for $\tan x$.

21. (Calculator) Use series and a calculator to estimate the integral

$$\int_0^{0.5} \cos x^2 \, dx$$

with an error of magnitude less than 0.0001 .

SOLUTIONS TO CHAPTER 16 SELF-TEST

1. (a) $a_n = \sqrt{n+1} - \sqrt{n} = \frac{(\sqrt{n+1} - \sqrt{n})(\sqrt{n+1} + \sqrt{n})}{(\sqrt{n+1} + \sqrt{n})} = \frac{(n+1) - n}{\sqrt{n+1} + \sqrt{n}}$

$= \frac{1}{\sqrt{n+1} + \sqrt{n}} \to 0$ as $n \to \infty$

(b) $\sqrt[n]{n} \to 1$, but $1 + (-1)^n$ alternates back and forth between 0 and 1 . Thus, for n large, a_n alternates between numbers very close to 1 and 0 ; hence the sequence diverges.

(c) $a_n = \left(\frac{n - 0.05}{n}\right)^n = \left(1 + \frac{-0.05}{n}\right)^n \to e^{-0.05} \approx 0.951$.

(d) $5^{3 + 1/n} = 125 \sqrt[n]{5} \to 125$, but $2^n \to +\infty$. Therefore, the sequence $\{a_n\}$ is unbounded and diverges.

2. (a) $\sum_{n=0}^{\infty} (-1)^n \frac{3}{5^n} = \sum_{n=0}^{\infty} 3\left(-\frac{1}{5}\right)^n = \frac{3}{1 + \frac{1}{5}} = \frac{5}{2}$.

(b) Using the partial fraction decomposition,

$\frac{2}{(4k-3)(4k+1)} = \frac{1}{2}\left(\frac{1}{4k-3}\right) - \frac{1}{2}\left(\frac{1}{4k+1}\right)$, we write the partial sum

$$s_k = \sum_{n=4}^{k} \frac{2}{(4n-3)(4n+1)}$$

as

$$s_k = \frac{1}{2}\left(\frac{1}{13} - \frac{1}{17}\right) + \frac{1}{2}\left(\frac{1}{17} - \frac{1}{21}\right) + \frac{1}{2}\left(\frac{1}{21} - \frac{1}{25}\right) + \ldots + \frac{1}{2}\left(\frac{1}{4k-3} - \frac{1}{4k+1}\right).$$

Thus,

$$s_k = \frac{1}{2}\left(\frac{1}{13} - \frac{1}{4k+1}\right) \rightarrow \frac{1}{26} \quad \text{as} \quad k \rightarrow \infty$$

so that

$$\sum_{n=4}^{\infty} \frac{2}{(4k-3)(4k+1)} = \frac{1}{26} .$$

(c) $$\sum_{n=0}^{\infty} \left(\frac{5}{3^n} - \frac{2}{7^n}\right) = \sum_{n=0}^{\infty} \frac{5}{3^n} - \sum_{n=0}^{\infty} \frac{2}{7^n} = \frac{5}{1 - \frac{1}{3}} - \frac{2}{1 - \frac{1}{7}} = \frac{31}{6} .$$

(d) $$\sum_{n=1}^{\infty} 127\left(\frac{1}{1000}\right)^n = \sum_{n=0}^{\infty} 127\left(\frac{1}{1000}\right)^n - 127 = \frac{127}{1 - \frac{1}{1000}} - 127$$

$$= \frac{127{,}000 - 126{,}873}{999} = \frac{127}{999} .$$

3. $\frac{\sqrt{n}}{n^2+3} < \frac{\sqrt{n}}{n^2} = \frac{1}{n^{3/2}}$ so that $\sum_{n=1}^{\infty} \frac{\sqrt{n}}{n^2+3}$ converges by comparison with the convergent p-series for $p = \frac{3}{2}$.

4. Using the ratio test, $\lim_{n \to \infty} \frac{(n+1)!\ 3^{n+1}}{10^{n+1}} \cdot \frac{10^n}{n!\ 3^n} = \lim_{n \to \infty} \frac{(n+1)3}{10} = \infty$.

Thus, $\sum_{n=1}^{\infty} \frac{n!\ 3^n}{10^n}$ diverges by the ratio test.

5. $\lim_{n \to \infty} \sin\left(\frac{n\pi - 2}{3n}\right) = \lim_{n \to \infty} \sin\left(\frac{\pi}{3} - \frac{2}{3n}\right) = \sin\frac{\pi}{3} = \frac{\sqrt{3}}{2} \neq 0$, so the series $\sum_{n=1}^{\infty} \sin\left(\frac{n\pi - 2}{3n}\right)$ diverges by the nth-term test for divergence.

6. If $a_n = \left(\frac{n}{2n+5}\right)^n$, then $\sqrt[n]{a_n} = \frac{n}{2n+5} \to \frac{1}{2}$. Thus, the series

$\sum_{n=1}^{\infty} \left(\frac{n}{2n+5}\right)^n$ converges by the root test.

7. $\frac{1}{n+\sqrt{n}} > \frac{1}{n+n} = \frac{1}{2n}$ so that $\sum_{n=1}^{\infty} \frac{1}{n+\sqrt{n}}$ diverges by comparison to the divergent series $\sum_{n=1}^{\infty} \frac{1}{2n}$.

8. $$\int_1^{\infty} \frac{\tan^{-1} x \, dx}{x^2+1} = \lim_{b \to \infty} \frac{1}{2}\left(\tan^{-1} x\right)^2\Big]_1^b = \lim_{b \to \infty} \frac{1}{2}\left(\tan^{-1} b\right)^2 - \frac{1}{2} \tan^{-1} 1$$

$$= \frac{1}{2}\left(\frac{\pi}{2}\right)^2 - \frac{1}{2}\left(\frac{\pi}{4}\right)$$

Therefore, the improper integral converges, so the original series

$\sum_{n=1}^{\infty} \frac{\tan^{-1} n}{n^2+1}$ converges by the integral test.

9. (a) If $|2x-1| < 1$, then $|2x-1|^{n!} < |2x-1|^n$, for $n \geq 1$.

If $|2x-1| \geq 1$, then $|2x-1|^{n!} \geq |2x-1|^n$, for $n \geq 1$.

Therefore, the series $\sum_{n=1}^{\infty} (2x-1)^{n!}$ converges absolutely for all values of x satisfying $|2x-1| < 1$, or $0 < x < 1$ by comparison with the convergent geometric series $\sum_{n=1}^{\infty} (2x-1)^n$. The series $\sum_{n=1}^{\infty} (2x-1)^{n!}$ diverges for all values of x satisfying $|2x-1| \geq 1$ since $(2x-1)^{n!} = |2x-1|^{n!} \geq |2x-1|^n$ if $n \geq 2$, and the geometric series $\sum_{n=1}^{\infty} (2x-1)^n$ diverges for $|2x-1| \geq 1$.

(b) Using the ratio test,

$$\lim_{n\to\infty} \frac{\ln(n+1)\,|x|^{n+1}}{(n+1)} \cdot \frac{n}{\ln n\,|x|^n} = \lim_{n\to\infty} \frac{\ln(n+1)}{\ln n} \cdot \frac{n+1}{n}|x| = |x| .$$

Thus, the given power series converges absolutely for $|x| < 1$ and diverges for $|x| > 1$. We test the end-points of the interval.

For $x = 1$, the power series is $\sum_{n=2}^{\infty} \frac{\ln n}{n}$. Now

$$\int_2^{\infty} \frac{\ln x}{x}\,dx = \lim_{b\to\infty} \frac{1}{2}(\ln x)^2\Big]_2^b = +\infty$$ diverges, so the series

$\sum_{n=2}^{\infty} \frac{\ln n}{n}$ is divergent by the integral test.

For $x = -1$, the power series is $\sum_{n=2}^{\infty} \frac{(-1)^n \ln n}{n}$. Since

$$0 \le \lim_{n\to\infty} \frac{\ln n}{n} \le \lim_{n\to\infty} \frac{\sqrt{n}}{n} = 0 ,$$ and $$\frac{d}{dx}\left(\frac{\ln x}{x}\right) = \frac{1 - \ln x}{x^2} < 0$$

for $x \ge 3$ implies that $\frac{\ln(n+1)}{n+1} < \frac{\ln n}{n}$, the alternating series

$\sum_{n=2}^{\infty} \frac{(-1)^n \ln n}{n}$ converges by Leibniz's Theorem. Therefore, the power

series $\sum_{n=2}^{\infty} \frac{\ln n}{n} x^n$ converges for all x satisfying $-1 \le x < 1$.

10. $\left|(-1)^{n+1} \frac{\sin n}{n^2+1}\right| \le \frac{1}{n^2+1}$, so the original series $\sum_{n=1}^{\infty} (-1)^{n+1} \frac{\sin n}{n^2+1}$

<u>converges absolutely</u> by the comparison test.

11. Since $\frac{1}{n^2} < \frac{1}{n+1} < \frac{1}{n}$, it follows that $\left(\frac{1}{\sqrt[n]{n}}\right)\left(\frac{1}{\sqrt[n]{n}}\right) < \frac{1}{\sqrt[n]{n+1}} < \frac{1}{\sqrt[n]{n}}$.

Thus, $\lim_{n\to\infty} \frac{1}{\sqrt[n]{n+1}} = 1$ so the original series

$\sum_{n=1}^{\infty} (-1)^{n+1} \frac{1}{(n+1)^{1/n}}$ <u>diverges</u> by the nth-term test.

12. $\lim_{n\to\infty} \dfrac{1}{(\ln n)^2} = 0$, and $\dfrac{1}{\left[\ln (n+1)\right]^2} < \dfrac{1}{(\ln n)^2}$ because $y = \ln x$ is an increasing function of x. Therefore the alternating series

$\sum_{n=2}^{\infty} (-1)^n \dfrac{1}{(\ln n)^2}$ converges by Leibniz's Theorem. However, since $\ln n < \sqrt{n}$ implies $\dfrac{1}{(\ln n)^2} > \dfrac{1}{n}$ if $n \geq 2$, the series of absolute values $\sum_{n=2}^{\infty} \dfrac{1}{(\ln n)^2}$ diverges by comparison with the divergent harmonic series. Therefore,

$\sum_{n=2}^{\infty} (-1)^n \dfrac{1}{(\ln n)^2}$ is conditionally convergent.

13. $\lim_{n\to\infty} \dfrac{n+1}{7n-2} = \dfrac{1}{7}$ so that $\sum_{n=1}^{\infty} (-1)^{n+1} \dfrac{n+1}{7n-2}$ diverges by the nth-term test.

14. $\sum_{n=1}^{\infty} (-1)^{n+1} \dfrac{2^n}{3^n} \approx \dfrac{2}{3} - \dfrac{4}{9} + \dfrac{8}{27} - \dfrac{16}{81} + \dfrac{32}{243} \approx 0.4527$ with an error of magnitude less than $2^6/3^6 < 0.0878$. Since the sign of the first unused term is negative, the sum 0.4527 overestimates the value of the series. In fact, the given geometric series sums to 0.4 .

15. We calculate the derivatives of $f(x) = \sqrt{x}$, and evaluate f and these derivatives at $a = 9$:

$f(x) = \sqrt{x}$	$f(9) = 3$
$f'(x) = \frac{1}{2} x^{-1/2}$	$f'(9) = \frac{1}{6}$
$f^{(2)}(x) = (-1)\left(\frac{1}{2}\right)\left(\frac{1}{2}\right) x^{-3/2}$	$f^{(2)}(9) = -\frac{1}{108}$
$f^{(3)}(x) = (-1)^2\left(\frac{1}{2}\right)\left(\frac{1}{2}\right)\left(\frac{3}{2}\right) x^{-5/2}$	$f^{(3)}(9) = \frac{1}{648}$
$f^{(4)}(x) = (-1)^3 \frac{3\cdot 5}{2^4} x^{-7/2}$	$f^{(4)}(9) = -\frac{5}{11664}$
$\vdots$	$\vdots$
$f^{(k)}(x) = (-1)^{k+1} \frac{3\cdot 5\cdots(2k-3)}{2^k} x^{-(2k-1)/2}$	$f^{(k)}(9) = (-1)^{k+1} \frac{3\cdot 5\cdots(2k-3)}{2^k\, 3^{2k-1}}$

Therefore, the Taylor series for $f(x) = \sqrt{x}$ at $a = 9$ is

$$3 + \frac{1}{6}(x-9) - \frac{1}{216}(x-9)^2 + \ldots + (-1)^{k+1} \frac{3\cdot 5\cdots(2k-3)}{2^k\, 3^{2k-1}\cdot k!}(x-9)^k + \ldots$$

16. $\ln(1+x) = x - \frac{x^2}{2} + \frac{x^3}{3} - \frac{x^4}{4} + \ldots, \quad -1 < x \le 1$

$\ln(1+x^2) = x^2 - \frac{x^4}{2} + \frac{x^6}{3} - \frac{x^8}{4} + \ldots, \quad -1 < x \le 1$

$x\ln(1+x^2) = x^3 - \frac{x^5}{2} + \frac{x^7}{3} - \frac{x^9}{4} + \ldots, \quad -1 < x \le 1$

or, in closed form, $x\ln(1+x^2) = \sum_{n=0}^{\infty} (-1)^n \frac{1}{n+1} x^{2n+3}$, valid for all x satisfying $-1 < x \le 1$.

17. $e^{-1/3} = 1 - \frac{1}{3} + \frac{(-1/3)^2}{2!} + \frac{(-1/3)^3}{3!} + \frac{(-1/3)^4}{4!} + \ldots$

By trial, $\frac{(-1/3)^4}{4!} < 0.00052$ and $\frac{(1/3)^3}{3!} > 0.001$.

Since the series is an alternating series,

$e^{-1/3} \approx 1 - \frac{1}{3} + \frac{1/9}{2!} - \frac{1/27}{3!} = 0.71605$ with an error in magnitude less than 0.00052.

18. Let $f(x,y) = xy^2 - 2xy + 2x^2 - 15x$. Then $f_x = y^2 - 2y + 4x - 15$ and $f_y = 2xy - 2x$. Thus, $f_y = 2x(y-1) = 0$, if $x = 0$ or $y = 1$. If $f_x = 0$ and $x = 0$, then $y^2 - 2y - 15 = 0$, or $(y-5)(y+3) = 0$. Thus, $(0,5)$ and $(0,-3)$ are critical points. If $f_x = 0$ and $y = 1$, then $1 - 2 + 4x - 15 = 0$, or $x = 4$. Thus, $(4,1)$ is a critical point. We test these three critical points.

$f_{xx} = 4$, $f_{xy} = 2y - 2$, and $f_{yy} = 2x$.

At $(0,-3)$: Let $A = f_{xx}(0,-3) = 4$, $B = f_{xy}(0,-3) = -8$, and $C = f_{yy}(0,-3) = 0$. Then, $AC - B^2 = -64 < 0$, so the surface has a saddle point at $(0,-3)$.

At $(0,5)$: Let $A = f_{xx}(0,5) = 4$, $B = f_{xy}(0,5) = 8$, and $C = f_{yy}(0,5) = 0$. Then, $AC - B^2 = -64 < 0$, so again the surface has a saddle point at $(0,5)$.

At $(4,1)$: Let $A = f_{xx}(4,1) = 4$, $B = f_{xy}(4,1) = 0$, and $C = f_{yy}(4,1) = 8$. Then, $AC - B^2 = 32 > 0$ and $A > 0$, so the surface has a relative minimum at $(4,1)$.

19. (a) The MacLaurin series for $\sin x$ and $x \cos x$, to terms in x^7, are

$$\sin x = x - \frac{x^3}{3!} + \frac{x^5}{5!} - \frac{x^7}{7!} + \ldots ,$$

$$x \cos x = x - \frac{x^3}{2!} + \frac{x^5}{4!} - \frac{x^7}{6!} + \ldots .$$ Hence

$$\sin x - x \cos x = x^3\left(\frac{1}{2!} - \frac{1}{3!}\right) + x^5\left(\frac{1}{5!} - \frac{1}{4!}\right) + x^7\left(\frac{1}{6!} - \frac{1}{7!}\right) + \ldots ,$$

and

$$\lim_{x \to 0} \frac{\sin x - x \cos x}{x^3} = \lim_{x \to 0} \left[\left(\frac{1}{2!} - \frac{1}{3!}\right) + x^2\left(\frac{1}{5!} - \frac{1}{4!}\right) + \ldots\right]$$

$$= \frac{1}{2} - \frac{1}{6} = \frac{1}{3} .$$

(b) The Maclaurin series for $\ln(1 - 2x)$ and $\tan \pi x$ are

$$\ln(1 - 2x) = -2x - \frac{(2x)^2}{2} - \frac{(2x)^3}{3} - \ldots ,$$

$$\tan \pi x = \pi x + \frac{(\pi x)^3}{3} + \frac{2(\pi x)^5}{15} + \ldots$$

Hence,

$$\frac{\ln(1 - 2x)}{\tan \pi x} = \frac{-x\left(2 + 2x + \frac{8x^2}{2} + \ldots\right)}{x\left(\pi + \frac{\pi^3 x^2}{3} + \frac{2\pi^5 x^4}{15} + \ldots\right)}$$

and

$$\lim_{x \to 0} \frac{\ln(1 - 2x)}{\tan \pi x} = \lim_{x \to 0} \frac{-(2 + 2x + \ldots)}{\left(\pi + \frac{\pi^3 x^2}{3} + \ldots\right)} = -\frac{2}{\pi} .$$

20. The Maclaurin series for $\tan x$, through the first three nonzero terms, is

$$\tan x = x + \frac{x^3}{3} + \frac{2x^3}{15} + \ldots$$

Hence, $\sec^2 x = \frac{d}{dx} \tan x = 1 + x^2 + \frac{2}{3} x^4 + \ldots .$

21. The Maclaurin series for $\cos x^2$ is

$$\cos x^2 = 1 - \frac{x^4}{2!} + \frac{x^8}{4!} - \frac{x^{12}}{6!} + \ldots + (-1)^k \frac{x^{4k}}{(2k)!} + \ldots$$

Hence,

$$\int_0^{0.5} \cos x^2 \, dx = x - \frac{x^5}{5 \cdot 2!} + \frac{x^9}{9 \cdot 4!} - \frac{x^{13}}{13 \cdot 6!} + \ldots \Big]_0^{0.5}$$

$$\approx 0.5 - 0.00313 + 0.0000090 - \ldots \approx 0.49687 ,$$

with an error in magnitude less than 0.000009 because the series is an alternating series.

CHAPTER 17 COMPLEX NUMBERS AND FUNCTIONS

17-1 Invented Number Systems.

OBJECTIVE: For given complex numbers, determine whether they are equal, find their sum, and find their product.

1. The complex numbers $a + ib$ and $c + id$ are equal if and only if _______ and _______ .

2. If $x + iy = 3 - 7i$, then $x =$ ___ and $y =$ ___ .

3. If $x + iy = -6$, then $x =$ ___ and $y =$ ___ .

4. $(-2 + 4i) + (7 - 5i) =$ _______ .

5. $(3,-2) + (-10,-6) =$ _______ .

6. $(4 + 5i) \cdot (-1 + i) =$ _______ .

7. $(-2,1)^2 =$ _______ .

8. $-3(5 - 3i) =$ _______ .

9. Let us solve the equation $4(x + iy) + (i - 1)^2 = x - iy$ for the real numbers x and y. Now $4(x + iy) = 4x + i4y$ and $(i - 1)^2 =$ ____ . Summing these last two complex numbers and equating the sum to $x - iy$ gives, $4x + i(4y - 2) = x - iy$ or, $4x =$ ___ and $4y - 2 =$ ___ . Therefore, $x =$ ___ and $y =$ ___ .

17-2 The Argand Diagram.

OBJECTIVE A: Express any rational combination of complex numbers as a single complex number in the form $a + ib$ (exclusive of division by the complex number $(0,0) = 0 + i0$). Express the answer in the form $r \text{ cis } \theta$, with $r \geq 0$ and $-\pi < \theta \leq \pi$.

10. In terms of the polar coordinates of x and y, we have $z = x + iy = r(\cos\theta + i\sin\theta)$, where $r =$ ____________ is the ______________ of the complex number z, and the polar angle θ is called the ____________ of z and written $\theta =$ _________ . The principal value of the argument is taken to be that value of θ for which __________ .

---

1. $a = c$, $b = d$ 2. 3, -7 3. -6, 0 4. $5 - i$

5. (-7,-8) or, $-7 - 8i$ 6. $-9 - i$ 7. (3,-4) or, $3 - 4i$ 8. $-15 + 9i$

9. $-2i$, x, $-y$, 0, 2/5 10. $\sqrt{x^2 + y^2}$, absolute value, argument, arg z, $-\pi < \theta \leq \pi$

11. We use the abbreviation $\text{cis}\ \theta$ = ____________________ . The complex-valued function $\text{cis}\ \theta$ satisfies the following properties:

$\text{cis}\ \theta_1 \cdot \text{cis}\ \theta_2$ = ________________ ,

$(\text{cis}\ \theta)^{-1}$ = _____________ , and $\dfrac{\text{cis}\ \theta_1}{\text{cis}\ \theta_2}$ = ____________________ .

12. Consider the rational expression $\dfrac{2+3i}{4-i}$. To reduce this to a single complex number in the form $x+iy$, multiply numerator and denominator by the complex conjugate, $4+i$, of the denominator, and simplify:

$$z = \frac{2+3i}{4-i} = \frac{(2+3i)(4+i)}{(4-i)(4+i)} = \frac{\_\_\_\_\_\_}{16+1} = \_\_\_\_\_\_\_\_\_\_\_\_ .$$

13. For $z = \frac{5}{17} + \frac{14}{17}i$, $|z| = \frac{1}{17}$ _____________ ≈ 0.874 , and

$\arg z = \tan^{-1} \frac{}{\_\_\_\_} \approx 1.23$ radians (or 70.3 degrees) .

Thus, in the form $r\ \text{cis}\ \theta$, $z \approx$ ___________________ .

OBJECTIVE B: Graph points $z = x+iy$ with an Argand diagram.

14. Representation of the complex number $z = x+iy$ as a point $P(x,y)$ in the xy-plane, or as the vector $\overrightarrow{OP}$ from the origin to P , is called an ________________ .

15. Sketch the complex number $z = -2+5i$ as a point in an Argand diagram.

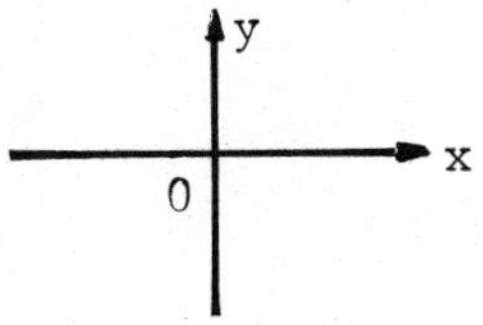

y
x
0

16. Sketch the complex number

$$z = 3\cos\left(-\frac{\pi}{4}\right) + i\,3\sin\left(-\frac{\pi}{4}\right)$$

as a vector in an Argand diagram.

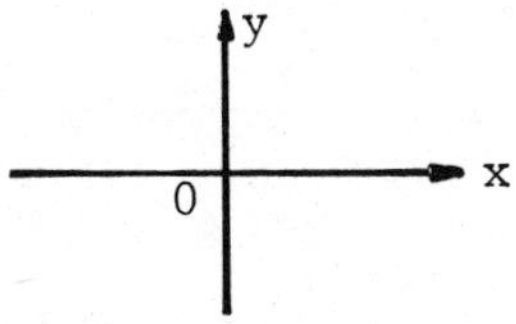

y
x
0

---

11. $\cos\theta + i\sin\theta$, $\text{cis}\,(\theta_1+\theta_2)$, $\text{cis}\,(-\theta)$, $\text{cis}\,(\theta_1-\theta_2)$

12. $5+14i$, $\frac{5}{17}+\frac{14}{17}i$

13. $\sqrt{25+196}$, $\frac{14}{5}$, $0.874\ \text{cis}\ 1.23$

14. Argand diagram

15.

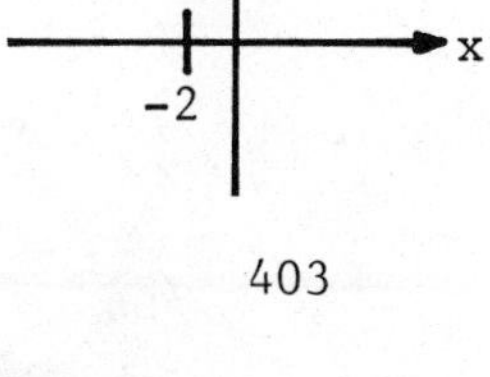

y
z
5
x
-2

16.

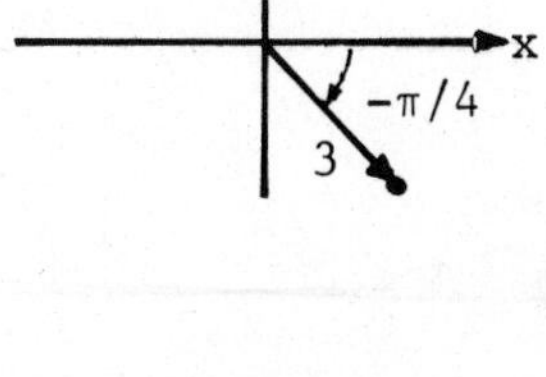

y
x
$-\pi/4$
3

17. The distance between two points z_1 and z_2 in an Argand diagram is equal to $|z_1 - z_2|$ (see Problem 8, page 838 in the Thomas/Finney text). Thus, the set of all points $z = x + iy$ that satisfy $|z - 2| = 3$ is a circle in an Argand diagram of radius ___ and centered at the point ________ . Sketch.

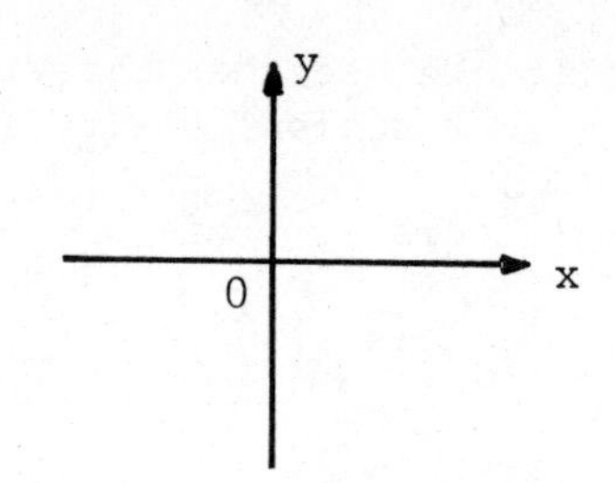

OBJECTIVE C: Use De Moivre's theorem to find powers, and roots, of complex numbers.

18. De Moivre's theorem gives $(\cos\theta + i \sin\theta)^n$ = ____________________ .

19. For $z = 2\left(-1 + i\sqrt{3}\right)$, we have $|z|$ = ___ , and

arg z = $\tan^{-1}$ ___ = $\frac{\quad}{\quad}$ radians. Thus, in polar notation,

$z = r \text{ cis } \theta$ = ____________ and z^6 = ____________ .

Expressed in the form $x + iy$, z^6 = ____________ .

Note that we replaced $\theta = 4\pi$ by $\theta = 0$ for the argument in z^6 because of our convention that $-\pi < \theta \leq \pi$ in this book.

20. Let us find the roots of $4\sqrt{2}\,(1 - i)$. The polar representation $r \text{ cis } \theta$ for the given complex number uses r = ___ and θ = ___ (you may wish to plot the given number in an Argand diagram). Thus,

$z = 4\sqrt{2}\,(1 - i) = 8 \text{ cis }\left(-\frac{\pi}{4}\right)$. One of the cube roots of z is

$2 \text{ cis }\left(-\frac{\pi}{12}\right)$. We obtain the others by successive additions of $\frac{2\pi}{\quad}$

to the argument of this first one. Hence,

$$\sqrt[3]{8 \text{ cis }\left(-\frac{\pi}{4}\right)} = 2 \text{ cis }\left(-\frac{\pi}{12}, \_\_\_, \_\_\_\right),$$

---

17. 3, (2,0)

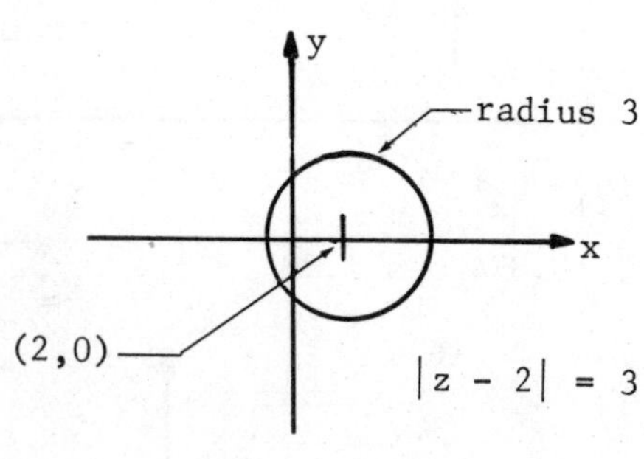

18. $\cos n\theta + i \sin n\theta$

19. 4, $-\sqrt{3}$, $\frac{2\pi}{3}$, $4 \text{ cis } \frac{2\pi}{3}$, $4096 \text{ cis } 0$, $4096 + i0$

20. 8, $-\frac{\pi}{4}$, 3, $\frac{7\pi}{12}$, $\frac{15\pi}{12}$ (replace by $-\frac{3\pi}{4}$)

and the three roots are

$$w_1 = 2\left[\cos\left(-\frac{\pi}{12}\right) + i\sin\left(-\frac{\pi}{12}\right)\right] = \frac{\sqrt{2}}{2}\left(\sqrt{3}+1\right) - i\frac{\sqrt{2}}{2}\left(\sqrt{3}-1\right),$$

$$w_2 = 2\left[\cos\left(\frac{7\pi}{12}\right) + i\sin\left(\frac{7\pi}{12}\right)\right] = -\frac{\sqrt{2}}{2}\left(\sqrt{3}-1\right) + i\frac{\sqrt{2}}{2}\left(\sqrt{3}+1\right),$$

$$w_3 = 2\left[\cos\left(-\frac{3\pi}{4}\right) + i\sin\left(-\frac{3\pi}{4}\right)\right] = -\sqrt{2} - i\sqrt{2} \quad .$$

17-3 Complex Variables.

OBJECTIVE A: In connection with the function $w = z^2$ discussed in the Thomas/Finney text on page 840, sketch the image in the w-plane of a specified set S of points in the z-plane. Use of polar coordinates may be helpful.

21. It is said that w is a function of the complex variable z on a domain S, written

______________________________ ,

if, to each ___ in the set S, there corresponds a complex number ___ = $u + iv$.

22. Let $w = z^2$ and consider the set S in the z-plane specified by $|z| = \sqrt{2}$ and $-\frac{\pi}{4} \leq \arg z \leq \frac{\pi}{4}$. Sketch the set S at the right.

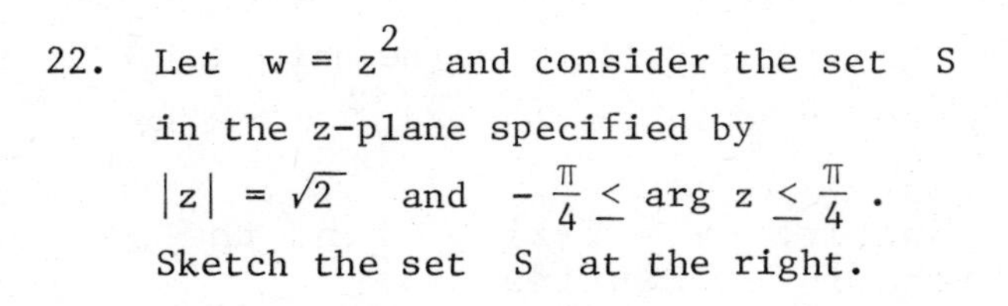

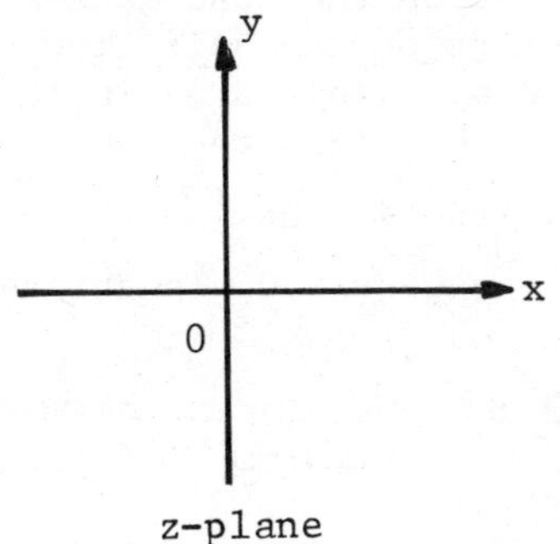

---

21. $w = f(z)$, z in S ; z, w

22.

S is arc of circle of radius $\sqrt{2}$ centered at 0

23. In polar coordinates,

$$w = z^2 = \underline{\qquad\qquad\qquad} .$$

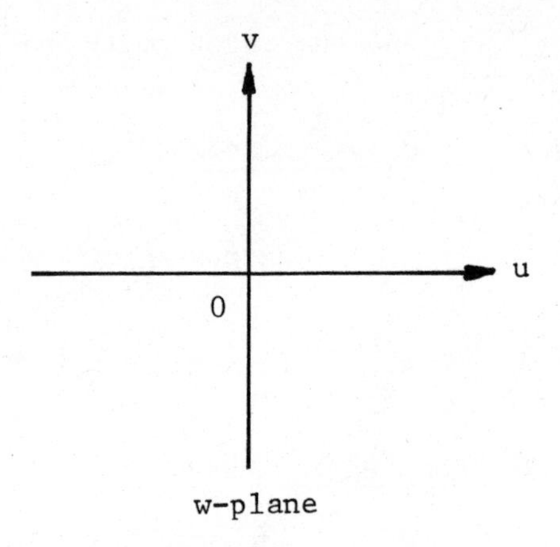

For the set S in Problem 22 , as $-\pi/4 \leq \arg z \leq \pi/4$, we find $\underline{\quad} \leq \arg w \leq \underline{\quad}$; and if $|z| = \sqrt{2}$, then $|w| = \underline{\quad}$. Therefore, the function $w = z^2$ maps a point $z = x + iy$ on the arc of the circle sketched in the z-plane in Problem 22 to a point on the arc of a circle in the w-plane. Sketch the arc in the w-plane.

24. Consider the line $y = x$ in the z-plane, so that S consists of all complex numbers of the form $z = x + ix$. For $w = z^2 = u + iv$, the text obtained (see page 840)

$$u = \underline{\qquad\qquad} \quad \text{and} \quad v = \underline{\qquad\qquad} .$$

Therefore, when $z = x + ix$, $u = \underline{\quad}$ and $v = \underline{\quad}$. That is, the function $w = z^2$ maps each point on the line $y = x$ in the z-plane to a point on the nonnegative v-axis in the w-plane. Notice, in fact, that the points on the <u>half-line</u> $y = x$, $x \geq 0$, in the z-plane map onto the nonnegative v-axis in the w-plane because the point z and the point $-z$ both map into the same point. Also, each point $(0,v)$, $v \geq 0$, in the w-plane is the image of the point $\left(\sqrt{v/2}\ ,\ \sqrt{v/2}\right)$ on the half-line $y = x$, $x \geq 0$ in the z-plane, since $v = 2x^2 = 2\left(\sqrt{v/2}\right)^2$ is true.

OBJECTIVE B: Given an elementary function $w = f(z)$, find the points $z = \alpha$ at which f is continuous.

25. A function $w = f(z)$ that is defined throughout some neighborhood of the point $z = \alpha$ is said to be <u>continuous</u> at α if

$$\underline{\qquad\qquad\qquad} \text{ as } |z - \alpha| \to 0 .$$

---

23. $r^2(\cos 2\theta + i \sin 2\theta)$,

$-\pi/2 \leq \arg w \leq \pi/2$, 2

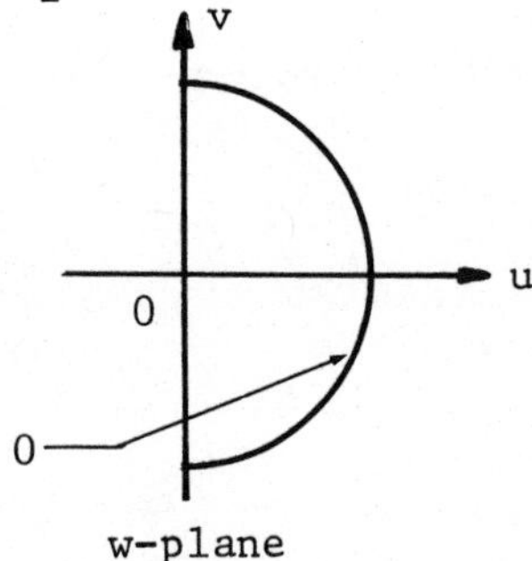

24. $x^2 - y^2$, $2xy$, 0, $2x^2$

25. $|f(z) - f(\alpha)| \to 0$

26. $f(z) = 1/z^2$ is continuous at any point $z = \alpha$ if $\alpha \neq 0$. For we have,

$$|f(z) - f(\alpha)| = \underline{\qquad\qquad} = \left|\frac{\alpha^2 - z^2}{\alpha^2 z^2}\right| = \frac{|\alpha - z| \cdot |\alpha + z|}{\alpha^2 z^2} .$$

Now, as $z \to \alpha$ we know $z^2 \to \alpha^2$ from the Example on page 841 of the Thomas/Finney text. Thus,

$$\lim_{z \to \alpha} |f(z) - f(\alpha)| = \lim_{z \to \alpha} \frac{|\alpha - z| \cdot |\alpha + z|}{\alpha^2 z^2} = \underline{\qquad} \lim_{z \to \alpha} |\alpha - z|$$

$$= \frac{|2\alpha|}{\alpha^4} \cdot \underline{\quad} = \underline{\quad} .$$

Therefore, the continuity condition in Problem 25 is satisfied.

17-4 Derivatives.

OBJECTIVE: Find the derivative with respect to z of a given function $w = f(z)$ at a specified point z_0 . Use the Δ-notation and calculate $\lim_{\Delta z \to 0} \Delta w/\Delta z$, or use the differential formulas on page 843 of the text.

27. We will use the differentiation formulas to calculate the derivative of $w = z^4 - 3z^2 - 2z + 7$ at $z_0 = i$. Now,

$$\frac{dw}{dz} = \frac{d}{dz}(z^4 - 3z^2 - 2z + 7)$$

$$= \frac{d}{dz}(z^4) + \frac{d}{dz}(-3z^2) + \frac{d}{dz}(\underline{\quad}) + \frac{d}{dz}(\underline{\quad})$$

$$= 4z^3 - 3\frac{d}{dz}(z^2) - 2\frac{d}{dz}(z) + \underline{\quad} = 4z^3 - (\underline{\qquad\qquad}) .$$

Therefore, at $z_0 = i$,

$$\left.\frac{dw}{dz}\right|_{z=i} = 4i^3 - 6i - 2 = \underline{\qquad\qquad} .$$

Notice that the calculation for the derivative with respect to z of the polynomial above in powers of the complex variable z produced the same result we obtain when differentiating a polynomial in x with respect to the real variable x . This result holds when we differentiate with

---

26. $\left|\frac{1}{z^2} - \frac{1}{\alpha^2}\right|$, $\frac{|2\alpha|}{\alpha^4}$, 0, 0 27. $-2z$, 7, 0, $6z + 2$, $-2 - 10i$

respect to z any polynomial in the complex variable z because of the differentiation formulas 1, 2, 3, and 6 on page 843: thus,

$$\frac{d}{dz}(a_n z^n + a_{n-1}z^{n-1} + \ldots + a_1 z + a_0) = na_n z^{n-1} + (n-1)a_{n-1}z^{n-1} + \ldots + a_1 .$$

28. Let us calculate $f'(\alpha)$ if $f(z) = 1/z^2$ and $\alpha \neq 0$ using the definition in Eq. (1), page 841 of the text. Then,

$$w + \Delta w = \frac{1}{(z+\Delta z)^2} \quad \text{so that}$$

$$\frac{\Delta w}{\Delta z} = \frac{1}{\Delta z}\left[\frac{1}{(z+\Delta z)^2} - \frac{1}{z^2}\right] = \frac{\_\_\_\_\_\_\_\_}{z^2(z+\Delta z)^2}$$

and

$$\lim_{\Delta z \to 0} \frac{\Delta w}{\Delta z} = \lim_{\Delta z \to 0} \frac{-2z}{z^2(z+\Delta z)^2} + \lim_{\Delta z \to 0} \frac{-\Delta z}{z^2(z+\Delta z)^2}$$

$$= \_\_\_\_\_\_\_\_\_\_\_\_ .$$

Notice that it makes no difference how Δz approaches zero in the limiting process. Thus,

$$f'(\alpha) = \_\_\_\_\_\_ \quad \text{whenever } \alpha \neq 0 .$$

17-5 The Cauchy-Riemann Equations.

OBJECTIVE A: Given a complex function $w = f(z)$, $z = x + iy$, find its real and imaginary parts $w = u + iv$, and determine if they satisfy the Cauchy-Riemann equations.

29. A necessary condition that $w = u + iv = f(z)$ be differentiable with respect to $z = x + iy$, is that the four partial derivatives of u and v with respect to x and y satisfy the following Cauchy-Riemann equations:

__________ and __________ .

30. If $w = u + iv = f(z)$ satisfies the Cauchy-Riemann equations, and, in addition, has __________ partial derivatives, u_x, u_y, v_x, v_y, then the function $w = u + iv$ is __________ with respect to ___ .

---

28. $-2z - \Delta z$, $\frac{-2z}{z^4} + 0$, $-\frac{2}{\alpha^3}$

29. $\frac{\partial u}{\partial x} = \frac{\partial v}{\partial y}$, $\frac{\partial u}{\partial y} = -\frac{\partial v}{\partial x}$

30. continuous, differentiable, z

31. Consider the complex function $w = \bar{z}$. Thus, for $z = x + iy$, $w =$ __________ $= u + iv$. Hence

$$\frac{\partial u}{\partial x} = \_\_\_ , \quad \frac{\partial v}{\partial x} = 0 , \quad \frac{\partial u}{\partial y} = \_\_\_ , \quad \frac{\partial v}{\partial y} = -1 .$$

Thus, $w = \bar{z}$ ____________ satisfy the Cauchy-Riemann equations. We conclde that $w = \bar{z}$ __________ differentiable in agreement with the result in Article 17-4 of the Thomas/Finney text.

32. Let $w = e^x(\cos y + i \sin y)$. Then,

$$\frac{\partial u}{\partial x} = \_\_\_\_\_\_\_\_\_\_ , \quad \frac{\partial v}{\partial x} = e^x \sin y$$

$$\frac{\partial u}{\partial y} = \_\_\_\_\_\_\_\_\_\_ , \quad \frac{\partial v}{\partial y} = e^x \cos y .$$

Thus, w _________ satisfy the Cauchy-Riemann equations. Since the four partial derivatives u_x, etc., are continuous, we conclude that $w = e^x(\cos y + i \sin y)$ _____ differentiable with respect to $z = x + iy$.

OBJECTIVE B: If the partial derivatives of first and second order of the real and imaginary parts of an analytic function $w = f(z) = u + iv$, $z = x + iy$, are continuous, verify that

$$\frac{\partial^2 u}{\partial x^2} + \frac{\partial^2 u}{\partial y^2} = 0 \quad \text{and} \quad \frac{\partial^2 v}{\partial x^2} + \frac{\partial^2 v}{\partial y^2} = 0 .$$

33. If a function $w = f(z)$ has a ______________ at every point of some region G in the z-plane, then the function is said to be ___________ in G .

34. For the function $w = e^x(\cos y + i \sin y)$ in Problem 32 above,

$$\frac{\partial^2 u}{\partial x^2} + \frac{\partial^2 u}{\partial y^2} = \_\_\_\_\_\_\_\_\_\_ + \_\_\_\_\_\_\_\_\_\_ = 0 ,$$

and

$$\frac{\partial^2 v}{\partial x^2} + \frac{\partial^2 v}{\partial y^2} = \_\_\_\_\_\_\_\_\_\_ + \_\_\_\_\_\_\_\_\_\_ = 0 .$$

---

31. $x - iy$, 1, 0, does not, is not

32. $e^x \cos y$, $-e^x \sin y$, does, is

33. derivative, analytic

34. $e^x \cos y + (-e^x \cos y)$, $e^x \sin y + (-e^x \sin y)$

17-6 Complex Series.

OBJECTIVE: Given a complex power series

$$\sum_{n=0}^{\infty} a_n z^n = a_0 + a_1 z + a_2 z^2 + \dots ,$$

find the region in the complex plane in which the series converges absolutely. (Use convergence tests studied for real series because $\sum |a_n z^n|$ is a series of nonnegative real numbers.)

35. Let $s_n = u_n(x,y) + i v_n(x,y)$ denote the nth-partial sum of the power series $\sum_{n=0}^{\infty} a_n z^n$. Then the power series converges at a point $z = x + iy$ to the value $w = u + iv$ if and only if

__________ and __________ at (x,y) .

36. If the series of absolute values $\sum_{n=0}^{\infty} |a_n z^n|$ converges, then the power series $\sum a_n z^n$ is said to ____________________ , and this implies convergence of the original power series (without absolute value signs).

37. Consider the complex series $\sum_{n=1}^{\infty} \frac{z^n}{n\,2^n}$. We apply the ratio test with

$$U_n = \left| \frac{z^n}{n\,2^n} \right| ,$$

and calculate

$$\frac{U_{n+1}}{U_n} = \_\_\_\_\_\_\_\_ \rightarrow \_\_\_\_ \quad \text{as} \quad n \rightarrow \infty .$$

---

35. $u_n \rightarrow u$, $v_n \rightarrow v$

36. converge absolutely

37. $\frac{n}{2(n+1)} |z|$, $\frac{|z|}{2}$, 2

This limit is less than unity if $|z| <$ ___ . Thus, the power series

$$\sum_{n=1}^{\infty} \frac{z^n}{n\,2^n}$$ converges absolutely for $|z| < 2$.

38. Consider the series $\sum_{n=0}^{\infty} \frac{(z-i)^n}{3^n}$. Let $U_n = \left|\frac{(z-i)^n}{3^n}\right|$, and apply the root test to U_n :

$$\sqrt[n]{U_n} = \_\_\_\_\_\_ \quad \text{and} \quad \lim_{n\to\infty} \sqrt[n]{U_n} = \_\_\_\_\_\_ .$$

This limit is less than unity for ________ < ___ . If $|z-i| = 3$, the series of absolute values is the divergent series

$$\sum_{n=0}^{\infty} 1 = 1 + 1 + 1 + \ldots$$. Therefore, the original power series

converges absolutely in the interior of the circle in the z-plane centered at the point $z_0 = i$ with radius $r = 3$.

17-7 Elementary Functions.

OBJECTIVE A: Establish elementary identities involving the five functions e^z , $\cos z$, $\sin z$, $\cosh z$, and $\sinh z$, by appealing either to their series definitions, or to their expressions in terms of exponentials. It may be required to use the exponential law $e^{z_1} \cdot e^{z_2} = e^{z_1+z_2}$ developed in the text.

39. We will establish that for any complex number z ,

$$\cos 2z = \cos^2 z - \sin^2 z .$$

Now, in terms of exponentials,

$$\cos z = \_\_\_\_\_\_ \quad \text{and} \quad \sin z = \_\_\_\_\_\_ .$$

---

38. $\frac{|z-i|}{3}$, $\frac{|z-i|}{3}$, $|z-i| < 3$

39. $\frac{1}{2}\left(e^{iz} + e^{-iz}\right)$, $\frac{1}{2i}\left(e^{iz} - e^{-iz}\right)$, $-\frac{1}{4}\left(e^{2iz} - 2e^0 + e^{-2iz}\right)$, $\frac{1}{2}\left(e^{2iz} + e^{-2iz}\right)$

Thus, squaring each side of these equations,

$$\cos^2 z = \frac{1}{4}\left(e^{2iz} + 2e^0 + e^{-2iz}\right), \quad \text{and} \quad \sin^2 z = \underline{\hspace{4cm}} .$$

It follows that

$$\cos^2 z - \sin^2 z = \underline{\hspace{4cm}} ,$$

and the right side of this last equation is precisely the exponential expression of $\cos 2z$ (when $2z$ is substituted for z in the above exponential expression of $\cos z$). Notice that we used $e^{z_1} \cdot e^{z_2} = e^{z_1 + z_2}$ when we squared the exponential expressions for $\cos z$ and $\sin z$.

40. In the text it was found that

$$\cos z = \cosh iz \quad \text{and} \quad i \sin z = \underline{\hspace{2cm}} .$$

Thus, using these equations and the exponential expressions for $\cosh z$ and $\sinh z$, we find

$$\cos iz = \cosh i^2 z = \underline{\hspace{3cm}} = \frac{1}{2}\left(e^{-z} + e^{-(-z)}\right) = \underline{\hspace{2cm}} ,$$

and

$$\sin iz = \frac{1}{i} \sinh i^2 z = \underline{\hspace{3cm}} = \frac{1}{2i}\left(e^{-z} - e^{-(-z)}\right)$$

$$= -\frac{1}{i} \sinh z = \underline{\hspace{3cm}} , \quad \text{since } i^2 = -1 \text{ implies } \frac{1}{i} = \underline{\hspace{1cm}} .$$

These relationships are very useful.

41. Euler's formula gives the identity

$$e^{iz} = \underline{\hspace{4cm}} ,$$

for all complex numbers z. It follows that, if $z = y$ is a real number, and if x is a real number, then

$$e^{x+iy} = e^x \cdot e^{iy} \quad \text{(by the exponential law)}$$

$$= e^x \left(\underline{\hspace{4cm}}\right) .$$

---

40. $\sinh iz$, $\cosh(-z)$, $\cosh z$, $\frac{1}{i}\sinh(-z)$, $i \sinh z$, $-i$

41. $\cos z + i \sin z$, $\cos y + i \sin y$, $e^x \cos y$, $e^x \sin y$

Therefore, the real part of the complex function $w = e^{x+iy}$ is

$u =$ ___________ and the imaginary part is $v =$ ___________ .

OBJECTIVE B: Calculate the derivatives of elementary complex functions by differentiating the appropriate power series term by term.

42. The series definition of the hyperbolic cosine function is

$\cosh z =$ ______________________________ ,

from which term by term differentiation yields,

$\frac{d}{dz} \cosh z =$ ______________________________ .

The power series on the right side of the previous equation is the series definition for __________ . Thus,

$\frac{d}{dz} \cosh z =$ ___________ .

17-8 Logarithms.

OBJECTIVE: Find all values of $\log z$ for a specified complex number $z \neq 0$.

43. In terms of polar coordinates, let $z = re^{i\theta}$. Then,

$\log z =$ ____ , $-\pi < \theta \leq \pi$.

44. Let us find all values of $\log\left(-2 + i\,2\sqrt{3}\right)$. In Problem 19 above we found that $z = -2 + i\,2\sqrt{3} = 4 \text{ cis } \frac{2\pi}{3}$. Thus,

$\log z =$ ____

$\approx 1.386 + 2\pi i\left(\frac{1+6n}{3}\right)$, $n = 0, \pm 1, \pm 2, \ldots$

---

42. $1 + \frac{z^2}{2!} + \frac{z^4}{4!} + \frac{z^6}{6!} + \ldots$, $z + \frac{z^3}{3!} + \frac{z^5}{5!} + \ldots$, $\sinh z$, $\sinh z$

43. $\ln r + i(\theta + 2n\pi)$, $n = 0, \pm 1, \pm 2, \ldots$

44. $\ln 4 + i\left(\frac{2\pi}{3} + 2n\pi\right)$, $n = 0, \pm 1, \pm 2, \ldots$

45. The principal value of $\log\left(-2 + i\,2\sqrt{3}\right)$ is

$\ln\left(-2 + i\,2\sqrt{3}\right)$ = ____________________ $\approx$ 1.386 + i 2.09 .

46. Consider $\log\left(\frac{1+2i}{2-i}\right)$. Here

$z = \frac{1+2i}{2-i} = i$ or, in polar coordinates, $z = \text{cis}\,\frac{\pi}{2}$. Thus,

$\log\left(\frac{1+2i}{2-i}\right)$ = ______________________________ .

CHAPTER 17 OBJECTIVE - PROBLEM KEY

Objective	Problems in Thomas/Finney Text	Objective	Problems in Thomas/Finney Text
17-1	p. 832, 1,2	17-5 A	p. 846, 1-4
17-2 A	p. 838, 14-17	B	p. 846, 5,6
B	p. 838, 1,2,6,8,9-13	17-6	p. 848, 1-8
C	p. 838, 18-22, 26	17-7 A	p. 853, 2,7-12,15,16
17-3 A	p. 841, 1	B	p. 853, 3
B	p. 841, 4,5	17-8	p. 855, 1,2
17-4	p. 844, 1-4		

CHAPTER 17 SELF-TEST

1. Express each of the following combinations as a single complex number in the form $x + iy$.

   (a) $(-8+5i) - 2(1-3i)$ (b) $(3-2i)(i-1)$

   (b) $\frac{1-4i}{i-2}$ (d) $(4+2i) + \frac{1}{i}\left(3-i^3\right)$

2. Express the given complex number in the form $r\,\text{cis}\,\theta$, with $r \geq 0$ and $-\pi < \theta \leq \pi$. Also, graph each point in an Argand diagram.

   (a) $z_1 = -1+i$ (b) $z_2 = \frac{1}{i}$ (c) $z_3 = -3 + i\sqrt{7}$

3. (a) Find $(1+i)^{20}$.

   (b) Find all complex z satisfying the equation $z^6 = 1 - i$.

---

45. $\ln 4 + i\,\frac{2\pi}{3}$ 46. $\ln 1 + i\left(\frac{\pi}{2} + 2n\pi\right)$, $n = 0, \pm 1, \pm 2, \ldots$

4. For the function $w = z^2$, sketch the images in the w-plane of the following sets in the z-plane. Use of polar coordinates may be helpful.

(a) $|z| < \sqrt{2}$, $-3\pi/8 < \arg z \leq \pi/4$ (b) $\arg z = \pi/8$

5. At which points in the z-plane is the function

$$f(z) = \frac{2z^3 - 3z + 1}{z^3 - 1}$$

continuous? Justify your answer.

6. Find the derivative with respect to z of the function $w = f(z)$ in Problem 5 above. Use the differentiation formulas.

7. Use the definition of the derivative in Eq. (1), Article 17-4 of the text, to find $f'(i)$ if $f(z) = z^2 + 1$.

8. Use the Cauchy-Riemann equations to determine whether the following functions are differentiable.

(a) $w = z + \bar{z}$ (b) $w = x + iy^3$

9. Find the region in the complex plane for which the series,

$$\sum_{n=1}^{\infty} (-1)^{n-1} \frac{(z+i)^n}{n\,5^n},$$

converges absolutely.

10. Show that $\sin(-z) = -\sin z$ and $\cos(-z) = \cos z$.

11. Show that the real and imaginary parts of

$$w = \cosh z = \cosh(x+iy) = \cosh x \cos y + i \sinh x \sin y,$$

satisfy the Cauchy-Riemann equations.

12. By differentiating an appropriate series term by term, show that

$$\frac{d}{dz}\left(e^{iz}\right) = ie^{iz}.$$

13. Find all values of $\log z$ for each of the following complex numbers z.

(a) $-\sqrt{3} + i$ (b) e^{1+i}

14. Show that $e^{\log z} = z$ for every complex number z.

SOLUTIONS TO CHAPTER 17 SELF-TEST

1. (a) $-10+11i$ (b) $-1+5i$

(c) $\dfrac{1-4i}{i-2} = \dfrac{(1-4i)(-2-i)}{(-2+i)(-2-i)} = \dfrac{-6+7i}{4+1} = -\dfrac{6}{5} + \dfrac{7}{5}i$

(d) $(4+2i) + \dfrac{1}{i}(3-i^3) = (4+2i) + (-i)(3+i) = 5-i$

2. (a) $|z_1| = \sqrt{(-1)^2+1^2} = \sqrt{2}$

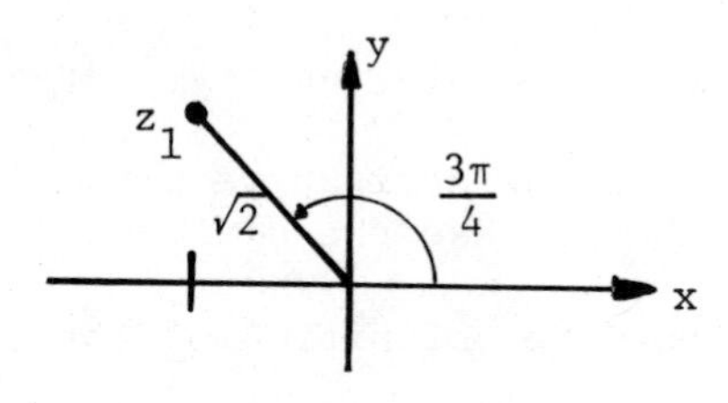

$\theta = \arg z_1 = \tan^{-1}\dfrac{1}{(-1)} = \dfrac{3\pi}{4}$

$z_1 = \sqrt{2}\ \text{cis}\ \dfrac{3\pi}{4}$

(b) $z_2 = \dfrac{1}{i} = -i$

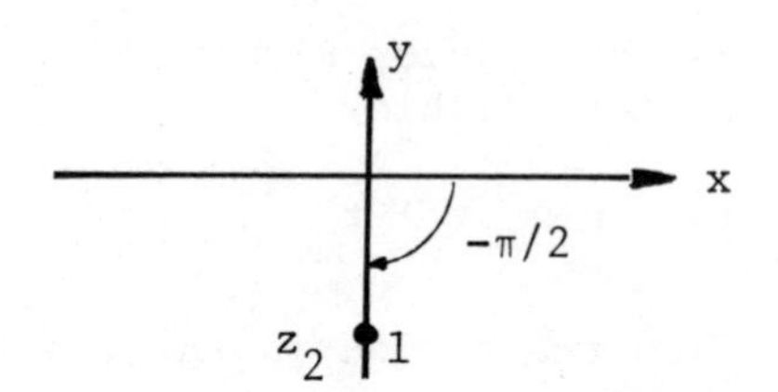

$|z_2| = 1$ and $\arg z_2 = -\dfrac{\pi}{2}$

$z_2 = \text{cis}\left(-\dfrac{\pi}{2}\right)$

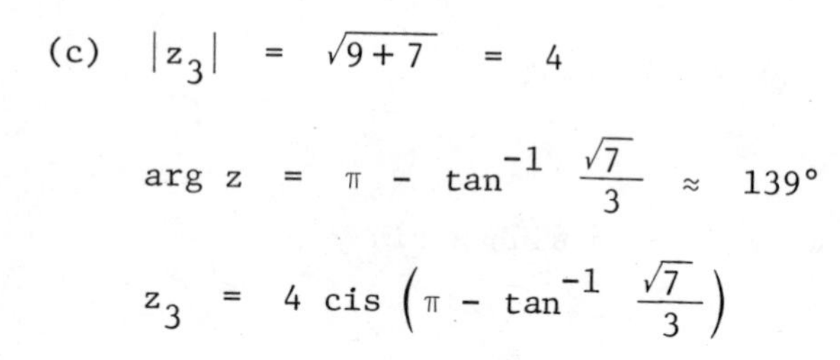

(c) $|z_3| = \sqrt{9+7} = 4$

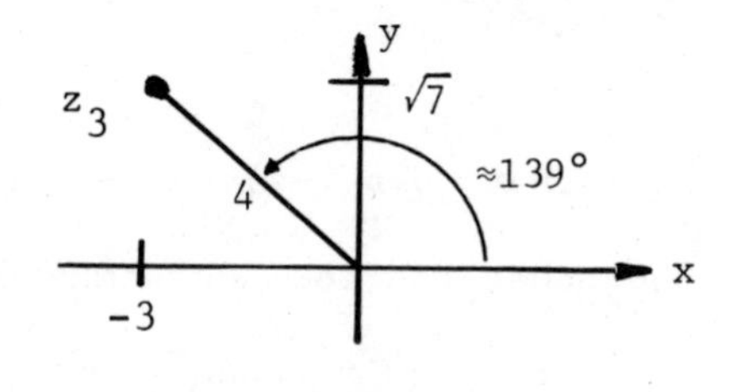

$\arg z = \pi - \tan^{-1}\dfrac{\sqrt{7}}{3} \approx 139°$

$z_3 = 4\ \text{cis}\left(\pi - \tan^{-1}\dfrac{\sqrt{7}}{3}\right)$

3. (a) $z = 1+i$ in polar form can be written $z = \sqrt{2}\ \text{cis}\ \dfrac{\pi}{4}$.

Thus, $z^{20} = \left(\sqrt{2}\right)^{20}\ \text{cis}\ 5\pi$ or, $z^{20} = -1024$.

(b) $z^6 = 1-i$ can be written $z^6 = \sqrt{2}\ \text{cis}\left(-\dfrac{\pi}{4}\right)$.

Thus, $z = 2^{1/12}\ \text{cis}\left(-\dfrac{\pi}{24} + k\dfrac{\pi}{3}\right)$, $k = -2,-1,0,1,2,3$.

Hence, the six roots are (using $2^{1/2} = r \approx 1.06$),

$$w_0 = r \text{ cis } \frac{7\pi}{24}, \quad w_1 = r \text{ cis } \frac{15\pi}{24}, \quad w_2 = r \text{ cis } \frac{23\pi}{24},$$

$$w_3 = r \text{ cis } \left(-\frac{\pi}{24}\right), \quad w_4 = r \text{ cis } \left(-\frac{9\pi}{24}\right), \quad w_5 = r \text{ cis } \left(-\frac{17\pi}{24}\right).$$

Remark: Our choice of the polar angles is in keeping with the polar notation $r \text{ cis } \theta$, where $-\pi < \theta \leq \pi$. However, the angles

$$-\frac{\pi}{24} + k\frac{\pi}{3}, \quad k = 0,1,2,3,4,5$$

would produce the same six points $w_0, w_1, w_2, w_3, w_4, w_5$, as would any six successive values for the integer k.

4. (a) (b)

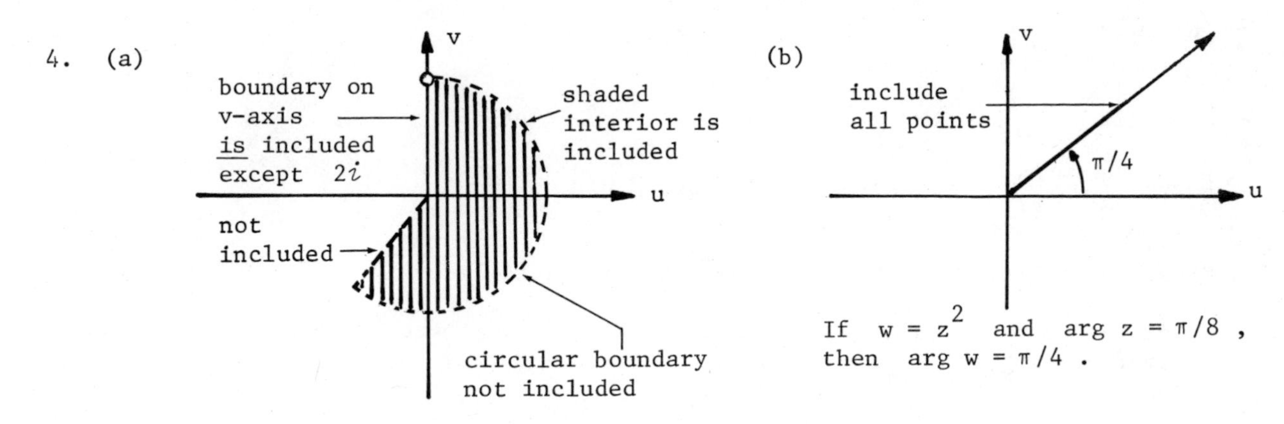

5. Since $w = f(z)$ is a rational function (i.e., a quotient of two complex polynomials), it is differentiable at every point except where $z^3 - 1 = 0$. Thus, $f(z)$ is continuous at every point in the z-plane except the cube roots of unity $z_1 = 1$, $z_2 = \left(-1 + i\sqrt{3}\right)/2$, and $z_3 = \left(-1 - i\sqrt{3}\right)/2$

6. $$\frac{d}{dz}\left(\frac{2z^3 - 3z + 1}{z^3 - 1}\right) = \frac{(z^3-1)(6z^2-3) - (2z^3-3z+1)(3z^2)}{\left(z^3-1\right)^2}$$

$$= \frac{6z^3 - 9z^2 + 3}{\left(z^3-1\right)^2}, \quad z^3 - 1 \neq 0.$$

7. Let $w = f(z) = z^2 + 1$. Then,

$$\frac{\Delta w}{\Delta z} = \frac{\left((z+\Delta z)^2 + 1\right) - (z^2+1)}{\Delta z} = \frac{2z\Delta z + \Delta z^2}{\Delta z} = 2z + \Delta z.$$

Also,

$$\lim_{\Delta z \to 0} \frac{\Delta w}{\Delta z} = 2z ,$$

since $|2z + \Delta z - 2z| = |\Delta z| \to 0$ as $\Delta z \to 0$,

independent of how Δz approaches zero.

Hence, $f'(i) = 2i$.

8. (a) $w = z + \bar{z} = 2x + i0$; $u = 2x$, $v = 0$

$$\frac{\partial u}{\partial x} = 2 , \quad \frac{\partial v}{\partial y} = 0 \qquad \text{so} \quad \frac{\partial u}{\partial x} \neq \frac{\partial v}{\partial y}$$

and the function is not differentiable for any z .

(b) $w = x + iy^3$; $u = x$, $v = y^3$

$$\frac{\partial u}{\partial x} = 1 , \quad \frac{\partial v}{\partial y} = 3y^2 , \quad \frac{\partial u}{\partial y} = 0 , \quad \frac{\partial v}{\partial x} = 0 .$$

Now, $\frac{\partial u}{\partial x} = \frac{\partial v}{\partial y}$ implies $3y^2 = 1$ or, $y = \pm \frac{1}{\sqrt{3}}$.

Thus, the function is differentiable at all points $z = x \pm i \frac{1}{\sqrt{3}}$,

since the Cauchy-Riemann equations are satisfied at these points and the partial derivatives u_x, u_y, v_x, v_y are all continuous there.

9. We apply the root test to $U_n = \left| \frac{(z+i)^n}{n\, 5^n} \right|$. Thus,

$$\sqrt[n]{U_n} = \frac{1}{5 \sqrt[n]{n}} |z+i| \to \frac{1}{5} |z+i| \quad \text{as} \quad n \to \infty .$$ Hence, $\frac{1}{5} |z+i| < 1$ implies

$|z+i| < 5$. The series converges absolutely for $|z+i| < 5$.

10. $$\sin(-z) = \frac{1}{2i} \left(e^{i(-z)} - e^{-i(-z)} \right) = - \frac{1}{2i} \left(e^{iz} - e^{-iz} \right) = - \sin z ,$$

$$\cos(-z) = \frac{1}{2} \left(e^{i(-z)} + e^{-i(-z)} \right) = \frac{1}{2} \left(e^{-iz} + e^{iz} \right) = \cos z .$$

11. $u = \cosh x \cos y$ and $v = \sinh x \sin y$

$$\frac{\partial u}{\partial x} = \sinh x \cos y = \frac{\partial v}{\partial y} \qquad \text{and} \qquad \frac{\partial v}{\partial x} = \cosh x \sin y = - \frac{\partial u}{\partial y}$$

so the Cauchy-Riemann equations are satisfied.

12. $$e^{iz} = 1 + iz + \frac{i^2z^2}{2!} + \frac{i^3z^3}{3!} + \frac{i^4z^4}{4!} + \cdots$$

$$\frac{d}{dz}(e^{iz}) = i + 2\frac{i^2z}{2!} + 3\frac{i^3z^2}{3!} + 4\frac{i^4z^3}{4!} + \cdots$$

$$= i\left(1 + iz + \frac{i^2z^2}{2!} + \frac{i^3z^3}{3!} + \cdots\right)$$

$$= ie^{iz} \quad \text{, as claimed.}$$

13. (a) The complex number $-\sqrt{3} + i$ has polar coordinates $r = 2$, $\theta_0 = 5\pi/6$. Hence

$$-\sqrt{3} + i = 2e^{i(5\pi/6 + 2n\pi)}$$

and

$$\log\left(-\sqrt{3} + i\right) = \ln 2 + i\left(\frac{5\pi}{6} + 2n\pi\right) \quad , \quad n = 0, \pm 1, \pm 2, \ldots .$$

(b) The complex number $e^{1+i} = e \cdot e^{i}$ has polar coordinates $r = e$, $\theta_0 = 1$. Hence

$$\log\left(e^{1+i}\right) = \ln e + i(1 + 2n\pi) \quad , \quad n = 0, \pm 1, \pm 2, \ldots$$

$$= 1 + i(1 + 2n\pi) .$$

Notice that the principal value of $\log\left(e^{1+i}\right)$ is $1+i$, which occurs here when $n = 0$.

14. Let us establish that $e^{\log z} = z$. In polar notation, let θ_0 denote the principle argument of z, $-\pi < \theta_0 \le \pi$, and

$$z = re^{i\theta_0} ,$$

Then, $$e^{\log z} = e^{\ln r + i(\theta_0 + 2n\pi)}$$

$$= e^{\ln r} \cdot e^{i(\theta_0 + 2n\pi)}$$

$$= r\left[\cos(\theta_0 + 2n\pi) + i \sin(\theta_0 + 2n\pi)\right]$$

$$= r\left(\cos\theta_0 + i\sin\theta_0\right) = z .$$

CHAPTER 18 DIFFERENTIAL EQUATIONS

18-1 Introduction.

1. A differential equation is an equation that involves one or more ________________________ , or ________________________ .

2. Differential equations are classified by ____________ , ____________ , and ____________ .

3. The order of a differential equation is that of the highest ______________ ____________ that occurs in the equation.

4. The degree of a differential equation is the ____________ of the highest power of the ___________________________________ , after the equation has been cleared of fractions and radicals in the dependent variable and its derivatives.

5. The differential equation

$$x \frac{d^3y}{dx^3} + \left(\frac{dy}{dx}\right)^2 - e^x y = 0$$

is an ________________________ differential equation of order ____________ and degree ____________ .

6. The differential equation

$$\frac{\partial u}{\partial t} = \left(h^2 \frac{\partial^2 u}{\partial x^2} + \frac{\partial^2 u}{\partial y^2}\right)$$

is a ________________________ differential equation of order ____________ and degree ____________ .

18-2 Solutions.

7. A function $y = f(x)$ is said to be a _______________ of a differential equation if the latter is satisfied when ______ and its __________________ are replaced throughout by ______ and its corresponding derivatives.

---

1. derivatives, differentials
2. type, order, degree
3. order derivative
4. exponent, highest order derivative
5. ordinary, three, one
6. partial, two, one
7. solution, y, derivatives, f(x),

8. If c_1 and c_2 are any constants, then for

$$y = c_1 \cos x + c_2 \sin x$$

we find

$$\frac{dy}{dx} = \underline{\hspace{6cm}},$$

and

$$\frac{d^2y}{dx^2} = \underline{\hspace{6cm}}.$$

Thus, the function $f(x) = c_1 \cos x + c_2 \sin x$ is a solution to the differential equation

$$\frac{d^2y}{dx^2} + y = \underline{\hspace{1.5cm}}.$$

OBJECTIVE: Show that a given function is a solution to a specified (ordinary) differential equation.

9. Consider the differential equation

$$3xy' - y = \ln x + 1, \; x > 0 ,$$

and the function $y = f(x) = Cx^{1/3} - \ln x - 4$, where C is any constant. Then,

$$\frac{dy}{dx} = f'(x) = \underline{\hspace{6cm}}, \text{ and}$$

$$3xy' = \underline{\hspace{5cm}}. \text{ Thus,}$$

$$3xy' - y = \left(Cx^{1/3} - 3\right) - \left(Cx^{1/3} - \ln x - 4\right)$$

$$= \underline{\hspace{4cm}}.$$

Therefore, the function $f(x) = Cx^{1/3} - \ln x - 4$, and its derivative, satisfy the differential equation. We have verified that $y = f(x)$ is a solution.

---

8. $-c_1 \sin x + c_2 \cos x$, $-c_1 \cos x - c_2 \sin x$, 0

9. $\frac{1}{3} Cx^{-2/3} - \frac{1}{x}$, $Cx^{1/3} - 3$, $\ln x + 1$

10. Consider the second order differential equation

$$2xy'' + (1 - 4x)y' + (2x - 1)y = e^x ,$$

and the function $y = f(x) = (c_1 + c_2\sqrt{x} + x)e^x$, $x > 0$, where c_1 and c_2 are any constants. Then,

$$y' = \Big(\_\_\_\_\_\_\_\_\_\_\_\_\_\_\_\_\_\_\_\_\Big)e^x , \text{ and}$$

$$y'' = \Big(\_\_\_\_\_\_\_\_\_\_\_\_\_\_\_\_\_\_\_\_\Big)e^x .$$

Then,

$$2xy'' = \left[(2c_1 + 4)x + 2x^2 - \frac{1}{2} c_2 x^{-1/2} + 2c_2 x^{1/2} + 2c_2 x^{3/2}\right]e^x ,$$

$$(1 - 4x)y' = \left[\_\_\_\_\_\_\_\_\_\_\_\_\_\_\_\_\_\_\_\_\_\_\_\_\right]e^x ,$$

$$(2x - 1)y = \left[-c_1 + (2c_1 - 1)x + 2x^2 - c_2 x^{1/2} + 2c_2 x^{3/2}\right] e^x .$$

Therefore,

$$2xy'' + (1 - 4x)y' + (2x - 1)y = \_\_\_\_\_\_\_\_\_\_ .$$

We conclude that $y = (c_1 + c_2\sqrt{x} + x)e^x$ __________ a solution to the differential equation.

18-3 First Order: Variables Separable.

OBJECTIVE: Solve first order differential equations in which the variables can be separated. If initial conditions are prescribed, determine the value of the constant of integration.

11. $(xy - x)dx + (xy + y)dy = 0$

We separate the variables and integrate:

$x(y-1)\ dx + y(x+1)\ dy = 0$, $\dfrac{y\ dy}{y-1} =$ ________ . Then,

$$\left(1 + \frac{1}{y - 1}\right) dy = \_\_\_\_\_\_\_\_\_\_\_\_$$

$$y + \ln|y - 1| = \_\_\_\_\_\_\_\_\_\_\_\_\_\_ + \ln C .$$

---

10. $c_1 + c_2\sqrt{x} + x + 1 + \frac{1}{2} c_2 x^{-1/2}$, $c_1 + c_2\sqrt{x} + x + 2 + c_2 x^{-1/2} - \frac{1}{4} c_2 x^{-3/2}$, $c_1 + 1 - (4c_1 + 3)x - 4x^2 + \frac{1}{2} c_2 x^{-1/2} - c_2 x^{1/2} - 4c_2 x^{3/2}$, e^x, is

11. $-\dfrac{x\ dx}{x + 1}$, $-\left(1 - \dfrac{1}{x + 1}\right)dx$, $- x + \ln|x + 1|$, $C|x + 1|$

We introduce $\ln C$, $C > 0$, as the constant of integration in order to simplify the form of the solution. Thus, by algebra,

$$x + y = \ln C \frac{|x + 1|}{|y - 1|} \quad \text{or,} \quad |y - 1|e^{x + y} = \underline{\qquad\qquad} .$$

12. $x^2 yy' = e^y$; when $x = 2$, $y = 0$

We change to differential form, separate the variables, and integrate:

$$x^2 \, y \, dy = e^y \, dx \quad \text{or,} \quad ye^{-y} \, dy = \underline{\qquad\qquad} .$$

We integrate the left side by parts, using $u = y$ and $dv = e^{-y} \, dy$:

$$\underline{\qquad\qquad\qquad\qquad} = -\frac{1}{x} + C$$

$$e^{-y}(\underline{\qquad}) = -\frac{1}{x} + C , \quad \text{or simplifying algebraically,}$$

$$x(y + 1) = \underline{\qquad\qquad} .$$

Using the initial condition $x = 2$ and $y = 0$ gives,

$$2(0 + 1) = \underline{\qquad} , \text{ or } C = \underline{\qquad} .$$

Thus, the solution is given by $x(y + 1) = \left(1 + \frac{x}{2}\right) e^y$.

18-4 First Order: Homogeneous.

OBJECTIVE A: Determine if a differential equation is homogeneous, and if it is, solve it.

13. If a differential equation can be put into the form $\frac{dy}{dx} = \underline{\qquad}$, then the equation is called homogeneous. The equation becomes separable in the variables x and v by defining $v = \underline{\qquad}$.

14. $3xy^2 \, dy = \left(4y^3 - x^3\right)dx$.

From the given equation, we have $\frac{dy}{dx} = \frac{4y^3 - x^3}{3xy^2} = \frac{1}{3}\left[4\left(\frac{y}{x}\right) - \underline{\qquad}\right]$.

The equation is homogeneous, and we let $v = y/x$, or $y = vx$.

---

12. $x^{-2} \, dx$, $-ye^{-y} + \int e^{-y} \, dy$, $y + 1$, $(1 - Cx)e^y$, $1 - 2C$, $-\frac{1}{2}$

13. $F\left(\frac{y}{x}\right)$, $\frac{y}{x}$

14. $(x/y)^2$, $v + x\frac{dv}{dx}$, $\frac{1}{3}\left(4v - v^{-2}\right)$, $v^3 - 1$, $\ln |v^3 - 1|$, $|x| = C\left|\frac{y^3}{x^3} - 1\right|$

Then $\frac{dy}{dx}$ = ____________ and the differential equation becomes,

$v + x\frac{dv}{dx}$ = ________________ , or separating the variables x and v ,

$$\frac{dx}{x} = 3\left(\frac{v^2\,dv}{\_\_\_\_\_\_}\right).$$

The solution of this is,

$$\ln|x| = \_\_\_\_\_\_\_\_\_\_ + \ln C$$

so that

$$|x| = C|v^3 - 1| .$$

In terms of x and y , the solution is

__________________________ .

OBJECTIVE B: Find the family of solutions of a given differential equation and the family of <u>orthogonal trajectories</u> (defined on page 864, in the Problem section).

15. Consider the differential equation $x + y\frac{dy}{dx} = 0$. Separating the variables, and integrating, gives

$x\,dx + y\,dy = 0$, and ___________ $= C_1$, where $C_1 > 0$.

Thus, the family of solutions is a family of ____________ centered at ______________ .

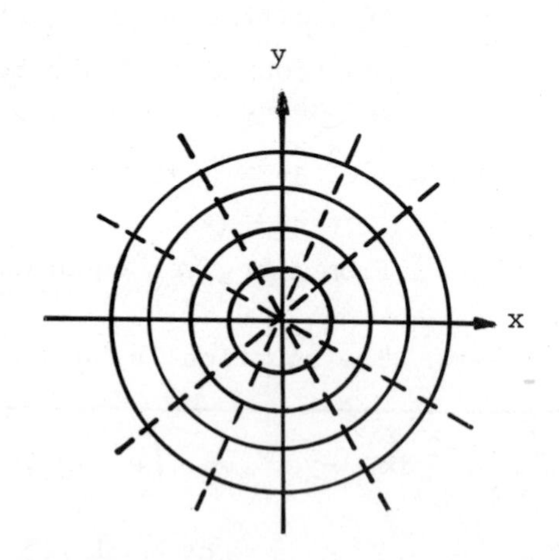

To find the family of orthogonal trajectories, we solve the differential equation __________________________ .
Separating the variables, and integrating, gives

$\frac{dy}{y}$ = ____ and $\ln|y|$ = _______ + $\ln C_2$,

or $y = \pm C_2 x$, $C_2 > 0$.

Notice that $y = 0$, the x-axis, also solves the differential equation for the orthogonal trajectories; so does $x = 0$, the y-axis. Thus, the orthogonal trajectories are the family of __________ passing through the origin. A sketch of the solution family and the family of orthogonal trajectories is given above.

---

15. $x^2 + y^2$, circles, the origin, $y\,dx - x\,dy = 0$, $\frac{dx}{x}$, $\ln|x|$, lines

18-5 First Order: Linear.

OBJECTIVE: Determine if a differential equation of first order is linear, and if it is, solve it.

16. A differential equation of first order, which is linear in the dependent variable y, can always be put in the standard form

_______________ ,

where P and Q are functions of x.

17. Assuming that P and Q are continuous functions of x, we can solve a linear differential equation $y' + Py = Q$ by finding an integrating factor,

$\rho =$ __________ ,

providing a solution $\rho y =$ _______________ .

18. A differential equation may be linear in x and dx/dy (se we are thinking of x as the dependent variable rather than y). In that case we can solve the equation by the technique outlined in Problem 17 after we ______________ the roles of x and y. A differential equation may be linear in one variable, but not in the other. It is also possible that it is linear in both variables.

19. Let us solve the equation $x\frac{dy}{dx} + (x-2)y = 3x^3e^{-x}$. In standard form,

$$\frac{dy}{dx} + \left(1 - \frac{2}{x}\right)y = 3x^2e^{-x}.$$

Here P = ________ and Q = __________ , and the differential equation is linear in y. An integrating factor is given by

$$\rho = e^{\int P\,dx} = e^{\int\left(1-\frac{2}{x}\right)dx} = e^{\_\_\_\_\_\_} = x^{\_\_} e^{x}.$$

Hence a solution is given by

$$x^{-2}e^{x}y = \int \_\_\_\_\_\_\_\_\_\_\_\_ \,dx + C = \int \_\_ \,dx + C$$

$$= \_\_\_\_\_\_\_\_ .$$

Thus,

$$y = \_\_\_\_\_\_\_\_\_\_\_\_ .$$

---

16. $\frac{dy}{dx} + Py = Q$

17. $e^{\int P\,dx}$, $\int \rho Q\,dx + C$

18. interchange

19. $1 - \frac{2}{x}$, $3x^2e^{-x}$, $x - 2\ln x$, -2, $x^{-2}e^{x}\cdot 3x^2e^{-x}$, 3, $3x + C$, $(3x^3 + Cx^2)e^{-x}$

20. Consider $y\,dx + (3x - xy + 2)\,dy = 0$.

Since the product $y\,dy$ occurs in the second term on the left side, the equation is not linear in y. It is, however, linear in x, and can be written in standard form as

$$\frac{dx}{dy} + \left(\frac{3}{y} - 1\right) x = \underline{\qquad} .$$

An integrating factor is found as,

$$\rho = e^{\underline{\qquad\qquad}} = e^{\underline{\qquad\qquad}} = y^3 e^{-y} .$$

Then a solution to the differential equation is,

$$\underline{\qquad\qquad} = \int y^3 e^{-y} \cdot \left(-\frac{2}{y}\right) dy + C$$

$$= -2 \int y^2 e^{-y}\, dy + C = 2y^2 e^{-y} + 4ye^{-y} + \underline{\qquad} + C,$$

or

$$xy^3 = 2y^2 + 4y + 4 + Ce^y .$$

21. The differential equation $y' = x - 4xy$ can be written in standard form as

$$y' + 4xy = x ,$$

so it is linear in the variable ___ . The equation may also be written in the form

$$\frac{dy}{1 - 4y} = \underline{\qquad} ,$$

so it is separable in the variables x and y. Thus we have a choice of methods of solution. As a separable equation, we integrate the last equation, and find

$$-\frac{1}{4} \ln |1 - 4y| = \frac{1}{2} x^2 + \ln C ,$$

or

$$|1 - 4y| = C_1 \underline{\qquad} , \quad \text{where } C_1 = C^{-4} .$$

If we consider the differential equation as linear in y, an integrating factor is

$$\rho = e^{\int 4x\, dx} = \underline{\qquad\qquad} ,$$

from which we get

---

20. $-\frac{2}{y}$, $\int \left(\frac{3}{y} - 1\right) dy$, $3 \ln y - y$, $xy^3 e^{-y}$, $4e^{-y}$

21. y, $x\,dx$, e^{-2x^2}, e^{2x^2}, xe^{2x^2}, $\frac{1}{4} e^{2x^2}$, $1 + 4C_2 e^{-2x^2}$

$$ye^{2x^2} = \int \_\_\_\_\_\_ \, dx + C_2 = \_\_\_\_\_\_ + C_2$$

or

$$4y = \_\_\_\_\_\_\_\_ .$$

If $1 - 4y < 0$, we choose $4C_2 = -C_1$, and if $1 - 4y \geq 0$, we choose $4C_2 = C_1$. Thus both solution forms agree.

18-6 First Order: Exact.

OBJECTIVE: Solve a differential equation that is exact. It might be necessary to make the given equation exact by multiplication by a suitable integrating factor $\rho(x,y)$.

22. $\cos y \, dx - (x \sin y - y^2) \, dy = 0$

$\frac{\partial}{\partial y}(\cos y) = -\sin y$ and $\frac{\partial}{\partial x}(-x \sin y + y^2) = \_\_\_\_\_\_$ so the equation is exact: that is, the left side is an exact differential $df(x,y)$.

Now, $\frac{\partial f}{\partial x} = \_\_\_\_$ so $f(x,y) = \_\_\_\_\_\_ + g(y)$.

Differentiating the last equation with respect to y with x held constant gives,

$\frac{\partial f}{\partial y} = \_\_\_\_\_\_\_\_\_\_$. Thus, $g'(y) = \_\_\_\_$. Hence,

$g(y) = \_\_\_\_ + C_1$ and $f(x,y) = -x \sin y + \frac{y^3}{3} + C_1$. Therefore,

$-x \sin y + \frac{y^3}{3} = C$, where C is an arbitrary constant, solves the differential equation.

23. Let us solve $3x^2y \, dx + (y^4 - x^3) \, dy = 0$.

Since $\frac{\partial}{\partial y}(3x^2y) = 3x^2 \neq -3x^2 = \frac{\partial}{\partial x}(y^4 - x^3)$, the equation, as it stands, is not exact. However, two terms in the coefficients of dx and dy are of degree three and the other coefficient is not of degree three, so we try regrouping the terms as

$$(3x^2y \, dx - x^3 \, dy) + y^4 \, dy = 0 .$$

If we rewrite this last equation as

$$\left[y \, d(x^3) - x^3 \, dy \right] + y^4 \, dy = 0 ,$$

the first two terms are suggestive of $d\left(\frac{u}{v}\right) = \_\_\_\_\_\_\_\_$.

---

22. $-\sin y$, $\cos y$, $x \cos y$, $-x \sin y + g'(y)$, y^2, $\frac{1}{3}y^3$

Therefore, we might divide the rewritten equation by ____ , obtaining

$$\frac{y\,d(x^3) - x^3\,dy}{y^2} + y^2\,dy = 0$$

or

$$d\left(\underline{\quad}\right) + y^2\,dy = 0\,, \quad \text{gives} \quad \underline{\qquad\qquad} = C\,.$$

18-7 Special Types of Second-Order Equations.

OBJECTIVE: Solve second-order differential equations in which the dependent variable, or the independent variable, is missing.

24. If a second-order equation has the special form

$$F\left(x,\ \frac{dy}{dx},\ \frac{d^2y}{dx^2}\right) = 0\,,$$

in which the ____________ variable is missing, we can reduce it to a first-order equation by substituting __________ and ____________ .

25. To solve $\frac{d^2y}{dx^2} = x\left(\frac{dy}{dx}\right)^3$, we substitute $p = \frac{dy}{dx}$ and $\frac{dp}{dx} = \frac{d^2y}{dx^2}$.

This gives, $\frac{dp}{dx} = $ ______ or $p^{-3}\,dp = $ _____ , $p^{-2} = -x^2 + C_1$,

$p = \frac{dy}{dx} = $ ___________ . Thus, $y + C_2 = \sin^{-1}\frac{x}{C_1}$ or $y + C_2 = $ ________ .

That is,

$$x = \underline{\qquad\qquad\qquad} \quad \text{or} \quad x = C_1\cos(y + C_2)\,.$$

However, since $\cos(y + C_2) = \sin\left(y + C_2 + \frac{\pi}{2}\right)$, and C_2 is an arbitrary constant, the second result is redundant, and we have $x = C_1\sin(y + C_2)$ as the general solution.

26. If a second-order equation has the special form

$$F\left(y,\ \frac{dy}{dx},\ \frac{d^2y}{dx^2}\right) = 0\,,$$

in which the _______________ variable is missing, the substitutions to use are

$$p = \underline{\quad} \quad \text{and} \quad \frac{d^2y}{dx^2} = \underline{\qquad}\,.$$

---

23. $\frac{v\,du - u\,dv}{v^2}$, y^2, $\frac{x^3}{y}$, $\frac{x^3}{y} + \frac{1}{3}y^3$ 24. dependent, $p = \frac{dy}{dx}$, $\frac{d^2y}{dx^2} = \frac{dp}{dx}$

25. xp^3, $x\,dx$, $\frac{\pm 1}{\sqrt{C_1^2 - x^2}}$, $\cos^{-1}\frac{x}{C_1}$, $C_1\sin(y + C_2)$ 26. independent, $\frac{dy}{dx}$, $p\frac{dy}{dx}$

27. Consider the equation $y \frac{d^2y}{dx^2} + \left(\frac{dy}{dx}\right)^2 + 1 = 0$. Let $\frac{dy}{dx} = p$, $\frac{d^2y}{dx^2} = p\frac{dp}{dy}$, and we have $\_\_\_\_\_\_ + p^2 + 1 = 0$.

We can separate the variables in the preceding equation, obtaining

$$\frac{p\,dp}{p^2+1} + \_\_\_\_ = 0 .$$

Integration gives, $\frac{1}{2}\ln(p^2+1) + \_\_\_\_\_\_ = \ln C_1$.

Solving the previous equation for p yields $\frac{dy}{dx} = p = \pm \_\_\_\_\_\_\_\_$.

Therefore,

$$\pm\, y\left(C_1^2 - y^2\right)^{-1/2} dy = dx ,$$

$$\mp \_\_\_\_\_\_\_\_ = x + C_2 , \quad \text{or} \quad (x+C_2)^2 = \_\_\_\_\_\_\_\_$$

is the general solution.

18-8 Linear Equations with Constant Coefficients.

28. A linear equation of order n can be written in the form

$\_\_\_\_\_\_\_\_\_\_\_\_\_\_\_\_\_\_\_\_\_\_\_\_\_\_\_\_\_\_\_\_\_\_\_\_\_\_\_\_$.

The coefficients $a_1, a_2, \ldots, a_n$ may be functions of x. If $F(x)$ is $\_\_\_\_\_\_\_\_\_\_$ the equation is said to be homogeneous; otherwise it is called $\_\_\_\_\_\_\_\_\_\_$.

29. The symbol D is introduced to represent differentiation with respect to $\_\_\_$, and powers of D mean taking successive derivatives:

$$D^2 f(x) = \_\_\_\_\_\_ \quad \text{and} \quad D^n f(x) = \_\_\_\_\_\_ , \quad n \geq 2 .$$

30. Let us find $(2D^2 - D + 3)(e^x + \sin x)$:

$$(2D^2 - D + 3)(e^x + \sin x) = 2D^2(e^x + \sin x) - D(\_\_\_\_\_\_\_\_) + 3(e^x + \sin x)$$

---

27. $yp\frac{dp}{dy}$, $\frac{dy}{y}$, $\ln|y|$, $\frac{\sqrt{C_1^2 - y^2}}{y}$, $\left(C_1^2 - y^2\right)^{1/2}$, $C_1^2 - y^2$

28. $\frac{d^ny}{dx^n} + a_1\frac{d^{n-1}y}{dx^{n-1}} + a_2\frac{d^{n-2}y}{dx^{n-2}} + \ldots + a_{n-1}\frac{dy}{dx} + a_n y = F(x)$, identically zero, nonhomogeneous

29. x, $\frac{d^2 f(x)}{dx^2}$, $\frac{d^n f(x)}{dx^n}$

$= 2D(\_\_\_\_\_\_\_\_\_\_) - (e^x + \cos x) + 3(e^x + \sin x)$

$= 2(\_\_\_\_\_\_\_\_\_\_) + 2e^x + 3 \sin x - \cos x$

$= \_\_\_\_\_\_\_\_\_\_\_\_\_\_\_\_\_\_\_\_.$

31. Linear differential operators that are polynomials in D with constant coefficients satisfy basic algebraic laws that make it possible to treat them as ____________________ so far as __________, multiplication, and __________ are concerned.

18-9 Linear, Second Order, Homogeneous Equations with Constant Coefficients.

OBJECTIVE: Solve linear, second-order, homogeneous equations with constant coefficients. The roots of the characteristic equation of the differential equation may be real and unequal, real and equal, or a pair of complex conjugate numbers.

32. $2 \frac{d^2y}{dx^2} + 5 \frac{dy}{dx} - 3y = 0$

The associated characteristic equation is given by ________________. This equation factors into $(2r - 1)(\_\_\_\_\_\_) = 0$, so the roots are $r_1 = \_\_\_\_$ and $r_2 = \_\_\_\_$. Since these roots are real and unequal, the solution of the differential equation is

$y = \_\_\_\_\_\_\_\_\_\_\_\_\_\_\_\_\_\_.$

33. $\frac{d^2y}{dx^2} - 4 \frac{dy}{dx} + 4y = 0$

The characteristic equation is ________________, and has roots $r_1 = \_\_\_\_$ and $r_2 = \_\_\_\_$. Since these roots are real and equal, the solution of the differential equation is

$y = \_\_\_\_\_\_\_\_\_\_\_\_\_\_\_\_.$

34. $\frac{d^2y}{dx^2} - 6 \frac{dy}{dx} + 13y = 0$

The characteristic equation is ________________, and has roots

$r_1 = \_\_\_\_\_\_$ and $r_2 = \_\_\_\_\_\_.$

---

30. $e^x + \sin x$, $e^x + \cos x$, $e^x - \sin x$, $4e^x + \sin x - \cos x$

31. ordinary polynomials, addition, factoring

32. $2r^2 + 5r - 3 = 0$, $r + 3$, $1/2$, -3, $C_1e^{x/2} + C_2e^{-3x}$

33. $r^2 - 4r + 4 = 0$, 2, 2, $(C_1 + C_2x)e^{2x}$

Thus, these roots are a pair of conjugate complex numbers with α = ___ and β = ___ . The solution of the differential equation is

$$y = \underline{\qquad\qquad\qquad\qquad} .$$

18-10 Linear, Second-Order, Nonhomogeneous Equations with Constant Coefficients.

OBJECTIVE: Use the method of variation of parameters to solve linear, second-order, nonhomogeneous equations with constant coefficients.

35. A method for solving the nonhomogeneous equation

$$\frac{d^2y}{dx^2} + 2a\frac{dy}{dx} + by = F(x) ,$$

is first to obtain the general solution y_h of the related homogeneous equation obtained by replacing ________________ . Let this solution be denoted by

$$y_h = C_1u_1(x) + C_2u_2(x) .$$

Next, determine two functions $v_1 = v_1(x)$ and $v_2 = v_2(x)$ in the following way: the derivatives v_1' and v_2' must satisfy the two equations

____________________ and ________________________ .

Solve this pair of simultaneous equations for v_1' and v_2', and integrate these functions to obtain the functions $v_1 = v_1(x)$ and $v_2 = v_2(x)$ (don't forget the constants of integration). Then, the general solution of the nonhomogeneous differential equation is given by

$$y = \underline{\qquad\qquad\qquad\qquad} .$$

The method described above is known as __________________________ .

36. Let us solve the equation $\frac{d^2y}{dx^2} - y = \frac{2}{e^x + 1}$ by variation of parameters. We first solve the associated homogeneous equation: $\frac{d^2y}{dx^2} - y = 0$ gives the characteristic equation __________ , which has the roots r_1 = ____ and r_2 = ___ . Thus, we find the solutions

$$u_1(x) = \underline{\quad} \quad\text{and}\quad u_2(x) = \underline{\quad}$$

to the homogeneous equation.

---

34. $r^2 - 6r + 13 = 0$, $3 + 2i$, $3 - 2i$, 3, 2, $e^{3x}(C_1 \cos 2x + C_2 \sin 2x)$

35. $F(x)$ by zero, $v_1'u_1 + v_2'u_2 = 0$, $v_1'u_1' + v_2'u_2' = F(x)$, $v_1(x)u_1(x) + v_2(x)u_2(x)$, variation of parameters

Next, we demand that the functions v_1 and v_2 satisfy the equations

$v_1'e^{-x} + v_2'e^{x} = 0$, and ________________ .

By Cramer's rule:

$$v_1' = \frac{\begin{vmatrix} 0 & e^x \\ \frac{2}{e^x+1} & e^x \end{vmatrix}}{\begin{vmatrix} e^{-x} & e^x \\ -e^{-x} & e^x \end{vmatrix}} = \frac{-e^x\,[2/(e^x+1)]}{\_\_\_} = \_\_\_\_\_\_ ,$$

$$v_2' = \frac{\begin{vmatrix} e^{-x} & 0 \\ -e^{-x} & \frac{2}{e^x+1} \end{vmatrix}}{\begin{vmatrix} e^{-x} & e^x \\ -e^{-x} & e^x \end{vmatrix}} = \frac{\_\_\_\_\_\_}{2} = \frac{e^{-x}}{e^x+1} .$$

Integration then gives:

$$v_1 = \int \frac{-e^x\,dx}{e^x+1} = \_\_\_\_\_\_ + C_1 ,$$

$$v_2 = \int \frac{e^{-x}\,dx}{e^x+1} = \int \frac{-u\,du}{u+1} \qquad (u = e^{-x})$$

$$= \int \left(-1 + \frac{1}{u+1}\right) du = \_\_\_\_\_\_ + C_2 .$$

Therefore, the general solution is given by

$$y = v_1u_1 + \_\_\_$$

$$= \left[-\ln(e^x+1) + C_1\right]\_\_\_ + \left[-e^{-x} + \ln(e^{-x}+1) + C_2\right]\_\_\_$$

$$= C_1e^{-x} + C_2e^x - 1 - e^{-x}\ln(e^x+1) + e^x\ln(e^{-x}+1) .$$

---

36. $r^2 - 1 = 0$, -1, 1, e^{-x}, e^x, $-v_1'e^{-x} + v_2'e^x = \frac{2}{e^x+1}$, 2, $\frac{-e^x}{e^x+1}$, $2e^{-x}/(e^x+1)$, $-\ln(e^x+1)$, $-e^{-x} + \ln(e^{-x}+1)$, v_2u_2, e^{-x}, e^x

18-11 Higher Order Linear Equations with Constant Coefficients.

OBJECTIVE: Solve higher order linear, homogeneous and nonhomogeneous, differential equations with constant coefficients. Employ solution methods that extend the methods of Articles 18-9 and 18-10.

37. Consider the differential equation $\left(D^3 - D^2 + D - 1\right)\left(D^2 + 4\right)^2 y = 0$.

The associated characteristic equation is $\left(r^3 - r^2 + r - 1\right)\left(r^2 + 4\right)^2 = 0$ whose roots are $r_1 = 1$, $r_2 = i$, $r_3 = -i$, $r_4 = -2i$, $r_5 = -2i$, $r_6 =$ ___, and $r_7 =$ ___. Thus, the general solution is

$y =$ ____ .

38. Let us find the general solution of the equation

$$\frac{d^3y}{dx^3} - \frac{d^2y}{dx^2} + \frac{dy}{dx} - y = x .$$

The characteristic equation of the associated homogeneous equation is

____

whose roots are $r_1 = 1$, $r_2 =$ ___, and $r_3 =$ ___. Thus, we find the three solutions $u_1(x) = e^x$, $u_2(x) = \cos x$, and $u_3(x) =$ ______ to the homogeneous equation.

Next, we demand that the three functions $v_1 = v_1(x)$, $v_2 = v_2(x)$, and $v_3 = v_3(x)$ satisfy the following three equations:

$$v_1' e^x + v_2' \cos x + v_3' \sin x = \_\_\_$$

$$v_1' e^x - v_2'' \sin x + v_3' \cos x = \_\_\_$$

$$\_\_\_\_\_\_\_\_\_\_\_\_\_\_\_\_\_\_\_\_ = x .$$

By Cramer's rule:

$$v_1' = \frac{\begin{vmatrix} 0 & \cos x & \sin x \\ 0 & -\sin x & \cos x \\ x & -\cos x & -\sin x \end{vmatrix}}{\begin{vmatrix} e^x & \cos x & \sin x \\ e^x & -\sin x & \cos x \\ e^x & -\cos x & -\sin x \end{vmatrix}} = \frac{x(\cos^2 x + \sin^2 x)}{2e^x} = \_\_\_\_\_\_ .$$

---

37. $2i$, $2i$, $C_1 e^x + C_2 \cos x + C_3 \sin x + (C_4 + C_5 x)\cos 2x + (C_6 + C_7 x)\sin 2x$

Similarly, $v_2' = \frac{1}{2} x(\sin x - \cos x)$, and $v_3' =$ ________________ .

Integrating each of these functions by parts, we find

$$v_1 = \int \frac{1}{2} x e^{-x}\, dx = \rule{4cm}{0.4pt} ,$$

$$v_2 = \int \frac{1}{2} x(\sin x - \cos x)\, dx = -\frac{1}{2}\left[x(\sin x + \cos x) - \sin x + \cos x\right] + C_2 ,$$

$$v_3 = \int -\frac{1}{2} x(\sin x + \cos x)\, dx = \frac{1}{2}\left[x(\cos x - \sin x) - \sin x - \cos x\right] + C_3 .$$

Therefore, the general solution to the nonhomogeneous differential equation is

$$y = v_1 u_1 + v_2 u_2 + v_3 u_3 = \rule{5cm}{0.4pt} .$$

18-12 Vibrations.

OBJECTIVE: Find an equation for a given vibratory motion specified by a second-order linear differential equation with constant coefficients. The motion may be damped or undamped.

39. An object weighing 2-lbs is suspended from the lower end of a spring, immersed in a medium, with its upper end attached to a rigid support. The object extends the spring 32 inches. After the object has come to rest in its equilibrium position, it is given an additional pull downward of 1 ft and released. As it moves up and down, its motion, taking into account the resistance of the medium, is described by the differential equation

$$m \frac{d^2x}{dt^2} = -kx - \frac{1}{2}\frac{dx}{dt} ,$$

where m is the mass of the object and k is the spring constant. Find its subsequent motion.

Solution. Let us first determine the spring constant k . Since 2 lbs stretches the spring 32 in = 8/3 ft , by Hooke's law $ks = mg$ we find,

_____ = 2 , or $k =$ ___ .

---

38. $r^3 - r^2 + r - 1 = 0$, $-i$, i, $\sin x$, 0, 0, $v_1' e^x - v_2' \cos x - v_3' \sin x$,

$\frac{1}{2} x e^{-x}$, $-\frac{1}{2} x(\sin x + \cos x)$, $-\frac{1}{2} e^{-x}(x+1) + C_1$,

$C_1 e^x + C_2 \cos x + C_3 \sin x - (x+1)$

39. $\frac{8}{3} k$, $\frac{3}{4}$, $\frac{2}{32}$, -2, -6, $C_1 e^{-2t} + C_2 e^{-6t}$, 1, 0, $0 = -2C_1 - 6C_2$, $\frac{3}{2}$, $-\frac{1}{2}$, overcritical

The mass m of the object is $m = \underline{\quad}$, and so the differential equation of the motion becomes

$$\frac{1}{16}\frac{d^2x}{dt^2} + \frac{1}{2}\frac{dx}{dt} + \frac{3}{4}x = 0, \quad \text{or}$$

$$\frac{d^2x}{dt^2} + 8\frac{dx}{dt} + 12x = 0 .$$

The characteristic equation is $r^2 + 8r + 12 = 0$, and it has the two roots $r_1 = \underline{\quad}$ and $r_2 = \underline{\quad}$. Thus, the general solution is

$$x = \underline{\qquad\qquad\qquad}, \quad \text{with} \quad \frac{dx}{dt} = -2C_1e^{-2t} - 6C_2e^{-6t} .$$

The initial conditions are $t = 0$, $x = \underline{\quad}$, and $\frac{dx}{dt} = \underline{\quad}$.

Substituting these values into the general solution and its derivative, we find

$$1 = C_1 + C_2 \quad \text{and} \quad \underline{\qquad\qquad\qquad} .$$

Solving these equations simultaneously gives $C_1 = \underline{\quad}$ and $C_2 = \underline{\quad}$.

Therefore, an equation of the motion is

$$x = \frac{3}{2}e^{-2t} - \frac{1}{2}e^{-6t} .$$

Notice that as $t \to +\infty$, $x \to 0$. The motion is nonoscillatory and we have ________________ damping.

40. For the same spring system as in Problem 39 above, suppose the resistance of the medium is such that the motion is described by the differential equation

$$m\frac{d^2x}{dt^2} = -kx - \frac{3}{8}\frac{dx}{dt} .$$

As before, $k = \frac{3}{4}$ and $m = \frac{1}{16}$ so the equation becomes

$$\frac{1}{16}\frac{d^2x}{dt^2} + \frac{3}{8}\frac{dx}{dt} + \frac{3}{4}x = 0, \quad \text{or}$$

$$\frac{d^2x}{dt^2} + 6\frac{dx}{dt} + 12x = 0 .$$

In this case the roots of the characteristic equation are $r_1 = \underline{\qquad}$ and $r_2 = \underline{\qquad}$. Thus, the general solution is

$$x = \underline{\qquad\qquad\qquad\qquad}, \quad \text{or} \quad x = Ce^{-3t}\sin\left(\sqrt{3}\,t + \phi\right) .$$

---

40. $-3 + \sqrt{3}\,i$, $-3 - \sqrt{3}\,i$, $e^{-3t}\left[C_1\cos\sqrt{3}\,t + C_2\sin\sqrt{3}\,t\right]$, $C\sin\phi$, $\sqrt{3}$, 2, undercritical, $\frac{2\pi}{\sqrt{3}}$

We will use the initial conditions $t = 0$, $x = 1$, and $\frac{dx}{dt} = 0$ to determine the constants C and ϕ.

Differentiation of the general solution to t yields

$$\frac{dx}{dt} = -3Ce^{-3t} \sin\left(\sqrt{3}\, t + \phi\right) + C\sqrt{3}\, e^{-3t} \cos\left(\sqrt{3}\, t + \phi\right),$$

and substitution of the values for the initial conditions into these equations for x and dx/dt gives,

$$1 = \underline{\qquad\qquad} \quad \text{and} \quad 0 = -3C \sin\phi + \sqrt{3}\, C \cos\phi .$$

Substituting $C = \csc\phi$ from the first of these equations into the second, we find

$$\cot\phi = \underline{\qquad}, \quad \text{so} \quad \phi = \frac{\pi}{6} \text{ or } \frac{7\pi}{6} .$$

Choosing $\phi = \frac{\pi}{6}$ and $\sin\phi = \frac{1}{2}$ gives $C =$ ___ .

Therefore, the equation of the motion becomes

$$x = 2e^{-3t} \sin\left(\sqrt{3}\, t + \frac{\pi}{6}\right).$$

As $t \to +\infty$, $x \to 0$. The motion is oscillatory and we have ______________ damping. The damped period of the motion is $T =$ _____ .

18-13 Poisson Probability Distribution.

OBJECTIVE: Assume that a random variable X, which takes on only nonnegative integer values $n = 0, 1, 2, \ldots$, is a Poisson random variable. For a specified nonnegative integer k, find the probability that $n = k$, or $n > k$, or $n \leq k$, etc. over a specified t interval. It may be required to calculate the mean proportionality factor λ.

41. Suppose that the average number of oil tankers arriving each day at a certain port city is known to be 5. The facilities at the port can handle at most 8 tankers per day. What is the probability that on a given day tankers will have to be sent away?

<u>Solution</u>. Let one day be the unit, $t = 1$. The mean number of arrivals per day is $\lambda =$ ___ . Assuming that the number of arrivals per day is (approximately) a Poisson distribution, then

$$P_n(t) = \text{probability of } n \text{ arrivals over } t \text{ days}$$

$$= \underline{\qquad\qquad\qquad}, \quad n = 0, 1, 2, \ldots$$

---

41. 5, $e^{-5t} \frac{(5t)^n}{n!}$, $\sum_{n=0}^{8} P_n(t)$, 0.0337, 0.0842

The probability of more than 8 arrivals on a single day is

$$\sum_{n=9}^{\infty} P_n(t) \text{ , with } t = 1 .$$

This is also equal to $1 -$ ________________ with $t = 1$, or

$$1 - e^{-5} \sum_{n=0}^{8} \frac{5^n}{n!} \approx 1 - (0.0067) + \_\_\_\_\_\_ + \_\_\_\_\_\_ + 0.1404 + 0.1755$$

$$+ 0.1755 + 0.1462 + 0.1044 + 0.0653)$$

$$\approx 1 - 0.9319 = 0.0681 .$$

Thus, the probability is about 7% that on a given day tankers will have to be sent away.

CHAPTER 18 OBJECTIVE - PROBLEM KEY

Objective	Problems in Thomas/Finney Text	Objective	Problems in Thomas/Finney Text
18-2	p. 860, 1-6	18-7	p. 870, 1-8
18-3	p. 861, 1-15	18-9	p. 874, 1-10
18-4 A	p. 864, 1-6	18-10	p. 877, 1-10
B	p. 864, 9,10,11	18-11	p. 878, 1-8
18-5	p. 866, 1-10	18-12	p. 882, 1,2,4-7
18-6	p. 868, 1-13	18-13	p. 889, 5,6,7,9

CHAPTER 18 SELF-TEST

1. Show that the function $f(x) = 4 + 2x + x^2 e^x$ is a solution to the differential equation

$$\frac{d^2y}{dx^2} - 2\frac{dy}{dx} + y = 2e^x + 2x .$$

In Problems 2 - 5, solve the given differential equation.

2. $\left[x \cos^2 (y/x) - y\right] dx + x\,dy = 0$

3. $2y(y^2 - x)\,dy = dx$

4. $e^x(y - 1)\,dx + 2(e^x + 4)\,dy = 0$

5. $y(y^3 - x)\,dx + x(y^3 + x)\,dy = 0$

6. Find the family of orthogonal trajectories to the family of curves given by

$$e^x + e^{-y} = C .$$

7. Solve the following second-order differential equations.

(a) $\dfrac{d^2y}{dx^2} = 2y\left(\dfrac{dy}{dx}\right)^3$

(b) $x\dfrac{d^2y}{dx^2} + \dfrac{dy}{dx} + x = 0$

In Problems 8 - 10, solve the given differential equation.

8. $\dfrac{d^4y}{dx^2} + 18\dfrac{d^2y}{dx^2} + 81\,y = 0$

9. $\dfrac{d^2y}{dx^2} + y = \sec^3 x$

10. $4\dfrac{d^3y}{dx^3} + 4\dfrac{d^2y}{dx^2} + \dfrac{dy}{dx} = 0$

11. A spring is stretched 4 inches by a 2-lb weight. After the weight has come to rest in its new equilibrium position, it is struck a sharp blow that starts it downward at a velocity of 4 ft/sec. If air resistance furnishes a retarding force of magnitude 0.02 of the velocity, find its subsequent motion.

12. The probability that a person dies from a certain respiratory infection is 0.002. Assuming a Poisson probability distribution, find the probability that fewer than 5 of the next 2000 people infected will die.

SOLUTIONS TO CHAPTER 18 SELF-TEST

1. $f(x) = 4 + 2x + x^2e^x$, $f'(x) = 2 + 2xe^x + x^2e^x$, and

$f''(x) = 2e^x + 4xe^x + x^2e^x$. Thus,

$$f'' - 2f' + f = (2+4x+x^2)e^x - (4+4xe^x+2x^2e^x) + (4+2x+x^2e^x) = 2e^x + 2x .$$

Thus, $y = f(x)$ satisfies the differential equation and is, by definition, a solution.

2. From the given equation, we have

$$\frac{dy}{dx} = \frac{y}{x} - \cos^2\left(\frac{y}{x}\right) = F\left(\frac{y}{x}\right), \quad \text{where } F(v) = v - \cos^2 v \quad \text{and} \quad v = \frac{y}{x} .$$

Thus, $\frac{dx}{x} + \frac{dv}{v - F(v)} = 0$ becomes $\frac{dx}{x} + \frac{dv}{\cos^2 v} = 0$.

The solution of this is

$$\ln|x| + \tan v = \ln C , \quad \text{or, in terms of } x \text{ and } y ,$$

$$\tan\left(\frac{y}{x}\right) = \ln\frac{C}{|x|} .$$

3. The differential equation is linear in x and can be written in the form

$$\frac{dx}{dy} + 2yx = 2y^3 .$$

An integrating factor is $\rho = e^{\int 2y\,dy} = e^{y^2}$, from which we get

$$xe^{y^2} = \int 2y^3 e^{y^2} dy + C$$

$$= \int ze^z dz + C \quad (z = y^2 , \ dz = 2y\,dy)$$

$$= e^{y^2}(y^2 - 1) + C \quad \text{or} \quad x = (y^2 - 1) + Ce^{-y^2} .$$

4. The variables are separable, and the differential equation can be written as

$$\frac{e^x}{e^x + 4}dx + \frac{2}{y-1}dy = 0 . \quad \text{Integration gives,}$$

$$\ln(e^x+4) + 2\ln|y-1| = \ln C , \quad \text{or} \quad (y-1)^2(e^x+4) = C .$$

5. $\frac{\partial}{\partial y}\left[y(y^3 - x)\right] = 4y^3 - x$ and $\frac{\partial}{\partial x}\left[x(y^3 + x)\right] = y^3 + 2x$,

so the equation is not exact. However, the equation can be rewritten as

$$x\left(\frac{x\,dy - y\,dx}{y^2}\right) + y(x\,dy + y\,dx) = 0,$$

or

$$\left(-\frac{x}{y}\right) d\left(\frac{x}{y}\right) + d(xy) = 0,$$

or

$$-u\,du + d(xy) = 0, \quad \text{where} \quad u = \frac{x}{y}.$$

Integration gives,

$$-\frac{1}{2}\left(\frac{x}{y}\right)^2 + xy = C, \quad \text{or} \quad 2xy^3 - x^2 = 2Cy^2.$$

6. The given family of curves satisfies the differential equation

$$e^x\,dx - e^{-y}\,dy = 0.$$

The family of orthogonal trajectories satisfies the differential equation

$$e^{-y}\,dx + e^x\,dy = 0, \quad \text{or} \quad e^{-x}\,dx + e^y\,dy = 0,$$

whose solution is

$$e^y - e^{-x} = C_1,$$

the family of orthogonal trajectories.

7. (a) The independent variable is missing. We substitute $p = \frac{dy}{dx}$ and $\frac{d^2y}{dx^2} = p\frac{dp}{dy}$, and obtain $p\frac{dp}{dy} = 2yp^3$, or $p^{-2}\,dp = 2y\,dy$.

Integration gives $-\frac{1}{p} = y^2 + C_1$, or $(y^2 + C_1)\,dy = -dx$.

Thus,

$$\frac{1}{3}y^3 + C_1 y = -x + C_2, \quad \text{or} \quad y^3 = 3(C_2 - x - C_1 y).$$

(b) The dependent variable is missing. We substitute $p = \frac{dy}{dx}$ and $\frac{dp}{dx} = \frac{d^2y}{dx^2}$, and obtain $x\frac{dp}{dx} + p + x = 0$, or $\frac{dp}{dx} + \frac{1}{x}p = -1$.

This last equation is linear in p, and an integrating factor is $\rho = e^{\int dx/x} = x$. Thus $px = \int -x\,dx + C_1$, or $x\frac{dy}{dx} = -\frac{1}{2}x^2 + C_1$.

The variables are separable, and we find

$$dy = \left(-\frac{1}{2}x + \frac{1}{x}C_1\right)dx \quad \text{gives} \quad y = -\frac{1}{4}x^2 + C_1 \ln|x| + C_2.$$

8. The characteristic equation is $r^4 + 18r^2 + 81 = 0$, or $\left(r^2+9\right)^2 = 0$. The pair of complex conjugate numbers $r_1 = -3i$ and $r_2 = 3i$ are each double roots, and the solution is

$$y = (c_1 + c_2 x)\cos 3x + (c_3 + c_4 x)\sin 3x .$$

9. The associated homogeneous equation has the solutions

$$u_1(x) = \cos x \quad \text{and} \quad u_2(x) = \sin x .$$

We then have

$$\begin{aligned} v_1' \cos x + v_2' \sin x &= 0 \\ -v_1' \sin x + v_2' \cos x &= \sec^3 x . \end{aligned}$$

By Cramer's rule:

$$v_1' = \frac{\begin{vmatrix} 0 & \sin x \\ \sec^3 x & \cos x \end{vmatrix}}{\begin{vmatrix} \cos x & \sin x \\ -\sin x & \cos x \end{vmatrix}} = -\frac{\sin x}{\cos^3 x} ,$$

$$v_2' = \frac{\begin{vmatrix} \cos x & 0 \\ -\sin x & \sec^3 x \end{vmatrix}}{\begin{vmatrix} \cos x & \sin x \\ -\sin x & \cos x \end{vmatrix}} = \sec^2 x .$$

Hence,

$$v_1 = \int \frac{-\sin x\, dx}{\cos^3 x} = -\frac{1}{2}\cos^{-2} x + C_1 = -\frac{1}{2}\sec^2 x + C_1 ,$$

$$v_2 = \int \sec^2 x\, dx = \tan x + C_2 ,$$

and

$$\begin{aligned} y &= v_1 u_1 + v_2 u_2 \\ &= -\frac{1}{2}\sec x + C_1 \cos x + \sin^2 x \sec x + C_2 \sin x \\ &= \sec x \left(\sin^2 x - \frac{1}{2}\right) + C_1 \cos x + C_2 \sin x \\ &= \sec x \left(\frac{1}{2} - \cos^2 x\right) + C_1 \cos x + C_2 \sin x \\ &= \frac{1}{2}\sec x + C_3 \cos x + C_2 \sin x . \end{aligned}$$

10. The characteristic equation is $4r^3 + 4r^2 + r = 0$, or $r(4r^2 + 4r + 1) = 0$. The roots are $r_1 = 0$, $r_2 = -\frac{1}{2}$, and $r_3 = -\frac{1}{2}$. The general solution is

$$y = C_1 + (C_2 + C_3 x)\, e^{-x/2} .$$

11. A differential equation of the motion is

$$m \frac{d^2x}{dt^2} = -kx - 0.02 \frac{dx}{dt} ,$$

where $\frac{1}{3} k = 2$ or $k = 6$ (by Hooke's law), and $m = \frac{2}{32}$.

Thus,

$$\frac{d^2x}{dt^2} + \frac{8}{25} \frac{dx}{dt} + 96\, x = 0 ,$$

is the differential equation of motion, and the general solution is (see page 881 of the Thomas/Finney text)

$$x = Ce^{-0.16t} \sin (9.8t + \phi) .$$

The derivative of the motion is

$$\frac{dx}{dt} = -0.16Ce^{-0.16t} \sin (9.8t + \phi) + 9.8Ce^{-0.16t} \cos (9.8t + \phi) .$$

From the initial conditions $t = 0$, $x = 0$, $dx/dt = 4$, we find

$$0 = C \sin \phi \quad \text{and} \quad 9.8C \cos \phi = 4$$

Thus, $\phi = 0$ and $C \approx 0.41$. Hence,

$$x = 0.41e^{-0.16t} \sin (9.8t)$$

describes the motion. The motion is oscillatory with undercritical damping.

12. We have $\lambda t = 0.002$ with $t = 1$, so $\lambda = 0.002$. The probability that fewer than 5 of the next 2000 so infected will die is given by

$$P = \sum_{n=0}^{4} P_n(t) = \sum_{n=0}^{4} e^{-\lambda t} \frac{(\lambda t)^n}{n!} , \quad \text{with } \lambda = 0.002 \text{ and } t = 2000 .$$

Thus,

$$P = e^{-4} \sum_{n=0}^{4} \frac{4^n}{n!} = e^{-4} \left(1 + 4 + 8 + \frac{32}{3} + \frac{32}{3}\right) \approx 0.6288 .$$